METHODS IN MOLECULAR BIOLOGY™

Series Editor
John M. Walker
School of Life Sciences
University of Hertfordshire
Hatfield, Hertfordshire, AL10 9AB, UK

For further volumes:
http://www.springer.com/series/7651

Epithelial Cell Culture Protocols

Second Edition

Edited by

Scott H. Randell and M. Leslie Fulcher

Cystic Fibrosis/Pulmonary Research and Treatment Center,
The University of North Carolina at Chapel Hill, Chapel Hill, NC, USA

Humana Press

Editors
Scott H. Randell
Cystic Fibrosis/Pulmonary Research and Treatment Center
The University of North Carolina at Chapel Hill
Chapel Hill, NC, USA

M. Leslie Fulcher
Cystic Fibrosis/Pulmonary Research and Treatment Center
The University of North Carolina at Chapel Hill
Chapel Hill, NC, USA

ISSN 1064-3745 ISSN 1940-6029 (electronic)
ISBN 978-1-4939-5679-1 ISBN 978-1-62703-125-7 (eBook)
DOI 10.1007/978-1-62703-125-7
Springer New York Heidelberg Dordrecht London

Printed on acid-free paper

Humana Press is a brand of Springer
Springer is part of Springer Science+Business Media (www.springer.com)

Preface

A lot of fascinating biology occurs at epithelial interfaces, whether between organism and environment or within body compartments. Many diseases inflicting huge personal and societal burdens result from dysfunction of epithelial systems, e.g., carcinomas. Bolstered by the breakthrough studies of George Gey that introduced HeLa cells to the world in 1953 (J Exp Med.; 97(5):695–710), epithelial cell cultures have been an integral and crucial part of the biomedical research enterprise, adding unique capabilities and enabling mechanistic approaches. In modern times, thousands of publications reporting studies using epithelial cell cultures are added to the peer-reviewed literature annually. The first edition of Epithelial Cell Culture Protocols hit the shelves in 2001, but since then there has been an ever-escalating series of research advances: directed differentiation of embryonic stem cells and induced pluripotent stem cells, robotic high-throughput screening, whole genome siRNA and shRNA libraries, massively parallel sequencing at low cost, identification of somatic stem cells in key organs, to name a few. There has been a quieter but steady change too. Depending on the question, many studies routinely done in the past with well-known and easily cultured cell lines on conventional plastic are no longer acceptable to reviewers and editors of the highest tier journals or to members of study sections making decisions for granting agencies. Common refrains are "will it hold up in primary cells?" or "needs to show in well-differentiated primary cells." Although still models, modern culture methods for primary cells that recapitulate the structure and function of the endogenous cells in vivo are often superior. In our own field of cystic fibrosis research, therapeutic validation in well-differentiated human airway epithelial air–liquid interface cultures appears better than heterologous screening assays for predicting ultimate success in clinical trials. However, there is a catch. Procurement of primary tissues requires approval by ethics committees and cooperation of pathologists or organ procurement agencies, modern state-of-the-art methods can be complex and costly, commercial cell suppliers use proprietary methods for expensive products, primary cells are unpredictable, variable, and difficult to transfect, and so on. Although it would be impossible to cover all epithelia, we sought to provide a cross-section of up-to-date culture protocols for the most heavily studied cell systems. Chapter by chapter, we start at the head and move down the body axis, focusing on human primary cells. In some cases of heavily used models or when no human counterpart is established, we include animal models. We also feature supporting technologies. Our goal was to provide the best possible information from outstanding investigators. For this, we are eternally grateful to the excellent scientists who responded positively to our solicitations. We are especially thankful to those who submitted well-formatted manuscripts in a timely fashion and those who filled in late stage gaps. We appreciate the patience and understanding of all our contributors and Professor Walker, the Series Editor. Our sincere hope is that the protocols herein will assist readers in their quest to advance biomedical science for the betterment of humankind.

Chapel Hill, NC, USA

M. Leslie Fulcher
Scott H. Randell

Contents

Preface . . . *v*
Contributors . . . *ix*

1 Progress Towards *Drosophila* Epithelial Cell Culture . . . 1
Amanda Simcox

2 Culture of Choroid Plexus Epithelial Cells and In Vitro Model of Blood–CSF Barrier . . . 13
Andrew D. Monnot and Wei Zheng

3 Methods for Culture of Human Corneal and Conjunctival Epithelia . . . 31
Sandra J. Spurr-Michaud and Ilene K. Gipson

4 The Culture and Maintenance of Functional Retinal Pigment Epithelial Monolayers from Adult Human Eye . . . 45
Timothy A. Blenkinsop, Enrique Salero, Jeffrey H. Stern, and Sally Temple

5 Oral and Pharyngeal Epithelial Keratinocyte Culture . . . 67
Kantima Leelahavanichkul and J. Silvio Gutkind

6 Primary Culture of the Human Olfactory Neuroepithelium . . . 81
Nancy E. Rawson and Mehmet Hakan Ozdener

7 Primary Culture of Mammalian Taste Epithelium . . . 95
Mehmet Hakan Ozdener and Nancy E. Rawson

8 Human Nasal and Tracheo-Bronchial Respiratory Epithelial Cell Culture . . . 109
M. Leslie Fulcher and Scott H. Randell

9 Culture and Differentiation of Mouse Tracheal Epithelial Cells . . . 123
Yingjian You and Steven L. Brody

10 Isolation and Culture of Alveolar Epithelial Type I and Type II Cells from Rat Lungs . . . 145
Robert F. Gonzalez and Leland G. Dobbs

11 Classical Human Epidermal Keratinocyte Cell Culture . . . 161
Cathy Rasmussen, Christina Thomas-Virnig, and B. Lynn Allen-Hoffmann

12 Mouse Epidermal Keratinocyte Culture . . . 177
Luowei Li

13 Three-Dimensional Culture of Human Breast Epithelial Cells: The How and the Why . . . 193
Pierre-Alexandre Vidi, Mina J. Bissell, and Sophie A. Lelièvre

14 Three-Dimensional Cultures of Mouse Mammary Epithelial Cells . . . 221
Rana Mroue and Mina J. Bissell

15 Isolation, Characterization, and Reaggregate Culture of Thymic Epithelial Cells 251
Natalie Seach, Maree Hammett, and Ann Chidgey

16 Enrichment and Clonal Culture of Hepatic Stem/Progenitor Cells During Mouse Liver Development 273
Akihide Kamiya and Hiromitsu Nakauchi

17 Long-Term Culture and Coculture of Primary Rat and Human Hepatocytes 287
Maria Shulman and Yaakov Nahmias

18 Tissue Culture Correlational Study of Genetic Cholangiopathy of Autosomal Recessive Polycystic Kidney Disease 303
Yasuni Nakanuma, Yasunori Sato, and Kenichi Harada

19 Primary Mouse Small Intestinal Epithelial Cell Cultures 319
Toshiro Sato and Hans Clevers

20 Isolation, Growth, and Characterization of Human Renal Epithelial Cells Using Traditional and 3D Methods 329
John J. Gildea, Helen E. McGrath, Robert E. Van Sciver, Dora Bigler Wang, and Robin A. Felder

21 Human Female Reproductive Tract Epithelial Cell Culture 347
Rachel W.S. Chan, Abby S.C. Mak, William S.B. Yeung, Kai-Fai Lee, Annie N.Y. Cheung, Hextan Y.S. Ngan, and Alice S.T. Wong

22 Primary Culture and Propagation of Human Prostate Epithelial Cells 365
Birunthi Niranjan, Mitchell G. Lawrence, Melissa M. Papargiris, Michelle G. Richards, Shirin Hussain, Mark Frydenberg, John Pedersen, Renea A. Taylor, and Gail P. Risbridger

23 Urothelial Cell Culture: Stratified Urothelial Sheet and Three-Dimensional Growth of Urothelial Structure 383
Yuanyuan Zhang and Anthony Atala

24 Analysis of Tissue Interactions in Ectodermal Organ Culture 401
Pauliina M. Munne, Katja Närhi, and Frederic Michon

25 Generation of Human Cell Lines Using Lentiviral-Mediated Genetic Engineering 417
Patrick Salmon

26 Viral Transformation of Epithelial Cells 449
Jennifer A. Regan and Laimonis A. Laimins

Index *467*

Contributors

B. LYNN ALLEN-HOFFMANN • *Department of Pathology and Laboratory Medicine, University of Wisconsin School of Medicine and Public Health, Madison, WI, USA*
ANTHONY ATALA • *Department of Urology, Wake Forest Institute for Regenerative Medicine, Wake Forest University School of Medicine, Winston-Salem, NC, USA*
DORA BIGLER WANG • *Department of Pathology, The University of Virginia, Charlottesville, VA, USA*
MINA J. BISSELL • *Lawrence Berkeley National Laboratory, Berkeley, CA, USA*
TIMOTHY A. BLENKINSOP • *Neural Stem Cell Institute, Rensselaer, NY, USA*
STEVEN L. BRODY • *Department of Internal Medicine, School of Medicine, Washington University in St. Louis, Saint Louis, MO, USA*
RACHEL W.S. CHAN • *Department of Pathology, School of Biological Sciences, University of Hong Kong, Hong Kong, China; Department of Obstetrics and Gynecology, School of Biological Sciences, University of Hong Kong, Hong Kong, China*
ANNIE N.Y. CHEUNG • *Department of Pathology, School of Biological Sciences, University of Hong Kong, Hong Kong, China; Department of Obstetrics and Gynecology, School of Biological Sciences, University of Hong Kong, Hong Kong, China*
ANN CHIDGEY • *Monash Immunology and Stem Cell Laboratories, Monash University, Melbourne, VIC, Australia*
HANS CLEVERS • *Hubrecht Institute, Utrecht, The Netherlands*
LELAND G. DOBBS • *Departments of Medicine and Pediatrics, The Cardiovascular Research Institute, University of California, San Francisco, CA, USA*
ROBIN A. FELDER • *Department of Pathology, The University of Virginia, Charlottesville, VA, USA*
MARK FRYDENBERG • *Prostate and Breast Cancer Research Program, Department of Anatomy and Developmental Biology, Monash University, Melbourne, VIC, Australia; Department of Surgery, Monash University, Melbourne, VIC, Australia*
M. LESLIE FULCHER • *Cystic Fibrosis/Pulmonary Research and Treatment Center, The University of North Carolina at Chapel Hill, Chapel Hill, NC, USA*
JOHN J. GILDEA • *Department of Pathology, The University of Virginia, Charlottesville, VA, USA*
ILENE K. GIPSON • *Department of Ophthalmology, Schepens Eye Research Institute, Harvard Medical School, Boston, MA, USA*
ROBERT F. GONZALEZ • *The Cardiovascular Research Institute, University of California, San Francisco, CA, USA*
J. SILVIO GUTKIND • *Oral and Pharyngeal Cancer Branch, National Institute of Dental and Craniofacial Research, Bethesda, MD, USA*

MAREE HAMMETT • *Monash Immunology and Stem Cell Laboratories, Monash University, Melbourne, VIC, Australia*
KENICHI HARADA • *Department of Human Pathology, Kanazawa University Graduate School of Medicine, Kanazawa, Japan*
SHIRIN HUSSAIN • *Prostate and Breast Cancer Research Program, Department of Anatomy and Developmental Biology, Monash University, Melbourne, VIC, Australia*
AKIHIDE KAMIYA • *Laboratory of Stem Cell Therapy, Tokai University Institute of Innovative Science and Technology, Tokyo, Japan*
LAIMONIS A. LAIMINS • *Department of Microbiology-Immunology, Feinberg School of Medicine, Northwestern University, Chicago, IL, USA*
MITCHELL G. LAWRENCE • *Prostate and Breast Cancer Research Program, Department of Anatomy and Developmental Biology, Monash University, Melbourne, VIC, Australia*
KAI-FAI LEE • *Department of Pathology, School of Biological Sciences, University of Hong Kong, Hong Kong, China; Department of Obstetrics and Gynecology, School of Biological Sciences, University of Hong Kong, Hong Kong, China*
KANTIMA LEELAHAVANICHKUL • *Oral and Pharyngeal Cancer Branch, National Institute of Dental and Craniofacial Research, Bethesda, MD, USA; Thammasat University, Pathumthani, Thailand*
SOPHIE A. LELIÈVRE • *Department of Basic Medical Sciences and Center for Cancer Research, Purdue University, West Lafayette, IN, USA*
LUOWEI LI • *Laboratory of Cancer Biology and Genetics, Center for Cancer Research, National Cancer Institute, Bethesda, MD, USA*
ABBY S.C. MAK • *Department of Pathology, School of Biological Sciences, University of Hong Kong, Hong Kong, China; Department of Obstetrics and Gynecology, School of Biological Sciences, University of Hong Kong, Hong Kong, China*
HELEN E. MCGRATH • *Department of Pathology, The University of Virginia, Charlottesville, VA, USA*
FREDERIC MICHON • *Institute of Biotechnology, University of Helsinki, Helsinki, Finland*
ANDREW D. MONNOT • *School of Health Sciences, Purdue University, West Lafayette, IN, USA*
RANA MROUE • *Lawrence Berkeley National Laboratory, Berkeley, CA, USA*
PAULIINA M. MUNNE • *Institute of Biotechnology, University of Helsinki, Helsinki, Finland*
YAAKOV NAHMIAS • *The Selim and Rachel Benin School of Engineering, The Hebrew University of Jerusalem, Jerusalem, Israel*
YASUNI NAKANUMA • *Department of Human Pathology, Kanazawa University Graduate School of Medicine, Kanazawa, Japan*
HIROMITSU NAKAUCHI • *Division of Stem Cell Therapy, The Institute of Medical Science, The University of Tokyo, Tokyo, Japan*
KATJA NÄRHI • *Institute of Biotechnology, University of Helsinki, Helsinki, Finland*

HEXTAN Y.S. NGAN • *Department of Pathology, School of Biological Sciences, University of Hong Kong, Hong Kong, China; Department of Obstetrics and Gynecology, School of Biological Sciences, University of Hong Kong, Hong Kong, China*
BIRUNTHI NIRANJAN • *Prostate and Breast Cancer Research Program, Department of Anatomy and Developmental Biology, Monash University, Melbourne, VIC, Australia*
MEHMET HAKAN OZDENER • *Monell Chemical Senses Center, Philadelphia, PA, USA*
MELISSA M. PAPARGIRIS • *Prostate and Breast Cancer Research Program, Department of Anatomy and Developmental Biology, Monash University, Melbourne, VIC, Australia*
JOHN PEDERSEN • *Prostate and Breast Cancer Research Program, Department of Anatomy and Developmental Biology, Monash University, Melbourne, VIC, Australia; TissuPath, Melbourne, VIC, Australia; Australian Prostate Cancer BioResource, Australian Prostate Cancer BioResource, Monash University, Melbourne, VIC, Australia*
SCOTT H. RANDELL • *Cystic Fibrosis/Pulmonary Research and Treatment Center, The University of North Carolina at Chapel Hill, Chapel Hill, NC, USA*
CATHY RASMUSSEN • *Department of Pathology and Laboratory Medicine, University of Wisconsin School of Medicine and Public Health, Madison, WI, USA*
NANCY E. RAWSON • *AFB International, Saint Charles, MO, USA*
JENNIFER A. REGAN • *Department of Microbiology-Immunology, Feinberg School of Medicine, Northwestern University, Chicago, IL, USA*
MICHELLE G. RICHARDS • *Prostate and Breast Cancer Research Program, Department of Anatomy and Developmental Biology, Monash University, Melbourne, VIC, Australia*
GAIL P. RISBRIDGER • *Prostate and Breast Cancer Research Program, Department of Anatomy and Developmental Biology, Monash University, Melbourne, VIC, Australia*
ENRIQUE SALERO • *Neural Stem Cell Institute, Rensselaer, NY, USA*
PATRICK SALMON • *Department of Neurosciences, Geneva School of Medicine (CMU), Geneva, Switzerland*
TOSHIRO SATO • *Department of Gastroenterology, School of Medicine, Keio University, Tokyo, Japan; Hubrecht Institute, Utrecht, The Netherlands*
YASUNORI SATO • *Department of Human Pathology, Kanazawa University Graduate School of Medicine, Kanazawa, Japan*
NATALIE SEACH • *Monash Immunology and Stem Cell Laboratories, Monash University, Melbourne, VIC, Australia*
MARIA SHULMAN • *The Selim and Rachel Benin School of Engineering, The Hebrew University of Jerusalem, Jerusalem, Israel*
AMANDA SIMCOX • *Department of Molecular Genetics, Ohio State University, Columbus, OH, USA*
SANDRA J. SPURR-MICHAUD • *Department of Ophthalmology, Schepens Eye Research Institute, Harvard Medical School, Boston, MA, USA*
JEFFREY H. STERN • *Neural Stem Cell Institute, Rensselaer, NY, USA*

Renea A. Taylor • *Prostate and Breast Cancer Research Program, Department of Anatomy and Developmental Biology, Monash University, Melbourne, VIC, Australia*
Sally Temple • *Neural Stem Cell Institute, Rensselaer, NY, USA*
Christina Thomas-Virnig • *Department of Pathology and Laboratory Medicine, University of Wisconsin School of Medicine and Public Health, Madison, WI, USA*
Robert E. Van Sciver • *Department of Pathology, The University of Virginia, Charlottesville, VA, USA*
Pierre-Alexandre Vidi • *Department of Basic Medical Sciences and Center for Cancer Research, Purdue University, West Lafayette, IN, USA*
Alice S.T. Wong • *Department of Pathology, School of Biological Sciences, University of Hong Kong, Hong Kong, China; Department of Obstetrics and Gynecology, School of Biological Sciences, University of Hong Kong, Hong Kong, China*
William S.B. Yeung • *Department of Pathology, School of Biological Sciences, University of Hong Kong, Hong Kong, China; Department of Obstetrics and Gynecology, School of Biological Sciences, University of Hong Kong, Hong Kong, China*
Yingjian You • *Department of Internal Medicine, School of Medicine, Washington University in St. Louis, Saint Louis, MO, USA*
Yuanyuan Zhang • *Department of Urology, Wake Forest Institute for Regenerative Medicine, Wake Forest University School of Medicine, Winston-Salem, NC, USA*
Wei Zheng • *School of Health Sciences, Purdue University, West Lafayette, IN, USA*

Chapter 1

Progress Towards *Drosophila* Epithelial Cell Culture

Amanda Simcox

Abstract

Drosophila epithelial research is at the forefront of the field; however, there are no well-characterized epithelial cell lines that could provide a complementary in vitro model for studies conducted in vivo. Here, a protocol is described that produces epithelial cell lines. The method uses genetic manipulation of oncogenes or tumor suppressors to induce embryonic primary culture cells to rapidly progress to permanent cell lines. It is, however, a general method and the type of cells that comprise a given line is not controlled experimentally. Indeed, only a small fraction of the lines produced are epithelial in character. For this reason, additional work needs to be done to develop a more robust epithelial cell-specific protocol. It is expected that *Drosophila* epithelial cell lines will have great utility for in vitro analysis of epithelial biology, particularly high-throughput analyses such as RNAi screens.

Key words: *Drosophila*, Ras, Tumor suppressor, GAL4/UAS system, Epithelial cell line

1. Introduction

The study of *Drosophila* epithelia including the imaginal discs, embryonic epidermis, larval midgut, and ovary has proved fertile ground for gene discovery and analysis to understand the mechanisms of cell fate, proliferation, polarity, and movement. What has been lacking in the analysis of *Drosophila* epithelia is the development of in vitro models. Cell lines have been derived from the imaginal discs but none of these has an overt epithelial character (1, 2). Recently we discovered a rapid way to generate *Drosophila* cell lines, which offers the promise of developing cell type-specific lines including those from epithelia. Using an approach that has been successful in mammalian tissue culture, we manipulated the expression of oncogenes and tumor suppressors to generate cell lines efficiently. Expression of oncogenic Ras, Ras^{V12}, has a profound effect on *Drosophila* primary cells in culture; they are protected from apoptosis and stimulated to proliferate so that they

Scott H. Randell and M. Leslie Fulcher (eds.), *Epithelial Cell Culture Protocols: Second Edition*, Methods in Molecular Biology, vol. 945, DOI 10.1007/978-1-62703-125-7_1, © Springer Science+Business Media, LLC 2012

rapidly give rise to continuous cell lines (3). A small fraction of these lines are epithelial. Blocking the Hippo/Warts tumor suppressor pathway has also produced epithelial cell lines ((3, 4); Simcox, unpublished).

In mammalian systems, there are widely used protocols to generate cell type-specific lines and lines corresponding to mutants. The *Drosophila* method has already been used to derive cells of a particular mutant strain (5). But its potential to generate cell type-specific cell lines has not been developed. In *Drosophila* the GAL4/UAS system can target gene expression in a cell type-specific manner (6). In theory, therefore, expressing oncogenic Ras, or inhibiting tumor suppressors, in epithelial cells should promote the generation of epithelial cell lines. It is likely, however, that the success of this will depend on two other developments: a system for conditional expression of the genetic agent and improved culture methods. The first could be achieved in a number of ways, mirroring those that work in mammalian cells, by using temperature-sensitive, drug-inducible, or recombination systems to regulate expression of the transgenes. The second will require the adaptation of existing culture strategies for *Drosophila* cells, such as surface treatments, strata, or matrices, that promote epithelial cell polarity and development of multicellular structures that more closely resemble their in vivo counterparts. There are certainly incentives for tailoring these methods to *Drosophila* where cell-based assays for gene discovery, such as RNAi screens, are cost effective and efficient because there are fewer gene redundancies than in vertebrates (7).

2. Materials

2.1. Fly Rearing and Egg Collection

1. *Drosophila* medium:

   Flies are raised in vials (e.g., Drosophila polystyrene vials, Genesee Scientific) on standard cornmeal and molasses food (http://flystocks.bio.indiana.edu/Fly_Work/media-recipes/media-recipes.htm). The food is supplemented with live yeast (e.g., Red Star active dry yeast, Lesaffre Yeast Corporation) to promote the health and fertility of the flies prior to the start of egg collections.

2. Egg laying medium (see Note 1):

   Solution A: 83 g dried bakers yeast (e.g., Red Star active dry yeast, Lesaffre Yeast Corporation), 75 mL water, 175 mL vinegar (apple cider, malt, or 4% acetic acid).

   Solution B: 17 g agar (e.g., molecular biology grade agar, TEKnova), 485 mL water.

Methylene blue (Sigma-Aldrich®).

Food dye (e.g., McCormick®).

3. Egg collection dishes: 60 mM Petri dishes.
4. Egg collection cages: Embryo collection cages—small (FlyStuff.com a division of Genesee Scientific).
5. Egg rinsing solution, TXN: NaCl (0.7%), Triton X (0.02%) in water.
6. Paintbrushes: Round, size 6.
7. Bleach: Diluted with water to 50%.
8. Sieve: Mesh, Nitex Nylon 120 μM (FlyStuff.com a division of Genesee Scientific).
9. Tubes for collecting embryos from sieve: 15 mL conical centrifuge tube.

2.2. Cell Culture

1. Tissue culture medium: 100 mL Schneider's insect medium (Sigma-Aldrich®) supplemented with 11.25 mL heat-inactivated fetal bovine serum (FBS) (56°C for 30 min) (e.g., Atlanta Biologicals®, see Note 2) and 1.25 mL antibiotic mix (Penicillin–Streptomycin liquid, Life Technologies™).
2. Trypsin–EDTA solution (1×) (Sigma-Aldrich®).
3. Freezing medium: 50% Schneider's medium:30% serum:20% dimethyl sulfoxide (DMSO).
4. Homogenizer: Glass 5 mL tissue grinder with Teflon pestle.
5. Tissue culture flasks, 25 cm^2.
6. Cryovials.

2.3. Antibody Staining

1. Primary antibody: Anti-DE-cadherin (dCad2, Developmental Studies Hybridoma Bank, University of Iowa).
2. Secondary antibody: Rodamine conjugated anti-rat (Jackson ImmunoResearch).
3. Tissue-culture coverslips: Thermenox™.
4. Chamber slide: Lab-Tek™ 4-well chamber slide.
5. 4% Paraformaldehyde: Dilute in PBS from 16% solution (Electron Microscopy Sciences).
6. PBS + 0.2% Triton™ X-100 (PBTX).
7. Block solution: 5% Normal Goat Serum (NGS) in PBS.
8. VectaShield® Mounting Medium (Vector Laboratories).

2.4. Transfection and RNAi

1. Transfection was performed using Effectene® transfection reagent (Qiagen).
2. A 500 bp dsRNA against the *argos* gene was generated using in vitro transcription (Megascript, Ambion®).

3. Methods

3.1. Modulation of Oncogenes and Tumor Suppressors to Establish Cell Lines

3.1.1. Expression of Oncogenic Ras^{V12} with the GAL4–UAS System

The GAL4/UAS system directs expression of *UAS-transgenes* under the control of the transcription factor GAL4 (6). There is a library of the so-called GAL4 driver lines in which GAL4 is expressed in a large number of different spatial and temporal patterns. In each strain the *GAL4* gene is under the control of different genomic enhancers. We have used the *Act5C-GAL4* strain in our experiments. *Act5C* is a broadly expressed cytoplasmic actin gene and therefore directs expression of GAL4 in many cell types. Flies from a genetic strain carrying a *UAS-*Ras^{V12} transgene are crossed to flies of an *Act5C-GAL4* strain and in response, Ras^{V12} is expressed in many cell types. The embryos are organism lethal but when homogenized and grown in vitro, the *Act5C-GAL4; UAS-*Ras^{V12} cells are viable and indeed outcompete cells of other genotypes and dominate the culture at confluence. For this reason there is no requirement to sort the embryos to select the desired genotype prior to setting up the primary cultures (see Note 3). The method can easily be adapted to generating mutant cell lines by using genetics to make strains in which only the homozygous mutant cells are also *Act5C-GAL4; UAS-*Ras^{V12} (5). While the method primarily gives rise to lines of cells with a spindle shape, two Ras^{V12}-expressing lines with epithelial character have been generated; one of these expresses both Ras^{V12} and wts^{dsRNA} (3).

3.1.2. Effect of Reduction of Tumor Suppressor Function

We assayed the effects of reducing the function of some tumor suppressors using mutants or RNAi. Blocking two pathways proved effective: PI3K/Pten or Warts (Wts) ((3); Justiniano and Simcox, unpublished). Three $Pten^{-}$ cell lines have been established; none of these has epithelial character. We tested the effect of knocking down *wts* with RNAi, wts^{dsRNA}, and mutation and found that both promoted the establishment of cell lines ((3); Simcox, unpublished). Two epithelial cell lines have been generated, one is mutant for *wts* (Simcox, unpublished) and, as mentioned above, one expresses both Ras^{V12} and wts^{dsRNA}.

3.2. Primary Cultures

3.2.1. Egg Collection and Processing

Two hundred males and 200 virgin females of the desired genotypes are crossed and kept in well-yeasted vials or bottles (at normal density) for 2–3 days (see Note 4). The flies are transferred to laying cages with sterile egg-collection plates (see Note 5). To make the plates, autoclave solutions A and B for 20 min and mix (see Subheading 2.1, step 2). Add 10 mg methylene blue or edible food-coloring drops to provide contrast for the eggs, and pour into 60 mm Petri dishes. Eggs are collected overnight in the dark at room temperature (22°C). The eggs are dislodged from the surface of the plate using a soft paintbrush and TXN (see Subheading 2.1, step 5). The suspended eggs are poured into a small sieve that retains eggs but allows smaller particles to pass through (see Note 6). The eggs are rinsed in the sieve with a stream of TXN from a squirt bottle to wash off any debris. The sieve is reversed over a 15 mL conical tube and the eggs are transferred into the tube by a stream of TXN. At this point further manipulations are done in a sterile hood. The TXN is removed and the eggs are dechorionated (the eggshell is removed) in 7 mL 50% bleach. The tube is inverted about three times to suspend the eggs in the bleach solution and then the eggs are allowed to settle. After 3 min, the bleach is removed and the eggs are washed with TXN (the success of the dechorionation step can be checked by examining the eggs under a dissecting microscope—they should lack the anterior appendages that are part of the eggshell and appear shiny due to the exposed vitelline membrane). The packed volume of the eggs is estimated at this point, as this will determine how many primary cultures to establish; 100 μL of packed eggs is sufficient for 2–3 primary cultures.

3.2.2. Homogenization and Plating

The eggs are transferred to a new sterile 15 mL conical tube by aspiration from the bottom of the first tube. This reduces the amount of bleach carried over. The eggs are rinsed in TXN three more times and transferred to a 5 mL glass homogenizer. The eggs are rinsed once in sterile water and once in cell culture medium (see Subheading 2.2, step 1). Care should be taken at these steps to avoid losing eggs, as without the detergent present in the TXN, they tend to stick together, float on the surface, and stick to the pipette. About 3 mL of culture medium is added and the pestle is inserted carefully to minimize trapped air at the interface with the medium. The embryos are homogenized with three gentle strokes keeping the pestle submerged even at the top of the upstroke, so as not to introduce air bubbles. The pestle is set aside (but kept sterile) and large clumps and unbroken embryos are allowed to settle. The supernatant is removed into a 15 mL conical tube. Fresh medium (3 mL) is added to the homogenizer, which is flicked to resuspend any clumps that are compressed at the bottom. The homogenization is repeated with slightly more vigorous strokes and a twist at the bottom of the tube to break up the pellet if necessary. The homogenate is pooled with the first fraction and the tube is spun in a benchtop clinical centrifuge for 2 min (room temperature at 1,380 × *g*). The supernatant is removed and the pellet is washed with two more changes of fresh medium. The washing removes yolk and other subcellular particles. The final pellet is resuspended in a small quantity of medium and divided into one or more T-flasks (25 cm^2), each with 3 mL of medium. The number of flasks depends on the starting number of embryos (see Subheading 3.2.1) (see Note 6). Flasks are placed in an incubator at 22°C (or 25°C for faster growth). No CO_2 is required for growing fly cells.

3.2.3. Care of Primary Cultures

Primary cultures are checked periodically and the medium is changed after 10 days for the first time and thereafter about every 10–14 days. Checking the cultures every few days initially is important as small fraction may be infected with yeast, mold, or bacteria and should be discarded. In our experience infections are not a major factor but additional surface sterilization of the eggs can be performed by rinsing with 70% EtOH after the bleaching step.

3.2.4. Epithelial and Other Cell Types Found in Primary Cultures

The development of *Drosophila* primary cultures has been described in detail (8–11). Briefly, many differentiated cell types appear including muscle, nerve, fat, epithelial, and spindle-shaped cells. The notable difference between cultures with cells expressing Ras^{V12} and wild-type cultures is that more of these latter two cell types, epithelial and spindle-shaped, are present. Figure 1a, b shows typical Ras^{V12}-expressing cultures with patches of these cell types.

3.3. Establishment of Lines

3.3.1. Subculturing

Once primary cultures reach about 70% confluence they can be subcultured. The lawn of cells covering the surface is rarely even and a judgment needs to be used about the best time to first passage a given culture. Once cultures reach confluence they typically deteriorate very rapidly; on the other hand, sparse subcultures do

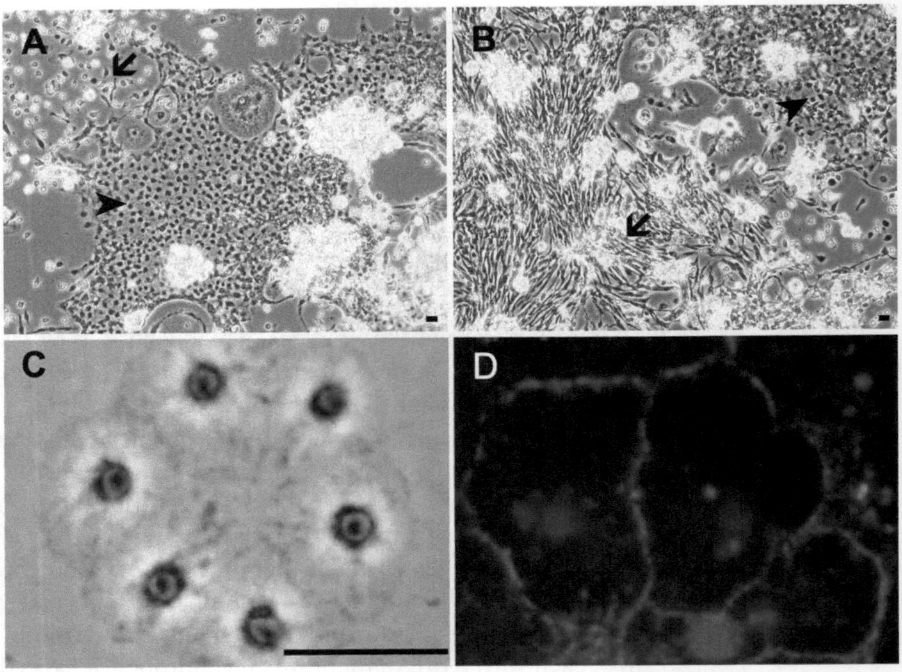

Fig. 1. *Drosophila* epithelial cells in primary cultures and established cell lines. (**a** and **b**) Primary cultures showing patches of epithelial (*arrow*) and spindle-shaped (*arrowhead*) cells. The cultures are about 70% confluent and ready for the first passage. (**c**) Cells of the Ras^{V12}; wts^{dsRNA} line are forming a small sheet in a new subculture soon after plating. (**d**) Cells of the *wts* line stained for E-cadherin. E-cadherin accumulates at the cell boundary. Bar in **a–c**, 10 μm.

not do well. Examples of primary cultures ready for subculturing are shown in Fig. 1a, b. In cultures with cells expressing *Ras*^{V12}, the first subcultures can be made starting at about 3 weeks (3). With wild-type cultures this is much longer and more variable (3).

1. To subculture, remove the medium (save in a labeled 15 mL conical tube).
2. Rinse the surface with 3 mL trypsin/EDTA, add a new aliquot of trypsin/EDTA, and incubate for 3 min at room temperature.
3. Squirt the trypsin/EDTA solution over the cells to help release them from the surface and dilute into the saved medium.
4. Spin the cells for 2 min in a centrifuge (room temperature at 1,380×*g*).
5. Plate half the cells in a new flask (i.e., a dilution of 1 in 2). At passage 10–15, cells can be diluted 1 in 4. On a case-by-case basis, cells can be split at higher dilutions.

3.3.2. Designation of a Line as a Continuous Cell Line

In our experience most cell cultures that have been passaged ten or more times can be propagated indefinitely with few exceptions. Two epithelial lines derived from *Ras*^{V12}; *wts*^{dsRNA}-expressing or *wts* mutant cells have been passaged more than 150 times, representing conservatively about 500 population doublings.

3.3.3. Frequency of Occurrence of Epithelial Like Cell Lines

In the primary cultures there are always patches of epithelial cells, but as the cultures are passaged these tend to be outcompeted by the spindle-shaped cells. Since the development of the method we have generated more than 50 cell lines and only three of these are epithelial. To increase the success rate, we are modifying the protocol to favor epithelial cells by using the GAL4/UAS system to induce gene expression only in epithelial cells and changing culture conditions.

3.4. Epithelial Cell Lines

3.4.1. Histochemical Analysis with E-Cadherin

Some cells in primary cultures (Fig. 1a, b) and continuous cell lines (Fig. 1c) have overt epithelial characteristics and form sheets of cells. These cells are positive for E-cadherin, the classical marker of epithelial cells. In *Drosophila* the gene encoding E-cadherin is called *shotgun* and there is a monoclonal antibody available that works well to detect the protein in cultured cells (Fig. 1d).

1. For antibody staining, cells are grown on tissue-culture coverslips or in tissue-culture chamber slides (see Subheading 2.3, step 4).
2. Cells are washed once in PBS and fixed for 20 min in 4% paraformaldehyde (see Subheading 2.3, step 5).
3. Cells are rinsed briefly in PBS and washed three times in PBS for 5 min.

4. PBTX (see Subheading 2.3, step 6) is used to permeabilize the cells. Cells are washed three times in PBS and blocked NGS (see Subheading 2.3, step 7) for 1 h and incubated with primary antibody (1:5) (see Subheading 2.3, step 1) and 5% NGS, overnight at 4°C.
5. Cells are washed three times in PBS and rhodamine-conjugated secondary antibodies (1:200) (see Subheading 2.3, step 2) are added and incubated for 30 min at room temperature.
6. Cells are washed three times in PBS and mounted using mounting medium (see Subheading 2.3, step 8). The image shown in Fig. 1d was captured with a Zeiss 510 META Laser Scanning Confocal microscope.

3.4.2. Transfection

The epithelial cells can be transfected using standard methods. We use Effectene® and the recommended procedure for adherent cells. A transfection efficiency of up to about 70% can be achieved for the *wts* cells.

3.4.3. RNAi

The epithelial cells derived from *wts* mutants have been subjected to RNAi. Microarray analysis and western analysis show that *wts* cells express components of the Egfr pathway (not shown), including the receptor and the three zygotic ligands. We tested for the effect of introducing dsRNAs against two secreted factors, the ligand Spitz and its inhibitor Argos. The dsRNA (15 μg) was delivered using transfection, as above. The expression of Argos was greatly reduced (see Fig. 2), whereas Spitz levels were not appreciably reduced (not shown). Variability in gene response is not

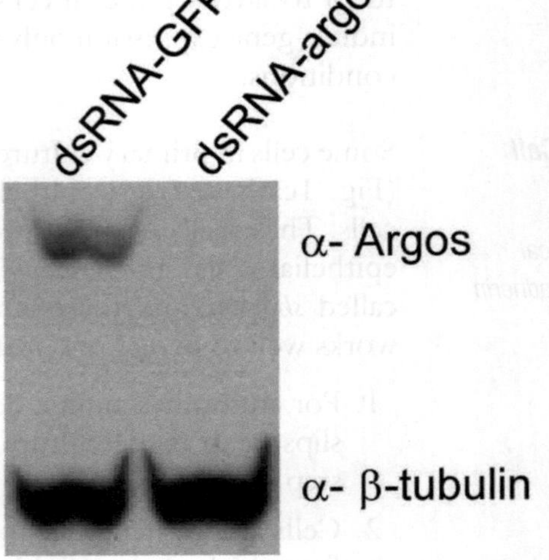

Fig. 2. Cells in the *wts* epithelial cell line are susceptible to RNAi. Treatment with a double-stranded RNA targeting the *argos* gene (dsRNA-argos), but not a control dsRNA targeting the *gfp* gene (dsRNA-GFP), reduces expression of Argos (β-tubulin is a loading control).

unexpected, and the success with *argos* shows that in principle the cells are susceptible to RNAi.

3.4.4. Freezing Cells for Long-Term Storage

We freeze aliquots of cells from the lines as they evolve, for example, at passages 5, 10, 20, and 30.

1. Cells are harvested from confluent T-flasks.
2. The cell pellet is suspended in a small volume of medium (50–100 μL remaining after the medium is removed) and then diluted into 1 mL of freezing medium (see Subheading 2.2, step 3).
3. This is divided into two aliquots in freezing tubes.
4. The tubes are placed at −80°C overnight for a "slow freeze" and then transferred to liquid nitrogen for long-term storage.
5. Multiple aliquots are frozen and one is tested after a few days for cell viability. The tube is defrosted in a beaker of room-temperature water for a "quick defrost."
6. To remove the freezing medium, the cells are diluted into 3 mL of cell-culture medium and spun and the pellet is plated in a single T-flask with 3 mL of fresh medium.
7. Cell viability is assessed by cell growth; typically some cells fail to adhere because they are inviable but many settle and start to proliferate.

4. Notes

1. We find that these plates, which incorporate some yeast into the medium, are more appealing to the flies and promote better egg laying than typical egg collection plates that are fruit-juice based. The fruit-juice plates have to be supplemented with killed yeast paste, and even this "sterile" paste can rapidly support the growth of microorganisms, and in our experience leads to more infections in the primary cultures.
2. We buy serum from a variety of sources based on cost. For the past 10 years we have *not* checked each new batch for whether it has adverse effects on *Drosophila* cells. There is certainly no harm in doing so, but in our experience we have never yet had to reject a batch.
3. Embryos of a desired genotype could be selected prior to setting up the cell cultures by using an embryo sorter (COPAS, Union Biometrica). Transgenes encoding fluorescent markers are used to select for, or against, particular genotypes.
4. Up to 2,000 flies can be housed in these small-size cages if a fluted sheaf of filter paper is inserted around the inside edge.

Also larger cages can be used. Plates should be checked for hatched larvae. These are sometimes present even though the collection period is shorter than the length of embryogenesis because females can retain eggs, some eggs are laid on the cage walls, and the hatched larvae crawl to the plate. It is important to remove these larvae, as their digestive track contents can cause yeast and other infections in the primary cultures. To remove larvae, we flood the plate with TXN and pick them out individually with forceps. Adding the TXN causes the larvae to move and makes them easier to spot. We find that it is well worth the investment of time (about 10 min) to include this step.

5. Small homemade sieves are constructed from the cutoff cap end of a 15 mL plastic centrifuge tube (about 2 cm of the tube). A small piece of mesh is stretched across the open end and held in place with the cap, which has its center cut out.
6. There is a limit to the number of embryos that can be processed successfully in a 5 mL glass homogenizer. Too many embryos tend to form a dense clump at the bottom of the homogenizer that cannot be disrupted without causing too much cell damage. Larger volume homogenizers can be used. In our experience, the 5 mL homogenizers work well and are convenient for the scale of work required for establishing cell lines. Three primary cultures can be set up in a typical procedure and as the efficiency of generating cells lines, especially with expression of *RasV12*, is good, a sufficient number of primary cultures can be set up in four runs of the procedure to guarantee generating 5–10 cell lines.

Acknowledgements

I thank Litty Paul for the RNAi experiment. The work was supported by NIH R01GM071856 and an ARRA supplement NIH 3R01GM071856-04S1.

References

1. Ui K, Ueda R, Miyake T (1987) Cell lines from imaginal discs of *Drosophila melanogaster*. In Vitro Cell Dev Biol 23:707–711
2. Peel DJ, Milner MJ (1990) The diversity of cell morphology in cloned cell lines derived from Drosophila imaginal discs. Roux's Arch Dev Biol 198:479–482
3. Simcox A, Mitra S, Truesdell S, Paul L, Chen T, Butchar JP, Justiniano S (2008) Efficient genetic method for establishing Drosophila cell lines unlocks the potential to create lines of specific genotypes. PLoS Genet 4:e1000142
4. Zhao B, Lei Q, Guan K-L (2010) The Hippo–YAP pathway in organ size control and tumorigenesis: an updated version. Genes Dev 24:862–874
5. Simcox AA, Austin CL, Jacobsen TL, Jafar-Nejad H (2008) Drosophila embryonic 'fibroblasts': extending mutant analysis in vitro. Fly (Austin) 2:306–309

6. Brand A, Perrimon N (1993) Targeted gene expression as a means of altering cell fates and generating dominant phenotypes. Development 118:401–415
7. Mohr S, Bakal C, Perrimon N (2010) Genomic screening with RNAi: results and challenges. Annu Rev Biochem 79:37–64
8. Shields G, Dubendorfer A, Sang JH (1975) Differentiation in vitro of larval cell types from early embryonic cells of *Drosophila melanogaster*. J Embryol Exp Morphol 33:159–175
9. Shields G, Sang JH (1970) Characteristics of five cell types appearing during in vitro culture of embryonic material from *Drosophila melanogaster*. J Embryol Exp Morphol 23: 53–69
10. Cross DP, Sang JH (1978) Cell culture of individual Drosophila embryos. I. Development of wild-type cultures. J Embryol Exp Morphol 45:161–172
11. Eschalier G (1997) Drosophila cells in culture. Academic, New York, NY

Chapter 2

Culture of Choroid Plexus Epithelial Cells and In Vitro Model of Blood–CSF Barrier

Andrew D. Monnot and Wei Zheng

Abstract

Chemical homeostasis in the extracellular fluid of the central nervous system (CNS) is maintained by two brain barrier systems, i.e., the blood–brain barrier (BBB) that separates the blood circulation from brain interstitial fluid and the blood–cerebrospinal fluid barrier (BCB) that separates the blood from the cerebrospinal fluid (CSF). The choroid plexus, where the BCB is located, is a polarized tissue, with the basolateral side of the choroidal epithelium facing the blood and the apical microvilli in direct contact with the CSF. The tissue plays a wide range of roles in brain development, aging, nutrient transport, endocrine regulation, and pathogenesis of certain neurodegenerative disorders. This chapter describes two in vitro cultures that have been well established to allow for study of the BCB structure and function. The primary choroidal epithelial cell culture can be established from rat choroid plexus tissue, and a similar immortalized murine choroidal epithelial cell culture known as Z310 cells has also been established. Both cultures display a dominant polygonal morphology, and immunochemical studies demonstrate the presence of transthyretin, a thyroxine transport protein known to be exclusively produced by the choroidal epithelia in the CNS. These cultures have been adapted for use on freely permeable Transwell® membranes sandwiched between two culture chambers, facilitating transport studies of various compounds across this barrier in vitro. These choroidal epithelia cultures with the Transwell system will perceivably assist blood–CSF barrier research.

Key words: Choroid plexus, Cerebrospinal fluid, Transthyretin, Z310 cells

1. Introduction

The brain barrier system that separates the systemic circulation from the cerebrospinal fluid (CSF) compartment is known as the blood–CSF barrier, which is primarily located in the choroid plexus. The function of the blood–CSF barrier is to restrict the access of substances from the blood to the CSF as well as remove substances from the CSF to the blood. The choroid plexus also actively produces and secretes CSF along with critical molecules such as transthyretin

Scott H. Randell and M. Leslie Fulcher (eds.), *Epithelial Cell Culture Protocols: Second Edition*, Methods in Molecular Biology, vol. 945, DOI 10.1007/978-1-62703-125-7_2, © Springer Science+Business Media, LLC 2012

(TTR) and transferrin into the brain. Under a microscope, the choroid plexus is composed of three cellular layers: (1) the apical epithelial cells facing the CSF, (2) the underlying supporting connective tissue, and (3) the inner layer of endothelial cells with immediate contact with the blood. Given that the endothelial cells of the choroid plexus are functionally leaky, the passage of substances across this barrier is essentially controlled by the apical layer of epithelial cells which are tightly connected with one another through tight junction proteins. These tight junctions constitute the structural basis for the blood–CSF barrier, through which only the selected materials may gain access to the CSF, while most of the water-soluble substances, proteins, ions, and macromolecules are impeded from the blood to the CSF.

Increasing research effort in understanding pharmacokinetics/toxicokinetics of active molecules in the CSF, a central milieu of the central nervous system (CNS), and in discovering the etiology and therapy of neurodegenerative diseases demands appropriate in vitro blood–CSF model systems that allow for characterizing the transport property of interested molecules across the blood–CSF barrier. During the past 15 years, we have successfully developed a standard protocol for primary culture of choroidal epithelial cells from rodents and routinely used the cultured cells to investigate the transport kinetics as well as molecular mechanisms of metals such as manganese (Mn), iron (Fe), and copper (Cu); chemicals such as thyroxin; and proteins such as β-amyloids (Aβ) and TTR (1–9). We have further immortalized rat choroidal epithelial cells and established a highly stable choroidal epithelial cell line known as the Z310 cell line (10, 11). Both model systems possess the essential morphology and characteristics of the blood–CSF barrier, and prove useful to study the structure and function of the blood–CSF barrier as well as examine transport properties of materials by this barrier.

In this chapter, we first describe the methods to conduct primary culture of rat choroidal epithelial cells. This is followed by the procedures to culture established Z310 cells. The methods to characterize the cells are addressed to allow the readers to typify the cells by their own hands. Finally, special notes are provided for advices learned from our own experiences and mistakes.

2. Material

2.1. Material for Primary Cell Culture

2.1.1. Coating of Culture Dishes

1. 0.1% Type 1 Collagen (Sigma-Aldrich®) solution: Diluted to 0.01% working solution with distilled-deionized water.
2. Transwell® culture chambers, 12 mm in diameter, 0.4 μm pore size (Corning Costar).
3. 35-mm tissue culture grade Petri dishes.

2.1.2. Tissue Isolation and Separation

1. Sprague-Dawley rats, male, 6–8 weeks old, 150–180 g (Harlan).
2. Ketamine/xylazine (75:10 mg/mL, 1 mL/kg body weight).
3. Dissection kit including scissors, bone cutter, and fine forceps.
4. 75% Ethanol.
5. Phosphate-buffered saline (PBS): To 800 mL of distilled-deionized water add 8.0 g NaCl, 0.2 g KCl, 1.44 g Na_2HPO_4, and 0.24 g KH_2PO_4. Adjust the pH to 7.4 with HCl, and bring the volume up to 1,000 mL. Autoclave and store at 4°C.

2.1.3. Primary Cell Culture

1. Digestion solution: 4 mg/mL pronase (Calbiochem-Novobiochem). Dissolve 6 mg in 1.5 mL HBSS. The solution should then be transferred to a syringe and passed through an attached 0.22 μm low-protein-binding filter unit (Millipore). The final stock solution (4 mg/mL) should be kept on ice in a culture hood until use. The solution must be made fresh on the day of the experiment.
2. 0.4% Trypan blue.
3. cis-4-Hydroxy-D-proline (Sigma-Aldrich®): Final concentration of 25 μg/mL in medium.
4. 0.25% Trypsin, 1 mol/L EDTA (Life Technologies™).
5. Dulbecco's Modified Eagle Medium (DMEM, Gibco®).
6. Hank's Buffered Salt Solution (HBSS, Gibco®).
7. Fetal bovine serum (FBS) (heat inactivated, sterile-filtered).
8. Penicillin–streptomycin (10,000 units/mL penicillin Na + 10,000 μg/mL streptomycin sulfate in 0.85% saline) (Gibco®).
9. Gentamicin (10 mg/mL) solution.
10. Mouse epidermal growth factor (EGF): Dissolve 0.1 mg in 1 mL of primary cell culture medium for a stock of 0.1 mg/mL.

2.2. Z310 Cell Culture

Like the culture of primary choroidal epithelial cells, the normal growth medium for Z310 cells consists of three major components in DMEM: (a) antibiotics to prevent infection, (b) FBS to provide nutrients, and (c) EGF to stimulate the growth of epithelia.

1. Growth medium: 450 mL DMEM supplemented with 50 mL FBS (10%), 5 mL penicillin–streptomycin (10× = 10%), penicillin–streptomycin (10,000 units/mL penicillin Na + 10,000 μg/mL streptomycin sulfate in 0.85% saline), 2 mL gentamicin solution (10 mg/mL), 50 μL of EGF (0.1 mg/mL stock) for a final concentration of 10 ng/mL.
2. Dexamethasone (only used in media during Transwell® studies): Dissolve in Z310 cell culture medium and dilute in medium for a final concentration of 1 μM.

3. Cryoprotective medium: 90–95% Culture medium supplemented with 5–10% dimethyl sulfoxide (DMSO).

2.3. Two-Chamber Transepithelial Model

1. Transwell® Permeable Supports (12 mm insert, Corning).
2. Epithelial voltohmmeter (model EVOM, World Precision Instruments).

2.4. Immuno-cytochemical Studies

1. Anti-TTR polyclonal antibody produced in the mouse (Sigma-Aldrich®).
2. 4% Paraformaldehyde in PBS.
3. 1% Permeabilization buffer: Triton™ X-100 in PBS 10 μL/mL.
4. 1% Bovine serum albumin (BSA) in PBS.
5. 1:250 Mouse anti-human TTR antibody, diluted in 1% BSA.
6. 1:2,000 Fluorescein-conjugated goat anti-mouse antibody, diluted in PBS.
7. Confocal microscope with fluorscein isothiocyanate (FITC) and phase contrast optics.

2.5. Reverse Transcription Polymerase Chain Reaction

1. TRIzol® reagent (Life Technologies™).
2. RNeasy Mini kit (Qiagen).
3. Chloroform.
4. Ethanol.
5. Isopropanol.
6. Primers specifically selected for rat TTR (synthesized by IDT).
7. DEPC-treated water: Prepare by adding 1 mL of DEPC to 1,000 mL of distilled, deionized water, standing overnight, and autoclaving prior to use.
8. RT buffer: DEPC water, 25 mM $MgCl_2$, 10× PCR buffer II, 10 mM dNTP mix, 50 μM Oligo dT, 20 U/μL Rnase inhibitor, MuLV reverse transcriptase (Applied Biosystems).
9. Master Mix: DEPC water, 25 mM $MgCl_2$, 10 mM dNTP mix, 5× GoTaq Flexi buffer, GoTaq polymerase.
10. 1.0% Agarose gels containing 0.5 μg/mL ethidium bromide.

3. Methods

3.1. Procedures for Cell Culture

3.1.1. Coating of Culture Dishes

1. To coat dishes or the membranes of Transwell® inner chambers with collagen, dilute the stock collagen (0.1%) 1:10 with DDI H_2O for a 0.01% working solution (see Subheading 2.1.1).
2. Add an aliquot of diluted collagen to 35-mm dishes (800 μL) or Transwell® inserts (100 μL). Swirl the dishes or inserts to ensure an even distribution of the coating solution.

3. Incubate at room temperature in the culture hood for 4–5 h to allow the protein to bind to the surface.
4. Remove the excess liquid and then allow the coated dishes to air-dry in the hood under UV light to avoid contamination.

3.1.2. Tissue Isolation and Separation

Figure 1 illustrates the flowchart of the general procedures pertaining to the primary culture of choroidal epithelial cells. Please keep in mind that all procedures must be operated in sterile condition; the surgical tools must be autoclaved; the skin where the insertion

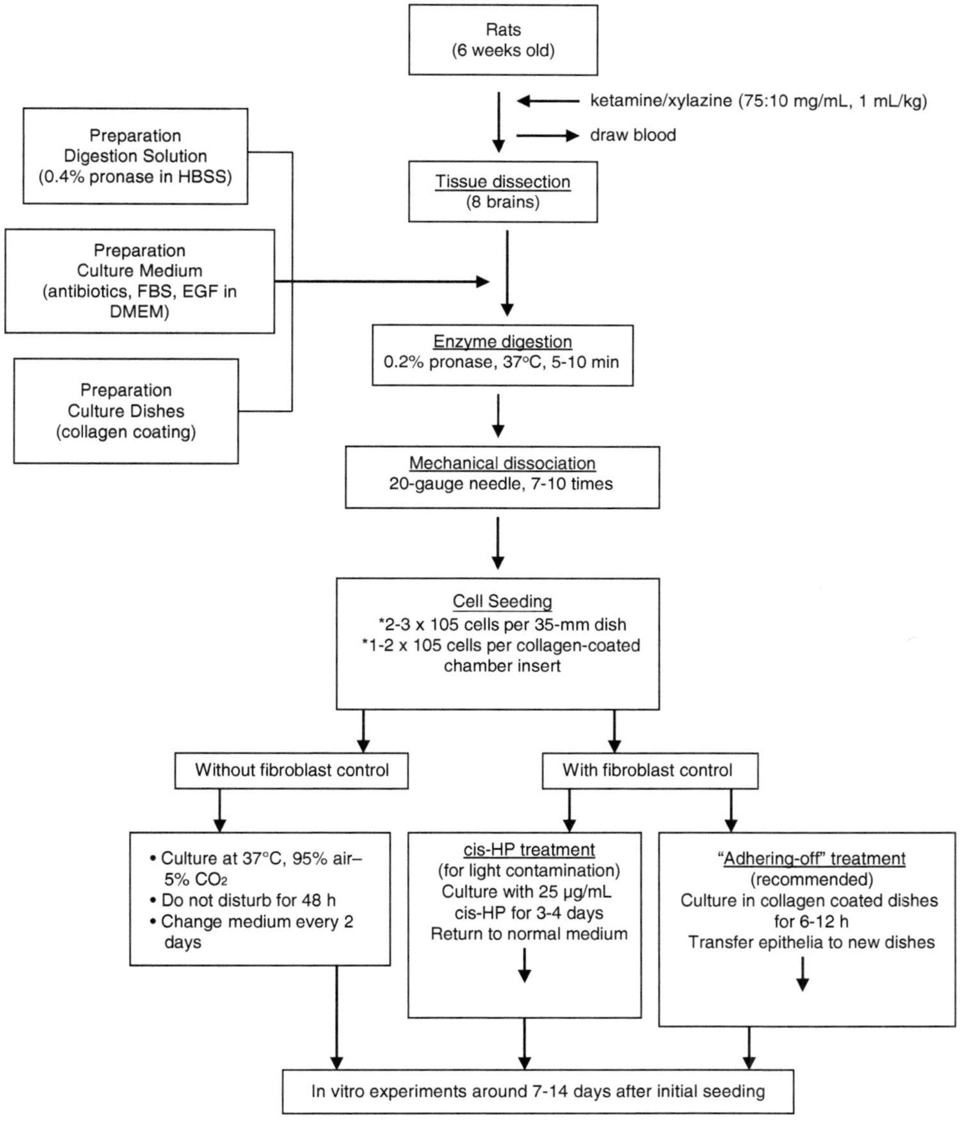

Fig. 1. Flowchart of procedures in establishing primary culture of choroidal epithelial cells.

made must be sterilized with 75% ethanol; and all glassware must also be autoclaved.

1. Anesthetize the rats with an i.p. injection of ketamine/xylazine.
2. To minimize the amount of blood present in choroid plexus tissues, draw as much blood as possible from the inferior vena cava using a syringe.
3. Remove the hair on the back of the head with a pair of scissors or electric trimmer.
4. Sterilize the exposed skin using cotton wool saturated with 75% ethanol.
5. Use a pair of scissors to cut the skin and to expose the skull. Use a bone cutter to remove the skull bone.
6. Remove the brain from the skull and place in a beaker containing PBS on ice to chill the tissue and wash off excess blood.
7. Once 5–8 brains have been collected, move them into a culture hood for dissection of the choroid plexuses.
8. Dissect the choroid plexuses from both the lateral and third ventricles and immerse them in 0.5 mL of HBSS at room temperature.
9. When all the choroid plexus tissues are collected, mince the plexus tissues with a pair of fine ophthalmologic scissors to roughly 1-mm cubes (~5 min of chopping).
10. Bring the total volume up to 1 mL by the addition of 0.5 mL HBSS.

3.2. Primary Cell Culture

3.2.1. Tissue Digestion

1. Add 1 mL of digestion solution (see Subheading 2.1.3, step 1) to the beaker to give a final pronase concentration of 4 mg/mL.
2. Swirl the beaker lightly by hand to allow a complete mixing of the digestion solution with the tissues.
3. Incubate at 37°C for 5–10 min.
4. Stop the digestion reaction by adding 4 mL of HBSS solution to the digestion mixture (see Note 1).
5. Centrifuge at 800 × *g* for 5 min at 4°C in a 15-mL sterile tube.
6. Discard the supernatant and wash the pellet once more with HBSS by resuspension and centrifugation. At this point the pellet should contain clumps of primary epithelial cells probably joined by tight junction proteins.
7. Resuspend the pellet in 2 mL of growth medium.
8. Mechanically dissociate the cells by 7–10 forced passages through a 20-gauge needle (see Note 1).

9. Remove an aliquot (0.1 mL) of cell suspension and mix with 0.1 mL of 0.4% trypan blue to count cell numbers and to assess the viability.
10. The procedure for cell isolation described here yields ~$0.8–1 \times 10^5$ epithelial cells per rat.

3.2.2. Culture of Epithelial Cells

1. Prior to cell seeding, dilute the cell preparations with growth medium to ~$1–2 \times 10^5$ cells/mL (see Note 2).
2. Plate the cells onto 35-mm coated Petri dishes ($2–3 \times 10^5$ cells per dish) and culture in a humidified incubator with 95% air/5% CO_2 at 37°C.
3. After 10 h in culture, remove unattached epithelial cells in culture medium and leave behind the attached fibroblast cells. This minimizes fibroblast contamination, a major problem in primary culture of epithelial cells. This "fibroblast adhering-off" method effectively leaves fibroblasts behind in the collagen-coated dishes, because fibroblasts usually attach to the collagen-coated surface much faster (6–10 h) than epithelial cells (16–24 h).
4. Replate the epithelial cells into new 35-mm plates and then leave cells undisturbed for at least 48 h.
5. Change the medium every 2–3 days thereafter for the duration of the culture.
6. Two days after the seeding, remove the culture medium, and replace with fresh medium containing cis-HP (see Subheading 2.1.3, step 3) to further control fibroblast contamination if necessary (see Note 3).
7. Usually the initial reseeding and treatment with cis-HP suffice for the purpose of inhibition of the growth of fibroblasts. Typical photographs of cultured choroidal epithelial cells under a phase contrast microscope are seen in Fig. 2.
8. After 3–5 days of culture with cis-HP, return the cells to normal growth medium without cis-HP, providing there are no visible fibroblasts under the microscope.
9. From our own experience, if the digestion procedure works well, the epithelia usually attach and grow rapidly. Therefore, the "fibroblast adhering-off" and cis-HP may not be necessary. However both methods greatly enhance the likelihood of a successful culture and are recommended.
10. To detach the cells for bioassays, incubate the culture with trypsin–EDTA in PBS at 37°C for 10 min.
11. Harvest the cells, centrifuge, and wash. They can then be used for further molecular studies or for Transwell® transport studies.

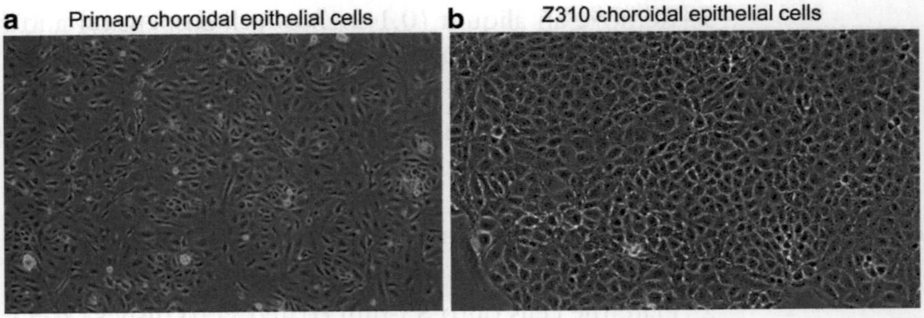

Fig. 2. Morphology of choroidal epithelial cells in culture. (**a**) Primary culture of choroidal epithelial cells after 5 days in culture (10×). Note the confluent layer of cells with a predominant polygonal cell type. The choroid plexus tissue was obtained from 6-week-old Sprague-Dawley rats. (**b**) Immortalized Z310 choroidal epithelial cells in culture (20×). Passage 86.

3.3. Two-Chamber Transepithelial Transport Model with Primary Choroidal Cells

1. The procedure for preparation of epithelial suspension is the same as described in Subheading 3.2.
2. Prior to seeding the cells in Transwell® chambers (inserts), coat the permeable membranes attached to the inserts with collagen as described in Subheading 3.1.1 (or purchase collagen-coated membranes).
3. Insert the inner chambers into the outer (basal) chambers, which should already contain 1.2 mL of growth medium (see Fig. 3).
4. Plate aliquots (0.8 mL) of cell suspension into the 12-mm collagen-coated culture wells (inserts) and place in incubator at 37°C.
5. Allow the cells to grow for 48 h.
6. Change the medium in both chambers every 2 days thereafter.
7. The formation of confluent impermeable cell monolayers is judged by three criteria: (1) the height of the culture medium in the inner chamber has to be at least 2 mm higher than that in the outer chamber for at least 24 h; (2) the appearance of a confluent monolayer on the insert under the microscope; and (3) the electrical resistance across the cell layer has to fall into the range of 65–80 $\Omega\,cm^2$ (see Note 4).
8. Transepithelial electrical resistance can be measured using an epithelial voltohmmeter after the cells have been cultured in the chambers for at least 4 days.
9. The net value of electrical resistance is calculated by subtracting the background (which is measured on collagen-coated cell-free chambers) from values of epithelial cell-seeded chambers.

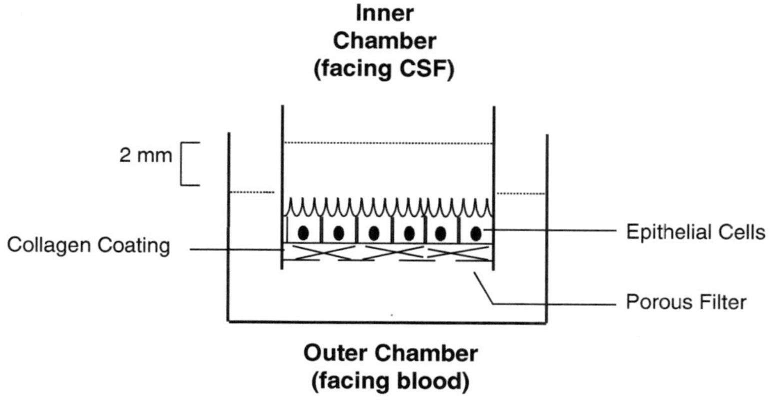

Fig. 3. Transepithelial model of the blood–CSF barrier used to study transepithelial transport. Epithelial cells are connected by tight junctions and form a barrier between fluids in the inner and outer chambers. Fluid in the inner chamber is in contact with the apical microvilli on the surface of the cells, while the fluid in the outer chamber has access to the basal surface of the cells.

3.4. Immuno-cytochemical Studies of the Marker of Choroidal Epithelia

A reliable method for the identification of choroidal epithelial cells is visualization of TTR, a unique marker for choroidal epithelial cells (see Notes 5 and 6). TTR is a 55,000-Da protein consisting of four identical subunits in tetrahedral symmetry. Per unit of weight, rat choroid plexus contains ten times more TTR mRNA than liver, and per gram of tissue, synthesizes TTR 13 times faster than the liver; the latter is the major organ in the body producing serum TTR (12, 13).

1. Isolate choroid plexus tissues from the lateral ventricles as previously described in Subheading 3.1.2.
2. Fix the tissue in 4% paraformaldehyde in PBS.
3. Wash three times with PBS.
4. Permeabilize the tissue by incubation in 0.1% permeabilization buffer (see Subheading 2.4, step 3) for 30 min.
5. Wash with PBS.
6. Block with 1% BSA for 60 min.
7. Incubate the cells with anti-TTR antibody (see Subheading 2.4, step 5) overnight at 4°C.
8. Wash with 1% BSA.
9. Incubate with secondary antibody for 1 h at 37°C.
10. Wash three times with PBS.
11. Examine the tissue using confocal microscopy (see Fig. 4).

3.5. RT-PCR Analysis

While immunohistochemical staining of TTR proteins is regarded as an acceptable approach to identify choroidal epithelial cells, it

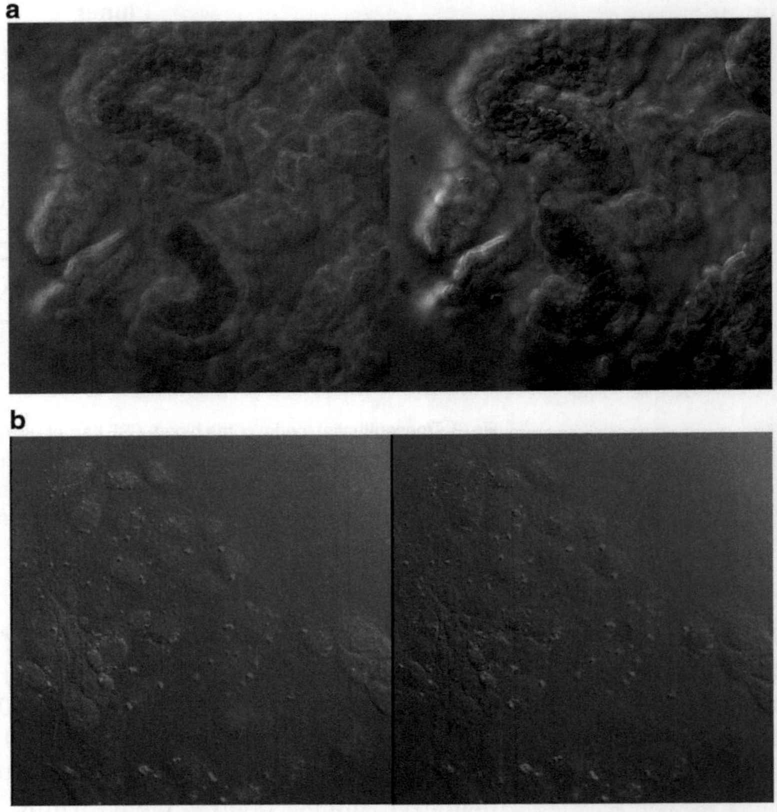

Fig. 4. Confocal microscopic depicting TTR staining in the choroid plexus tissue (**a**) and cultured Z310 cells (**b**). Tissue was treated with anti-TTR primary antibody followed by secondary antibody conjugated with fluorescein. Note the positive staining primarily along the basolateral side of the choroid plexus tissue. A Nikon C1 series modular confocal microscope was used to view the tissue through a 60× oil immersion objective with a 488 nm laser line for excitation.

may also be used in conjunction with reverse transcription polymerase chain reaction (RT-PCR) to validate expression of the mRNA encoding the protein.

1. Extract total RNA from the cultured cells or rat liver (as a positive control) using the TRIzol® method followed by RNA cleanup using an RNeasy mini kit.
2. Carry out the RT on 1 μg of total RNA, using MuLV reverse transcriptase with oligo dT primers.
3. The reaction should be carried out as follows: 25°C 10 min, 48°C 60 min, 95°C 5 min, and 4°C hold.
4. For PCR amplification, one set of specific oligonucleotide pairs should be incubated with the new cDNA synthesized in the above reaction mixture.

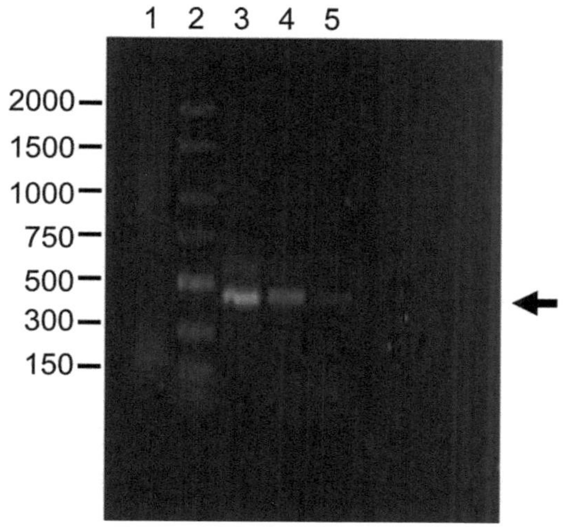

Fig. 5. Expression of TTR mRNA in primary choroidal epithelial cells and in choroidal Z310 cells by RT-PCR analysis. All samples underwent RT-PCR unless otherwise stated. *Arrow* indicates bands corresponding to TTR mRNA. Lane ID: *Lane 1* total RNA for PCR without RT, *lane 2* base pair ladder, *lane 3* liver mRNA with selected primer, *lane 4* cultured primary plexus cells mRNA with selected primer, *lane 5* Z310 cells mRNA with selected marker.

5. Add 80 μL Master Mix (see Subheading 2.5, step 9), containing 2.5 U Taq DNA polymerase, giving rise to a total volume of 100 μL.
6. The cycle parameters are 5 min at 94°C for initial denaturation, 0.5 min at 57°C for annealing, and 0.5 min at 72°C for extension. The subsequent cycle is as follows: 1 min at 94°C, 0.5 min at 57°C, and 0.5 min at 72°C for 40 cycles. Follow this with a 5-min incubation at 72°C.
7. An aliquot (5 μL) of each reaction mixture should be analyzed by electrophoresis on 1.0% agarose gels containing 0.5 μg/mL ethidium bromide (see Fig. 5).
8. The primers (custom-synthesized) designed by us specifically for rat TTR consist of:

   Primer sense: 5′-TTCCCTTCGCCTGTTCTTCTT-3′.

   Primer antisense: 5′-TTCTGGGGGTTAACTGACGACA-3′.

   This amplifies a product of 443 bp covering the mature TTR peptide from rats.

3.6. Immortalized Z310 Choroidal Epithelial Cells from the Murine Choroid Plexus

To facilitate the in vitro study of the blood–CSF barrier we have used a gene transfection technique to immortalize murine choroidal epithelial cells, known as Z310 cells. The cells display the same polygonal epithelial morphology and characterization as primary choroidal epithelial cells (see Fig. 2).

3.6.1. Transfer/Passage

1. Pre-warm PBS and growth medium in the water bath at 37°C.
2. Aspirate all media from the plate(s).
3. Add 5 mL of PBS to the plates, swirl slightly, and then aspirate PBS (repeat 2×).
4. Add 200 μL of trypsin–EDTA to detach the cells.
5. Incubate the plate(s) at 37°C for 10 min.
6. Add 8–10 mL of medium to each of the new plates.
7. After 10-min digestion, remove the cells from the incubator and observe under the microscope. If the cells look like "round balls," the digestion is complete and cells are detached.
8. Add 3 mL of cell culture medium into the old plate for a 1/12 split or 4 mL for a 1/16 split.
9. Mix well by using a pipette to break up all the cell clusters.
10. Pipette 0.25 mL of cells from the old plate into a new plate which has already had 8–10 mL of new medium.
11. Label the new passage and place in a 37°C incubator.
12. Passage cells every 2–3 days, when reaching roughly 80% confluence.

3.7. Freezing and Thawing Z310 Cells

3.7.1. Freezing Z310 Cells

1. After cells are digested and resuspended in growth medium (see Subheading 2.2, step 1), pellet the cells (200×*g* for 5–10 min, 4°C).
2. Aspirate supernatant above the cell pellet.
3. Slowly add (drop by drop) cryoprotective medium (see Subheading 2.2, step 3) to the cell pellet on ice.
4. Resuspend the cell pellet thoroughly in the cryoprotective medium.
5. Aliquot ~1 mL of cells (density ~1–2 × 10^7) to the appropriate tubes for storage.
6. Leave the aliquots on ice for 5–10 min.
7. Place aliquots in an insulated box (styrofoam) and place at −80°C overnight.
8. The next day remove the insulated container from the freezer, and place the aliquots in appropriate container for storage in liquid nitrogen.

3.7.2. Thawing Z310 Cells

1. Remove the vial from liquid nitrogen.
2. Thaw in a water bath at 37°C for 2–5 min.
3. Remove the vial from the bath, and clean with 70% ethanol.
4. Under the hood remove the cells from the vial and place in 8–10 mL of culture medium.

5. Use this medium to resuspend the cells and wash.
6. Pellet cells and resuspend in new culture medium.
7. Plate the cells on uncoated dishes and place into an incubator at 37°C for 1–2 days.
8. Pass cells upon 85–90% confluence.

3.8. Two-Chamber Transepithelial Transport Model with Z310 Cells

1. The procedure for preparation of epithelial suspension is the same as described in Subheading 3.6.1.
2. Prior to seeding the cells in Transwell® chambers (inserts), coat the permeable membranes attached to the inserts with collagen as described in Subheading 3.1.1 (or purchase collagen-coated membranes).
3. Insert the inner chambers into the outer (basal) chambers, which should already contain 1.2 mL of growth medium (see Fig. 3). For Z310 cells add 1 μM dexamethasone (see Subheading 2.2, step 2) in the growth medium for Transwell® studies in order to improve the tightness of the Z310 barrier.
4. Plate aliquots (0.8 mL) of cell suspension into the 12-mm collagen-coated culture wells (inserts).
5. Typical seeding numbers of Z310 cells can range from 5.0×10^4 to 2.0×10^6 cells per well. Depending on the seeding density, cells can reach confluence anywhere from 2 to 6 days.
6. Change the medium every 2 days thereafter.
7. The formation of confluent impermeable cell monolayers is judged by three criteria: (1) the height of the culture medium in the inner chamber has to be at least 2 mm higher than that in the outer chamber for at least 24 h; (2) the appearance of a confluent monolayer on the insert under the microscope; and (3) the electrical resistance across the cell layer has to fall into the range of 85–100 $\Omega \, cm^2$.
8. Transepithelial electrical resistance can be measured using an epithelial voltohmmeter after the cells have been cultured in the chambers for 1 day.
9. The net value of electrical resistance is calculated by subtracting the background (which is measured on collagen-coated cell-free chambers) from values of epithelial cell-seeded chambers.

4. Notes

1. Isolation of epithelial cells from the choroid plexus. To help ensure a large yield of epithelial cells from the primary culture, the digestion procedure is critical. Both collagenase (2 mg/mL)

and pronase (2 mg/mL) have been used to digest the tissues in our laboratory. We have found that pronase is the most effective at releasing the epithelial cells. It is important that the concentration and duration of pronase digestion be well monitored. The ideal digestion time with pronase varies depending on the tissue mass. The general rule of thumb is to carefully watch the color change of the medium. With a complete digestion, the medium usually changes from a light red to a yellow-orange, and from transparent to slightly cloudy. Digestion with pronase should not exceed 10 min; prolonged digestion reduces cell attachment down the line. The other vital procedure is the mechanical digestion of the cells. The epithelial clumps present after digestion normally attach to the surface of the dish, but will form fewer colonies if they are not dissociated from one another. Mechanical digestion through the needle effectively dissociates the cell clumps to produce a maximal yield of epithelia and greater plate efficiency.

2. Density of cells for initial seeding: An effective primary culture of choroidal epithelia requires a sufficient number of cells at the initial seeding. When cells are plated at a density of $<10^4$/mL, the cell proliferation can be very slow or virtually nonexistent. We recommend seeding the cells at a density of $2–4 \times 10^5$ cells/35-mm dish, which is roughly 2 mL of $1–2 \times 10^5$ cells/mL after digestion. One further noteworthy detail pertains to the transfer of the cells from centrifuge tubes to culture dishes. It is important to pre-wet the tips of the glass pipettes prior to transferring the cells. This minimizes the number of cells adhering to the dry tip of the glass pipette.
3. Control of fibroblasts: Arguably the most significant challenge of this procedure is the control of fibroblast contamination in the choroidal epithelial cell culture. Fibroblasts are visible in the culture under the light microscope and are typically elongated with their nuclei condensed. Fibroblasts usually rapidly spread between the epithelial clusters. We have tried several methods to inhibit fibroblast contamination. From our experience the most successful strategy is a two-step approach. The first step is known as the "fibroblast adhering-off" approach. This technique takes advantage of the higher affinity of the fibroblast cells to collagen-coated surfaces in the early cell selection stage. A relatively complete fibroblast adherence will occur 6–12 h after the initial seeding. At this time remove all the media carefully and plate onto freshly coated 35-mm dishes. After 48 h add new medium that contains 25 μg/mL of cis-HP, which can be withdrawn from the medium a few days (3–5) later. We have found that both concentration of cis-HP as well as time of its addition to the medium are critical to its success. The earlier the cis-HP treatment the more effective it

is. The caveat for cis-HP treatment is that the chemical also kills epithelial cells. Thus, we recommend adding the cis-HP no earlier than 48 h after the initial seeding. From our experience this process effectively reduces fibroblast contamination and gives us the purest epithelial culture.

4. Culture on Transwell® chambers: The procedure for culturing primary cells on Transwell® membranes is identical to the standard primary culture. Cells grown on the inner chamber membranes display the same morphology as those grown in culture dishes. The epithelial cells are connected by tight junctions and once they grow to confluence they form an impermeable barrier between the medium in the inner and outer chambers. The net electrical resistance across this barrier in our studies is 65–80 $\Omega\,cm^2$ for the primary culture and 85–100 $\Omega\,cm^2$ in Z310 cells (10). It is worthy to note that many factors can influence the determination of electrical resistance such as temperature, pH (physiological solutions vs. culture media), age of tissues or cultures, and freshness of the culture medium. From our experience, a higher pH, colder culture medium, and fresher culture medium usually result in a higher resistance reading. We have used this in vitro blood–CSF barrier model to study protein and metal transport and have demonstrated that the choroid plexus can remove Aβ, a vital peptide in the etiology of Alzheimer's disease, from the CSF (2). In addition this model has been used to show that manganese (Mn) exposure alters iron (Fe) flux across the blood–CSF barrier (7). Both Z310 and primary culture in vitro Transwell® models have been used to demonstrate the effect of lead (Pb) exposure on transport of thyroxin and Aβ at the blood–CSF barrier. Pb has been shown to hinder the transepithelial transport of thyroxine as well as decrease the blood–CSF barrier's permeability (1, 14, 15). The decrease in thyroxine transport appears to be due to the inhibitory effect of Pb on the production and secretion of TTR by the choroid plexus (5, 6). The decrease in blood–CSF barrier permeability is the result of early Pb exposure selectively inhibiting the expression of essential tight junction proteins (15). Recently, we have also applied this model to study copper transport at the blood–CSF barrier (data not shown). From these studies we have observed that the transepithelial transport of metal ions is a slow process and may take up to 24 h to reach equilibrium. Short time course studies may not be adequate in assessing the steady-state transport of substances across the blood–CSF barrier; so patience is vital for these experiments.

5. Use of immortalized choroidal epithelial Z310 cells: While primary culture serves as an ideal model for the in vitro study of the blood–CSF barrier, it has many limitations making it

undesirable to use on a daily basis. The disadvantages include a lack of cell abundance, reliance on a large number of animals, difficulty in obtaining pure cultures, and a short life span in culture. Therefore, the development of an immortalized choroidal epithelial line for in vitro studies has been a long, overdue summon in brain barrier transport research. Z310 cells possess many of the same phenotypic characteristics of primary choroidal cells, including a polygonal epithelial type as well as the presence of TTR. These cells have been proven an acceptable model for in vitro blood–CSF study and have been widely used (9, 16).

6. Characterization of primary and Z310 choroidal epithelial cells: Immunohistochemical staining with an anti-TTR antibody is a reliable way to distinguish choroidal epithelia from other types of cells. The presence of TTR has been repeatedly demonstrated in the choroidal epithelia of the brain (17, 18). In addition to immunochemical staining, RT-PCR is an acceptable methodology to validate the presence of TTR mRNA transcripts in choroidal epithelia. The sequence of TTR mRNA has been reported by Dickson et al. (13). We have designed a set of primers in order to run RT-PCR for TTR mRNA from the rat. The methods presented in this article display reliable evidence for TTR expression in our cultured Z310 cells, freshly isolated choroid plexus tissues, as well as the liver. The establishment of choroidal epithelial cell cultures from various sources has been well documented in the literature. These include cells from mouse (19, 20), rat (20–22), rabbit (12), sheep (23), and cow (24), each having its own particular set of advantages and disadvantages.

References

1. Zheng W, Zhao Q, Graziano JH (1998) Primary culture of rat choroidal epithelial cells: a model for in vitro study of the blood–cerebrospinal fluid barrier. In Vitro Cell Biol Dev 34:40–45
2. Crossgrove JS, Li GJ, Zheng W (2005) The choroid plexus removes beta-amyloid from the cerebrospinal fluid. Exp Biol Med 230(10):771–776
3. Deane R, Zheng W, Zlokovic BV (2004) Brain capillary endothelium and choroid plexus epithelium regulate transport of transferrin-bound and free iron into the rat brain. J Neurochem 88:813–820
4. Zheng W, Deane R, Redzic Z, Preston JE, Segal MB (2003) Transport of L-(125I)thyroxine by in-situ perfused ovine choroid plexus: inhibition by lead exposure. J Toxicol Env Health 66:435–451
5. Zheng W, Perry DF, Nelson DL, Aposhian HV (1991) Protection of cerebrospinal fluid against toxic metals by the choroid plexus. FASEB J 5:2188–2193
6. Zheng W, Shen H, Blaner WS, Zhao Q, Ren X, Graziano JH (1996) Chronic lead exposure alters transthyretin concentration in rat cerebrospinal fluid: the role of the choroid plexus. Toxicol Appl Pharmacol 139:445–450
7. Wang XQ, Li GJ, Zheng W (2008) Efflux of iron from the cerebrospinal fluid to the blood at the blood–CSF barrier: effect of manganese exposure. Exp Biol Med 233:1561–1571
8. Choi BS, Zheng W (2009) Copper transport to the brain by the blood brain barrier and blood–CSF barrier. Brain Res 1248:14–21

9. Zheng W (2001) Toxicology of choroid plexus: special reference to metal-induced neurotoxicities. Microsc Res Tech 52:89–103
10. Zheng W, Zhao Q (2002) Establishment and characterization of an immortalized Z310 choroidal epithelial cell line from murine choroid plexus. Brain Res 958:371–380
11. Shi LZ, Wang SZ, Li GJ, Zheng W (2008) Use of Z310 cells as an in vitro blood–cerebrospinal fluid barrier model: tight junction proteins and transport properties. Toxicol In Vitro 22: 190–199
12. Mayer SE, Sanders-Bush E (1993) Sodium-dependent antiporters in choroid plexus epithelial cultures from rabbit. J Neurochem 60:1304-13–1304-16
13. Dickson PW, Howlett GJ, Schreiber G (1985) Rat transthyretin (prealbumin): molecular cloning, nucleotide sequence, and gene expression in liver and brain. J Biol Chem 260:8214–8219
14. Zheng W, Blaner WS, Zhao Q (1999) Inhibition by Pb of production and secretion of transthyretin in the choroid plexus: its relationship to thyroxine transport at the blood–CSF barrier. Toxicol Appl Pharmacol 155:24–31
15. Shi LZ, Zheng W (2007) Early lead exposure increases the leakage of the blood–cerebrospinal fluid barrier, in-vitro. Hum Exp Toxicol 26:159–167
16. Shi LZ, Zheng W (2005) Establishment of an in vitro brain barrier epithelial transport system for pharmacological and toxicological study. Brain Res 1057:37–48
17. Aldred AR, Brack CM, Schreiber G (1995) The cerebral expression of plasma protein genes in different species. Comp Biochem Physiol 111B:1–15
18. Herbert J, Wilcox JN, Pham KC (1986) Transthyretin: a choroid plexus-specific transport protein in human brain. Neurology 36:900–911
19. Bouille C, Mesnil M, Barriere H, Gabrion J (1991) Gap junctional intercellular communication between cultured ependymal cells, revealed by Lucifer yellow CH transfer and freeze-fracture. Glia 4:25–36
20. Peraldi-Roux S, Nguyen-Than Dao B, Hirn M, Gabrion J (1990) Choroidal ependymocytes in culture: expression of markers of polarity and function. Int J Dev Neurosci 8:575–588
21. Strazielle N, Ghersi-Egea JF (1999) Demonstration of a coupled metabolism-efflux process at the choroid plexus as a mechanism of brain protection toward xenobiotics. J Neurosci 19:6275–6289
22. Tsutsumi M, Skinner MK, Sanders-Bush E (1989) Transferrin gene expression and synthesis by cultured choroid plexus epithelial cells. J Biol Chem 264:9626–9631
23. Harter DH, Hsu KC, Rose HM (1967) Immunofluorescence and cyto-chemical studies of visna virus in cell culture. J Virol 1:1265–1270
24. Whittico MT, Hui AC, Giacomini KM (1991) Preparation of brush border membrane vesicles from bovine choroid plexus. J Pharmacol Methods 25:215–227

Chapter 3

Methods for Culture of Human Corneal and Conjunctival Epithelia

Sandra J. Spurr-Michaud and Ilene K. Gipson

Abstract

The surface of the eye is exposed to the outside world and is, thus, subject to surface abrasion, infections, and drying, cicatrizing diseases. Availability of in vitro methods for culture of the human corneal and conjunctival epithelia, which cover the ocular surface, is therefore important in understanding the biology of these epithelia and their response to disease/infections, as well as for providing human-relevant models for preclinical testing of potential therapeutic agents. The ensuing chapter describes several methods for primary culture of both corneal and conjunctival epithelia and culture of immortalized cell lines, and methods employed to induce differentiation in the cultured epithelia.

Key words: Corneal epithelium, Conjunctival epithelium, Corneal epithelial cell culture, Conjunctival epithelial cell culture, Epithelial differentiation in culture, Ocular surface epithelia

1. Introduction

The surface of the eye is covered by a wet-surfaced mucosa, the epithelium of which is stratified, squamous, and non-keratinized. The ocular surface epithelium is responsible for maintaining and protecting the cornea and vision itself (1). The corneal epithelium over the translucent, avascular cornea is specialized at its apical surface to refract light. At the same time, the epithelium must be transparent to light and protect the cornea from noxious substances, injury, and pathogen invasion. It is an epithelium of 5–7 cell layers that turns over in about a week. The adult stem cells of the epithelium are located at the corneal periphery in the limbus, and their progeny move centripetally over time toward the center of the cornea to effect cell renewal.

Peripheral to the cornea and limbus, the epithelium changes character as it covers the bulbar and tarsal (under the eyelid)

Scott H. Randell and M. Leslie Fulcher (eds.), *Epithelial Cell Culture Protocols: Second Edition*, Methods in Molecular Biology, vol. 945, DOI 10.1007/978-1-62703-125-7_3, © Springer Science+Business Media, LLC 2012

conjunctiva. This epithelium, while maintaining the same wet-surfaced squamous epithelial phenotype of the cornea, has goblet cells intercalated within it, requires a vascularized connective tissue bed, as compared to the avascular cornea, and expresses a different keratin gene profile.

Culture of the two types of epithelia facilitates study of protective mechanisms of the epithelia, their resistance and response to pathogen invasion, toxicity of cosmetics to the epithelia, and, importantly, transport of pharmaceuticals, such as those used to treat glaucoma, across the epithelium. Culture of human epithelia is of particular importance due to species variation in differentiation markers (for example, keratin and membrane mucin gene expression varies between human and mouse) and relevance to development of human in vitro preclinical models.

The major technical problem encountered in culture of ocular surface epithelia is access to sufficient human tissue for primary culture. Recent efforts to develop immortalized human cell lines that differentiate and show characteristics of the native epithelium have facilitated research. The ensuing chapter describes several methods of primary culture of both corneal and conjunctival epithelia and culture of immortalized cell lines, and methods employed to induce differentiation in the cultured epithelia.

2. Materials

2.1. Tissue

1. Cornea: Donor rims from corneas used for transplant surgery (see Notes 1 and 2).
2. Conjunctiva: Biopsies or discarded conjunctiva obtained by ophthalmic surgeons (see Note 2).

2.2. Culture Media and Supplies

1. Keratinocyte Serum Free Growth Medium (k-SFM, Life Technologies™): Includes a vial of bovine pituitary extract (BPE) and recombinant human epidermal growth factor (EGF); reduce BPE to 25 μg/mL (½ of supplied vial) and EGF to 0.2 ng/mL (100 ng per 500 mL bottle), add additional $CaCl_2$ ($CaCl_2 \cdot 2H_2O$, Sigma-Aldrich®) to a final concentration of 0.4 mM and 1× penicillin–streptomycin (Life Technologies™) for growth to confluence (see Note 3).
2. DMEM/F12 Growth Medium 1: DMEM/F12 with L-glutamine, 15 mM HEPES, 1.0 mM $CaCl_2$ supplemented with 10% newborn calf serum (Hyclone Laboratories), 10 ng/mL human recombinant EGF (Life Technologies™), and 1× penicillin–streptomycin to induce stratification and differentiation (see Note 3).

3. Fetal bovine serum (FBS).
4. 0.05% Trypsin–EDTA (Sigma-Aldrich®) for cell dissociation.
5. DMEM/F12 Growth Medium 2: DMEM/F12 with L-glutamine, 15 mM HEPES, 1.0 mM $CaCl_2$ supplemented with 10% FBS, 10 ng/mL EGF, 5 μg/mL insulin (Sigma-Aldrich®), 24 μg/mL adenine (Sigma-Aldrich®), 0.4 μg/mL hydrocortisone, 0.1 nM cholera toxin (Sigma-Aldrich®), 2 nM Triiodo-L-thyronine (Sigma-Aldrich®), 1× penicillin–streptomycin (Life Technologies™).
6. Bronchial Epithelial Growth Medium (BEGM®) Bullet Kit (Lonza, includes all supplements): With 0.15 mg/mL bovine serum albumin (BSA, Sigma-Aldrich®) and 50 ng/mL amphotericin B.
7. Dispase® II (Roche): Dissolve 10 U/mL of Dispase II in 50:50 Hank's Balanced Salt Solution (HBSS, Life Technologies™): k-SFM and sterile filter.
8. Protease (Sigma-Aldrich®): Dissolve 0.1% Protease (w/v) in DMEM/F12 with 1× penicillin–streptomycin and sterile filter.
9. 3T3 Fibroblasts (CCL-92™, ATCC™) or irradiated 3T3 fibroblasts (Waisman Biomanufacturing).
10. DMEM/F12 Growth Medium 3: DMEM/F12 with L-glutamine, 15 mM HEPES, 1.0 mM $CaCl_2$ supplemented with 10% FBS, 5 μg/mL insulin, 24 μg/mL adenine, 0.4 μg/mL hydrocortisone, 0.1 nM cholera toxin, 2 nM Triiodo-L-thyronine, 1× penicillin–streptomycin.
11. Keratinocyte Growth Medium (KGM): KGM (Thermo Scientific®) supplemented with 10 ng/mL human EGF, 5 μg/mL insulin, 0.5 μg/mL hydrocortisone, 30 μg/mL BPE, 50 μg/mL gentamicin, 50 ng/mL amphotericin B. KGM contains 0.15 mM $CaCl_2$—supplement with additional $CaCl_2$ (Sigma-Aldrich®) to reach 0.9 mM KGM.
12. All-trans-retinoic acid (RA, Sigma-Aldrich®): Dissolve RA in high-purity DMSO and dilute to 100 nM in DMEM/F12 and sterile filter.
13. Phosphate-buffered saline (PBS), Ca^{2+}/Mg^{2+} free.
14. Number ten scalpel blade, sterile forceps, and surgical scissors.
15. 2× Freezing medium: 20% (v/v) high-purity DMSO and 20% (v/v) newborn calf serum in DMEM/F12 medium (see Subheading 2.2, step 5, Note 4).
16. Mr. Frosty (Nalgene®) for slow freezing of cells for cryostorage.
17. Culture flasks/dishes/plates; 1.8 mL cryovials (Corning).
18. Locator Cryobiological Storage System (Thermo Scientific®).

19. BD BioCoat Collagen I, Rat Tail Multiwell Plates (BD Biosciences).
20. Transwell® clear, collagen- or Matrigel™-coated PTFE Membrane Inserts for 6- or 12-well plates (Corning).
21. Bright-Line™ Hemacytometer for cell counting.
22. Water-jacketed CO_2 incubator with humidity tray (see Note 5).
23. Fyrite® Gas Analyzer (Bacharach) (see Note 5).
24. Inverted microscope equipped with phase contrast optics.
25. Transport medium: DMEM/F12 with 50 U/mL penicillin, 50 μg/mL streptomycin, and 2.5 μg/mL amphotericin B.
26. Conjunctival Growth Medium 2: Transport medium (see Subheading 2.2, step 25) additionally supplemented with 10% FBS, 1 μg/mL insulin, 2 ng/mL EGF, 0.1 μg/mL cholera toxin, and 5 μg/mL hydrocortisone.
27. 0.02% (w/v) EDTA: Dissolve in unsupplemented DMEM/F12 and sterile filter.

2.3. Reagents Needed to Prove Epithelial Identity and Differentiation

1. Cytokeratin 3 and 12 antibodies (corneal epithelium; Santa Cruz Biotechnology®).
2. Cytokeratin 4 and 19 antibodies (conjunctival epithelium; Santa Cruz Biotechnology®).
3. Antibodies to the membrane-tethered mucins, MUCs 1, 4, and 16, which are differentiation markers (commercially available from various suppliers, see www.biocompare.com to search and compare available antibodies).
4. ZO-1 antibody (Life Technologies™) to identify tight junctions.
5. Rose bengal dye (Sigma-Aldrich®).

3. Methods

3.1. Primary Culture

Several methods are successfully used for primary culture of cornea and conjunctival epithelia. For primary culture of corneal epithelium, the first method is to explant cultures without feeder layer (2, 3). The second method is to dissociate cell culture with a feeder layer (4).

When working with primary culture of conjunctival epithelium, there are four methods that will be shared. The first is to dissociate cells with feeder layer (5), followed by the second method, which is to explant culture without feeder layer (6). Method 3 of 4 is to explant culture without feeder layer or serum (7). And the final method is to dissociate cells without a feeder layer (8).

3.1.1. Primary Culture of Corneal Epithelium, Method 1 of 2: Explant Cultures Without Feeder Layer

1. Aseptically remove donor rim from vial.
2. Pick up cornea by edge and place it epithelial side up in a sterile 100 mm Petri dish.
3. Hold onto sclera to stabilize the cornea and cut the rim into 6–12 equal pieces with a size ten sterile scalpel blade.
4. Put epithelium side down in Biocoat wells (see Subheading 2.2, step 19), three wells per cornea if a 6-well plate.
5. Gently press each tissue slice lightly with the blunt end of the forceps and let sit uncovered in hood for ~20 min to allow tissue to adhere to plate (see Note 6).
6. Add one drop of k-SFM (see Subheading 2.2, step 1) on top of each tissue piece and incubate overnight at 37°C, 5% CO_2. Fibroblasts will not grow in k-SFM.
7. The next day add 1 mL of k-SFM to each well (day 1).
8. Feed twice weekly with 1 mL of k-SFM per well.
9. On day 5, remove the tissue slices with sterile forceps and feed each well with 3 mL of fresh k-SFM (see Notes 7–9).
10. By day 14, harvest cultures with trypsin–EDTA (see Subheading 2.2, step 4) and expand. Expect $\sim 3 \times 10^5$ cells per cornea (see Notes 10 and 11).
11. Alternatively, isolate the epithelial cells from the limbal region by incubation in Dispase II: 10 U/mL of Dispase II in 50:50 HBSS: k-SFM (see Subheading 2.2, step 7) at 4°C for 24–48 h; remove epithelium by gentle scrape of tissue with a scalpel blade; gently pipette the cells to generate a cell suspension; wash in DMEM/F12 with 10% FBS; centrifuge at low speed; and resuspend in k-SFM. Plate all cells obtained on plastic or collagen-coated 60-mm or 6-well culture plates (3) (see Note 12).

3.1.2. Primary Culture of Corneal Epithelium, Method 2 of 2: Dissociated Cell Culture with Feeder Layer

1. Aseptically remove donor rim from vial.
2. Mince specimens and treat them in a flask with 0.05% trypsin and 0.01% EDTA.
3. Harvest cells at 1 and 2 h and plate them in tissue culture-grade plastic dishes that were previously plated with irradiated 3T3-fibroblasts (see Subheading 2.2, step 9, Note 13).
4. Culture in DMEM/F12 Growth Medium 2 (see Subheading 2.2, step 5).
5. After 8–10 days, rinse cells with unsupplemented DMEM/F12 and briefly incubate in 0.02% EDTA.
6. Vigorously pipette the cell suspension to remove fibroblast feeder cells and any contaminating human fibroblasts.
7. Prepare single cell suspension by incubating in 0.05% trypsin–EDTA (see Subheading 2.2, step 4).
8. Plate cells on tissue culture plastic and feed with k-SFM (see Subheading 2.2, step 1) to expand (see Notes 7–9).

3.1.3. Primary Culture of Conjunctival Epithelium, Method 1 of 4: Dissociated Cells with Feeder Layer

1. Obtain conjunctival biopsies from surgery on patients with undamaged ocular surface.
2. Treat samples with 0.05% trypsin–EDTA at 37°C for 80 min, harvest solution, store at 4°C, and replace with fresh trypsin–EDTA solution every 20 min. Pool collections and centrifuge at 5,000 × *g* for 4 min to collect all cells.
3. Plate cells on lethally irradiated 3T3 cells (see Subheading 2.2, step 9, Note 13) and culture at 37°C, 5% CO_2 in DMEM/F12 Growth Medium 3 (see Subheading 2.2, step 10).
4. Three days after plating, feed cells with above medium with the addition of 10 ng/mL EGF.
5. Feed cells every other day thereafter (see Notes 7, 8, and 14).
6. Passage cells prior to their reaching confluence (see Notes 11 and 15).

3.1.4. Primary Culture of Conjunctival Epithelium, Method 2 of 4: Explant Culture Without Feeder Layer

1. Obtain conjunctival biopsy from surgery on patients with undamaged ocular surface.
2. Transport specimen in DMEM/F12 (see Subheading 2.2, step 25).
3. Carefully remove connective tissue with surgical scissors under a microscope.
4. Place explants epithelial side up in a 35 mm culture dish.
5. Incubate at 37°C, 5% CO_2 in Conjunctival Growth Medium 2 (see Subheading 2.2, step 26).
6. Feed cultures every 2–3 days (see Notes 7, 8, and 15).
7. Passage prior to reaching confluence (see Notes 11 and 15).

3.1.5. Primary Culture of Conjunctival Epithelium, Method 3 of 4: Explant Culture Without Feeder Layer or Serum

1. Two formulations of KGM are used in this protocol (see Note 16).
2. Obtain conjunctival biopsy from surgery on patients with undamaged ocular surface.
3. Separate the conjunctiva from the underlying Tenon's capsule.
4. Dissect the biopsy into 0.5–1 mm pieces (if biopsy is larger than 1 mm). Place pieces in a 35 mm tissue culture dish.
5. Slowly cover with enough 0.9 mM $CaCl_2$ KGM (see Subheading 2.2, step 11, Note 16) to submerge the explants (~500 μL), taking care to prevent the explants from floating off the bottom of the dish.
6. Culture at 37°C, 5% CO_2 and monitor cell outgrowth on inverted microscope.
7. The next day, add stock KGM—with the standard $CaCl_2$ concentration of 0.15 mM (see Subheading 2.2, step 11, Note 16) to completely submerge explants.

8. Feed cells every 2–3 days with fresh KGM (see Notes 7, 8, and 14).
9. Passage cells at 70–80% confluence for future experiments (see Notes 10 and 11).

3.1.6. Primary Culture of Conjunctival Epithelium, Method 4 of 4: Dissociated Cells Without Feeder Layer

1. Obtain conjunctival biopsies from surgery on patients with undamaged ocular surface.
2. Transport specimen in DMEM/F12 (see Subheading 2.2, step 25).
3. Incubate biopsies for 16–20 h at 4°C, in 0.1% protease in DMEM/F12 with penicillin–streptomycin (see Subheading 2.2, step 8).
4. Scrape the loosened cells with a pipette, wash 3× with DMEM/F12, and resuspend in DMEM/F12 with 10% FBS and penicillin–streptomycin.
5. Preplate cells on 60 mm tissue culture dishes for 1 h at 37°C, 5% CO_2 to eliminate fibroblasts by differential attachment.
6. After 1 h, plate suspended epithelial cells at 3×10^4 cells per 60 mm tissue culture dish in fully supplemented BEGM (see Subheading 2.2, step 6) to which 50 ng/mL amphotericin B and 0.15 mg/mL BSA have been added.
7. Feed 1 day after plating, followed by every other day feeding regimen (see Notes 7, 8, and 14).
8. Passage cells at 60–70% confluence for future experiments (see Notes 10 and 11).
9. Grow passaged cells in a mixture of 1 part supplemented BEGM:1 part DMEM/F12, reducing the EGF supplement in BEGM to 0.5 ng/mL EGF.

3.1.7. Determine Cell Origin of Corneal and Conjunctival Cells

1. Cytokeratin expression patterns, as determined by immunohistochemistry, western blot, and/or polymerase chain reaction (PCR) of cell cultures, should match those of the native tissue (see Subheading 3.3).
2. Expression of differentiation markers, such as MUC1, -4, -16, or -5AC, by immunohistochemistry, western blot, and/or PCR should match those of native tissue (9).

3.2. Culture of Immortalized Cell Lines

Immortalized human corneal and conjunctival epithelial cells have been successfully cultured using a two-step process: growth to confluence in low-calcium, serum-free medium, followed by induction of stratification using high-calcium, serum-containing medium (9).

3.2.1. Growth to Confluence (Proliferation Permissive)

1. Plate cells on tissue culture plastic in k-SFM (see Subheading 2.2, step 1), supplemented with $CaCl_2$ to 0.4 mM (low calcium is proliferation permissive), 0.2 ng/mL EGF, 1/2 supplied bovine pituitary extract (25 μg/mL), and 1× penicillin–streptomycin.

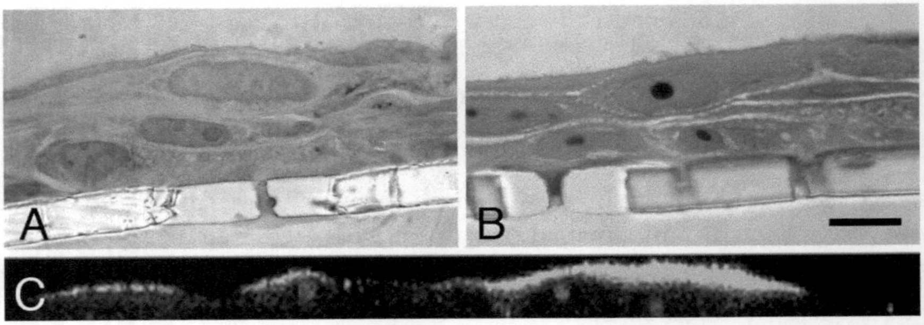

Fig. 1. Micrographs of immortalized human corneal (**a** and **c**) and conjunctival (**b**) epithelial cells cultured on collagen-coated Transwell inserts to induce stratification. The cells differentiate to express the apical surface membrane mucin MUC16 (shown by immunofluorescence in **c**).

2. Feed cells every 2 days with k-SFM and grow to ~1/2 confluence, if you plan to subculture cells for experimentation or to freeze down a stock of cells (see Subheading 3.4). It usually takes about 1 week to reach this stage, but need to check cells regularly to be sure not to overgrow. If you plan to induce stratification, grow the cells to confluence. As cell density reaches and exceeds 1/2 confluence, the nutrients in k-SFM become a limiting factor, so it may be necessary to feed more frequently (see Notes 7–9 and 14).

3.2.2. Stratification Induction

1. At confluence, switch cells to DMEM/F12 growth Medium 1 (see Subheading 2.2, step 2). The switch to high calcium (1.0 mM)-containing medium promotes stratification and differentiation (9, 10).
2. Feed cells every 2 days for 7 days.

3.3. Induction of Differentiation

3.3.1. Differentiation Markers

1. Cytokeratins: The cytokeratin pairs CK3/CK12 are specific for corneal epithelium, and CK4/CK19 for conjunctival epithelium (7).
2. Stratification: The ocular surface in vivo is a stratified, nonkeratinizing epithelium; therefore, achieving true stratification in vitro is one hallmark of a differentiated culture (see Fig. 1a or b).
3. Mucins: MUCs 1, 4, and 16 are markers of differentiated corneal and conjunctival epithelium, as they are expressed by apical cells of stratified epithelium; MUC5AC is specific for the goblet cells of conjunctival epithelium (11). Figure 1c illustrates the apical localization of the membrane mucin MUC16.
4. Tight junctions: ZO-1 antibody is used to help identify tight junctions which play a role in the barrier function of the ocular surface, and their presence in the apical layer of stratified ocular surface epithelia is characteristic of a mature ocular surface epithelium (12, 13).

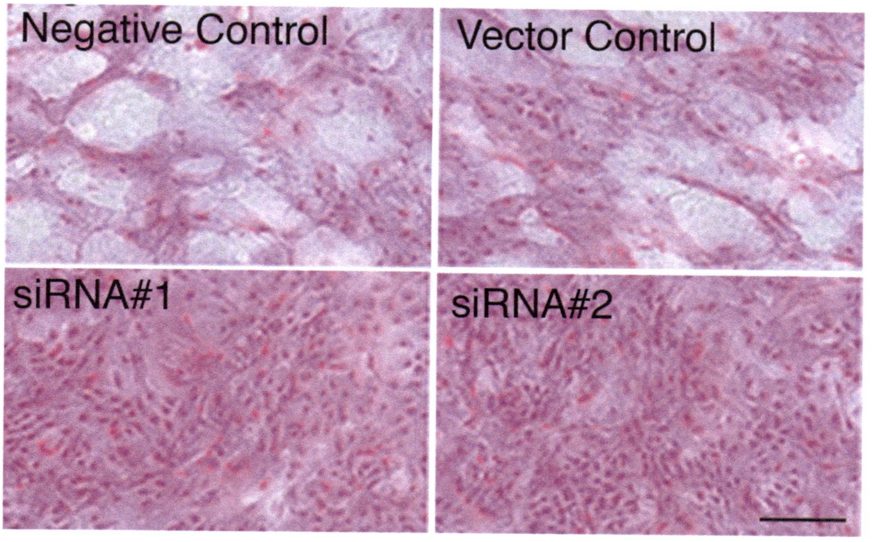

Fig. 2. Example of use of cultured, differentiated human corneal epithelial (HCLE) cells to determine functions of the membrane mucin MUC16 in barrier function. Micrographs demonstrate rose bengal dye penetrance into cultured stratified HCLE cells with or without knockdown of MUC16, using siRNA methods. The nontransfected (negative) control and vector-only control show islands of cells that exclude rose bengal dye, whereas cells with MUC16 knockdown, either siRNA sequence v1 or 2, do not exclude the dye. These data demonstrate that the membrane-tethered mucin MUC16 is involved in barrier function. Original data in Blalock et al. (15). Bar = 50 mm.

5. Rose bengal dye exclusion: Exclusion of the anionic dye rose bengal is also a characteristic of intact ocular surface epithelium; the membrane-associated mucins (MUCs 1, 4, 16) play a role in establishing this aspect of the barrier function of the ocular surface (14–16). Figure 2 illustrates the use of rose bengal dye exclusion for study of the role of the membrane mucin MUC16 in the barrier function of the ocular surface.

3.3.2. Airlift Culture

1. Plate 5×10^4 cells/mm^2 cells on 12-mm diameter Transwell® clear or collagen-coated inserts in 12-well plates (see Subheading 2.2, step 20) and grown in k-SFM (see Subheading 2.2, step 1) to confluence as a submerged culture.
2. Next, remove medium from the inserts, and feed the cultures from below to allow for culture of the epithelial surface at an air–liquid interface (17). This is especially helpful in inducing stratification (17) and with airway epithelia to obtain a normal phenotype (18).

3.3.3. Transwell® Culture

1. Transwell® inserts are available with different pore sizes (0.4 or 3.0 µm) and the option of being uncoated ("clear"), or collagen or Matrigel™ coated.
2. Culture using Transwell® inserts provides a means to coculture 2 cell types without their being in physical contact (9), and to culture at an air–liquid interface by feeding the cells from

below (17). They are useful for studies of transepithelial barrier function (18, 19).

3.3.4. Serum Culture and Stratification

1. Addition of serum in combination with increasing the concentration of calcium in the medium induces stratification of ocular surface epithelia (9).
2. Culture in the presence of serum is necessary for expression of the membrane-associated mucin MUC4, a marker of a differentiated culture (9).
3. Some degree of stratification is also seen when cells are cultured in the presence of retinoic acid, which is normally present in serum, and is required for maintenance of the wet-surfaced phenotype of the ocular surface (20).

3.3.5. Coculture with Fibroblasts and Neurons

1. Plate cells in Transwell® inserts with either 3T3 fibroblasts or sensory neurons from the trigeminal ganglion in the lower chamber to study the effect of factors produced by these cell types on the epithelial cells (9, 21).
2. Alternatively, plate cells in the same chamber with neurons to examine the effect of direct contact of 2 cell types as occurs in vivo (22).

3.4. Harvesting and Storage of Cells

1. Passage cells by aspirating medium from culture vessel, add 0.08 mL of trypsin/EDTA solution (trypsin concentration 0.05–0.2%, EDTA concentration 0.01–0.02% in PBS per cm^2 growth area to flask (see Notes 10, 11, and 15)), and incubate at 37°C for 5–12 min.
2. Check culture using an inverted microscope; cells will round up and detach spontaneously (see Note 11).
3. After detachment, rapidly add an equal volume of serum-containing medium to culture vessel to neutralize trypsin activity.
4. Pipette cells up by gently washing the surface of culture vessel and transfer to a centrifuge tube.
5. Spin cells down at $5{,}000 \times g$ for 4 min, saving an aliquot of the cells to count using a hemacytometer.
6. Aspirate off medium and resuspend in appropriate volume of culture medium (volume depends on cell density and what one wants to do with cells).
7. Prepare cell stocks by plating 10^4, 3×10^4, and 10^5 cells per 75 cm^2 flasks.
8. Feed cultures every 2 days and use as needed for experiments. Exponentially growing cultures 5–8 days old are used as the source of cells for the next passage (see Notes 7–9 and 14).
9. It is recommended that cells be cryopreserved for future use; add an equal volume of room-temperature 2× freezing medium (see Subheading 2.2, step 15) to cells already suspended to

appropriate density in culture medium. Pipette up and down to mix (see Notes 17 and 18).

10. Promptly distribute to cryovials, slow-freeze in Mr. Frosty (~1–3°C/min cooling rate) to <–50°C, and put on BOTTOM shelf of –80°C freezer; freezes to the lowest temperature in ~1.5 h; can leave overnight (see Subheading 2.2, step 16).
11. Store in liquid N_2 cryobiological storage system (see Subheading 2.2, step 18).

4. Notes

1. Limbal epithelial cells have the best proliferative capacity; central corneal epithelial cells senesce after 1 or 2 passages (4).
2. Specimens obtained from younger donors have better proliferative capacity than those obtained from older donors.
3. All media is light sensitive and should be protected from light during storage.
4. Add high-purity DMSO to chilled, 20% calf-serum-supplemented DMEM/F12 medium so that the final DMSO concentration is 20% (v/v). Mix well and let stand for 2 h or overnight at 4°C. Filter-sterilize with a 0.2 μm filter. Aliquot into plastic tubes. Place in a –80°C freezer to freeze solid, after which tubes can be stored at –20°C, if desired.
5. A reliable water-jacketed CO_2 incubator is imperative for good cell culture. Before first use, autoclave trays, and pan and calibrate temperature and CO_2 concentration. Verify temperature at 37°C and 5% CO_2 concentration (using a Bacharach Fyrite® gas analyzer) on a regular basis. Maintain the water level in stainless steel humidification tray with fresh water and clean with ethanol on a regular basis.
6. When preparing explant cultures, it is critical to gently press each tissue slice down onto the bottom of the culture dish and let it sit uncovered in the hood for ~20 min to adhere tissue to plastic to allow growth of epithelial cells out onto plastic culture dish. Tissue slice will float if full volume of medium is added immediately.
7. A healthy, proliferative cell population should contain small cells that have a population doubling time of ~24–40 h.
8. Medium to large epithelial cells are post-mitotic, and many of these will express the terminal differentiation protein involucrin (23, 24).
9. Sister cells of growing colonies of corneal epithelia remain adherent to one another in k-SFM, forming discrete colonies.

10. No need to rinse culture prior to trypsinization if grown in serum-free, low-Ca^{2+} medium.
11. Epithelial cells are fragile during trypsinization; allow the cells to round up and lift off the surface on their own. Do not hit the bottom of the flask or attempt to spray the cells off using a stream of fluid as this will break the cells.
12. The use of plastic or collagen-coated plates is dependent on experimental goals. Collagen-coated plates provide a substrate closer to the native condition and, therefore, may be useful for inducing differentiation of the cells to a more native state.
13. Cells are irradiated with 6,000 rad from a 60 Co source, either as a trypsinized cell suspension or already plated in the culture flask. Alternatively, irradiated 3T3 feeder cells may be purchased from Waisman Biomanufacturing.
14. Sister cells of growing colonies of conjunctival epithelia separate from each other and grow as single cells until they start to touch as the culture approaches confluence. Conjunctival epithelia are very sensitive to CO_2 concentration and will not grow well if CO_2 levels vary from 5% by ±2–3% (see Note 5).
15. If grown in serum-containing medium, rinse cells with sterile Ca^{2+}, Mg^{2+}-free PBS or trypsin–EDTA prior to incubation with trypsin–EDTA.
16. Two formulations of KGM are used in the protocol in Subheading 3.1.5 (see Subheading 2.2, step 11) (7). For initial incubation, use KGM supplemented with 10 ng/mL human EGF, 5 μg/mL insulin, 0.5 μg/mL hydrocortisone, 30 μg/mL BPE, 50 μg/mL gentamicin, 50 ng/mL amphotericin B, and 0.9 mM $CaCl_2$. After initial outgrowth of cells from the explant (day 2 of incubation), reduce the calcium concentration in the above supplemented KGM to 0.15 mM to promote cell proliferation.
17. It is good cell culture practice to produce a number of vials of cryopreserved cells as soon as possible after obtaining a cell line. This provides a uniform stock for use in future experiments.
18. Cells can be frozen at any concentration, from ten cells to 3×10^6 cells/mL; typical freezing density is ~10^6 cells per vial. Be sure to label tubes with cell name, passage number, and concentration before fill and freeze.

Acknowledgment

The authors thank Ann S. Tisdale for preparation of the figures.

References

1. Gipson IK (2007) The ocular surface: the challenge to enable and protect vision: the Friedenwald lecture Invest Ophthalmol Vis Sci 48:4390–4398
2. Kahn CR, E Young, IH Lee, and JS Rhim (1993) Human corneal epithelial primary cultures and cell lines with extended life span: in vitro model for ocular studies Invest Ophthalmol Vis Sci 34:3429–3441
3. Offord EA, NA Sharif, K Mace, Y Tromvoukis, EA Spillare, O Avanti, WE Howe, and AM Pfeifer (1999) Immortalized human corneal epithelial cells for ocular toxicity and inflammation studies Invest Ophthalmol Vis Sci 40:1091–1101
4. Lindberg K, ME Brown, HV Chaves, KR Kenyon, and JG Rheinwald (1993) In vitro propagation of human ocular surface epithelial cells for transplantation Invest Ophthalmol Vis Sci 34:2672–2679
5. Pellegrini G, O Golisano, P Paterna, A Lambiase, S Bonini, P Rama, and M De Luca (1999) Location and clonal analysis of stem cells and their differentiated progeny in the human ocular surface J Cell Biol 145:769–782
6. Diebold Y, M Calonge, A Enriquez de Salamanca, S Callejo, RM Corrales, V Saez, KF Siemasko, and ME Stern (2003) Characterization of a spontaneously immortalized cell line (IOBA-NHC) from normal human conjunctiva Invest Ophthalmol Vis Sci 44:4263–4274
7. Ang LP, DT Tan, TT Phan, J Li, R Beuerman, and RM Lavker (2004) The in vitro and in vivo proliferative capacity of serum-free cultivated human conjunctival epithelial cells Curr Eye Res 28:307–317
8. Chung SH, JH Lee, JH Yoon, HK Lee, and KY Seo (2007) Multi-layered culture of primary human conjunctival epithelial cells producing MUC5AC Exp Eye Res 85:226–233
9. Gipson IK, S Spurr-Michaud, P Argueso, A Tisdale, TF Ng, and CL Russo (2003) Mucin gene expression in immortalized human corneal-limbal and conjunctival epithelial cell lines Invest Ophthalmol Vis Sci 44: 2496–2506
10. Boyce ST, and RG Ham (1983) Calcium-regulated differentiation of normal human epidermal keratinocytes in chemically defined clonal culture and serum-free serial culture J Invest Dermatol 81:33s–40s
11. Gipson IK (2004) Distribution of mucins at the ocular surface Exp Eye Res 78:379–388
12. Ban Y, A Dota, LJ Cooper, NJ Fullwood, T Nakamura, M Tsuzuki, C Mochida, and S Kinoshita (2003) Tight junction-related protein expression and distribution in human corneal epithelium Exp Eye Res 76:663–669
13. Sugrue SP, and JD Zieske (1997) ZO1 in corneal epithelium: association to the zonula occludens and adherens junctions Exp Eye Res 64:11–20
14. Argueso P, A Tisdale, S Spurr-Michaud, M Sumiyoshi, and IK Gipson (2006) Mucin characteristics of human corneal-limbal epithelial cells that exclude the rose bengal anionic dye Invest Ophthalmol Vis Sci 47:113–119
15. Blalock T, S Spurr-Michaud, A Tisdale, S Heimer, M Gilmore, V Ramesh, and I Gipson (2007) Functions of MUC16 in corneal epithelial cells. Invest Ophthalmol Vis Sci 48: 4509–4518
16. Blalock TD, SJ Spurr-Michaud, AS Tisdale, and IK Gipson (2008) Release of membrane-associated mucins from ocular surface epithelia Invest Ophthalmol Vis Sci 49:1864–1871
17. Robertson DM, L Li, S Fisher, VP Pearce, JW Shay, WE Wright, HD Cavanagh, and JV Jester (2005) Characterization of growth and differentiation in a telomerase-immortalized human corneal epithelial cell line Invest Ophthalmol Vis Sci 46:470–478
18. Fulcher M, S Gabriel, K Burns, J Yankaskas, and S Randell. (2004) Well-differentiated human airway epithelial cell cultures. In: Picot, J. (ed). Methods in Molecular Medicine. Human Cell Culture Protocols, Second Ed., Humana Press, Inc., Totowa, NJ; 183–206
19. Yi X, Y Wang, and FS Yu (2000) Corneal epithelial tight junctions and their response to lipopolysaccharide challenge Invest Ophthalmol Vis Sci 41:4093–4100
20. Hori Y, SJ Spurr-Michaud, CL Russo, P Argueso, and IK Gipson (2005) Effect of retinoic acid on gene expression in human conjunctival epithelium: secretory phospholipase A2 mediates retinoic acid induction of MUC16 Invest Ophthalmol Vis Sci 46:4050–4061
21. Garcia-Hirschfeld J, LG Lopez-Briones, and C Belmonte (1994) Neurotrophic influences on corneal epithelial cells Exp Eye Res 59:597–605
22. Suuronen EJ, M Nakamura, MA Watsky, PK Stys, LJ Muller, R Munger, N Shinozaki, and M Griffith (2004) Innervated human corneal equivalents as in vitro models for nerve-target cell interactions FASEB J 18:170–172
23. Banks-Schlegel S, and H Green (1981) Involucrin synthesis and tissue assembly by keratinocytes in natural and cultured human epithelia J Cell Biol 90:732–737
24. Watt FM (1983) Involucrin and other markers of keratinocyte terminal differentiation J Invest Dermatol 81:100s–103s

Chapter 4

The Culture and Maintenance of Functional Retinal Pigment Epithelial Monolayers from Adult Human Eye

Timothy A. Blenkinsop, Enrique Salero, Jeffrey H. Stern, and Sally Temple

Abstract

The retinal pigment epithelium (RPE) is implicated in many eye diseases, including age-related macular degeneration, and therefore isolating and culturing these cells from recently deceased adult human donors is the ideal source for disease studies. Adult RPE could also be used as a cell source for transplantation therapy for RPE degenerative disease, likely requiring first in vitro expansion of the cells obtained from a patient. Previous protocols have successfully extracted RPE from adult donors; however improvements in yield, cell survival, and functionality are needed. We describe here a protocol optimized for adult human tissue that yields expanded cultures of RPE with morphological, phenotypic, and functional characteristics similar to freshly isolated RPE. These cells can be expanded and cultured for several months without senescence, gross cell death, or undergoing morphological changes. The protocol takes around a month to obtain functional RPE monolayers with accurate morphological characteristics and normal protein expression, as shown through immunohistochemistry analysis, RNA expression profiles via quantitative PCR (qPCR), and transepithelial resistance (TER) measurements. Included in this chapter are steps used to extract RPE from human adult globes, cell culture, cell splitting, cell bleaching, immunohistochemistry, and qPCR for RPE markers, and TER measurements as functional test.

Key words: Human adult retinal pigment epithelium, Epithelial mesenchymal transition, RPE culture, RPE expansion, Characterization of RPE, Age-related macular degeneration

1. Introduction

The retinal pigment epithelium (RPE) is involved in many eye diseases, including age-related macular degeneration (AMD), the leading cause of blindness for those 55 years of age and older. To study the role of RPE in disease, an efficient protocol for dissection, isolation, and culturing of these cells is highly valuable. Much work has been done to culture RPE cells with varying success. Recently optimized culture protocols recognize the difficulties in growing adult RPE in culture and therefore focus on human fetal

Scott H. Randell and M. Leslie Fulcher (eds.), *Epithelial Cell Culture Protocols: Second Edition*, Methods in Molecular Biology, vol. 945, DOI 10.1007/978-1-62703-125-7_4, © Springer Science+Business Media, LLC 2012

RPE as their source (1–4). However, to directly study RPE in its most clinically relevant form, adult RPE is the gold standard. The main obstacle to studying RPE from adult humans is that RPE cells have a high probability of changing morphology when cultured in vitro. In the past, the capacity of RPE to change during culture has been given various names, including in vitro aging (5), dedifferentiation (6), and transdifferentiation (7). New evidence suggests that the change is due to the RPE undergoing an epithelial to mesenchymal transition (EMT) (8–16). EMT is a process by which epithelial cells lose their epithelial characteristics and acquire a mesenchymal like phenotype (17). This phenotypic change, the basis of which is still largely unknown, often begins with the loss of cadherin-mediated cell–cell adhesion and results in a cascade of changes that deconstruct the epithelial state, preventing its reformation. Here we describe a protocol that takes into consideration the burgeoning understanding of RPE physiology in vitro, minimizes the factors that lead to EMT, and increases the likelihood of maintaining the in vivo morphological RPE state.

In the eye, the RPE performs numerous roles that support surrounding tissues, which include blood–retina barrier (18, 19), light absorption and photooxidation protection (20), reisomerization of 11-*cis* retinal (21, 22), phagocytosis of photoreceptor outer segments (ROS) (23), and control of volume composition in the subretinal space (see Fig. 1) (24–27). Loss of normal RPE function can result in the photoreceptors being unable to transmit light signals, thereby leading to vision loss (28). When cultured, RPE should mimic these essential features as much as possible in order to be useful for translational research. However, once cultured, the adult human RPE cells typically lose their hexagonal morphology along with other key functions over time. As seen with other epithelial cells, when RPE cells are extracted from donors and stimulated to proliferate in vitro, an array of morphologies is often observed, including hexagonal/cobblestone, fusiform/fibroblastic, elongated, polygonal, and epithelioid shapes (29). Cultured RPE cells that change to fusiform morphology lose their ability to phagocytose ROS (30). As RPE cells divide, their pigment becomes diffuse and often is released from cultured cells over time (31, 32). Associated with assumption of the fusiform morphology is loss of adherens junction-related proteins, known to be essential for actin filament-based epithelial phenotype (33–35).

Much work has been done to optimize human fetal RPE (hfRPE) culture, since these cells seem to more easily regain epithelial morphology after in vitro culture and passage (1–4, 36–41). As good as these protocols are for hfRPE cell culture, they do not translate to the adult human RPE. In comparison to hfRPE, adult human RPE seems less able to undergo the cytoskeletal and cell–cell contact changes that occur during proliferation without being irreversibly altered, thereby resulting in effectively permanent EMT.

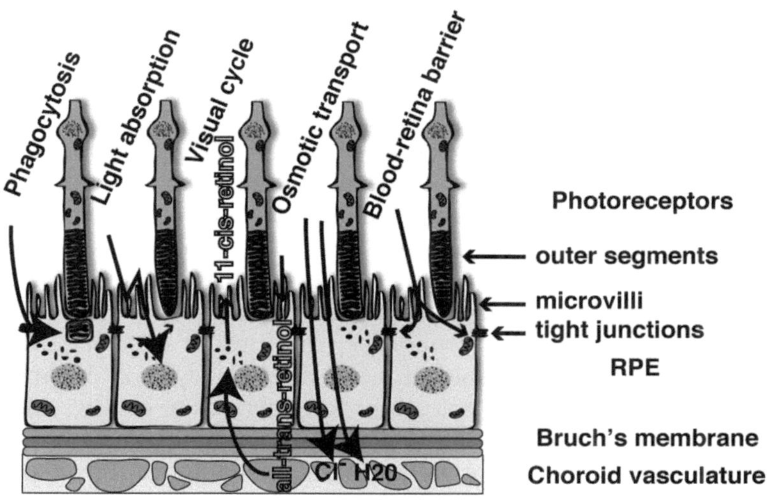

Fig. 1. Diagram depicting the functional relationships between the RPE, photoreceptors, and choroidal vasculature. Disruption of any of these relationships leads to dysfunction of photoreceptor light signal transmission and loss of vision. Among the diseases associated with photoreceptor-pigment epithelial malfunction is age-related macular degeneration, a leading cause of blindness in the elderly.

Prior studies have made some progress towards obtaining cultured adult human RPE cells that maintain in vivo morphology over long-term culture (42–44). Calcium has been found to be very important for proper cell–cell junctions. For example, McKay and Burke (1994) show that when RPE cells are extracted from the adult eye and incubated in high calcium (1.8 mM) with gentle agitation, calcium-dependant cell–cell contact improves and a more epithelioid physiology results (44). More recently, Rak and McKay (2006) developed a technique using adult pig RPE, in which they can trigger a change from fusiform to epithelioid morphology by priming the RPE with low-calcium medium and then switching to a high-calcium medium (42). These data clearly show that calcium-dependent cell–cell contact is required by the cultured RPE to recreate an epithelium.

Since EMT by RPE is likely to be involved in proliferative vitreoretinopathy, some attempts have been made from a clinical perspective to prevent RPE from making this change. For example, blocking the Smad3 pathway reduces the likelihood of RPE from undergoing EMT, as shown by the fact that Smad3 null mice have reduced incidence of PVR and EMT marker expression after retinal detachment (10). Targeting the TGF-β pathway has also been shown to reduce the expression of the mesenchymal marker αSMA in cultured porcine RPE (15). Lithium has been recognized to block the accumulation of αSMA in lens epithelium, preserving polarized cobblestone morphology (45). And finally, a hyperactive

TGF-β pathway resulted in the loss of epithelial marker E-cadherin and hypermethylation of its promoter region in a human breast cancer cell line (46). Conversely, disrupting TGF-β signaling resulted in demethylation of E-cadherin gene promoter region, re-expression of the protein, reacquisition of epithelial morphology, and reduced migration in wound healing assay (46). Future attempts to optimize RPE culture protocols will likely include suppression of the signaling pathways involved in EMT.

1.1. Method Outline

Our approach to obtaining RPE with characteristic in vivo morphology relies on our observation that pieces of RPE sheets isolated intact from the eye will maintain their cobblestone morphology for long periods of culture. Moreover, the cells included in these fragments of epithelium can proliferate and grow, albeit slowly, generating daughter cells that appear also of epithelial morphology. Thus we have found that maintaining cell-to-cell contacts between neighboring RPE cells results in a "community effect" that can prevent them from undergoing the changes that lead to EMT. To this end, a primary focus of our culture strategy is to extract RPE in a way that preserves the junctional bonds resulting in fragments of the epithelial sheet. This strategy is incorporated into many facets of the culturing method, from the type of dissection to the length and temperature of enzyme incubation. We have optimized the protocol so that after a month of culture, we routinely obtain confluent monolayers of hexagonal, cobblestone RPE.

1.2. Eye Dissection

During dissection, the anterior chamber is removed, along with the vitreous and retina and the eyecup is maintained as a hemisphere during enzymatic digestion (see Fig. 2). Maintaining the eyecup preserves an unbroken Bruch's membrane and reduces the risk of contaminating the RPE with choroidal cells during this procedure. We pay careful attention not to scrape the RPE/choroid membrane to prevent pigmented mesenchymal cells from spilling into the inner side of the Bruch's membrane. These pigmented mesenchymal cells look very similar (albeit smaller in size) to RPE and, if present as a contaminant, will proliferate more rapidly and contribute to the culture. These pigmented cells with non-epithelial morphology could well contribute to the heterogeneous morphologies identified in RPE cultures by others.

We have taken particular care to choose which instrument to use when scraping the RPE off the Bruch's membrane. Since we are removing the RPE cells while they are still in the eyecup, we need an instrument that is rounded so as not to poke through the Bruch's membrane. The spoon blade is angled and double beveled so that it easily removes sheets of RPE without threatening the integrity of the Bruch's membrane. Additionally, of all the enzyme digestions we have surveyed, we have found that dispase is

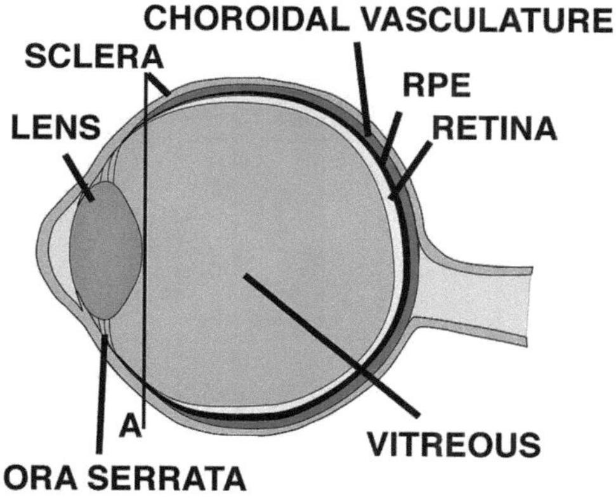

Fig. 2. Diagram of the human eye. During dissection a scalpel is used to make an incision below the ora serrata of the eye as designated by *line A*. The anterior structures—cornea, lens, and iris—are removed. The vitreous and retina are peeled off the underlying RPE/choroid layers.

preferable, since it breaks the RPE cell–basement membrane associations while leaving cell–cell junctions intact. A major concern when extracting RPE is that it will begin to secrete extracellular matrix (ECM) if too much shear stress is applied to cell body. For this reason we conducted an array of tests varying temperature, dispase exposure time, and agitation during the enzyme digestion, and thus we have identified an optimal treatment for easy RPE removal. Finally, we try to gather RPE sheet fragments that are similar in size and shape so that we have better control of sheet fragment spacing when plating. When done correctly, the wells will grow into an epithelial monolayer after 1 month in culture (see Fig. 3). Included are protocols used during RPE cell extraction, culture, and confirmation of functionality of the resulting RPE epithelial monolayers.

1.3. Separating RPE Sheets Out from Single Cells Using a Sucrose Gradient

Due to the lack of cell–cell contacts, individual cells have an increased probability of undergoing EMT. Therefore, efforts to reduce the percentage of single cells while isolating RPE sheet cultures improve the probability of culturing homogenous epithelial cell cultures. Single cell/sheet separation can be conducted by using a sucrose gradient, which allows cell fractionation according to size (44).

1.4. Bleaching RPE Monolayers for Immunostaining

To characterize the RPE monolayers, immunostaining with RPE-specific antibodies is performed (47). Before proceeding with the staining, it is necessary to bleach the monolayers as the pigment in RPE cells autofluoresces. Our lab developed a protocol whereby pigmented cells can be bleached, thereby minimizing autofluorescence when using fluorescent antibodies.

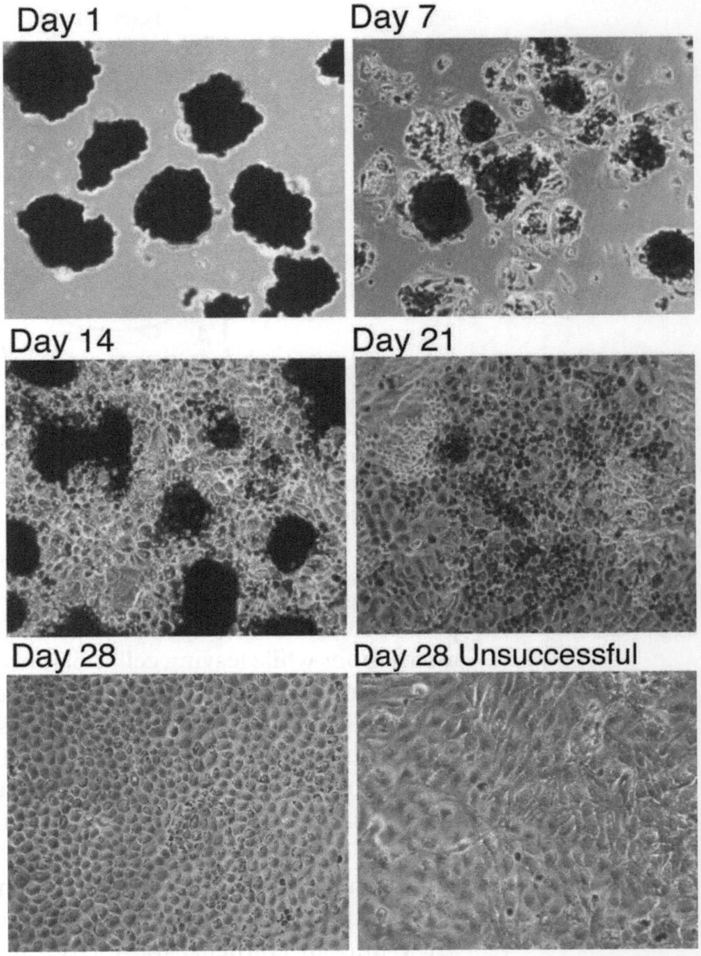

Fig. 3. RPE culture timeline. Light microscope images of RPE taken during the culture period. Please note that representative examples were chosen for this sequence of pictures, which do not represent the same exact field. Day 1 illustrates sheets recently attached. Often, after dissection from the eye, RPE sheets curl up before laying flat. On Day 7 some cells break down cell–cell contact and migrate away from sheets. Day 14 cell cultures become confluent, with some cells maintaining hexagonal shape whereas others becoming rounded. By Day 21 the cells have spaced out and are becoming more homogenous morphologically. Around Day 28 expect the RPE monolayer to become much more uniform and hexagonal, mimicking the in vivo cobblestone appearance. An example of an unsuccessful functional epithelium shows cells with more fusiform morphology.

1.5. Characterization of Human RPE Monolayers Using Immunohistochemistry

One of the best ways to determine whether functional RPE cells have been successfully cultured is by staining for proteins integral to RPE function (see Fig. 4). MITF is a transcription factor that is essential for pigment cell survival and regulates genes involved in pigmentation (48). OCCLUDIN is a functional component of tight junctions integral to RPE function as a blood–retina barrier (49). RPE65 is a protein involved in the production of 11-*cis* retinal of the visual cycle (50). ZO-1 is a protein that connects tight junctions to the

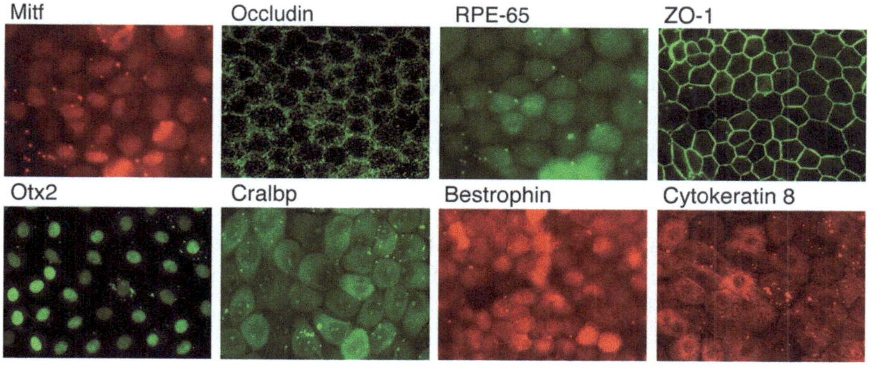

Fig. 4. Immunostaining of confluent RPE cultures after 1 month. Immunostaining shows expression of markers unique to the RPE. *Punctate spots* are unquenched autofluorescence.

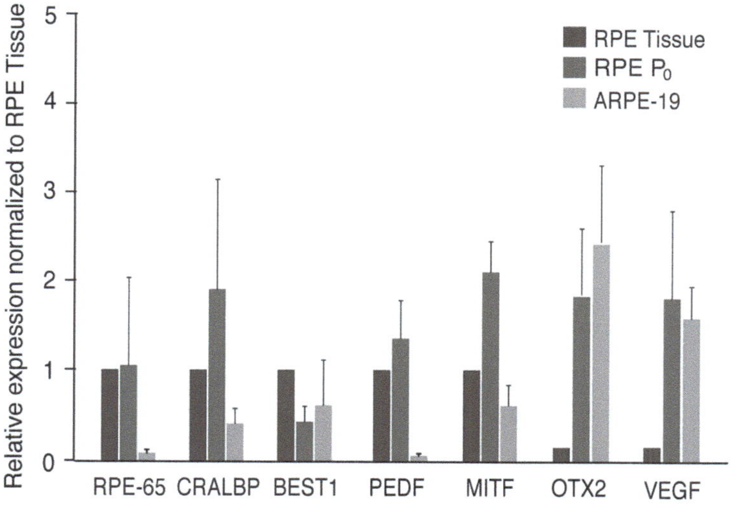

Fig. 5. Gene expression comparison between cultured RPE, freshly isolated RPE, and immortalized RPE cell line ARPE-19. Long-term RPE cultures express RPE genes, at a level similar to freshly isolated RPE. Measurements were normalized to expression level of freshly isolated RPE tissue.

cytoskeleton (51). OTX2 is a transcription factor that activates the specific genetic expression of RPE (52). CRALBP is involved in the reisomerization of 11-*cis*-retinal in the visual cycle (53). BESTROPHIN is a Ca^{2+}-activated Cl^- channel involved in osmotic transport (54). CYTOKERATIN 8 is an intermediate filament protein specifically expressed in epithelium and is expressed in RPE (55).

1.6. Gene Expression Profiling of RPE Monolayer Cultures

Real-time (quantitative) PCR provides a way to compare RNA transcript expression between various conditions. We found that long-term cultured RPE cells express markers similarly to RPE freshly isolated from the eye, as shown by comparing gene expression between RPE tissue and primary RPE cultures by qPCR (see Fig. 5). Since immortalized ARPE-19 cells are unable to generate a functional epithelial monolayer, their confluent cell cultures were used as a negative control of the primary cultured RPE (56). RNA isolation was performed by following the protocol included in the RNAeasy plus Mini Kit (Qiagen). The relative differences (*n*-fold) in gene expression of a characteristic marker were determined by using cyclophilin G as an internal control (housekeeping gene) for each reaction set. Data is represented as the mean relative change ± SEM of the *n* separate qPCR amplification reactions. The primer sequences, annealing temperatures, and product size are shown in Table 1.

1.7. Transepithelial Resistance

Functional epithelial monolayers are characterized by a high transepithelial resistance (TER), which depends on the formation of tight junctions between cells. Only after confluent epithelial cells form tight junctions does the resistance of current passing from one side of a Transwell® to the other become high. A functional epithelial monolayer possesses a resistance above ~200 $\Omega\,cm^2$. To measure TER, freshly isolated RPE sheets were plated in Matrigel™-treated Transwells®. Resistance was measured using the EVOM² every 5 days for 30 days (see Fig. 6).

2. Materials

2.1. Dissection

1. Adult human donor eyes were obtained from the Eye-Bank for Sight Restoration, Inc., New York, NY, and the National Disease Research Interchange, Philadelphia, PA.
2. Scalpel, #11 style.
3. Scissors, bent, pointed, 25.4 mm.
4. Microscissors, straight tips, long handle 19.05 mm.
5. Forceps, Style #5, 11.2 cm (Miltex®).
6. Disposable inoculating loop.
7. Sterile gauze sponges.
8. Tissue culture plates, 24 well.
9. Transwell® culture plates, 24 well (Corning).
10. Conical tube, 15 mL.
11. Conical tube, 50 mL.
12. Tissue culture-treated dishes, 10 cm.
13. Microsurgical angled, double beveled spoon blade, 3.0 mm (Sharpoint®).
14. Matrigel™ Basement Membrane Matrix (BD Biosciences).
15. Dispase, 1 mg/mL (Sigma-Aldrich®): Dissolve the sterile lyophilized enzyme in sterile double-distilled water for a stock

Table 1
List of primers used for real-time PCR on adult human RPE

Human gene	Forward 5′–3′	Reverse 3′–5′	Product size (bp)	T_{ann} (°C)	Gene Bank ID
RPE65	TACAGAAAGCACTGAGTTGAGC	CCATTTAGTAAGTCCACATTCATTTCC	153	55	NM_000329
RLBP1 (CRALBP)	CAAAGCCATCCACTTCATCCACCA	AAGTCAGAGGGCAGGATGTTCTCA	161	60	NM_000326
BEST1 (BESTROPHIN)	CCTTTATGGGCTCCACCTTCAACATC	CAGTAGTTTGGTCCTTGAGTTTGCCC	166	65	NM_004183
PEDF (SERPINF1)	TTATGAAGGCGAAGTCACCAAGTCCC	CATCCTCGTTCCACTCAAAGCCA	142	60	NM_002615
VEGFA	TTCGTCCAACTTCTGGGCTGTTCT	TCTCCTCTTCCTTCTCTTCTTCCTCCTC	180	60	NM_003376
OTX2	CCATGACCTATACTCAGGCTTCAGG	GAAGCTCCATATCCCTGGGTGGAAAG	211	62	NM_021728
MITF	TTGTCCATCTGCCTCTGAGTAG	CCTATGTATGACCAGGTTGCTTG	87	55	NM_198178
ZO-1 (TJP1)	CCA GAA TCT CGG AA A AGT GC	ACC GTG TAA TGG CAG A CT CC	396	52	NM_003257
GAPDH	CCCCTTCATTGACCTCAACTACA	TTGCTGATGATCTTGAGGCTGT	342	56	NM_002046
CYCLOPHILLIN G	CTTGTCAATGGCCAACAGAGG	GCCCATCTAAATGAGGAGTTGGT	82	60	NM_004792

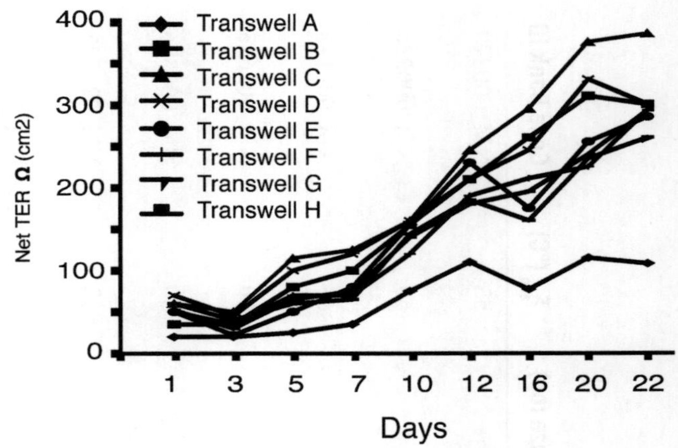

Fig. 6. TER measurements from adult RPE. As RPE culture grows confluent and tight junctions become more functional, resistance rises. *Each line* represents recordings from separate Transwell® insert cultures.

concentration of 5 mg/mL. Further dilute with sterile PBS (Ca^{2+}/Mg^{2+} free) for a final concentration of 1 mg/mL.

16. Dulbecco's PBS (D-PBS) Ca^{2+} and Mg^{2+} free.

17. DNase (Sigma-Aldrich®): Supplied as 4 mg/mL of glycine pH 5.0, 5 mM calcium acetate, and 50% glycerol. Use at 12.2 μg/mL.

18. RPE Medium 1: DMEM/F12, high glucose (Life Technologies™), penicillin/streptomycin 1% (Life Technologies™), N1 supplement 1% (Sigma-Aldrich®), fetal bovine serum (FBS) 10%, non-essential amino acids (10 mM) (Life Technologies™), L-glutamine 1%, taurine 0.25 mg/mL (Sigma-Aldrich®), hydrocortisone 0.02 μg/mL (Sigma-Aldrich®), triiodothyronine 0.013 ng/mL (Sigma-Aldrich®).

19. RPE Medium 2: DMEM/F12, high glucose, penicillin/streptomycin 1%, N1 supplement 1%, FBS 5%, nonessential amino acids (10 mM), L-glutamine 1%, taurine 0.25 mg/mL, hydrocortisone 0.02 μg/mL, triiodothyronine 0.013 ng/mL, human basic fibroblast growth factor 20 ng/mL (R&D™), human epidermal growth factor 1 ng/mL (Life Technologies™).

20. Betadine, povidone-iodine, 10%.

21. Sterile transfer pipette.

22. 10% Sucrose medium: Add 5 g of sucrose to 50 mL of RPE medium 1. Sterile filter using 0.2 μm. Aliqout into 5 mL/tube and store at −20°C.

2.2. Immunohistochemistry

1. 4′,6-Diamidino-2-phenylindole, dihydrochloride, DAPI (Life Technologies™): 5 mg/mL in PBS.
2. D-PBS Ca^{2+} and Mg^{2+} free.
3. Glass coverslips, 12-mm diameter.
4. Normal Goat Serum (NGS, Vector Laboratories): Serum was collected from normal, healthy animals, heat-treated at 56°C for 2 h, kept at 4°C for several weeks, centrifuged to remove precipitates, and filtered through a 0.45 μm filter.
5. 4% Paraformaldehyde (PFA) in PBS.
6. 1× PBS (GIBCO®).
7. ProLong®Gold antifade reagent (Life Technologies™).
8. Triton™ X-100 (Sigma-Aldrich®): Prepare a 0.2% PBST solution by adding 1 mg/500 mL of PBS; for a 2% PBST solution, add 10 mg/500 mL of PBS.
9. Vectashield® Mounting Medium (Vector Laboratories).
10. 0.1% Solution of $KMnO_4$: Add 0.1 mg $KMnO_4$ powder into 10 mL of PBS.
11. 1% Solution of oxalic acid: Add 0.1 mg of oxalic acid powder into 10 mL of PBS.
12. Hydrogen peroxide (H_2O_2): 0.3% (v/v) in methanol 100% (v/v).

2.3. RNA

1. RNeasy Plus Mini Kit (Qiagen).
2. Ethanol absolute for molecular biology.
3. SuperScript®III First—Strand Synthesis System for RT-PCR (Life Technologies™).
4. Power SYBR® Green PCR Master Mix (Applied Biosystems™).
5. Oligonucleotide primers for q-PCR (see Table 1).

2.4. Equipment

1. Agilent, 2100 Bioanalyzer.
2. Applied Biosystems™ 7500 Real-Time PCR System.
3. EVOM^2 Epithelial Voltohmmeter (World Precision Instruments).
4. Cell scrapers, PE blade, PS handle (Nalgene® Nunc™ International).
5. Glass Pasteur pipette.
6. MiniSpin plus® (Eppendorf).
7. Nanodrop ND-1000 Spectrophotometer PCR thermocycler (DNA Engine®, BioRad).
8. PCR thermocycler (DNA Engine®, BioRad).

9. Phase contrast microscope (Carl Zeiss).
10. Plastic disposable transfer pipettes (5, 10, 25, and 50 mL).
11. Tabletop centrifuge.
12. Zeiss Axiovert 200 inverted microscope fitted with a Zeiss AxioCam MRm digital camera image capture system and analyzed with AxioVision 4.72 software (Carl Zeiss).

3. Methods

3.1. RPE Dissection

1. The eyes are received in individual moist chambers. Upon receiving the eyes from the eye bank, place one globe at 4°C while working on the other. Then repeat the steps with the second globe.
2. Place the globe in Betadine (see Subheading 2.1, step 20) for 5 min.
3. Wash the globe in sterile D-PBS for 5 min. Rinse the globe in PBS twice.
4. Place the eye into a 10 cm Petri dish on sterile gauze for stabilization.
5. Make a small incision with a scalpel just below the ora serrata and using the scissors, make a circular cut to remove the anterior chamber (see Fig. 2a).
6. Tip the posterior eyecup on its side so that the vitreous partially pours out. Using a sterile inoculating loop, gently tease out the rest of the vitreous (see Note 1).
7. Using forceps gently pull the rest of the retina off the RPE from the initial opening cut.
8. Once the retina is separated, using curved microscissors, cut the retina from the optic disc.
9. Rinse the eyecup with sterile PBS and then incubate the posterior eyecup in 5 mL sterile-filtered dispase (1 mg/mL) containing DNAse (12.2 μg/mL) (see Subheading 2.1, step 17) for 45 min to 1 h at 37°C (see Note 2).
10. After incubation, place the eyecup in a 10 cm Petri dish onto a piece of sterile gauze for stabilization (see Note 3).
11. Gently vacuum out the dispase and fill the eyecup with 5 mL DMEM/F12 media containing 20% FBS.
12. Using a dulled angled, double-bevel spoon blade (3.0 mm), gently scrape small sheets (1 mm^2) of RPE off the Bruch's membrane.

13. Using a sterile transfer pipette, remove the 5 mL media containing RPE sheet fragments in suspension from the eye-cup and transfer these into a 15 mL conical tube.
14. Separate the cell sheet fragments from single cells by sucrose gradient (see Subheading 3.2, and Note 4).
15. Plate into matrigel-pretreated tissue culture plates (see Subheading 3.3) so that approximately half of the well is covered by the sheets as shown in Fig. 3, Day 1 in RPE medium 1 (see Subheading 2.1, step 18 and Note 5).

3.2. Sucrose Gradient

1. Thaw an aliquot of 10% sucrose medium (see Subheading 2.1, step 22) and bring to room temperature.
2. After RPE extraction, the RPE sheet fragments are suspended in 5 mL of DMEM/F12 media containing 20% FBS per eye. Centrifuge at 259 × *g* for 5 min at 4°C. Remove the supernatant and resuspend the pellet in 2 mL of RPE medium 1.
3. Layer the 2 mL of suspended RPE sheet fragments onto 2 mL 10% sucrose medium.
4. Let sit for 15–20 min during which time the sheets will separate from single cells based on size (the upper half of the medium will contain the single cells while the lower half will contain RPE sheets).
5. Using a transfer pipette, take the lower half and transfer it into a new 15 mL conical tube. Centrifuge at 259 × *g* for 5 min at 4°C.
6. Discard the supernatant and resuspend the cell pellet in RPE medium 2 (see Subheading 2.1, step 19) by gently pipetting up and down with a transfer pipette.
7. Examine the resulting RPE sheet fragments by light microscopy. If single cells are still observed (more than 5% of total cell number), repeat the sucrose gradient by centrifuging the cells at 259 × *g* for 5 min at 4°C, resuspending pellet in RPE medium 1 and re-layering suspension onto a fresh, equal volume of 10% sucrose medium.
8. Plate the RPE sheet fragments so that around half of the well is covered as shown in Day 1 of Fig. 3. The RPE sheet fragments are plated onto Matrigel™-treated tissue culture quality plates containing basic RPE medium 2. Two eyes on average yield enough sheet fragments to cover six wells in a 24-well plate.
9. Plate RPE sheet fragments into 24 well Transwell® in RPE medium 1 (covering half of the surface of the Transwell®) for functional testing of TER (see Subheading 3.8).
10. Incubate the plated cells at 37°C and 5% CO_2.

11. After 24 h, switch to RPE medium 2.
12. Change the medium 2–3 times per week and monitor the culture morphology by light microscopy.

3.3. Thin Matrigel™ Coating Preparation

1. Thaw an aliquot of Matrigel™ Basement Membrane Matrix as recommended by the manufacturer. Using pipettes cooled on ice, mix the Matrigel™ to homogeneity. Store Matrigel™ on ice while preparing the 24-well tissue culture plate.
2. Put the tissue culture plate on ice for 5 min.
3. Keeping the culture plate on ice add enough Matrigel™ to cover the bottom of each well. Remove the excess.
4. Place the plates at 37°C for 30 min. The plates are now ready to use.

3.4. Splitting Confluent RPE Epithelial Monolayers

1. Once a phenotypically homogenous confluent monolayer is obtained it must be passaged in order to expand the cells. However, we found that using trypsin to lift the cells from the plate jeopardizes the likelihood of regaining epithelial functionality. For this reason, we chose dispase to split confluent RPE monolayers.
2. Rinse the RPE monolayers twice with sterile Ca^{2+}-, Mg^{2+}-free PBS.
3. Using a P200 pipette tip, scrape around the edge of the well, and then scrape in a 1 mm^2 grid. This divides the monolayer into equal sized sheets of RPE.
4. Incubate with dispase (1 mg/mL) containing DNAse (12.2 μg/mL) for 20 min at 37°C, 5% CO_2. Check to see if sheets of RPE are lifting off the bottom of the well. Gently tap a few times on the side to dislodge sheets. If sheets are not lifting, leave in the incubator and check at 5-min intervals until the sheets become unattached.
5. Centrifuge at 115 × g for 5 min, remove supernatant, resuspend the cell pellet in RPE medium 1, and plate in Matrigel™-treated plates at 50% confluency. Usually we split one well of a 24-well plate into two wells of a 24-well plate.

3.5. Bleaching RPE Monolayers for Immunostaining

1. Remove the medium and rinse the cells twice with PBS.
2. Fix the human RPE cell monolayers by adding 4% (wt/v) PFA for 10 min at room temperature (RT) followed by two rinses with D-PBS (see Note 6).
3. Permeabilize the cells with 0.2% (v/v) PBST (see Subheading 2.2, step 8) for 5 min at RT followed by two washes with 0.5 mL of the solution. After permeabilization, this solution is discarded.

4. Incubate with $KMnO_4$ from 0.1% (see Subheading 2.2, step 10) to 0.25% (wt/v) for 30 min at RT followed by one wash with 0.5 mL of PBST for 5 min (see Note 7).
5. Incubate with oxalic acid/PBS 1% (wt/v) solution (see Subheading 2.2, step 11) for 10 min at RT until cells are clear.
6. Wash with PBST for 5 min at RT. Check the monolayers with fluorescence microscopy to determine whether autofluorescence is quenched (see Notes 8 and 9).
7. Incubate with H_2O_2 0.3% (v/v) in methanol 100% (v/v) for 30 min at RT to block endogenous peroxidase activity, followed by three washes of PBS for 2 min each (see Note 10).
8. At this point the cells are ready for immunostaining.

3.6. Characterization of Human RPE Monolayers Using Immuno-histochemistry

1. Permeabilize the PFA-fixed cells by adding 0.5 mL of 0.2% (v/v) PBST (see Subheading 2.2, step 8) into each 24-well for 30 min at room temperature.
2. Remove the permeabilizing solution and rinse each well twice with PBS at room temperature.
3. Block the cells by adding 0.5 mL 5% (wt/v) NGS/PBS solution for 1 h at room temperature.
4. Incubate with antibodies diluted in 5% (wt/v) NGS/PBS overnight at 4°C. Antibodies should be diluted as described in Table 2.
5. Remove the primary antibody, wash three times with PBS for 5 min each, and then incubate with the corresponding fluorescent-labeled secondary antibody for 45 min at room temperature in the dark (see Table 3). Remove the secondary

Table 2
Primary antibodies for detection of RPE markers by immunofluorescence

Antibody	Isotype	Host	Supplier	Dilution	Description
RPE65	IgG1	Rabbit	Dr. T. Michael Redmond, NEI®	1:250	Cytoplasmic marker
CRALBP	IgG2a	Mouse	ABCAM®	1:1,000	Cytoplasmic marker
Bestrophin	IgG1	Mouse	Novus Biologicals®	1:100	Basolateral membrane marker
Mitf	IgG1	Mouse	NeoMarkers®	1:50	Nuclear marker
Cytokeratin 8	IgG1	Mouse	Sigma-Aldrich®	1:800	Epithelial marker
Otx2	IgG1	Rabbit	Chemicon Millipore®	1:1,000	Nuclear marker
ZO-1	IgG1	Rabbit	Life Technologies™	1:100	Tight junction marker
Occludin	IgG1	Rabbit	Life Technologies™	1:50	Tight junction marker

Table 3
Secondary antibodies for detection of RPE markers by immunofluorescence

Antibody	Catalog	Supplier	Dilution
Anti-mouse Alexa 488	A-11001	Molecular Probes Invitrogen	1:1,500
Anti-mouse Alexa 546	A-11030	Molecular Probes Invitrogen	1:1,500
Anti-rabbit Alexa 488	A-11008	Molecular Probes Invitrogen	1:1,500
Anti-rabbit Alexa 546	A-11035	Molecular Probes Invitrogen	1:1,500

antibodies and wash the cells three times with PBS. Counterstain the nuclei by adding DAPI (see Subheading 2.2, step 1) for 10 min at room temperature, in the dark, for example in a closed drawer.

6. Wash cells three times in PBS for 10 min each at room temperature, in the dark.
7. Immediately mount the cells using 0.5 mL of Vectashield® or ProLong® Gold mounting medium per well, sealing each well over the top with a coverslip (12 mm). Gently press down on the coverslip, removing air bubbles or excess mounting medium. Leave at room temperature protecting from light for a few minutes. The preparations can be maintained at 4°C for 1 week or at −20°C for 4 weeks protected from light.
8. View preparations under a fluorescence microscope. Images can be taken with a fluorescence microscope equipped with a digital camera (or a similar equipment) at a magnification of at least 400×. A confocal microscope can be used for more detailed examination (see Notes 11–13).

3.7. Gene Expression Profiling of RPE Monolayer Cultures

1. RNA is extracted and purified from RPE using the RNeasy Plus Mini Kit.
2. RNA is reverse-transcribed by following the protocol provided with the purchase of the Superscript® III Reverse Transcriptase kit. QPCR was performed using the Power SYBR® Green PCR Master Mix.
3. The real-time PCR cycling program used on all qPCR experiments is described in Table 4.

3.8. Measuring Transepithelial Resistance

1. Plate RPE sheet fragments into 24 well Matrigel™ pre-treated Transwells® covering half of the surface of the Transwell®. Using all RPE from two globes, sheets can be plated in approximately 12 Transwells®.
2. Every few days during epithelial culture, measure the TER of the Transwell® RPE cell cultures (see Fig. 6).

Table 4
Three-step qPCR cycling conditions for SYBR Green I assay

Step number	Denature	Anneal	Extend
1 cycle	95°C for 10 min		
40 cycles	95°C for 15s	60°C for 1 min	72°C for 45s
Dissociation step	95°C, 15s 60°C, 1 min 95°C, 15s 60°C, 1 min		

3. Rinse the electrodes of EVOM^2 Epithelial Voltohmmeter with 70% ethanol and dry. Measurements are made ideally at ~32°C and so try to record soon after removing from the 37°C incubator.
4. Place the electrode pair in the Transwell® in a way that the long side enters and touches the bottom of the well and the short electrode enters into the Transwell®.
5. Using the EVOM^2 or equivalent, set the function switch to Ohms and measure the resistance of a control, for example a Matrigel™-treated Transwell® containing only media.
6. Subtract all the following measurements from control measurement to obtain Net TER. Functional epithelial layers are acquired when the Net TER is greater than ~200 $\Omega\,cm^2$.

4. Notes

1. Clean detachment of retina from RPE: When dissecting out the vitreous, try to also detach the retina, allowing it to follow the vitreous out of the eyecup. Often the retina is attached well enough to the vitreous that if some tension is made between it and the RPE layer, the retina will easily pull out with the vitreous. If the retina remains, very gently, at the edge of the rim where the circular incision was made, tease a small area of retina away from the RPE without detaching the RPE/choroid membrane from the scleral wall. Then retry to tip the posterior eyecup and the retina should follow the vitreous.
2. RPE/choroid detachment: When dissecting out the vitreous and retina pay close attention to the tension put upon the RPE/choroid so that they do not detach from the sclera. If this happens, however, as can be the case with elderly donors after long

death-to-enucleation times, use syringe needles (30 gauge) to pin up the RPE/choroid back onto the scleral wall.

3. RPE expelling ECM: During dispase digestion, some protocols add agitation (shaker at 80 rpm) to the 37°C incubation step. In our experience, we find that the RPE expels either ECM or DNA. When plating, the cells are trapped in this ECM and do not attach to the tissue culture plate. The cells that do end up attaching are the more migratory type, resulting in less homogenous cultures.
4. Separating RPE sheet fragments from single cells: After enzymatic digestion, and sheet scraping, single cells are typically intermixed with the single sheets. In our experience, single cells are more likely to undergo EMT and therefore, we actively remove them. To reduce single cell contamination, separate them using the sucrose gradient as described. If single cells represent more than 5% of the total RPE population (evaluate a small sample in a well using a light microscope), increase the sucrose to 20%, or wait less time (for example, 10 min) to remove the lower layer.
5. Heterogeneous culture: If after isolation of RPE from donor eyes multiple cell phenotypes are observed, one emergency strategy is to manually select under the microscope the cells with the desired phenotype and plate them into a new well. To do this, draw a circle around the desired subpopulation using a syringe needle. Then, using a P200 pipette, scrape the well, tracing the circle, and scrape the cells until they lift off the plate. Using the same pipette, suck up the removed cells and replate into a 96-well plate pretreated with Matrigel™. In our experience, this method works approximately 20% of the time to produce a confluent, homogenous epithelial culture.
6. PFA is highly toxic. Work in a fume cupboard. Wearing a protective mask, glasses, and gloves is mandatory.
7. Bleaching: Make this solution fresh every time and do not store it. As a starting point use a 0.15% $KMnO_4$ solution; however the optimal working solution concentration must be determined by each user.
8. Autofluorescence remains after bleaching. After bleaching, using a fluorescence microscope, check the RPE layer at the 488, 568 and 633 nm wavelengths to make sure that RPE cells have been bleached adequately. If autofluorescence remains, increase the $KMnO_4$ concentration, or increase the length of the $KMnO_4$ incubation time.
9. Cell damage after bleaching: $KMnO_4$, oxalic acid, and H_2O_2 can damage cells if exposed for too long. It is best to use shorter incubation times initially and make changes from there.

10. Blocking endogenous peroxidase activity: Let the methanol solution stand for about 10 min at room temperature before use.
11. Poor staining: The fixation procedure used must be compatible with the primary antibody. If using a different fixation method, ensure that the antibody will work with any changes.
12. Cells removed from the well during staining or washing procedures: Gentle washing is strongly recommended to avoid cell detachment. Use transfer pipettes and gentle aspirate. Do not dispense solutions directly on top of the cells but squirt gently down the inside of the well to avoid cell disruption.
13. Low signal: Fluorescent-labeled antibodies are highly light sensitive. It is strongly recommended to avoid direct light during incubations and storage.

References

1. Sonoda S, Spee C, Barron E, Ryan SJ, Kannan R, Hinton DR (2009) A protocol for the culture and differentiation of highly polarized human retinal pigment epithelial cells. Nat Protoc 4:662–673
2. Gamm DM, Melvan JN, Shearer RL, Pinilla I, Sabat G, Svendsen CN, Wright LS (2008) A novel serum-free method for culturing human prenatal retinal pigment epithelial cells. Invest Ophthalmol Vis Sci 49:788–799
3. Geisen P, McColm JR, King BM, Hartnett ME (2006) Characterization of barrier properties and inducible VEGF expression of several types of retinal pigment epithelium in medium-term culture. Curr Eye Res 31:739–748
4. Maminishkis A, Chen S, Jalickee S, Banzon T, Shi G, Wang FE, Ehalt T, Hammer JA, Miller SS (2006) Confluent monolayers of cultured human fetal retinal pigment epithelium exhibit morphology and physiology of native tissue. Invest Ophthalmol Vis Sci 47:3612–3624
5. Flood MT, Gouras P (1981) The organization of human retinal pigment epithelium in vitro. Vis Res 21:119–126
6. Vielkind U, Crawford BJ (1988) Evaluation of different procedures for the dissociation of retinal pigmented epithelium into single viable cells. Pigment Cell Res 1:419–433
7. Lopashov GV (1983) Transdifferentiation of pigmented epithelium induced by the influence of lens epithelium in frogs. Differentiation 24:27–32
8. Grisanti S, Guidry C (1995) Transdifferentiation of retinal pigment epithelial cells from epithelial to mesenchymal phenotype. Invest Ophthalmol Vis Sci 36:391–405
9. Casaroli-Marano RP, Pagan R, Vilaro S (1999) Epithelial–mesenchymal transition in proliferative vitreoretinopathy: intermediate filament protein expression in retinal pigment epithelial cells. Invest Ophthalmol Vis Sci 40: 2062–2072
10. Saika S, Kono-Saika S, Tanaka T, Yamanaka O, Ohnishi Y, Sato M, Muragaki Y, Ooshima A, Yoo J, Flanders KC, Roberts AB (2004) Smad3 is required for dedifferentiation of retinal pigment epithelium following retinal detachment in mice. Lab Invest 84:1245–1258
11. Lee H, O'Meara SJ, O'Brien C, Kane R (2007) The role of gremlin, a BMP antagonist, and epithelial-to-mesenchymal transition in proliferative vitreoretinopathy. Invest Ophthalmol Vis Sci 48:4291–4299
12. Garcia S, Lopez E, Lopez-Colome AM (2008) Glutamate accelerates RPE cell proliferation through ERK1/2 activation via distinct receptor-specific mechanisms. J Cell Biochem 104:377–390
13. Kim JW, Kang KH, Burrola P, Mak TW, Lemke G (2008) Retinal degeneration triggered by inactivation of PTEN in the retinal pigment epithelium. Genes Dev 22:3147–3157
14. Liu Y, Ye F, Li Q, Tamiya S, Darling DS, Kaplan HJ, Dean DC (2009) Zeb1 represses Mitf and regulates pigment synthesis, cell proliferation, and epithelial morphology. Invest Ophthalmol Vis Sci 50:5080–5088
15. Tamiya S, Liu L, Kaplan HJ (2009) Loss of cell–cell contact initiates epithelial–mesenchymal transition and proliferation of retinal pigment epithelial cells. Invest Ophthalmol Vis Sci 50(11):5080–5088

16. Liu Y, Xin Y, Ye F, Wang W, Lu Q, Kaplan HJ, Dean DC (2010) Taz-Tead1 Links Cell-Cell Contact to Zeb1 Expression, Proliferation and Dedifferen-tiation in Retinal Pigment Epithelial Cells. Invest Ophthalmol Vis Sci 51(7): 3372–3378
17. Hay ED (1995) An overview of epithelio-mesenchymal transformation. Acta Anat (Basel) 154:8–20
18. Tso MO, Cunha-Vaz JG, Shih CY, Jones CW (1980) Clinicopathologic study of blood-retinal barrier in experimental diabetes mellitus. Arch Ophthalmol 98:2032–2040
19. Caldwell RB, McLaughlin RJ, Boykins LG (1982) Intramembrane changes in retinal pigment epithelial cell junctions of the dystrophic rat retina. Invest Ophthalmol Vis Sci 23:305–318
20. Noell WK, Albrecht R (1971) Irreversible effects on visible light on the retina: role of vitamin A. Science 172:76–79
21. Bridges CD (1976) Vitamin A and the role of the pigment epithelium during bleaching and regeneration of rhodopsin in the frog eye. Exp Eye Res 22:435–455
22. Saari JC, Bredberg L, Garwin GG (1982) Identification of the endogenous retinoids associated with three cellular retinoid-binding proteins from bovine retina and retinal pigment epithelium. J Biol Chem 257:13329–13333
23. Edwards RB, Szamier RB (1977) Defective phagocytosis of isolated rod outer segments by RCS rat retinal pigment epithelium in culture. Science 197:1001–1003
24. Noell WK, Crapper DR, Paganelli CV (1965) Transretinal currents and ion fluxes. In: Snell FM, Noell WK (eds) Transcellular membrane potentials and ion fluxes. Gordon and Breach, New York, NY, pp 92–130
25. Steinberg RH, Miller S (1973) Aspects of electrolyte transport in frog pigment epithelium. Exp Eye Res 16:365–372
26. Miller SS, Steinberg RH (1977) Active transport of ions across frog retinal pigment epithelium. Exp Eye Res 25:235–248
27. Miller SS, Steinberg RH (1977) Passive ionic properties of frog retinal pigment epithelium. J Membr Biol 36:337–372
28. Eagle RC Jr (1984) Mechanisms of maculopathy. Ophthalmology 91:613–625
29. Burke JM, Skumatz CM, Irving PE, McKay BS (1996) Phenotypic heterogeneity of retinal pigment epithelial cells in vitro and in situ. Exp Eye Res 62:63–73
30. Feng W, Zheng JJ, Lutz DA, McLaughlin BJ (2003) Loss of RPE phenotype affects phagocytic function. Graefes Arch Clin Exp Ophthalmol 241:232–240
31. Whittaker JR (1967) Loss of melanotic phenotype in vitro by differentiated retinal pigment cells: demonstration of mechanisms involved. Dev Biol 15:553–574
32. Flood MT, Gouras P, Kjeldbye H (1980) Growth characteristics and ultrastructure of human retinal pigment epithelium in vitro. Invest Ophthalmol Vis Sci 19:1309–1320
33. Rodriguez-Boulan E, Nelson WJ (1989) Morphogenesis of the polarized epithelial cell phenotype. Science 245:718–725
34. Luna EJ, Hitt AL (1992) Cytoskeleton–plasma membrane interactions. Science 258:955–964
35. Hitt AL, Luna EJ (1994) Membrane interactions with the actin cytoskeleton. Curr Opin Cell Biol 6:120–130
36. Turksen K, Opas M, Kalnins VI (1989) Cytoskeleton, adhesion, and extracellular matrix of fetal human retinal pigmented epithelial cells in culture. Ophthalmic Res 21:56–66
37. Song MK, Lui GM (1990) Propagation of fetal human RPE cells: preservation of original culture morphology after serial passage. J Cell Physiol 143:196–203
38. Gouras P, Cao H, Sheng Y, Tanabe T, Efremova Y, Kjeldbye H (1994) Patch culturing and transfer of human fetal retinal epithelium. Graefes Arch Clin Exp Ophthalmol 232: 599–607
39. Castillo BV Jr, Little CW, del Cerro C, del Cerro M (1995) An improved method of isolating fetal human retinal pigment epithelium. Curr Eye Res 14:677–683
40. Zhu M, Provis JM, Penfold PL (1998) Isolation, culture and characteristics of human foetal and adult retinal pigment epithelium. Aust N Z J Ophthalmol 26(Suppl 1):S50–S52
41. Hu J, Bok D (2001) A cell culture medium that supports the differentiation of human retinal pigment epithelium into functionally polarized monolayers. Mol Vis 7:14–19
42. Rak DJ, Hardy KM, Jaffe GJ, McKay BS (2006) Ca++-switch induction of RPE differentiation. Exp Eye Res 82:648–656
43. Tezel TH, Del Priore LV (1998) Serum-free media for culturing and serial-passaging of adult human retinal pigment epithelium. Exp Eye Res 66:807–815
44. McKay BS, Burke JM (1994) Separation of phenotypically distinct subpopulations of cultured human retinal pigment epithelial cells. Exp Cell Res 213:85–92
45. Stump RJ, Lovicu FJ, Ang SL, Pandey SK, McAvoy JW (2006) Lithium stabilizes the

polarized lens epithelial phenotype and inhibits proliferation, migration, and epithelial mesenchymal transition. J Pathol 210:249–257

46. Papageorgis P, Lambert AW, Ozturk S, Gao F, Pan H, Manne U, Alekseyev YO, Thiagalingam A, Abdolmaleky HM, Lenburg M, Thiagalingam S (2010) Smad signaling is required to maintain epigenetic silencing during breast cancer progression. Cancer Res 70:968–978
47. De S, Rabin DM, Salero E, Lederman PL, Temple S, Stern JH (2007) Human retinal pigment epithelium cell changes and expression of alphaB-crystallin: a biomarker for retinal pigment epithelium cell change in age-related macular degeneration. Arch Ophthalmol 125:641–645
48. Hemesath TJ, Steingrimsson E, McGill G, Hansen MJ, Vaught J, Hodgkinson CA, Arnheiter H, Copeland NG, Jenkins NA, Fisher DE (1994) Microphthalmia, a critical factor in melanocyte development, defines a discrete transcription factor family. Genes Dev 8:2770–2780
49. McCarthy KM, Skare IB, Stankewich MC, Furuse M, Tsukita S, Rogers RA, Lynch RD, Schneeberger EE (1996) Occludin is a functional component of the tight junction. J Cell Sci 109(Pt 9):2287–2298
50. Redmond TM, Yu S, Lee E, Bok D, Hamasaki D, Chen N, Goletz P, Ma JX, Crouch RK, Pfeifer K (1998) Rpe65 is necessary for production of 11-cis-vitamin A in the retinal visual cycle. Nat Genet 20:344–351
51. Denker BM, Nigam SK (1998) Molecular structure and assembly of the tight junction. Am J Physiol 274:F1–F9
52. Martinez-Morales JR, Dolez V, Rodrigo I, Zaccarini R, Leconte L, Bovolenta P, Saule S (2003) OTX2 activates the molecular network underlying retina pigment epithelium differentiation. J Biol Chem 278:21721–21731
53. Saari JC, Nawrot M, Kennedy BN, Garwin GG, Hurley JB, Huang J, Possin DE, Crabb JW (2001) Visual cycle impairment in cellular retinaldehyde binding protein (CRALBP) knockout mice results in delayed dark adaptation. Neuron 29:739–748
54. Fischmeister R, Hartzell HC (2005) Volume sensitivity of the bestrophin family of chloride channels. J Physiol 562:477–491
55. Hunt RC, Davis AA (1990) Altered expression of keratin and vimentin in human retinal pigment epithelial cells in vivo and in vitro. J Cell Physiol 145:187–199
56. Dunn KC, Aotaki-Keen AE, Putkey FR, Hjelmeland LM (1996) ARPE-19, a human retinal pigment epithelial cell line with differentiated properties. Exp Eye Res 62: 155–169

Chapter 5

Oral and Pharyngeal Epithelial Keratinocyte Culture

Kantima Leelahavanichkul and J. Silvio Gutkind

Abstract

Primary human oral epithelial cells are readily available and have been recently employed for tissue engineering. These cells are currently being widely utilized in multiple research efforts, ranging from the study of oral biology, mucosal immunity, and carcinogenesis to stem cell biology and tissue engineering. This chapter describes step-by-step protocols for the successful isolation and culture of human oral epithelial cells and fibroblasts, and techniques for their use in two-dimensional and three-dimensional culture systems. The described methods will enable to generate reconstituted tissues that resemble epithelial like structures in vitro, which can recapitulate some of the key features of the oral epithelium in vivo.

Key words: Keratinocytes, Oral mucosa, gingival, pharyngeal mucosa, Cell culture, Three-dimensional organotypic cell culture, raft, epithelial like construct

1. Introduction

The differentiation patterns of the superficial layers of the oral mucosa and the skin epidermis are quite distinct. The oral epithelium can be classified into three types, masticatory, lining, and specialized, which include keratinized, parakeratinized, and non-keratinized stratified epithelium (1). In general, though, a parakeratinized stratified epithelium forms the lining of most of the oral mucosa, thus providing a barrier against mechanical, physical, and pathological injury (2). Specifically, the main morphological difference between the buccal mucosa and the skin epidermises is the differentiation pattern found in the superficial layers of the epithelium. The oral mucosa lacks the granulous and cornified layers, but instead presents parakeratin as the last step in the differentiation process. In addition, emerging evidence suggests that the epithelial stem cells maintaining the homeostasis of the oral mucosa and skin might be distinct (3). In the skin, both follicular and interfollicular epithelial stem cells can contribute to the replacement of the

Scott H. Randell and M. Leslie Fulcher (eds.), *Epithelial Cell Culture Protocols: Second Edition*, Methods in Molecular Biology, vol. 945, DOI 10.1007/978-1-62703-125-7_5,

exfoliated cells and to regeneration of the tissue upon injury (4, 5). In this regard, an emerging body of information suggests that the interfollicular (IF) stem cells and their derived transient amplifying cells may play a primary role in epidermal homeostasis and tissue repair, and that in some cases IF stem cells may be alone sufficient for the healing of cutaneous wounds (5–8). The oral mucosa, however, represents an epidermal anatomical site that presents similar architecture to that of the skin but devoid of any hair follicle structures. Thus, the oral epithelium is expected to be maintained by a single type of adult epithelial stem cell possessing self-renewal capacity.

Recent studies addressing the contribution of the small GTPase Rac1 to the healing of oral and dermal wounds suggest that oral epithelial stem cells resemble the IF skin stem cells regarding their homeostatic function. However, distinct signaling mechanisms may be deployed in these two adult stem cell populations during wound healing (3). While the distinct roles of these stem cells in tissue maintenance and their behavior during tissue regeneration represent areas of active investigation, we expect that the availability of primary human oral epithelial cell culture systems may aid in dissecting their multiple biological properties and underlying molecular mechanisms. In general, we use for these studies two-dimensional culture systems that enable to study multiple aspects of cellular behavior, such as proliferation, senescence, death, and motility, as well as epithelial cell-to-cell interactions, and cell interactions with defined cellular and extracellular components of their microenvironment, such as basement membrane, fibroblasts, immune cells, and cytokines (see Fig. 1a). In addition, the development and availability of three-dimensional cultures that aim at recreating the epithelial like environment has opened new avenues for research, as it offers powerful tools to study epithelial cell biology in the context of the complex interplay among cellular and extracellular components of the skin and oral mucosa.

Oral epithelial cell culture methods have been developed for many years. Indeed, there are multiple different culture methods available, ranging from explant outgrowth or the use of enzymatically digested tissues, which rely on distinct media and supplements, and coating the culture surface with an ever-increasing number of extracellular matrix components. This chapter describes step by step the successfully simplified methods that have been used in our laboratory to culture primary oral epithelial cells in monolayer without fibroblast as a seeding layer. It also describes in detail simplified culture methods for the establishment of three-dimensional organotypic (epithelial constructed) oral keratinocyte cultures. This chapter describes how to obtain oral keratinocytes and fibroblasts from human tissue biopsies and culture them in both monolayers and three-dimensional cultures. This will be accomplished by first describing our current methods for the preparation of oral keratinocytes and fibroblasts in monolayers, which can be accomplished in less than a week, and will be followed by

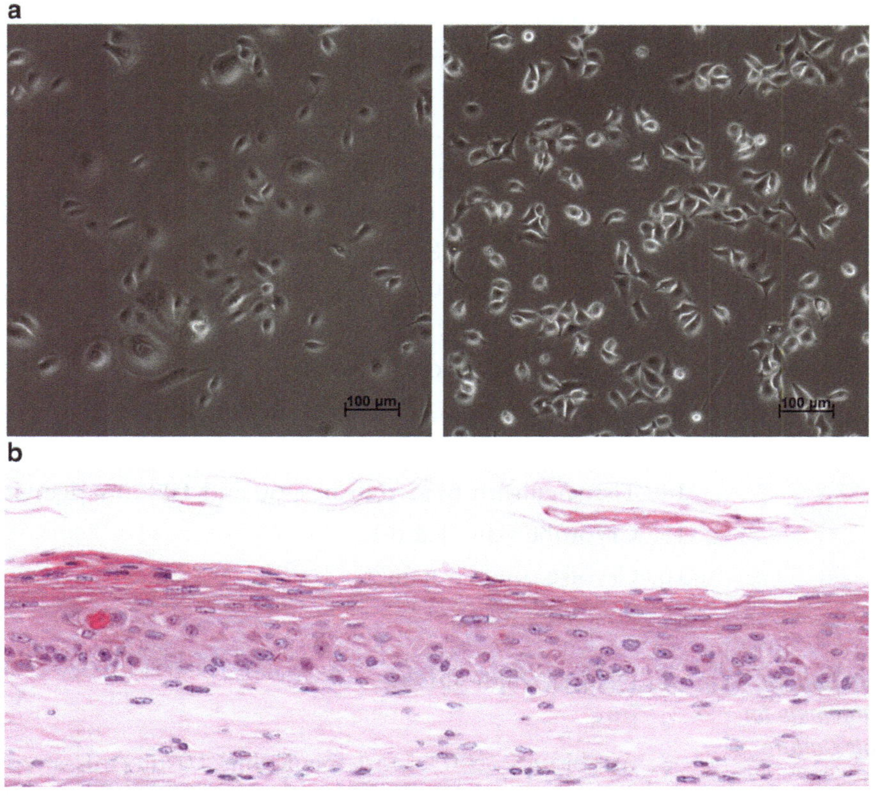

Fig. 1. Monolayer cell morphology of oral keratinocytes from gingiva. (**a**) The primary keratinocytes (*left*) are not homogenous in size and shape compared to immortalized keratinocytes (*right*). Size of the cells varies from 25 μm to over 100 μm. (**b**) Three-dimensional organotypic culture of primary oral keratinocytes and oral fibroblasts. H&E staining shows the different layers from the basal proliferative cells to the parakeratinized (differentiated) keratinocytes in the upper layers.

methods that describe the incorporation of these constituent cells into three-dimensional organotypic cultures, which often takes approximately 2–3 weeks (see Fig. 1b). In every case, multiple manufacturers and companies have available reagents that may be comparable or even superior to the ones that we describe. We have included the information of each of the reagents used in our laboratory only as a reference point to initiate the new investigator in this still highly evolving field.

2. Materials

2.1. Primary Culture of Oropharyngeal Keratinocytes

1. Defined Keratinocyte Serum-Free Media (K-SFM) (Life Technologies™).
2. Dulbecco's Modified Eagle's Medium (DMEM), high glucose.
3. Dulbecco's phosphate-buffered saline (PBS).

4. Trypsin (0.25%) solution.
5. Trypsin (0.05%)–EDTA (0.02%) solution.
6. Antibiotics: 100 U/mL penicillin, 10 μg/mL streptomycin, and 0.25 μg/mL amphotericin B, added to the medium.
7. Fetal bovine serum (FBS).
8. Collagen type I (rat tail): Dilute in 3% acetic acid at final concentration 0.3 mg/mL and store at 4°C.
9. Dimethyl sulfoxide (DMSO).
10. Scalpels, No. 21.
11. Forceps, fine, two pairs.
12. Centrifuge tubes, 15 and 50 mL, conical.
13. Cell strainer, 100 μm.
14. Tissue culture-treated dishes, 60 and 100 mm diameters.
15. Cryotube vials, 1.8 mL.
16. Cryofreezing container.
17. 70% Ethanol.

2.2. Primary Culture of Oral Fibroblasts for Organotypic Culture

1. DMEM, high glucose.
2. FBS.
3. Antibiotics: 100 U/mL penicillin, 10 μg/mL streptomycin, and 0.25 μg/mL amphotericin B, added to the medium.
4. Trypsin (0.05%)–EDTA (0.02%) solution.
5. Scalpels, No. 21.
6. Forceps, fine, two pairs.
7. Centrifuge tubes, 15 mL conical.
8. Tissue culture-treated dishes, 60 and 100 mm diameters.

2.3. Three-Dimensional Organotypic Culture

1. Bovine collagen, type I (organogenesis) at a final concentration of 0.78–1.01 mg/mL (see Note 1).
2. K-SFM (Life Technologies™) with bovine pituitary extract (BPE) and human epidermal growth factor (EGF).
3. Eagle's Minimum Essential Medium (EMEM), 10×.
4. L-Glutamine, 200 mM.
5. Dialyzed Fetal Bovine Serum (dFBS).
6. Sodium bicarbonate, 7.5%: Dissolve 7.5 g in 100 mL of ddH_2O and filter with pore size 0.22 μm.
7. $CaCl_2$: Final concentration is 2.4 mM. Make stock concentration of 500× by dissolving 3.55 g in 20 mL of ddH_2O and filter with pore size 0.22 μm. Aliquot into 2 mL aliquots and store at −20°C.

8. Hank's Balance Salt Solution (HBSS) without calcium and magnesium.
9. Formaldehyde solution, 4% in buffered zinc (Z-fix®, Anatech).
10. Optimal cutting temperature (O.C.T.™) compound.
11. Sucrose solution, 50%: To prepare 50% sucrose solution, dilute 50 g of sucrose in 100 mL of dH_2O.
12. Sucrose solution, 2 M: To prepare 2 M of sucrose solution, dilute 68.46 g of sucrose in 100 mL of dH_2O until dissolved.
13. Tray and cover sets with transwells (organogenesis) (see Note 2).
14. Biopsy papers.
15. Tissue cassettes.
16. Scalpels, No. 21.
17. Plastic cryomold.
18. Forceps, fine, two pairs.
19. Kimwipes®.
20. Centrifuge tubes, 15 and 50 mL, conical.

3. Methods

3.1. Primary Oral and Pharyngeal Keratinocytes

3.1.1. Receipt of Tissues

1. Healthy gingival can be obtained from healthy volunteers from third molar impact surgery or biopsy (see Note 3). To achieve high yield of cells, donor's age in general is best to be limited to 60 years or under (9).
2. Place tissue in a 15 mL tube filled with cold K-SFM enough to cover the tissue.
3. Transfer the tissue from the clinical operating room to the research laboratory on ice and proceed to the next step within 2–4 h after biopsy excision.

3.1.2. Primary Culture of Oral Keratinocytes

1. Transfer to a new 60 mm dish and add enough trypsin (0.25%) solution to cover the tissue. Incubate overnight at 4°C (10).
2. Coat 60 mm dish with 1 mL of collagen solution (see Subheading 2.1, step 8) for 1 h, and wash with PBS three times.
3. Hold the tissue using a pair of forceps, and then peel and scrape the epithelium with another pair of forceps placing it into the trypsin solution. Keep the remaining tissue in trypsin solution for culturing the oral fibroblast (see Subheading 3.2.1). Finely mince the epithelium with scalpels.
4. Pipette up and down the tissue and cell suspension several times in a 5 mL pipette to disrupt the epithelium, and pass through cell strainer into a 50 mL tube.

5. Rinse the dish with DMEM with 10% FBS and add this rinsing solution to the cell suspension in 50 mL tube. It is not necessary to determine the number of cells at this time. If tissue size is less than 10 × 10 mm, keratinocytes can be seeded on 60 mm dish.
6. Pellet the cells at 125 × *g*, for 5 min at 4°C.
7. Carefully remove the supernatant, resuspend the cells in K-SFM, and then seed the cells on a collagen-coated dish (see Subheading 3.1.2, step 2). Incubate at 37°C with 5% CO_2.
8. After 24 h, wash with PBS twice to remove cell debris and any remaining red blood cells, and replace with fresh medium. Change the culture medium every 2–3 days.

3.1.3. Subculture of Oral Keratinocytes

1. Wash the cells twice with PBS, when cells reach 70–80% subconfluence.
2. Add enough trypsin–EDTA (see Subheading 2.1, step 5) to cover the cells.
3. Incubate at 37°C until the cells round up and/or detach from the dish. Monitor this process under the microscope, checking the dish every 5 min for up to 15 min.
4. Do not tap the tissue culture dish. Add DMEM with 10% FBS to inactivate trypsin (see Note 4), pipette the cell suspension multiple times over the surface of the tissue culture dish to detach the cells mechanically and disrupt the cell clumps, and transfer to 15 mL tubes.
5. Take an aliquot of the cell suspension to determine the number of cells by hemocytometer.
6. Pellet the cells at 125 × *g* for 5 min at 4°C.
7. Coat new 100 mm dishes with 2 mL of collagen for 1 h, and wash with PBS three times.
8. Carefully remove the supernatant and resuspend the pellet in K-SFM.
9. Seed the cells into a collagen-coated dish, 100 mm, at 2.5–5 × 10^5 cells.
10. Incubate the cells at 37°C, 5% CO_2. Change the medium every 2–3 days.

3.1.4. Cryopreservation of Oral Keratinocytes in Liquid Nitrogen

1. After the culture reaches 70–80% confluence level, detach the cells using trypsin–EDTA and monitor under a microscope.
2. After cells round up and detach from the culture dish, add DMEM with 10% FBS to inactivate trypsin, pipette the cell suspension multiple times over the surface of the tissue culture dish to detach the cells mechanically and disrupt the cell clumps, and transfer to 15 mL tubes. Take an aliquot of the cell suspension to count the number of cells.
3. Pellet the cells at 125 × *g* for 5 min at 4°C.

4. Carefully remove the supernatant and resuspend the cells at 5–8 × 10^5 cells/mL of K-SFM containing 10% DMSO.
5. Transfer 0.5 mL of cell suspension into each cryotube vial, and then place the vials into a cryofreezing container. Store at −80°C for at least 4 h, and transfer the vials to the designated liquid nitrogen container.

3.1.5. Thawing Cryopreserved Oral Keratinocytes

1. Add 10 mL of K-SFM into a 15 mL tube.
2. Thaw the frozen cells by briefly placing the vial into a water bath at 37°C.
3. As soon as the cell suspension has thawed, wipe the vial with 70% ethanol and transfer the cells into prepared 15 mL tube.
4. Pellet the cells at 125 ×*g* for 5 min at 4°C.
5. Prepare a 100 mm collagen-coated dish.
6. Carefully remove the supernatant, resuspend the pellet in 10 mL of K-SFM, and transfer the cell suspension into collagen-coated dish.
7. Carefully swirl and shake the dish to homogeneously distribute the cells.
8. Incubate the cells at 37°C, 5% CO_2.
9. After 24 h, replace the medium with fresh K-SFM to get rid of unattached cells. Change the medium every 2–3 days.

3.2. Preparation of Primary Oral Fibroblasts for Organotypic Culture

3.2.1. Primary Culture of Oral Fibroblasts

1. The isolation of oral fibroblasts can be conducted concomitant with the isolation of oral keratinocytes. Start this process from Subheading 3.1.2. Mince down the remaining tissue after keratinocyte isolation using scalpels.
2. Disrupt the cell suspension by pipetting up and down multiple times and transfer the cells to a new 60 mm tissue culture-treated dish containing 3 mL of DMEM with 10% FBS to allow tissue to attach well to the dish.
3. Incubate at 37°C in 5% CO_2.
4. After 24 h, add 2 mL of fresh medium. Change the medium every 2–3 days.

3.2.2. Subculture of Oral Fibroblasts

1. After the culture has reached 80–90% confluency level, wash the cells twice with PBS.
2. Add enough trypsin–EDTA to cover the cells.
3. Incubate at 37°C until the cells round up and/or detach from the dish, about 5 min.
4. Add DMEM with 10% FBS to inactivate trypsin, disrupt by passing several times through a pipette, and transfer to a new 100 mm dish.
5. Incubate the cells at 37°C, 5% CO_2. Change the medium every 2–3 days.

3.2.3. Cryopreservation of Oral Fibroblasts in Liquid Nitrogen

1. After reaching 80–90% confluency level, detach the cells using trypsin–EDTA.
2. Count the number of cells.
3. Pellet cells at 125 × *g* for 5 min at 4°C.
4. Resuspend the cells in DMEM with 10% FBS and 10% DMSO at 2–2.5 × 10^6 cells/mL.
5. Transfer 0.5 mL of cell suspension into each cryotube vial, transfer the vials to a cryofreezing container, store at −80°C for at least 4 h, and then move the vials to designated liquid nitrogen tank.

3.2.4. Thawing Cryopreserved Oral Fibroblasts

1. Prepare a 15 mL tube with 10 mL of DMEM with 10% FBS.
2. Thaw the frozen cells by briefly placing the vial in a water bath at 37°C.
3. As soon as the cell suspension has thawed, transfer the cell suspension to the prepared tube.
4. Pellet the cells at 125 × *g* for 5 min at 4°C.
5. Aspirate supernatant and resuspend the pellet in 10 mL of DMEM with 10% FBS and transfer to a 100 mm dish.
6. Carefully swirl and shake the dish to homogeneously distribute the cells.
7. Incubate the cells at 37°C, 5% CO_2.
8. After 24 h, replace the medium with fresh medium to get rid of unattached cells, and replace it every 2–3 days.

3.3. Three-Dimensional Oral and Pharyngeal Organotypic Culture

3.3.1. Preparation of Collagen Matrix

1. Prepare acellular collagen matrix (see Table 1). Each insert requires 1 mL of collagen matrix. Keeping all components on ice, gently mix the components in a 50 mL tube on ice in the order, avoiding bubbles.
2. The color of mixed collagen matrix should be from straw-yellow to light pink; otherwise the collagen will not solidify.

Table 1
Components for preparing acellular matrix

Substance	Volume for six gels (mL)
MEM w/Earle's salts, 10×	0.59
L-Glutamine	0.05
Fetal bovine serum	0.60
Sodium bicarbonate	0.12
Bovine collagen, type I	4.60

Table 2
Components for preparing cellular matrix

Substance	Volume for six inserts (mL)
MEM w/Earle's salts, 10×	1.65
L-Glutamine	0.15
Fetal bovine serum	1.85
Sodium bicarbonate	0.35
Bovine collagen	14.00
Oral fibroblasts (75,000 cells/insert)	4.5×10^5 in 1.5 mL of DMEM with 10% FBS (3×10^5/mL)

If the color is bright yellow, add more sodium bicarbonate drop by drop (see Note 5).

3. Add 1 mL of collagen matrix in each insert, avoiding bubbles. Make sure that collagen covers the culture surface. Let the plate sit for 15–30 min in the hood without disturbing. The polymerization leads to deeper pink color.
4. Prepare cellular matrix (see Table 2). Each insert requires 3 mL of cellular matrix. Gently mix the components by pipette in a 50 mL tube on ice in the indicated order, avoiding bubbles.
5. The color of mixed cellular collagen should be from straw-yellow to pink before adding the oral fibroblasts (see Note 6). If the color is bright yellow, add more sodium bicarbonate drop by drop.
6. Add 3 mL of cellular collagen matrix into each insert, on top of the solidified acellular collagen, avoiding bubbles. Make sure that the collagen covers the culture surface.
7. Let the tray sit for 1 h in 37°C, 5% CO_2 incubator without disturbing. The polymerization leads to deeper pink color.
8. When the collagen matrix turns pink and firm, add 10–12 mL of DMEM with 10% FBS into the outside of each insert and 2 mL into the inside of each insert.
9. Incubate the collagen matrix in incubator for 5–7 days until it contracts. Within 7 days, the collagen matrix should be contracted away from the side of inserts and form a central raised area.

3.3.2. Growth of Oral Keratinocytes in Organotypic Culture

1. Before seeding keratinocytes, carefully aspirate medium and change to HBSS with 1% dFBS to wash off the serum. Incubate at 37°C, 5% CO_2 for 30 min.

Table 3
Epithelial reconstruct media for primary oral keratinocyte culture (see Note 7)

Substance	Reconstruct media I (day 1–2)	Reconstruct media II (day 3–4)	Reconstruct media III (day 5)
Keratinocyte serum-free medium	98%		
Dialyzed fetal bovine serum	2%		
Bovine pituitary extract (BPE)	60 μg/mL		
Epidermal growth factor (EGF)	1 ng/mL	0.2 ng/mL	None
$CaCl_2$	None	None	2.4 mM

2. Change the medium again and incubate for another 30 min.
3. Each insert requires 500,000 cells. Trypsinize the keratinocytes with trypsin–EDTA and count cells. Transfer the appropriate amount of cell suspension to a 15 mL tube, and spin them down at 150 × *g* for 5 min at 4°C.
4. While waiting, aspirate the washing media from inserts and add 1.5 mL of reconstruct media I into the inserts and 10–12 mL outside of inserts (see Table 3 and Note 7).
5. Add 100 μL of reconstruct medium I per 500,000 cells and gently resuspend the cell pellet with a 200 μL pipette.
6. Transfer 100 μL of keratinocyte suspension into the center of collagen matrix drop by drop. Do not move the tray for 15 min to allow cells to attach to collagen matrix.
7. Carefully place the tray in incubator at 37°C, 5% CO_2.
8. On day 3, carefully remove media both inside and outside inserts and add 2 mL of reconstruct media II inside of inserts and 10–12 mL outside of inserts (see Table 3).
9. On day 5, completely aspirate all media and replace with 7.5 mL of reconstruct media III only outside of inserts, allowing the epithelial layer to be exposed to the air (see Table 3). Incubate in 37°C, 5% CO_2 incubator.
10. Replace reconstruct media III every other day until day 14. Every time that media are changed, make sure to completely remove media inside of the inserts.

3.3.3. Harvesting the Three-Dimensional Organotypic Culture

1. Aspirate media from wells both outside and inside.
2. Use a pair of forceps to remove inserts from well. Cut the organotypic culture and filter by tracing a circle close to the edge with a scalpel.

3. Cut the organotypic culture in half on a hard surface. Depending on the purpose of the study, number of tissues and method of preserving tissues might vary (see Note 8).
4. For paraffin-embedded tissue, place one half of the culture between wet biopsy papers and put it into a histology cassette. Immediately submerge the culture in Z-fix® for 4–6 h and replace with 70% ethanol and store at 4°C until paraffin processing.
5. For frozen tissue, place the other half of the culture in 50% sucrose (see Subheading 2.3, step 11) for 1–2 h at 4°C and replace with 2 M of sucrose solution (see Subheading 2.3, step 12) for another 1–2 h at 4°C.
6. Fill a plastic cryomold with O.C.T.™ compound about ½ full; gently remove the tissue from the sucrose solution using forceps to grab the edge and a thin spatula underneath the membrane. Do not disturb the tissue side. Get rid of some excess sucrose solution by gently touching the membrane side to Kimwipes®.
7. Transfer the tissue using forceps into a plastic cryomold filled with O.C.T.™ compound. Add more O.C.T.™ compound on top of the tissue, avoiding bubbles.
8. Place the cryomold on evenly crushed dry ice and allow the O.C.T.™ compound to completely freeze until it turns white.

4. Notes

1. Other sources of collagen can be used to culture as well, but the ready-to-use collagen from organogenesis gives more consistent results.
2. Other sources of culture trays can be used to grow the three-dimensional organotypic culture as well. However, the trays from organogenesis give us reproducible contracting of collagen matrix. In addition, the unique shape of the plates is able to hold up to 7.5 mL of media in the air-lift stage.
3. Although this method is for culture of oral keratinocytes and fibroblasts from gingival tissue, it can be modified to grow tissues from other oropharyngeal squamous epithelia such as oral mucosa (11), peritonsillar mucosa (12), and pharyngeal mucosa (13).
4. Although high calcium concentration in DMEM with FBS will stimulate the keratinocytes to differentiate, we found that it can increase the number of cells that can be centrifuged down when compared with PBS. So, the cells should be briefly resuspended in DMEM with FBS, and after spinning down, media

should be removed as much as possible with minimum disruption on the pellet.

5. The pH of the collagen is critical for the appropriate polymerization. It is important to mix the solution well by pipetting before deciding to adjust the pH of solution. The provided recipe gives consistent polymerization of collagen gels.
6. We have found some variations in the ability of fibroblasts to support oral keratinocyte growth depending on the fibroblast origin. Primary oral fibroblasts provide the best oral keratinocyte growth and differentiation when compared to foreskin fibroblasts or murine NIH3T3 cells.
7. This three-dimensional organotypic culture is modified from a reproducible and successful skin organotypic culture method described by the Herlyn (14). Although the method in this chapter has been developed and adapted for oral keratinocytes, it can be modified for other oral cells as well. For example, to culture head and neck squamous cell carcinomas (HNSCC), the majority of which are derived from the malignant transformation of oral epithelial cells, the cells will need more supplement nutrients in their culture media that allow optimal growth in epithelial like culture as described (15, 16).
8. The three-dimensional organotypic culture is a robust method that can be used in various approaches, such as the detailed study of cell-to-cell interactions (17), biomaterial compatibility (18), wound healing (19), cell migration (20), and tumor progression (21), among others.

Funding

This research was supported by the Intramural Research Program of the National Institutes of Health, National Institute of Dental and Craniofacial Research, project Z01DE00558. The funders had no role in study design, data collection and analysis, decision to publish, or preparation of the manuscript.

References

1. Nanci A, Ten Cate R (2008) Ten Cate's oral histology: development, structure, and function. Mosby Elsevier Health Sciences, St. Louis, MO
2. Wertz PW, Squier CA (1991) Cellular and molecular basis of barrier function in oral epithelium. Crit Rev Ther Drug Carrier Syst 8:237–269
3. Castilho RM, Squarize CH, Leelahavanichkul K, Zheng Y, Bugge T, Gutkind JS (2010) Rac1 is required for epithelial stem cell function during dermal and oral mucosal wound healing but not for tissue homeostasis in mice. PLoS One 5:e10503
4. Xu X, Lyle S, Liu Y, Solky B, Cotsarelis G (2003) Differential expression of cyclin D1 in

the human hair follicle. Am J Pathol 163: 969–978

5. Levy V, Lindon C, Zheng Y, Harfe BD, Morgan BA (2007) Epidermal stem cells arise from the hair follicle after wounding. FASEB J 21:1358–1366
6. Langton AK, Herrick SE, Headon DJ (2008) An extended epidermal response heals cutaneous wounds in the absence of a hair follicle stem cell contribution. J Invest Dermatol 128: 1311–1318
7. Ito M, Liu Y, Yang Z, Nguyen J, Liang F, Morris RJ, Cotsarelis G (2005) Stem cells in the hair follicle bulge contribute to wound repair but not to homeostasis of the epidermis. Nat Med 11:1351–1354
8. Levy V, Lindon C, Harfe BD, Morgan BA (2005) Distinct stem cell populations regenerate the follicle and interfollicular epidermis. Dev Cell 9:855–861
9. Grossman N, Slovik Y, Bodner L (2004) Effect of donor age on cultivation of human oral mucosal keratinocytes. Arch Gerontol Geriatr 38:114–122
10. Kedjarune U, Pongprerachok S, Arpornmaeklong P, Ungkusonmongkhon K (2001) Culturing primary human gingival epithelial cells: comparison of two isolation techniques. J Craniomaxillofac Surg 29:224–231
11. Sundqvist K, Liu Y, Arvidson K, Ormstad K, Nilsson L, Toftgard R, Grafstrom RC (1991) Growth regulation of serum-free cultures of epithelial cells from normal human buccal mucosa. In Vitro Cell Dev Biol 27A:562–568
12. Neugebauer P, Bonnekoh B, Wevers A, Michel O, Mahrle G, Krieg T, Stennert E (1996) Human keratinocyte culture from the peritonsillar mucosa. Eur Arch Otorhinolaryngol 253: 245–251
13. Rickert D, Franke RP, Fernandez CA, Kilroy S, Yan L, Moses MA (2007) Establishment and biochemical characterization of primary cells of the upper aerodigestive tract. Clin Hemorheol Microcirc 36:47–64
14. Meier F, Nesbit M, Hsu MY, Martin B, Van BP, Elder DE, Schaumburg-Lever G, Garbe C, Walz TM, Donatien P et al (2000) Human melanoma progression in skin reconstructs: biological significance of bFGF. Am J Pathol 156:193–200
15. Margulis A, Zhang W, Garlick JA (2005) In vitro fabrication of engineered human skin. Methods Mol Biol 289:61–70
16. Formanek M, Millesi W, Willheim M, Scheiner O, Kornfehl J (1996) Optimized growth medium for primary culture of human oral keratinocytes. Int J Oral Maxillofac Surg 25:157–160
17. Schaller M, Weindl G (2009) Models of oral and vaginal candidiasis based on in vitro reconstituted human epithelia for the study of host-pathogen interactions. Methods Mol Biol 470:327–345
18. Tomakidi P, Schuster G, Breitkreutz D, Kohl A, Ottl P, Komposch G (2000) Organotypic cultures of gingival cells: an epithelial model to assess putative local effects of orthodontic plate and occlusal splint materials under more tissue-like conditions. Biomaterials 21:1549–1559
19. Egles C, Garlick JA, Shamis Y (2010) Three-dimensional human tissue models of wounded skin. Methods Mol Biol 585:345–359
20. Nystrom ML, Thomas GJ, Stone M, MacKenzie IC, Hart IR, Marshall JF (2005) Development of a quantitative method to analyse tumour cell invasion in organotypic culture. J Pathol 205:468–475
21. Yoo GH, Washington J, Piechocki M, Ensley J, Shibuya T, Oda D, Wei WZ (2000) Progression of head and neck cancer in an in vitro model. Arch Otolaryngol Head Neck Surg 126: 1313–1318

Chapter 6

Primary Culture of the Human Olfactory Neuroepithelium

Nancy E. Rawson and Mehmet Hakan Ozdener

Abstract

The central cell type involved in the initial perception of odors and transduction of the sensory signal are the olfactory receptor neurons (ORNs) located in the olfactory neuroepithelium of the nasal cavities. The olfactory epithelium is a unique system similar to the neuroepithelium of the embryonic neural tube, in which new neurons are continually generated throughout adult life. Olfactory neurons are derived from precursor cells that lie adjacent to the basal lamina of the olfactory epithelium; these precursor cells divide several times and their progeny differentiate into mature sensory neurons throughout life. Thus, the human olfactory epithelium has the potential to be used as a tool to examine certain human disorders resulting from abnormal development of the nervous system. This chapter presents methods for primary culture of human ORNs, which have been used successfully by multiple investigators. The protocol provides a consistent, heterogeneous cell population, which demonstrates functional responses to odorant mixtures and exhibits a complex neuronal phenotype, encompassing receptors and signaling pathways pertinent to both olfaction and other aspects of CNS function. These cultured neural cells exhibit neurotransmitter pathways important in a number of neuropsychiatric disorders, and the ability to culture cells from living human subjects provides a tool for assessing cellular neuropathology at the individual patient level.

Key words: Olfactory, Smell, Odor, Neuroregeneration, Neuropsychiatric disorders

1. Introduction

Our understanding of olfactory cell function has arisen largely from studies of dissociated cells and tissues. In humans, the olfactory neuroepithelium is accessible via biopsy using relatively simple procedures, and olfactory epithelial biopsies have been established as a viable approach to obtain fresh sensory epithelial tissue for physiological and molecular studies (1, 2). Such tissue represents a uniquely accessible neuroepithelium available from living subjects, and studies utilizing this tissue have provided novel insights into the neuropathology of psychiatric disorders such as bipolar depression (3), schizophrenia (4, 5), and Rett's syndrome (6). While ex vivo preparations provide the ability to examine the native cells,

Scott H. Randell and M. Leslie Fulcher (eds.), *Epithelial Cell Culture Protocols: Second Edition*, Methods in Molecular Biology, vol. 945, DOI 10.1007/978-1-62703-125-7_6, © Springer Science+Business Media, LLC 2012

they are subject to the effects of dissociation and tissue extraction trauma, which may compromise and alter their cellular physiology. In addition, the availability of fresh tissue may be limited due to the need to obtain a regular pool of willing subjects and to engage the services of an otorhinolaryngologist trained in the procedure to extract the biopsies. As a regenerative tissue, the olfactory epithelium contains a population of progenitor cells capable of replicating and differentiating to replace mature neurons as they are lost due to injury, infection, or normal senescence (7). In view of this, several laboratories have now developed protocols to establish primary cultures of the human olfactory epithelium (hOE) using biopsies from the sensory epithelium as a starting point. Olfactory-specific functional and molecular properties are retained by cells in these preparations, although the classic bipolar morphology is generally lost, particularly with extended passaging.

These primary cultured cells provide the means to examine stimulus–receptor cell interactions, intracellular signaling pathways, and processes involved in modulation of responses in a system amenable to higher throughput assay protocols. Further the cultures can be used to study the pathways involved in proliferation and differentiation in a controlled environment that supports cells which retain key features characteristic of their native counterparts. Accordingly, the method provides a tool for examination of cellular lineages, and to assess the impact of growth factors and xenobiotics on cellular proliferation, differentiation, and survival. Most importantly, the ability to obtain cultures from individual, living human subjects represents a unique opportunity to correlate phenotype with individual genotype with respect to the metabolic, neurophysiological, and pharmacological responses of a neuronal cell population. This approach could one day provide the basis for personalized therapeutics based on the neurophysiological properties of the individual.

Molecular markers are typically used to define cellular phenotype, but may not fully predict functional characteristics. Accordingly, both molecular and functional approaches should be used to establish the neuronal cell characteristics in these primary cultures. A number of neuronal and non-neuronal markers identified in hOE have been used for culture characterization (7). Key among these are olfactory marker protein, neuron-specific tubulin, neural cell adhesion molecule, and adenylate cyclase III (ACIII), an element of the olfactory transduction pathway (8–12). Olfactory marker protein is a prototypical and specific marker for mature olfactory receptor (OMP) neurons (13) while neuron-specific tubulin and neural cell adhesion molecule are found in the immature and mature olfactory neurons in vivo and in vitro (1, 10, 14). Adenylate cyclase and other components of the signal transduction pathway are key markers of function both for odorant receptors and neurotransmitter response. Unfortunately, no reliable antibodies for human odorant receptors (ORs) are presently available, so molecular methods are required to verify the presence and population of ORs that are expressed in vitro. Cells generated using the protocol described here have been shown to express several ORs using both PCR and in situ hybridization as well as relevant transduction and neurotransmission elements (1).

The methods described here provide for the generation, maintenance, and characterization of primary hOE cultures as used in our laboratory. Factors contributing to variability and success are noted.

1.1. Anatomy of the Human Nasal Cavity and Olfactory Epithelium

The sensory epithelium, roughly the area of a postage stamp, occupies the olfactory cleft and projects onto the dorsal portion of the superior turbinate. In humans, the extent of the neuroepithelial sheet varies within and among individuals, in relation to age and disease, becoming patchy and interspersed with respiratory or fibrotic tissue (15–17).

The olfactory neuroepithelium is a pseudostratified epithelium comprising multiple cell types. Mature olfactory sensory neurons (OSNs) are bipolar cells classically identified by the expression of olfactory marker protein (18). These cells are derived from a population of proliferative basal cells that produce precursors which differentiate into immature neurons expressing Gap-43 and neuron-specific tubulin (17, 19). In addition, supporting cells including microvillar and sustentacular cells provide trophic and protective support, expressing detoxification enzymes such as cytochrome P450 (20). These differentiated cell types sit on a layer of horizontal and globose basal cells. Horizontal basal cells may represent a function comparable to a blood–brain barrier; although little is known about their physiological properties, they are molecularly defined by the expression of specific cytokeratins (e.g., CK-18) and in human (but not rodent), OE, express p75-nerve growth factor receptor (21, 22).

Unlike the very orderly layers of OMP-ir OSNs observed in rodent olfactory epithelium, the frequency of OMP-ir OSNs in the human middle turbinate tends to be sparse and scattered. This accounts for the approximately 60% success rate in obtaining sensory epithelium from biopsies obtained from this region (23–25).

2. Materials

2.1. Human Olfactory Epithelial Biopsy

The main sensory region is located within the olfactory cleft, separated from the cranial cavity by a thin, porous bone called the cribriform plate, and the fragility of this has led current investigators to obtain tissue from the superior aspect of the middle turbinate and apposed septum rather than the olfactory cleft. Detection thresholds for one or more odorants may be obtained unilaterally to insure that the individual's sense of smell is normal on the side from which the biopsy is obtained (2). However, it must be noted that there is only a minimal association between the likelihood of obtaining OSNs from a biopsy and that individual's olfactory performance, given that the biopsy represents only a tiny fraction of the total epithelial sheet. Over 700 biopsies from healthy subjects and a variety of patient populations have been obtained using this technique without severe adverse events (refs. 2–7, 10, 14, 23–28 and unpublished records from the Monell-Jefferson Taste and Smell Center, Philadelphia, PA).

1. 0.5% Phenylephrine hydrochloride spray and tetracaine hydrochloride, 0.5%, to facilitate visualization and anesthetization.
2. Small sponge plaget such as a Merocel® Otic Wick.
3. Xylocaine (e.g., Lidocaine, 4%).
4. Sterile Giraffe forceps or a similar instrument.

2.2. Cell Culture

1. T25 (25 cm^2) and T75 (75 cm^2) tissue culture flasks.
2. PBS—phosphate-buffered saline with calcium and magnesium.
3. Trypsin/EDTA—0.05% Trypsin 0.53 mM EDTA (Gibco®/BRL).
4. Iscove's Modified Eagle's Medium (IMEM) (Gibco®/BRL) supplemented with 10% fetal bovine serum (FBS, HyClone) and a triple cocktail of antibiotics (100 U/mL/100 μg/mL, penicillin/streptomycin, and 0.5 μg/mL amphotericin B).
5. Preparation of coverslips: Prior to use, treat coverslips with 2 M NaOH for 1 h and leave overnight in 70% nitric acid (HNO_3). Wash coverslips with 9 M HCl acid for 1 h, autoclave in water, rinse with 70% ethanol and 100% ethanol, and then air-dry.
6. Isolation solution (Hank's buffer): 145 mM NaCl, 5 mM KCl, 2 mM EDTA, 1 mM Na-Pyruvate, 20 mM HEPES, 100 μg/mL gentamicin.
7. 15 mL conical tubes.
8. Cryovials.
9. Freezer containers, e.g., Nalgene Nunc Cryo 1°C Mr. Frosty Freezing Container.
10. Freezing medium for cryopreservation: Add DMSO into FBS to make final volume 5%.

2.3. Immunocytochemistry

1. 4% Paraformaldehyde in PBS.
2. Peroxidase blocking solution: 4 mL PBS, 0.5 mL 100% methanol (4°C), 0.5 mL 30% H_2O_2 (4°C).

3. Blocking and antibody diluting solution: 3% Normal goat serum, 3% bovine serum albumin, 0.3% Triton™ X-100 in PBS.
4. Primary antibodies: Polyclonal antibody against olfactory marker protein (OMP) made in goat (Wako), 1:100 dilution; polyclonal antibody against the tyrosine receptor kinase B made in rabbit (Trk-B; Santa Cruz Biotechnology®), 1:250 dilution; and polyclonal anti-adenylate cyclase III (ACIII) made in rabbit (Santa Cruz Biotechnology®) at 1:200 dilution.
5. Secondary antibodies were Alexa Fluor® 488-conjugated anti-goat IgG made in donkey (Molecular Probes®) or Alexa Fluor® 633-conjugated anti-rabbit IgG made in goat (Molecular Probes®).
6. Mounting medium and nuclear stain: VectaShield® with 4,6-diamidino-2-phenylindole (DAPI) (Vector Labs).

2.4. Calcium Imaging

1. Mammalian Ringer's solution: 145 mM NaCl, 5 mM KCl, 1 mM $CaCl_2$, 1 mM $MgCl_2$, 1 mM sodium pyruvate, and 20 mM Na–HEPES; adjust pH to 7.1–7.2 and osmolarity to 300–310 mOsmol/L using 5 M NaCl, then filter sterilize.
2. For high K^+ stimulation, 130 mM of NaCl is replaced with KCl. For Ca^{2+}-free Ringer's, $CaCl_2$ is removed and replaced with 1 mM ethyleneglycol tetraacetic acid (EGTA).
3. For the isolation solution, $CaCl_2$ and $MgCl_2$ are omitted from the mammalian Ringer's and replaced with 2 mM ethylenediamine tetraacetic acid (EDTA).
4. Cell loading solution: 1 mM Fura-2 AM (Molecular Probes®) in 10 mg/mL Pluronic F127 (Molecular Probes®) in Ringer's solution.
5. Odors: A variety of odor stimuli have been used in functional assays. Odors should be the purest available and stored in amber, airtight vials, frozen and/or under liquid nitrogen to minimize deterioration. Some odors exhibit autofluorescence and negative controls should always be used to avoid artifacts. For general culture characterization, a mixture of odors is typically used to increase the likelihood of finding responsive cells. In our studies, Odor Mix A consists of hedione, geraniol, phenylethyl alcohol, citralva, citronellal, eugenol, and menthone, while Odor mix B contained lyral, lilial, triethylamine, ethylvanillin, isovaleric acid, and phenylethyl amine, freshly dissolved at 100 μM in Ringer's Solution with sonication. Oxidation products may exhibit an odor different from the fresh compound, so experiments aiming to relate molecular features to a particular odorant response require assurance of the odor identity and purity using an appropriate analytical technique.

3. Methods

3.1. Harvest, Isolation, Culture, and Maintenance of Human Olfactory Cells

1. Spray the anterior portion of the superior turbinate and apposed nasal septum (see Note 1) with 0.5% phenylephrine hydrochloride spray and tetracaine hydrochloride, 0.5%, to facilitate visualization and anesthetization (see Note 2).
2. Wait 10–15 min for anesthesia of the biopsy site. After anesthetization, biopsies of approximately 1 mm^3 are obtained using a giraffe forceps or a similar instrument from the turbinate and apposed septum and immediately transferred to culture medium (see Subheading 2.2, step 4) for transport to the laboratory (see Note 3).
3. Subjects are observed for 10–15 min after the procedure in the event of nosebleed, and are asked to refrain from strenuous activity or blowing their nose for several hours after the procedure.
4. Biopsy specimens are transferred into isolation solution (see Subheading 2.2, step 6) and minced finely with iridectomy scissors.
5. The minced tissue is incubated for ~30 min at room temperature inside cell culture hood.
6. Cells are dissociated by trituration with a fire-polished pipette and centrifugation at 600 × *g* for 5 min at room temperature.
7. The pellet is resuspended in 3 mL culture medium and transferred to a 25 cm^2 culture flask.
8. Flasks are maintained in a humidified incubator (37°C, 5% CO_2).
9. Allow to grow for 2–4 weeks until cell growth is sufficient for transfer to continuous culture.

3.2. Propagation of Human Olfactory Cells

1. Replace 1/3 of the culture medium every 6–7 days until cultured cells have reached 80–90% confluence (see Fig. 1a, b; Note 4).
2. To passage cells, wash cells once with sterile PBS, then trypsinize cells using 0.25% (w/v) trypsin/EDTA for 2–3 min at 36°C, and gently collect detached cells into a centrifuge tube.
3. After centrifugation (see Subheading 3.1, step 7), resuspend cells in Iscove's medium and transfer to fresh T75 flasks (passage 1).
4. Replace 1/3 of medium every 6–7 days.
5. Repeat steps 2 and 3 when cells have reached 80–90% confluence.
6. Cells are now ready to be processed for immunohistochemistry, RNA isolation, calcium imaging, or other applications.
7. For growth on coverslips for functional assays or immunocytochemistry, transfer approximately 2,000–4,000 of suspended

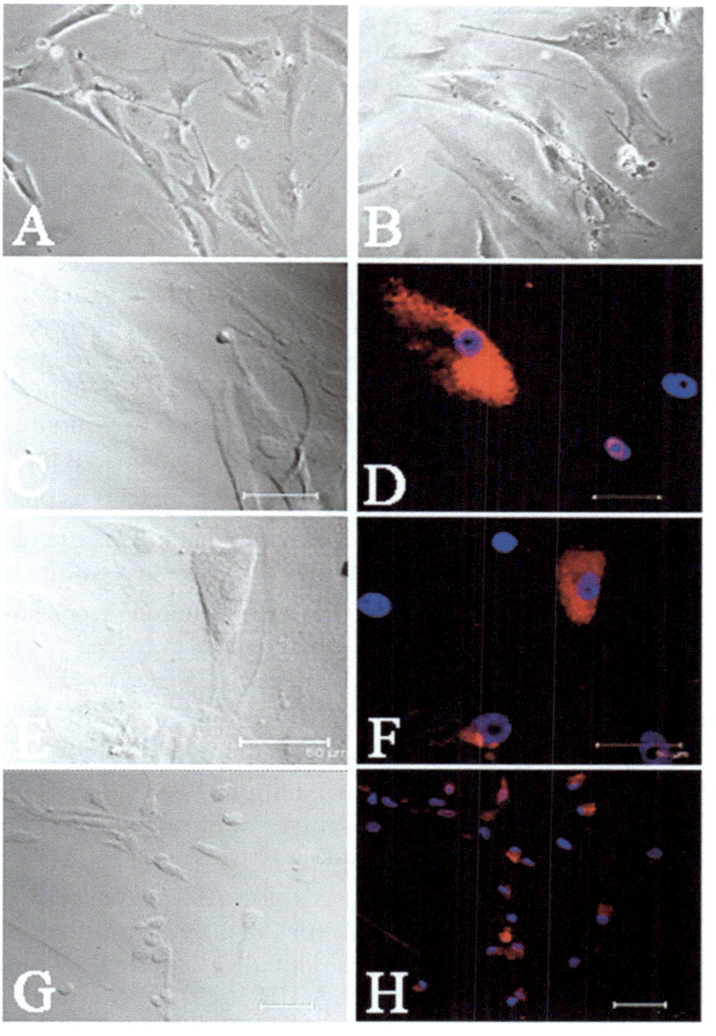

Fig. 1. Morphology and immunostaining of cultured human olfactory cells. Human olfactory cells were cultured and maintained without losing their immunologic and physiological properties. (**a** and **b**) Primary olfactory cell cultures were grown and maintained for up to 4–6 months. Immunostaining of cultured human olfactory cells showed the presence of specific biomarkers. (**c**, **e**, and **g**) Transmission images of corresponding fields are shown on the *left*. (**d**) Immunoreactivity was observed for olfactory marker protein in cultured human olfactory cells. (**f** and **h**) Immunoreactivity for Trk-B and adenylate cyclase III was also observed in cultured human olfactory cells. Nuclei of cells were stained blue with DAPI. Scale bars = 50 μm.

cells to cleaned, sterilized uncoated coverslips and maintain in culture media for up to 3–5 days (90% confluence).

8. Split cells at a 1:4 dilution in a T75 flask for maintaining adequate growth of the cells over time (see Note 5). At this stage, aliquots may be frozen for future use or used for experimental analysis.

3.3. Freezing and Thawing Cultured Human Olfactory Cells

1. To freeze stocks of primary olfactory cells, after trypsinization, add complete Iscove's medium and transfer cells to sterile 15 mL conical centrifuge tubes. Centrifuge at $600 \times g$ for 5 min at room temperature.
2. Carefully discard the supernatant and gently resuspend cells with appropriate volume of freezing medium (see Subheading 2.2, step 10).
3. Transfer cells to labeled, sterile cryovials, cap tightly, and place in a freezing container containing isopropanol. Place into a −80°C freezer for at least 1 day prior to transferring indefinitely to liquid nitrogen or −80°C freezer. We did not observe significant differences in viability of cells stored at −80°C or liquid nitrogen for up to 6 months.
4. For thawing, transfer cryovial directly to 37°C water bath and swirl vial to thaw as quickly as possible.
5. Spray vial with ethanol. Transfer cells into a sterile T25 or T75 cell culture flask with cell culture medium.
6. Continue to culture cells according to Subheading 3.2.

3.4. Confocal Immunofluorescence for Olfactory Cell Markers

1. Human olfactory cells grown on sterile, uncoated coverslips for 3–5 days were fixed with 4% paraformaldehyde in 0.1 M phosphate buffer (pH 7.2) for 10 min at room temperature.
2. After washing in PBS, the cells were treated with peroxidase solution (see Subheading 2.3, step 2) to remove endogenous peroxidase activity.
3. Cells were blocked with blocking solution (see Subheading 2.3, step 3) for 30–60 min and then incubated with primary antibodies diluted in blocking solution (OMP at 1:100; TrkB at 1:250; ACIII at 1:200) overnight at 4°C.
4. After washing with PBS, cells were then incubated with the appropriate secondary antibody diluted 1:500 in blocking buffer for 30 h at room temperature.
5. After washing in PBS (3×15 min) and water (3×20 min), coverslips were mounted with Vectashield® with DAPI (Vector Laboratories).
6. Fluorescent images were captured with a Leica TCS SP2 Spectral Confocal Microscope (Leica Microsystems) (see Fig. 1c–h, Note 6).

3.5. Calcium Imaging for Functional Assays

1. Aliquots of 2,000–4,000 suspended cells are transferred onto sterile, uncoated coverslips in Petri dishes and grown for 2–4 days in a humidified incubator to achieve 80–90% confluence.
2. Cells are loaded calcium-sensitive dye by applying ~0.5 mL loading solution and incubating for 1 h at 36°C.

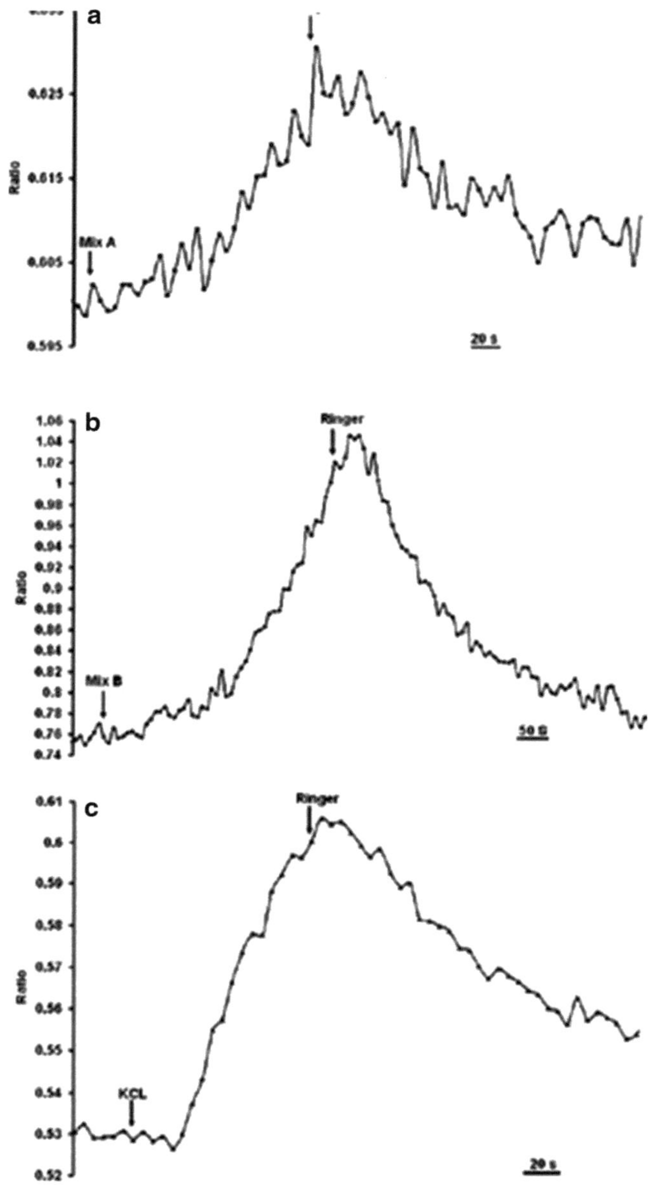

Fig. 2. Cultured human olfactory cells respond to odor stimuli. Changes in intracellular calcium levels ($[Ca^{2+}]_i$) in cultured human olfactory cell were measured using Fura-2. Graphs illustrate representative changes in $[Ca^{+2}]_i$ levels in individual cells during exposure to (**a**) Mix A, (**b**) Mix B, and (**c**) potassium chloride (KCl 40 mM).

3. Cells were first superfused with Ringer's solution, then exposed to odorant stimuli (see Subheading 2.4) via superfusion, with each stimulus applied for 30–60 s, and washed with Ringer's solution for at least 2 min (see Fig. 2 and Note 7).
4. Tracking of changes in cytosolic fluorescence is performed using standard imaging equipment and data analysis procedures (see refs. 28, 33 and Note 8).

5. Intracellular calcium may either increase or decrease in response to odor stimulation, although the mechanism and role for odorant-induced calcium decreases in odorant signaling are unclear (2). Following a response, calcium levels should return to baseline. Odor stimulation should be organized to minimize repetitive exposures to the extent possible to minimize adaptation. Individual cells rarely respond to multiple odors and the proportion of odor-responsive cells may vary across cultures and days in culture (see Note 8).

4. Notes

1. This region contains olfactory epithelium in approximately 60% of healthy subjects (27). Subjects are asked to refrain from taking anticoagulants such as aspirin prior to a biopsy. Medical history is obtained to insure that other health criteria are met. These medical criteria may vary depending on the purpose of the study, and viable olfactory neurons have been obtained from individuals ranging in age from 12 to 84, smokers and nonsmokers, and subjects with various psychiatric or neurological disorders (2–7, 10, 12, 14, 15, 23–33).
2. Alternately, following the epinephrine/tetracaine spray, additional anesthetic may be applied by placing a small sponge plaget such as a Merocel® Otic Wick soaked in xylocaine between the middle turbinate and nasal septum. This procedure reduces the degree of pressure sensed by the subject during the procedure, but also increases the risk of tissue damage from the sponge plaget and potential adverse effects of the xylocaine on the OE cells (34). Care must be taken by the surgeon to inspect the nasal cavity for any signs of infection or inflammation, and for adequate access to the region to be biopsied. We have observed that samples containing blood tend to be less viable. Nosebleed constitutes the most likely difficulty, and has been readily addressed with standard office procedures.
3. In our hands, this process typically takes about 20–30 min, but longer delays of up to 2 h were not found to significantly impact viability.
4. Cell growth rates vary, likely in relation to the proportion of proliferating precursors in the biopsy sample. In some cases, 2 weeks may be sufficient to attain 80% confluence, while other samples grow more slowly and 4–5 weeks was required (see Fig. 1a, b). Either culture type could generate odorant-responsive cells, and could be propagated further, although the impact of this difference on the proportion of cells expressing OMP or responding to odor stimuli has not be directly compared.

5. The culture protocol described here takes advantage of the mixed cell population and the use of serum to promote growth and differentiation. The full complement of growth factors required for this process remains unknown, but a variety of conditions and factors have been investigated to determine whether a more defined medium may be achieved. Among these are basic fibroblast growth factor, nerve growth factor, and dopamine (9, 31, 32). While some features of growth or differentiation are influenced modestly by these factors, a sufficiently comprehensive analysis of the complex OE environment governing proliferation and differentiation has yet to be achieved to construct a defined media that may be considered a true reflection of that environment.
6. The proportion of cells expressing proteins for neuronal markers varies among biopsies and across passages. The proliferation rate generally declines across passages and days post plating, but the specific rates of proliferation vary greatly across cultures.
7. As activation of odorant receptors triggers the influx of extracellular calcium (10), calcium imaging has been widely used to evaluate odorant responsiveness. Ratiometric imaging has the advantage of minimizing effects due to leaching of dye. Values obtained from such experiments are not absolute intracellular calcium concentrations unless a calibration is performed for each field of cells imaged. Components and considerations for experimental design and equipment necessary for conducting such experiments are described in ref. (33).
8. The prevalence of odorant-responsive cells typically peaks between day 2 and 4 of culture when cells are plated on coverslips, and declines thereafter. In contrast, the prevalence of cells exhibiting voltage-sensitive calcium responses is typically lower initially and increases during days 5–6 post plating (unpublished data). It is critical to characterize the cultures using both molecular and functional measures to insure that cells expressing the phenotype of interest are present. Rather than attempting to achieve homogeneity in these parameters, one may exploit the individuality of each culture to provide insight into the donor.

References

1. Borgmann-Winter KE, Rawson NE, Wang HY, Wang H, Macdonald ML, Ozdener MH, Yee KK, Gomez G, Xu J, Bryant B et al (2009) Human olfactory epithelial cells generated in vitro express diverse neuronal characteristics. Neuroscience 158:642–653
2. Rawson NE, Gomez G, Cowart B, Brand JG, Lowry LD, Pribitkin EA, Restrepo D (1997) Selectivity and response characteristics of human olfactory neurons. J Neurophysiol 77:1606–1613
3. Hahn CG, Gomez G, Restrepo D, Friedman E, Josiassen R, Pribitkin EA, Lowry LD, Gallop RJ, Rawson NE (2005) Aberrant intracellular calcium signaling in olfactory neurons from patients with bipolar disorder. Am J Psychiatry 162:616–618
4. Matigian N, Abrahamsen G, Sutharsan R, Cook AL, Vitale AM, Nouwens A, Bellette B, An J,

Anderson M, Beckhouse AG et al (2010) Disease-specific, neurosphere-derived cells as models for brain disorders. Dis Model Mech 3:785–798
5. McCurdy RD, Feron F, Perry C, Chant DC, McLean D, Matigian N, Hayward NK, McGrath JJ, Mackay-Sim A (2006) Cell cycle alterations in biopsied olfactory neuroepithelium in schizophrenia and bipolar I disorder using cell culture and gene expression analyses. Schizophr Res 82:163–173
6. Ronnett GV, Leopold D, Cai X, Hoffbuhr KC, Moses L, Hoffman EP, Naidu S (2003) Olfactory biopsies demonstrate a defect in neuronal development in Rett's syndrome. Ann Neurol 54:206–218
7. Hahn CG, Han LY, Rawson NE, Mirza N, Borgmann-Winter K, Lenox RH, Arnold SE (2005) In vivo and in vitro neurogenesis in human olfactory epithelium. J Comp Neurol 483:154–163
8. Baker H, Grillo M, Margolis FL (1989) Biochemical and immunocytochemical characterization of olfactory marker protein in the rodent central nervous system. J Comp Neurol 285:246–261
9. Ensoli F, Fiorelli V, Vannelli B, Barni T, De Cristofaro M, Ensoli B, Thiele CJ (1998) Basic fibroblast growth factor supports human olfactory neurogenesis by autocrine/paracrine mechanisms. Neuroscience 86:881–893
10. Gomez G, Rawson NE, Hahn CG, Michaels R, Restrepo D (2000) Characteristics of odorant elicited calcium changes in cultured human olfactory neurons. J Neurosci Res 62:737–749
11. Vannelli GB, Ensoli F, Zonefrati R, Kubota Y, Arcangeli A, Becchetti A, Camici G, Barni T, Thiele CJ, Balboni GC (1995) Neuroblast long-term cell cultures from human fetal olfactory epithelium respond to odors. J Neurosci 15:4382–4394
12. Wolozin B, Sunderland T, Zheng BB, Resau J, Dufy B, Barker J, Swerdlow R, Coon H (1992) Continuous culture of neuronal cells from adult human olfactory epithelium. J Mol Neurosci 3:137–146
13. Danciger E, Mettling C, Vidal M, Morris R, Margolis F (1989) Olfactory marker protein gene: its structure and olfactory neuron-specific expression in transgenic mice. Proc Natl Acad Sci U S A 86:8565–8569
14. Yee KK, Pribitkin EA, Cowart BJ, Vainius AA, Klock CT, Rosen D, Feng P, McLean J, Hahn CG, Rawson NE (2010) Neuropathology of the olfactory mucosa in chronic rhinosinusitis. Am J Rhinol Allergy 24:110–120
15. Paik SI, Lehman MN, Seiden AM, Duncan HJ, Smith DV (1992) Human olfactory biopsy. The influence of age and receptor distribution. Arch Otolaryngol Head Neck Surg 118:731–738
16. Trojanowski JQ, Newman PD, Hill WD, Lee VM (1991) Human olfactory epithelium in normal aging, Alzheimer's disease, and other neurodegenerative disorders. J Comp Neurol 310:365–376
17. Leopold DA, Hummel T, Schwob JE, Hong SC, Knecht M, Kobal G (2000) Anterior distribution of human olfactory epithelium. Laryngoscope 110:417–421
18. Kream RM, Margolis FL (1984) Olfactory marker protein: turnover and transport in normal and regenerating neurons. J Neurosci 4:868–879
19. Monti-Graziadei GA, Margolis FL, Harding JW, Graziadei PP (1977) Immunocytochemistry of the olfactory marker protein. J Histochem Cytochem 25:1311–1316
20. Ling G, Gu J, Genter MB, Zhuo X, Ding X (2004) Regulation of cytochrome P450 gene expression in the olfactory mucosa. Chem Biol Interact 147:247–258
21. Schlage WK, Bulles H, Friedrichs D, Kuhn M, Teredesai A (1998) Cytokeratin expression patterns in the rat respiratory tract as markers of epithelial differentiation in inhalation toxicology. I. Determination of normal cytokeratin expression patterns in nose, larynx, trachea, and lung. Toxicol Pathol 26:324–343
22. Feron F, Bianco J, Ferguson I, Mackay-Sim A (2008) Neurotrophin expression in the adult olfactory epithelium. Brain Res 1196:13–21
23. Lane AP, Gomez G, Dankulich T, Wang H, Bolger WE, Rawson NE (2002) The superior turbinate as a source of functional human olfactory receptor neurons. Laryngoscope 112:1183–1189
24. Lovell MA, Jafek BW, Moran DT, Rowley JC III (1982) Biopsy of human olfactory mucosa. An instrument and a technique. Arch Otolaryngol 108:247–249
25. Lanza DC, Moran DT, Doty RL, Trojanowski JQ, Lee JH, Rowley JC III, Crawford D, Kennedy DW (1993) Endoscopic human olfactory biopsy technique: a preliminary report. Laryngoscope 103:815–819
26. Rawson NE, Brand JG, Cowart BJ, Lowry LD, Pribitkin EA, Rao VM, Restrepo D (1995) Functionally mature olfactory neurons from two anosmic patients with Kallmann syndrome. Brain Res 681:58–64
27. Féron F, Perry C, McGrath JJ, Mackay-Sim A (1998) New techniques for biopsy and culture of human olfactory epithelial neurons. Arch Otolaryngol Head Neck Surg 124(8): 861–866

28. Restrepo D, Okada Y, Teeter JH, Lowry LD, Cowart B, Brand JG (1993) Human olfactory neurons respond to odor stimuli with an increase in cytoplasmic Ca2+. Biophys J 64:1961–1966
29. Johnson GS, Basaric-Keys J, Ghanbari HA, Lebovics RS, Lesch KP, Merril CR, Sunderland T, Wolozin B (1994) Protein alterations in olfactory neuroblasts from Alzheimer donors. Neurobiol Aging 15:675–680
30. Wolozin B, Lesch P, Lebovics R, Sunderland T (1993) A.E. Bennett Research Award 1993. Olfactory neuroblasts from Alzheimer donors: studies on APP processing and cell regulation. Biol Psychiatry 34:824–838
31. Feron F, Vincent A, Mackay-Sim A (1999) Dopamine promotes differentiation of olfactory neuron in vitro. Brain Res 845:252–259
32. Vawter MP, Basaric-Keys J, Li Y, Lester DS, Lebovics RS, Lesch KP, Kulaga H, Freed WJ, Sunderland T, Wolozin B (1996) Human olfactory neuroepithelial cells: tyrosine phosphorylation and process extension are increased by the combination of IL-1beta, IL-6, NGF, and bFGF. Exp Neurol 142:179–194
33. Restrepo D, Zviman M, Rawson NE (1995) The measurement of intracellular calcium in chemosensory cells. In: Spielman A, Brand J (eds) Methods in chemosensory research. CRC Press, Boca Raton, FL, pp 387–398
34. Sakaguchi M, Yoshihiro Kuroda Y, Hirose M (2006) The antiproliferative effect of lidocaine on human tongue cancer cells with inhibition of the activity of epidermal growth factor receptor. Anesth Analg 102:1103–1107

Chapter 7

Primary Culture of Mammalian Taste Epithelium

Mehmet Hakan Ozdener and Nancy E. Rawson

Abstract

Establishment of primary and immortalized cultures of many cell types has facilitated efforts to understand the signals involved in proliferation and differentiation and yielded tools to rapidly assay new molecules targeting specific receptor pathways. Taste cells are specialized sensory epithelial cells which reside within taste buds on the lingual epithelium. Only recently have successful culturing protocols been developed which maintain essential molecular and functional characteristics. These protocols provide a tractable tool to examine the molecular, regenerative, and functional properties of these unique sensory cells within a controlled environment. The method involves an enzymatic isolation procedure and standardized culture conditions, and may be applied to either dissected rodent tissue or human fungiform papillae obtained by biopsy. Human fungiform cells can be maintained in culture for more than seven passages, without loss of viability and with retention of the molecular and biochemical properties of acutely isolated taste cells. Cultured primary human fungiform papillae cells also exhibit functional responses to taste stimuli indicating the presence of taste receptors and at least some relevant signaling pathways. While the loss of the three-dimensional structure of the intact taste bud must be taken into consideration in interpreting results obtained with these cells, this culture protocol provides a useful model for molecular studies of the proliferation, differentiation, and physiological function of mammalian taste receptor cells.

Key words: Fungiform, Culture, Taste, Gustducin, Taste receptor, Regeneration, Sweet

1. Introduction

Taste receptor cells are specialized epithelial cells with unique histological, molecular, and physiological characteristics that permit detection of a wide range of both simple and structurally complex molecules. In mammals, taste buds are associated with fungiform papillae on the anterior two-thirds of the tongue, and circumvallate and foliate papillae are on the posterior third of the tongue. Taste buds are also present in the epithelium of the soft palate and pharynx (1). Taste receptor molecules and their downstream signaling components have also been found in enteroendocrine cells throughout the lining of the gastrointestinal tract (2).

Scott H. Randell and M. Leslie Fulcher (eds.), *Epithelial Cell Culture Protocols: Second Edition*, Methods in Molecular Biology, vol. 945, DOI 10.1007/978-1-62703-125-7_7, © Springer Science+Business Media, LLC 2012

Human fungiform papillae contain from zero to over 25 taste buds, with over half having no taste buds and the rest having an average of three or four buds (3). Each taste bud contains 50–100 cells of four morphologically and functionally distinct types, which exhibit properties of both neuronal and epithelial cells (4). About half of the cells in the taste bud are spindle-shaped type I (dark) cells, which appear to have a glial like function because they surround other cell types and express molecules involved in neurotransmitter inactivation (5). An additional 25% of the taste bud cells are type II (light) cells, which express several proteins including the G-protein alpha-gustducin, phospholipase C-β2 (PLC-β2), inositol 1,4,5-trisphosphate receptor type 3 (IP3R3), and transient receptor potential channel M5 (TRPM5), which have been implicated in transduction of sweet, umami, and bitter taste responses (6, 7). Current evidence indicates that G-protein-coupled receptors (GPCRs) implicated in sweet and umami taste (T1Rs) and bitter taste (T2Rs) are expressed in nonoverlapping subsets of type II cells (8, 9). Cells mediating sour (acid) taste are likely to be a subset of type III cells (10), which comprise an additional 15% of the taste cells. The cells mediating salty taste have not yet been identified. Importantly, taste buds are one of the four very few truly regenerative organs in the human body, with taste cells having an average life span of about 12 days (11). A small number of type IV or basal cells in each bud are generally considered to be the stem cells giving rise to the other cell types (12, 13). Properties of taste cells often vary among species, and it is presently not known how maturational stage or lineage relates to the stimulus–response properties of these cells (12).

To date, no immortalized taste cell lines or long-term (i.e., months) taste culture methods have been published. Because of the unavailability of a human taste cell culture model, investigators are now dependent on freshly isolated primary cells, explant cultures, or nonhuman species to study taste cell development and physiological properties and require the use of a large number of experimental animals (14, 15). Importantly, these methods are not conducive to screening large number of chemicals for taste activity. Although heterologous systems expressing known specific taste receptors are used routinely to screen the activity of putative taste compounds, results may fail to reflect the actual more complicated taste detection processes. Functional studies of taste cells have been done using freshly isolated cells in primary culture, explant cultures from rodents, or semi-intact taste buds in tissue slices (14, 15). While each of these preparations has advantages, the development of long-term cultures would have provided significant benefits, particularly for studies of taste cell proliferation and differentiation. Most attempts to culture taste cells have reported limited viability, with cells typically not lasting beyond 3–5 days (16–19). We recently reported on a method for the extended culture of rodent taste cells (20). Here, we describe the establishment of long-term primary cultures of cells from human fungiform taste papillae that have molecular and physiological properties consistent with both developing and mature taste cells. In addition, these cultures were amenable to the use of moderate throughput screening (MTS) to examine responses to individual or combinations of taste stimuli. The establishment of this human fungiform cell culture protocol provides an important in vitro model to study intracellular signaling mechanisms, the impact of trophic or toxic agents on taste cell growth and function, and the assessment of potential taste stimuli and taste modifiers.

2. Materials

2.1. Human Fungiform Taste Papillae Biopsy

1. Biopsy instruments: Small fine-tip forceps and extra fine spring scissors (Fine Science Tools), surgical razor.
2. Taste cell isolation solution: 26 mM $NaHCO_3$, 2.5 mM NaH_2PO_4, 20 mM glucose, 65 mM NaCl, 20 mM KCl, and 1 mM EDTA dissolved in nuclease-free water and filter sterilize.
3. Enzyme mixture: Mix collagenase type 1 (550 U/mL, Worthington), elastase (10 U/mL, Worthington), and soy bean trypsin inhibitor (0.9 mg/mL, Worthington) in calcium-free Ringer solution just before use.
4. Enzyme mixture (for alternative route): Mix pronase E (1.5 mg/mL, Sigma-Aldrich®), elastase (1 mg/mL, Sigma-Aldrich®) in calcium-free Ringer solution just before use.

2.2. Cell Culture

1. Tissue culture plastics: T25 (25 cm^2), T75 (75 cm^2) tissue culture flasks and other tissue culture plastics (i.e., conical tubes, 96-well plates).
2. Iscove's Modified Dulbecco's Medium (IMDM) (Gibco®/BRL) supplemented with 10% fetal bovine serum (FBS, HyClone).
3. MCDB 153 medium (Sigma-Aldrich®): Dissolve a package of MCDB 153 (17.6 g) completely in 1 L of tissue culture grade water, supplemented with 1.18 g/L sodium bicarbonate according to the manufacturer's instructions. pH 7.0–7.1 (see Note 1).
4. Taste cell culture medium: IMDM containing 10% FBS, 1:5 ratio of MCDB 153, and a triple cocktail of antibiotics (100 U/mL/100 μg/mL, penicillin/streptomycin, 2.5 μg/mL gentamicin, and 0.5 μg/mL fungizone) (see Note 1).

5. Preparation of coverslips: Prior to use, treat coverslips with 2 M NaOH for 1 h and leave overnight in 70% nitric acid (HNO_3). Wash coverslips with 9 M HCl acid for 1 h, autoclave in water, rinse with 70% ethanol and 100% ethanol, and then air-dry.
6. Coating coverslips with rat tail collagen type-1: Dilute rat tail collagen type-1 (3.96 mg/mL, BD Sciences) with sterile nuclease-free water at 1:4 ratio. Add 0.5–1 mL of rat tail collagen type-1 onto coverslips into a 12-well tissue culture plate for 15 min at room temperature. Remove rat tail collagen type-1 and let coverslip air-dry for 10–20 min.
7. Cloning cylinder 10–12 mm in diameter, sterilized sealed onto coverslip by using autoclaved grease coated on glass piece.
8. 0.25% (w/v) trypsin/EDTA (Life Technologies™).
9. PBS with calcium and magnesium.
10. Cloning cylinder, 10–12 mm in diameter.

2.3. Cryopreservation of Cultured Cells

1. Freezing medium: Add DMSO into FBS to make final volume 5%.
2. Sterile cryovials, 2 mL.
3. Freezing container (e.g., Mr. Frosty, NUNC).
4. Isopropanol.
5. 15 mL conical tubes.

2.4. Immunocytochemistry

1. 4% Paraformaldehyde in PBS.
2. Deperoxidase solution: 4 mL PBS, 0.5 mL 100% methanol (4°C), 0.5 mL 30% H_2O_2 (4°C).
3. Blocking and antibody dilution solution: 3% Normal goat serum, 3% bovine serum albumin, and 0.3% Triton X-100 in PBS.
4. Primary antibodies (see Table 1).
5. Secondary antibodies (see Table 1).

Table 1
Antibodies used for detecting expression of specific molecules

Primary antibody	Source	Host	Dilution	Secondary antibody	Source	Host	Dilution
Gustducin	SantaCruz	Rabbit	1:500	Anti-rabbit IgG Alexa 633	Molecular Probes	Goat	1:500
PLC-β_2	SantaCruz	Rabbit	1:500	Anti-rabbit IgG Alexa 633	Molecular Probes	Goat	1:500

6. Mounting medium and nuclear stain: Vecta® Shield with 4,6-diamidino-2-phenylindole (DAPI) (Vector Laboratories).
7. Leica TCS SP2 Spectral Confocal Microscope (Leica Microsystems Inc.).

2.5. Reverse Transcription-Polymerase Chain Reaction

1. Total RNA from cultured human fungiform cells.
2. Superscript First Strand Synthesis System for RT-PCR® (Life Technologies™).
3. 1× AmpliTaq® Gold PCR buffer (Applied Biosystems™).
4. 2.5 mM $MgCl_2$: Add 0.0050 g $MgCl_2 \cdot 6H_2O$ into 10 mL ddH_2O. Autoclave solution.
5. 1 mM deoxynucleoside triphosphates: From 100 mM stock of dTTP, dATP, dGTP, dCTP, add 10 μL of each to 60 μL of ddH_2O to make a 100 μL mix of dNTPs.
6. 0.4 μM of each primer: For an amount of 38.5 nM of lyophilized primer provided by the primer supplier, 385 μL of PCR grade sterile water were added to get 100 μM/L stock solution. From this stock solution, to prepare 10 μM/L (or 10 picom/μL) as a PCR working solution, dilute stock solution 1:100 to give a concentration of 10 μM.
7. AmpliTaq® Gold polymerases (Applied Biosystems™).
8. 1× TBE buffer: Dissolve 108 g Tris base, 55 g boric acid, 9.3 g EDTA in 800 mL H_2O and adjust volume to 1 L with additional distilled dH_2O.
9. 2% Agarose gels in 1× TBE buffer.
10. 0.2 μg/mL of ethidium bromide (10 mg/mL): Dissolve 0.2 g ethidium bromide to 20 mL water. Mix well and store at 4°C in the dark.

2.6. Moderate Throughput Screening

1. Modified Ringer solution (pH 7.1–7.2, 300–310 mOsmol): Dissolve 4.67 g NaCl, 0.373 g KCl, 0.203 g $MgCl_2$, 0.147 g $CaCl_2$, 0.110 g Na-pyruvate, and 4.76 g HEPES-Na in 1 L water and adjust pH to 7.1–7.2 and osmolarity to 300–310 mOsm/L by 5 M NaCl. Filter sterilize (0.2 μm).
2. Cell loading solution: 1 mM Fura-2 AM (Molecular Probes Inc.) in 10 mg/mL Pluronic F127 (Molecular Probes Inc.) in modified Ringer solution.
3. Tastant dissolved in modified Ringer solution: 1 mM denatonium; 0.446 g of denatonium benzoate dissolved in 50 mL of modified Ringer solution. 1 mM sucralose; 0.019 g of sucralose dissolved in 50 mL of modified Ringer solution. 250 ppm AceK; 25 mg of AceK was dissolved in 50 mL modified Ringer solution. 3 mM mono potassium glutamate (MPG); 0.025 g of MPG dissolved in 50 mL of modified Ringer solution. All solutions are filter sterilized.

3. Methods

3.1. Isolation, Culture, and Maintenance of Human Fungiform Cells

1. Remove 4–8 human fungiform taste papillae from the dorsal surface of the anterior portion of the tongue using curved spring microscissors.
2. Place immediately taste papillae into the taste isolation solution (see Subheading 2.1, step 2).
3. Digest fungiform papillae with collagenase, elastase, and soy bean trypsin inhibitor (see Subheading 2.1, step 3) under O_2 bubbling in water bath with circulation for 30 min (see Note 2).
4. For alternative digestion protocol, please refer to Note 2.
5. Remove taste isolation solution and add 1 mL of taste cell culture medium (see Subheading 2.1, step 4).
6. Transfer digested fungiform papillae into glass dish.
7. Dissect fungiform papillae with surgical razor. Dissect gently to dissociate tissue pieces.
8. Add 250 μL of dissected papillae into cloning cylinder onto rat tail collagen type-1-coated coverslip (see Note 3).
9. Add 1 mL of taste cell culture medium into each well.
10. Incubate plate at 36°C (see Note 4) in a humidified incubator containing 5% CO_2.
11. Place in an incubator undisturbed for 2 days prior to the first change of complete medium. Remove cloning cylinder from plate and aspirate medium completely. Add 1 mL of taste cell medium into each well (see Fig. 1a).
12. Replace 1/3 of medium every 6–7 days.

3.2. Passage of Cultured Human Fungiform Cells

1. Once 40–50% of the site of cloning cylinder is covered in expanding taste cells (see Fig. 1b), trypsinize cells using 0.25% (w/v) trypsin/EDTA for 2–3 min at 36°C.
2. Remove cells from well(s) and add into a 15 mL conical tube. Add 3 volumes of taste cell culture medium followed by centrifugation at 1,500 × *g* for 5 min at room temperature.
3. Remove supernatant and resuspend cells with 1 mL of taste cell medium.

3.3. Propagation of Human Fungiform Taste Cells

1. Transfer cells into T25 flask and add 4 mL of taste cell medium. Maintain cells at 36°C (see Note 5) in a humidified incubator containing 5% CO_2 (see Fig. 1c).
2. Replace 1/3 of medium every 6–7 days until cultured taste cells have reached 100% confluence (see Fig. 1d).

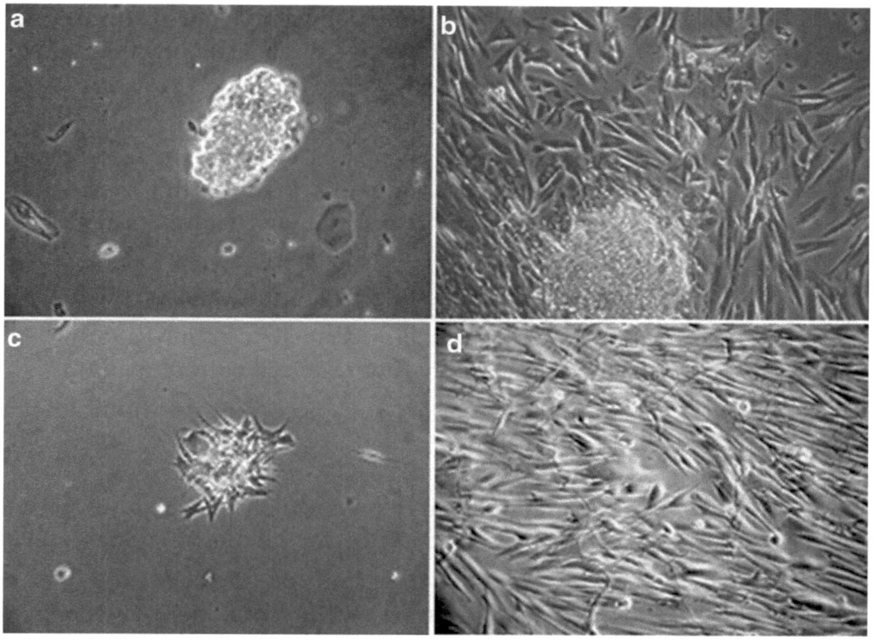

Fig. 1. Attachment and morphology of cultured human fungiform taste cells. (**a**) Primary cell cultures grown on collagen type-1-coated plates were imaged after 2 days. (**b**) Cells from human fungiform papillae grew for up to 2–4 weeks under attached cell clusters. (**c** and **d**) Represent day 2 and 4 weeks after harvesting, respectively. During this period we did not observe growth of cells with the appearance of non-taste epithelial cells.

3. To passage taste cells, wash cells once with sterile PBS and then trypsinize cells using 0.25% (w/v) trypsin/EDTA for 2–3 min at 36°C.
4. After centrifugation as described above, resuspend in complete taste cell medium and transfer cells to fresh T-75 flasks (passage 1).
5. Replace 1/3 of medium every 6–7 days.
6. Repeat steps 3 and 4 when cells have reached close to 100% confluence.
7. Cells are now ready to be processed for immunohistochemistry, RNA isolation, calcium imaging, or other applications.
8. Split cells at a highest 1:4 dilution in a T75 flask for maintaining adequate growth of the cells over time. We then choose whether to proceed with freezing numerous vials of passage-1 cells for archival purposes.

3.4. Freezing and Thawing Cultured Human Fungiform Cells

1. To freeze stocks of primary taste cells, after trypsinization (see Subheading 3.2, step 1), add complete taste cell medium and transfer cells to sterile 15 mL conical centrifuge tubes. Centrifuge at 1,000 × g for 5 min at room temperature.

2. Carefully discard the supernatant and gently resuspend cells with appropriate volume of freezing medium (see Subheading 2.3, step 1).
3. Transfer cells to labeled, sterile cryovials, cap tightly, and place in a freezing container containing isopropanol (see Subheading 2.3). Place into a −80°C freezer for at least 1 day prior to transferring indefinitely to liquid nitrogen.
4. To thaw a cryovial, transfer a vial of cells directly to 37°C water bath and swirl vial to thaw as quickly as possible.
5. Spray vial with ethanol. Transfer cells into a sterile T25 or T75 cell culture flask having taste cell culture medium.
6. Continue to culture cells according to Subheadings 3.1–3.6 (see Note 5).

3.5. Confocal Immunofluorescence for Taste Cell Markers

1. Human fungiform cells grown on coverslips were fixed with 4% paraformaldehyde in PBS for 10 min at room temperature.
2. After washing coverslips in PBS, the cells were treated with 3% deperoxidase solution (see Subheading 2.4, step 2) to remove endogenous peroxidase activity.
3. Cells were blocked with blocking solution (see Subheading 2.4, step 3) for 30–60 min and then incubated with primary antibodies (see Table 1) overnight at 4°C.
4. After washing coverslips with PBS, cells were then incubated with secondary antibodies (see Table 1) diluted in blocking buffer for 30 h at room temperature.
5. After washing in PBS (3 × 15 min) and water (3 × 20 min), coverslips were mounted with Vectashield® mounting solution with DAPI (see Note 6).
6. Fluorescent images were captured with a Leica TCS SP2 Spectral Confocal Microscope (see Fig. 2).

3.6. Reverse Transcription-Polymerase Chain Reaction for Determining Taste Cell Markers

1. Total RNA (0.5 μg) was reverse transcribed for 50 min at 42°C using the Superscript® First Strand Synthesis System for RT-PCR (see Note 7).
2. PCR amplification of cDNA for each RT reaction was performed in a final volume of 25 μL containing 1 μL of cDNA, 1× AmpliTaq Gold PCR buffer, 2.5 mM $MgCl_2$, 1 mM deoxynucleoside triphosphates, 0.4 μM of each primer (see Table 2), and 0.25 U/μL of AmpliTaq Gold polymerases.
3. PCR products were separated on 2% agarose gels and stained with 0.2 μg/mL of ethidium bromide (see Subheading 2.5, step 10) to verify their expected size (see Fig. 3).

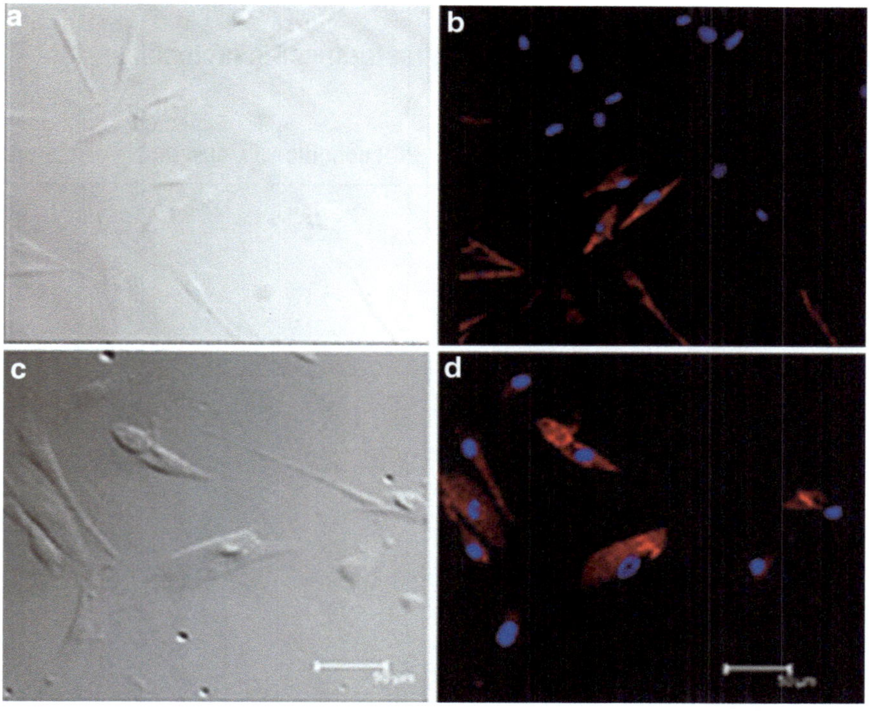

Fig. 2. Immunostaining of cultured human fungiform papillae cells showed the presence of taste cell-specific biomarkers. Images were acquired with a Leica TCS-SP2 confocal laser scanning microscope. (**a** and **c**) Transmission images of corresponding fields are shown on the *left*. (**b**) Immunoreactivity was observed for gustducin in about 60% of cultured cells. (**d**) Immunoreactivity for PLC-β_2 was also observed in about 25% of cultured cells (h). Nuclei of cells were stained as *blue* with DAPI. For controls, immunostaining with antibody-specific immunoglobulin demonstrated the absence of nonspecific immunoreactivity (data not shown). Scale bars = 50 μm.

3.7. Moderate Throughput Screening

1. Cultured human fungiform cells were grown for 2–4 days in 96-well plates.
2. Cells were loaded with loading solution in modified Ringer solution (see Subheading 2.6) for 1 h at 36°C (see Note 8).
3. Cells were first superfused with bath solution, then exposed to taste stimuli (see Table 3) via superfusion, with each stimulus applied for 30–60 s, and washed with modified Ringer for at least 2 min (see Fig. 4).

4. Notes

1. Store at 4°C and protect from light by covering medium bottle with aluminum foil. MCDB 153 should be colorless after preparation; any change of color is indication of expiration. MCDB 153 medium expires in 3 months.

Table 2
Primers and conditions used for detecting taste-specific molecules

Gene	Sequence	PCR condition		Expected size (bp)	Reference
β-Actin	GGACTTCGAGCAAGAGATGG AGCACTGTGTTGGCGTACAG	95 94 53 72 72	7 min 45 s 45 s 45 s 7 min	234	NM_001101.3
Gustducin	TCTGGGTATGTGCCAAATGA GGCCCAGTGTATTCTGGAAA	95 94 53 72 72	7 min 45 s 45 s 45 s 7 min	386	NM_001102386
PLC-β2	GTCACCTGAAGGCATGGTCT TTAAAGGCGCTTTCTGCAAT	95 94 53 72 72	3 min 30 s 30 s 60 s 7 min	333	NM_004573
T2R5	TATGGTTTGCCACCTTCCTC AAGGACTTCAGCGCAGTGAT	95 94 53 72 72	7 min 45 s 45 s 45 s 7 min	394	NM_01898
T1R3	CTTTTGTGGCCAGGATGAGT TGCAGGAAGAGTGTGCTCAG	95 94 56 72 72	4 min 45 s 45 s 50 s 10 min	345	NM_152228
TRPM5	TGGTAGAGCGCATGATGAAG ACCAACAGGAAGGTGACCAG	95 94 63 72 72	10 min 20 s 30 s 45 s 7 min	301	NM_001101

2. Alternative protocol: Digest fungiform papillae with pronase and elastase mixed in isolation solution at room temperature for 30–45 min and then follow step 5 of Subheading 3.1. These two protocol steps give the same result.
3. Taste cells will eventually attach to the coated coverslip, though it may not be clearly visible in 1–3 days, and you may observe some cubical cells which may be distinct from majority of cells which do not show any sign of proliferation. These cells will be eliminated after the first passaging.
4. The exact temperature (36°C) of incubator was found to be critical for growth of cultured taste papillae cells.

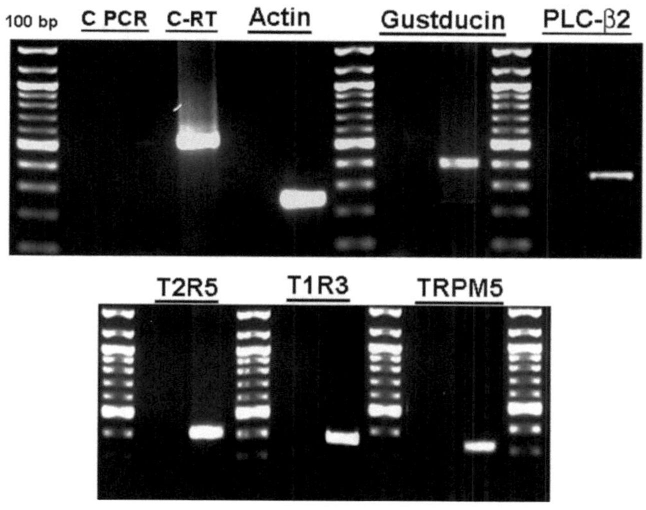

Fig. 3. The expression taste cell marker mRNAs were demonstrated by RT-PCR. The expression taste cell markers of β-actin, gustducin, PLC β_2, T1R3, T2R5, and TRPM5 mRNA were shown in cultured human fungiform taste cells. The cDNA transcribed from total RNA was amplified with intron spanning-specific primers (Table 2). PCR products were found at expected size and confirmed by sequencing. Specific mRNA was not detected in control experiments without reverse transcriptase indicating no genomic DNA contamination. M = marker (100-bp division).

Table 3
Response frequency of cultured human fungiform taste cells to different chemical stimuli

Stimuli	Response frequency (%)
Denatonium 2 mM	15
Sucralose 1 mM	2
AceK 250 ppm	2
MPG 3 mM	1

5. Experiments are typically performed using cells at passage-2 through -7; however, cells may be used beyond that passage number.
6. Controls for immunofluorescence consisted of omitting the primary antibody or substituting the primary antibody with the host IgG from which the antibody was generated. In all cases these controls revealed no artifactual labeling. Immunoreactive cells were counted in at least three sampling fields.
7. As a control to check genomic DNA contamination and nonspecific amplification, samples of RNA were treated in

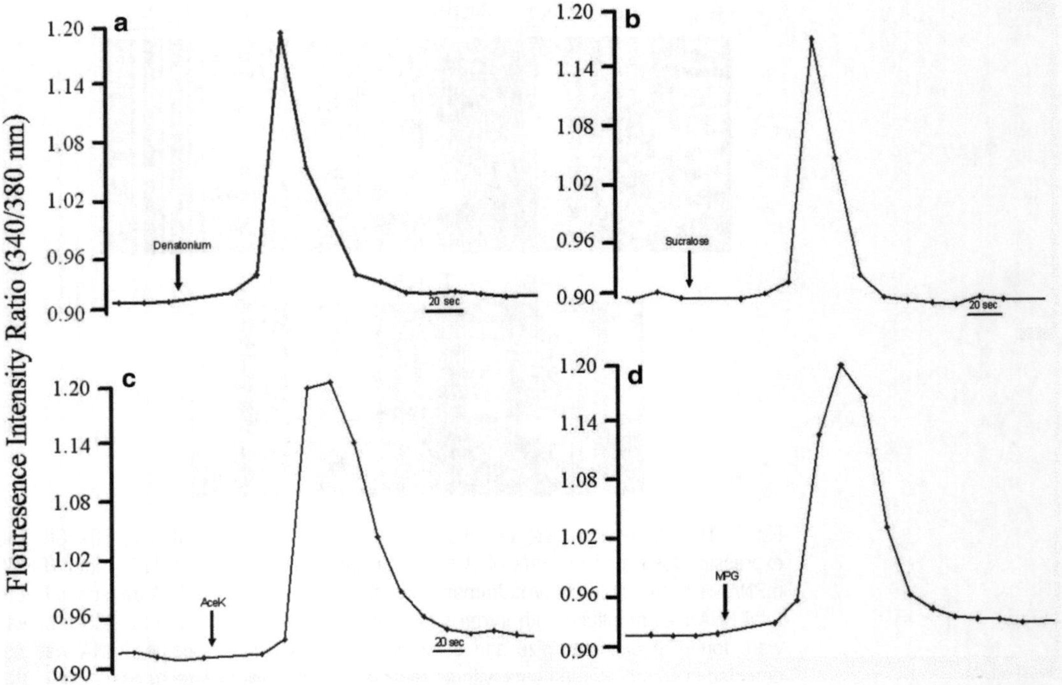

Fig. 4. Cultured taste cells responded to sweet, umami, and bitter stimuli. Changes in intracellular calcium levels ($[Ca^{2+}]_i$) in cultured human fungiform taste cell were measured in 96-well plates using fura-2, a medium-throughput system. Stimuli were dissolved in modified Ringer solution and adjusted for pH and osmolarity. Graphs illustrate representative changes in $[Ca^{2+}]_i$ levels in individual cells during exposure to (**a**) denatonium benzoate (2 mM), (**b**) sucralose (1 mM), (**c**) Acesulfame-K (AceK, 250 ppm), (**d**) mono potassium glutamate (MPG, 3 mM).

parallel in the presence and absence of reverse transcriptase and used for PCR by amplifying with primers designed for detection of β-actin, gustducin, PLC-β_2, T2R5, T1R3, and Trpm5 (see Table 2). Primers were chosen to span one or more introns to exclude confusion with amplified fragments from genomic DNA.

8. Changes in intracellular calcium levels ($[Ca^{2+}]_i$) in response to taste stimuli were measured in a 96-well format using a Discovery-1 imaging station and Metamorph software® (Molecular Devices). The cells in each well were visualized with an inverted fluorescence microscope using excitation wavelengths of 340 and 380 nm and an emission wavelength set by a band-pass filter centered at 510 nm. A series of image pairs was acquired for each well at 20× with a CCD camera (Photometrics®) as follows: four baseline, 12 poststimulus, and four baseline, for each of the three stimuli. A 2-min washout period followed each poststimulus block of images. Stimulus selection and delivery, focusing, and image acquisition, as well as plate movement were controlled by custom-designed Discovery-1 software (Molecular Devices).

Acknowledgement

We acknowledge the technical assistance of Esi Quayson, Aimee Myers, Linda Wysocki, and Valerie Audige. This work was supported in part by P50-DC006760, National Science Foundation (NSF) 0216310, and Givaudan Inc. grant.

References

1. Arvidson K (1979) Location and variation in number of taste buds in human fungiform papillae. Scand J Dent Res 87:435–442
2. Barlow LA, Northcutt RG (1994) Analysis of the embryonic lineage of vertebrate taste buds. Chem Senses 19:715–724
3. Beidler LM, Smallman RL (1965) Renewal of cells within taste buds. J Cell Biol 27:263–272
4. Bezencon C, Le Coutre J, Damak S (2007) Taste-signaling proteins are coexpressed in solitary intestinal epithelial cells. Chem Senses 32:41–49
5. Caicedo A, Jafri MS, Roper SD (2000) In situ Ca2+ imaging reveals neurotransmitter receptors for glutamate in taste receptor cells. J Neurosci 20:7978–7985
6. Chandrashekar J, Hoon MA, Ryba NJ, Zuker CS (2006) The receptors and cells for mammalian taste. Nature 444:288–294
7. Finger TE (2005) Cell types and lineages in taste buds. Chem Senses 30(Suppl 1):i54–i55
8. Gilbertson TA, Damak S, Margolskee RF (2000) The molecular physiology of taste transduction. Curr Opin Neurobiol 10:519–527
9. Huang YA, Maruyama Y, Stimac R, Roper SD (2008) Presynaptic (Type III) cells in mouse taste buds sense sour (acid) taste. J Physiol 586:2903–2912
10. Kishi M, Emori Y, Tsukamoto Y, Abe K (2001) Primary culture of rat taste bud cells that retain molecular markers for taste buds and permit functional expression of foreign genes. Neuroscience 106:217–225
11. Mbiene JP, Maccallum DK, Mistretta CM (1997) Organ cultures of embryonic rat tongue support tongue and gustatory papilla morphogenesis in vitro without intact sensory ganglia. J Comp Neurol 377:324–340
12. Ozdener H, Yee KK, Cao J, Brand JG, Teeter JH, Rawson NE (2006) Characterization and long-term maintenance of rat taste cells in culture. Chem Senses 31:279–290
13. Paran N, Mattern CF, Henkin RI (1975) Ultrastructure of the taste bud of the human fungiform papilla. Cell Tissue Res 161:1–10
14. Perez CA, Huang L, Rong M, Kozak JA, Preuss AK, Zhang H, Max M, Margolskee RF (2002) A transient receptor potential channel expressed in taste receptor cells. Nat Neurosci 5:1169–1176
15. Ruiz CJ, Stone LM, McPheeters M, Ogura T, Bottger B, Lasher RS, Finger TE, Kinnamon SC (2001) Maintenance of rat taste buds in primary culture. Chem Senses 26:861–873
16. Spielman AI, Mody I, Brand JG, Whitney G, MacDonald JF, Salter MW (1989) A method for isolating and patch-clamping single mammalian taste receptor cells. Brain Res 503:326–329
17. Stone LM, Wilcox CL, Kinnamon SC (2002) Virus-mediated transfer of foreign DNA into taste receptor cells. Chem Senses 27:779–787
18. Stone LM, Tan SS, Tam PP, Finger TE (2002) Analysis of cell lineage relationships in taste buds. J Neurosci 22:4522–4529
19. Zhao GQ, Zhang Y, Hoon MA, Chandrashekar J, Erlenbach I, Ryba NJ, Zuker CS (2003) The receptors for mammalian sweet and umami taste. Cell 115:255–266
20. Travers SP, Nicklas K (1990) Taste bud distribution in the rat pharynx and larynx. Anat Rec 227:373–379

Chapter 8

Human Nasal and Tracheo-Bronchial Respiratory Epithelial Cell Culture

M. Leslie Fulcher and Scott H. Randell

Abstract

Human airway epithelial (hAE) cell cultures are instrumental for studying basic and applied aspects of respiratory tract biology, disease, and therapy. When primary epithelial cells from the human nasal passages or tracheo-bronchial airways are grown on porous supports at an air-liquid interface (ALI) they undergo mucociliary differentiation, reproducing both the *in vivo* morphology and key physiologic processes. These cultures are useful for studying basic biology, disease pathogenesis, gene therapy and aerosol administration of drugs. This chapter gives detailed protocols for tissue procurement, cell isolation, production of complex media, and cell culture initiation and maintenance needed for hAE cell ALI cultures with non-proprietary reagents.

Key words: Epithelial cell, Respiratory tract, Differentiation, Physiology, Pathogenesis

1. Introduction to Primary Human Airway Epithelial Cell Cultures

The epithelial cells lining the nasal cavity, trachea, and bronchi play an important role to protect the host from a variety of injurious stimuli including chemical and particulate pollutants and pathogens. Foreign particles deposited on the mucosal surface are trapped in mucus and are removed by coordinated cilia beating and/or coughing. Human airway epithelial (hAE) cell cultures are vital for basic and applied studies of airway biology, disease, and therapy related to respiratory tract diseases.

Primary hAE cells grown as well-differentiated cultures on porous supports at an air–liquid interface (ALI) recapitulate the characteristic pseudostratified mucociliary morphology and key physiologic functions and are a quantum leap towards the in vivo biology. These cultures serve as a critical milestone test of biological relevance. Verification of efficacy in this model is a rational step

Scott H. Randell and M. Leslie Fulcher (eds.), *Epithelial Cell Culture Protocols: Second Edition*, Methods in Molecular Biology, vol. 945, DOI 10.1007/978-1-62703-125-7_8, © Springer Science+Business Media, LLC 2012

for advancement of potential therapies, and peer reviewers for scientific journals and granting agencies often require its use.

Although primary hAE cultures have been created for over 25 years (1) and have been used for numerous studies, the expense, technical complexity, and experimental limitations inhibit their full application. From 1984 to 2012, The University of North Carolina Cystic Fibrosis Center Tissue Procurement and Cell Culture Core has prepared cells from more than 7,200 human tissue specimens, adopting new technologies and extending research capabilities. The current chapter distills and updates our prior detailed description (2), enabling others to employ this relevant cell culture model. The procedures detailed below are based on the original methods of Lechner and LaVeck (3), and Gray et al. (4), as currently employed in our laboratory.

2. Materials

2.1. Tissue Procurement

Airway epithelial cells can be extracted from excess surgical pathology or autopsy specimens procured through cooperating surgeons and pathologists using protocols in accordance with relevant regulations. These include nasal turbinates or polyps not requiring histopathologic examination; lung tissue after lobectomy, pneumonectomy, or transplantation (surgical pathology); and trachea/lungs (autopsy) after examination and release by a pathologist. A useful source of normal tissue is the donor lower trachea, carina, and mainstem bronchi left over after lung transplantation. These are transported to the laboratory in an appropriate container on wet ice in a physiologic solution (sterile saline, PBS, Lactated Ringer's solution, or tissue culture medium). Lungs from potential organ donors are frequently unsuitable for transplantation but are useful for research. These can be obtained via establishing protocols with the agencies that normally oversee organ donation or from nonprofit organizations that provide human biomaterials for research (e.g., National Disease Research Interchange, www.ndri.com). Criteria for specimen acceptability are discussed in Note 1. Finally, hAE cells are available from commercial suppliers, circumventing the need for tissue procurement.

2.2. Isolating Primary hAE Cells

1. Absorbent bench covering.
2. Large plastic sterile drape (3 M Health Care).
3. Lactated Ringer's (LR) solution.
4. Sterile specimen cups.
5. Instrument sterilizer or previously autoclaved instruments.

6. Suggested tools include the following: Curve-tipped scissors, delicate 4.5″; heavy scissors, straight, sharp, 11.5 cm; forceps, blunt-pointed, straight, 15 cm; rat-tooth forceps 1 × 2, 15.5 cm; scalpels, #10; sterile 4″ × 4″ cover sponges.
7. Platform rocker.
8. Tissue wash medium: Joklik Minimum Essential Medium (JMEM) plus desired antibiotics (see Subheading 2.6).
9. Tissue soak solution: Dithiothreitol (DTT, 0.5 mg/mL, Sigma-Aldrich®), and DNase (10 μg/mL, Sigma-Aldrich®) plus supplemental antibiotics (see Subheading 2.6).

2.3. Harvesting Cells

1. 15 and 50 mL conical tubes.
2. Fetal bovine serum (FBS).
3. Sterile scalpels, #10.
4. 1× PBS, Ca^{2+} and Mg^{2+} free.
5. 150 mm tissue culture dish.
6. Declump solution: 2 mM EDTA, 0.05 mg/mL DTT, 0.25 mg/mL collagenase, 0.75 mg/mL calcium chloride, 1 mg/mL magnesium chloride, and 10 μg/mL DNase in PBS. Sterile filter and store at −20°C.
7. Ham's F12 medium.

2.4. Porous Supports

There are multiple porous support options for ALI cultures. Ideal supports are optically clear, facilitate attachment and long-term growth, and are amenable to downstream analyses. However, in our experience there have been problems with membrane consistency and quality control. We strongly recommend coating all porous supports with human type IV placental collagen (see Subheading 3.5).

1. Recommended: Millipore, Millicell Cell Culture insert, 12 mm hydrophilic PTFE, 0.4 μM pore size.
2. Other options: Transwell-Clear®, 0.4 μM pore size or Snapwell® (Corning, Inc.) membranes.

2.5. Media

Two closely related medias are employed. Bronchial epithelial growth medium (BEGM) is used when plating initial cell harvests on type I/III collagen-coated plastic dishes (see Subheading 3.4) or to expand passaged cells on plastic. ALI medium is used to support growth and differentiation on porous supports. BEGM and ALI composition is given in Table 1 and the differences between BEGM and ALI are illustrated in Table 2. The base medias (LHC Basal, Life Technologies™ and DMEM-H) can be purchased commercially and additives are made as specified below.

Table 1
BEGM and ALI composition

Additive		Final concentration in media	Company	Cat. #
Bovine serum albumin		0.5 mg/mL	Sigma-Aldrich	A7638
Bovine pituitary extract		10 μg/mL	Sigma-Aldrich	P1476
Insulin		0.87 μM	Sigma-Aldrich	I6634
Transferrin		0.125 μM	Sigma-Aldrich	T0665
Hydrocortisone		0.21 μM	Sigma-Aldrich	H0396
Triiodothyronine		0.01 μM	Sigma-Aldrich	T6397
Epinephrine		2.7 μM	Sigma-Aldrich	E4250
Epidermal growth factor		25 ng/mL—BEGM 0.50 ng/mL—ALI	Invitrogen	PHG0313
Retinoic acid		5×10^{-8} M	Sigma-Aldrich	R2625
Phosphorylethanolamine		0.5 μM	Sigma-Aldrich	P0503
Ethanolamine		0.5 μM	Sigma-Aldrich	E0135
Zinc sulfate		3.0 μM	Sigma-Aldrich	Z0251
Penicillin G sulfate		100 U/mL	Sigma-Aldrich	P3032
Streptomycin sulfate		100 μg/mL	Sigma-Aldrich	S9137
Gentamicin[a]		50 μg/mL	Sigma-Aldrich	G1397
Amphotericin[a]		0.25 μg/mL	Sigma-Aldrich	A2942
Stock 4	Ferrous sulfate	1.5×10^{-6} M	Sigma-Aldrich	F8048
	Magnesium chloride	6×10^{-4} M	J.T Baker	2444
	Calcium chloride	1.1×10^{-4} M	Sigma-Aldrich	C3881
Trace Elements	Selenium	30 nM	Sigma-Aldrich	S5261
	Manganese	1 nM	Sigma-Aldrich	M5005
	Silicone	500 nM	Sigma-Aldrich	S5904
	Molybdenum	1 nM	Sigma-Aldrich	M1019
	Vanadium	5 nM	Sigma-Aldrich	398128
	Nickel sulfate	1 nM	Sigma-Aldrich	N4882
	Tin	0.5 nM	Sigma-Aldrich	S9262

[a]Not in ALI

2.5.1. Stock Additives for BEGM and ALI Media

Additives are 0.2 μM filtered (unless all components are sterile) and aliquots are stored at −20°C for up to 3 months unless specified otherwise.

1. Bovine serum albumin (BSA) 300× (150 mg/mL): Add PBS to BSA (essentially globulin free, ≥99%) at a concentration of >150 mg/mL, gently rock or stir at 4°C for 2–3 h until dissolved, and adjust volume to yield 150 mg/mL.

Table 2
Differences between ALI and BEGM medium

	ALI	BEGM
Base media	LHC Basal:DMEM-H 50:50	LHC Basal 100%
Base antibiotics	Pen/Strep (100 U/mL/100 μg/mL)	Pen/Strep (100 U/mL/100 μg/mL) Gentamicin 50 μg/mL Amphotericin 0.25 μg/mL
EGF	0.50 ng/mL	25 ng/mL
$CaCl_2$	1.0 mM	0.11 mM

2. Bovine pituitary extract (BPE) (dilution depends on lot, typically 125×): BPE is available from Sigma-Aldrich® and is used at a final concentration of 10 μg/mL. Check the protein concentration per mL of the specific lot to determine the dilution factor.
3. Insulin 1,000× (5 mg/mL; 0.87 mM): Dissolve bovine insulin in 0.9 N HCl.
4. Transferrin 1,000× (10 mg/mL; 0.125 mM): Reconstitute human-holo transferrin in PBS.
5. Hydrocortisone 1,000× (0.072 mg/mL; 0.21 mM): Reconstitute hydrocortisone in distilled water (dH_2O).
6. Triiodothyronine 1,000× (0.0067 mg/mL; 0.01 mM): Dissolve triiodothyronine in 0.001 M NaOH.
7. Epinephrine 1,000× (0.5 mg/mL; 2.7 mM): Dissolve epinephrine in 0.01 N HCl.
8. Epidermal growth factor 1,000× for BEGM, 50,000× for ALI (25 μg/mL; 4 μM): Dissolve human recombinant, culture-grade EGF (Life Technologies™) in PBS.
9. Retinoic acid (concentrated stock = 1×10^{-3} M in absolute ethanol, 1,000× stock = 5×10^{-5} M in PBS with 1% BSA): Retinoic acid (RA) is soluble in ethanol and is light sensitive. Dissolve 0.3125 mg of RA (Sigma-Aldrich®) per mL in 100% ethanol. Store in foil-wrapped tubes at −70°C for up to 2 weeks. To prepare the 1,000× stock, first confirm the RA concentration of the ethanol stock by diluting it 1:100 in absolute ethanol. Read the absorbance at 350 nm using a spectrophotometer and a 1 cm light path quartz cuvette, blanked on 100% ethanol. The molar extinction coefficient of RA in ethanol equals 44,300 at 350 nm. Thus, the absorbance of the diluted stock should equal 0.44. RA with absorbance readings below 0.18 should be discarded. If the absorbance equals 0.44, add 3 mL of 1×10^{-3} M ethanol stock solution to 53 mL PBS and add

4.0 mL of BSA 150 mg/mL stock (see Subheading 2.5.1, step 1). For absorbance values <0.44, calculate the needed volume of ethanol stock as 1.35/absorbance and adjust the PBS volume appropriately.

10. Phosphorylethanolamine 1,000× (0.07 mg/mL; 0.5 mM): Phosphorylethanolamine is dissolved in PBS.
11. Ethanolamine 1,000× (0.03 μL/mL; 0.5 mM): Dilute ethanolamine in PBS.
12. Stock 11 1,000× (0.863 mg/mL; 3 mM): Dissolve zinc sulfate in dH_2O. Store at room temperature.
13. Penicillin–streptomycin 1,000× (100,000 U/mL and 100 mg/mL): Dissolve penicillin-G sodium and streptomycin sulfate in dH_2O for a final concentration of 100,000 U/mL and 100 mg/mL, respectively.
14. Gentamicin 1,000× (50 mg/mL): Store at 4°C.
15. Amphotericin B 1,000× (250 μg/mL).
16. Stock 4 1,000×: Combine 0.42 g ferrous sulfate, 122.0 g magnesium chloride, 16.17 g calcium chloride-dihydrate, and 800 mL dH_2O in a volumetric flask. Add 5.0 mL concentrated HCl. Stir to dissolve and bring volume to 1 L.
17. Trace elements 1,000×: Prepare seven separate 100 mL stock solutions (see Table 3). Fill a 1 L volumetric flask to the 1 L mark with dH_2O. Remove 8 mL of dH_2O. Add 1.0 mL of each stock solution and 1.0 mL of concentrated HCl. Store at room temperature.

2.5.2. BEGM and ALI Media

We describe here production of 500 mL or 1 L batches, which are assembled in the reservoir of a 0.2 μm bottle top filter. Larger

Table 3
Stock solutions for trace elements

Component	Sigma-Aldrich Cat. #	Amount/100 mL	Molarity
Selenium ($NaSeO_3$)	S5261	520 mg	30.0 mM
Manganese ($MnCl_2 \cdot 4H_2O$)	M5005	20.0 mg	1.0 mM
Silicone ($Na_2SiO_3 \cdot 9H_2O$)	S5904	14.2 g	500 mM
Molybdenum [$(NH4)_6Mo_7O_{24} \cdot 4H_2O$]	M1019	124.0 mg	1.0 mM
Vanadium (NH_4VO_3)	398128	59.0 mg	5.0 mM
Nickel ($NiSO_4 \cdot 6H_2O$)	N4882	26.0 mg	1.0 mM
Tin ($SnCl_2 \cdot 2H_2O$)	S9262	11.0 mg	500 μM

quantities (e.g., >6 L) can be prepared in a volumetric flask and sterilized by peristaltic pumping (e.g., Masterflex pump, Cole-Parmer Instruments) through a cartridge filter (Pall). To clean tubing, rinse with dH_2O, and then ethanol followed again by dH_2O.

1. BEGM Medium: Dispense thawed additives into 100% LHC basal medium (Life Technologies™) in a bottle top filter unit. Note that some additives are not 1,000×. Add amphotericin after filtering. Store media at 4°C.
2. ALI Medium: The ALI base is 50:50 DMEM-H and LHC basal. Additives are thawed and dispensed as above. Note that some additives are not 1,000×. ALI medium contains low EGF and omits gentamicin and amphotericin. To prepare low endotoxin medium, use low endotoxin BSA.

2.6. Antibiotics

Primary human tissues frequently contain yeast, fungi, or bacteria and media for passage 0 cultures should be supplemented with at least gentamicin (50 μg/mL) and amphotericin (0.25 μg/mL) for the first 3–5 days. Less contamination will result by increasing the amphotericin concentration to 1.25 μg/mL and adding ceftazidime (100 μg/mL), tobramycin (80 μg/mL), and vancomycin (100 μg/mL). When processing tissues from cystic fibrosis (CF) patients, additional antibiotics are used as described in a prior publication (5). If no clinical microbiology information is available, and assuming *P. aeruginosa* contamination, consider adding ciprofloxacin (20 μg/mL), meropenem (100 μg/mL), and colymycin (5 μg/mL). CF lungs infected with *A. xylosoxidans*, *Burkholderia* sp., or *S. maltophilia* may require a different spectrum of antibiotics including sulfamethoxazole/trimethoprim (80 μg/mL, 16 μg/mL, respectively), chloramphenicol (5 μg/mL), minocycline (4 μg/mL), tigecycline (2 μg/mL), or moxifloxacin (20 μg/mL). For fungus or yeast contamination, Nystatin (100 U/mL) and Diflucan (25 μg/mL) can be added. Sterile liquids for injection may be added directly to media, whereas powders contain a given amount of antibiotic and unknown quantities of salts and buffers—purity of powders is determined by comparing the total vial powder weight to the designated antibiotic content and adjusting the μg/mL accordingly. A 25× concentrated antibiotic cocktail can be stored at 4°C, and used within 1–2 days. Note that nystatin and amphotericin are suspensions and cannot be filter-sterilized.

2.7. Assorted Reagents and Solutions

All non-sterile solutions are filter-sterilized and stored at −20°C unless otherwise noted.

1. Ham's F-12 Medium with 1 mM l-glutamine. Store at 4°C.
2. 1% Protease XIV with 0.01% DNase (10× stock): Dissolve Protease XIV (Sigma-Aldrich®) and DNase (Sigma-Aldrich®)

in desired volume of PBS and stir. A 1:9 dilution in JMEM (see Subheading 2.7, step 5) is used for cell dissociation.

3. Soybean trypsin inhibitor (STI, 1 mg/mL): Dissolve STI in Ham's F-12. Store at 4°C.
4. 0.1% Trypsin with 1 mM EDTA in PBS: Dissolve trypsin type III powder in PBS. Add EDTA from concentrated stock for a final concentration of 0.1% trypsin with 1 mM EDTA. pH solution to 7.2–7.4.
5. JMEM (Sigma-Aldrich®). Store at 4°C.
6. Type I/III collagen: Purecol® (Advanced BioMatrix). Store at 4°C.
7. Collagen type IV, from human placenta (Sigma-Aldrich®): Resuspend 10 mg of collagen powder in 20 mL dH_2O and add 50 μL of concentrated acetic acid. Incubate for 30 min at 37°C until dissolved.

2.8. Cryopreservation of Cells

1. Cell freezing solution: Combine 2 mL of 1.5 M HEPES, pH 7.2, 10 mL of FBS, and 78 mL Ham's F-12. Gradually add 10 mL DMSO.
2. Cryovials.
3. Nalgene Cryo Freezing Container (Nalgene® Labware).

3. Methods

3.1. Primary Human Airway Epithelial Cell Cultures, Overview

Primary hAE cells can be obtained from nasal, trachea, or lung tissue specimens and can be seeded directly onto porous supports for passage 0 ALI cultures or can be first grown on plastic for cryopreservation and/or subculture of passage 1 or passage 2 cells to porous supports (see Fig. 1).

3.2. Isolating hAE Cells

Primary hAE cells originate from nasal turbinates, nasal polyps, trachea, and bronchi. When handling human tissues always follow locally prescribed safety precautions to prevent potential bloodborne pathogen exposure. Tissue is transported to the laboratory in sterile containers containing sterile chilled LR solution, JMEM, F12, or another physiologic solution. Nasal tissue samples are usually processed without further dissection but whole lungs require dissection as described below.

1. Assemble material in a laminar flow hood:
   (a) Absorbent bench covering.
   (b) Large plastic sterile drape.

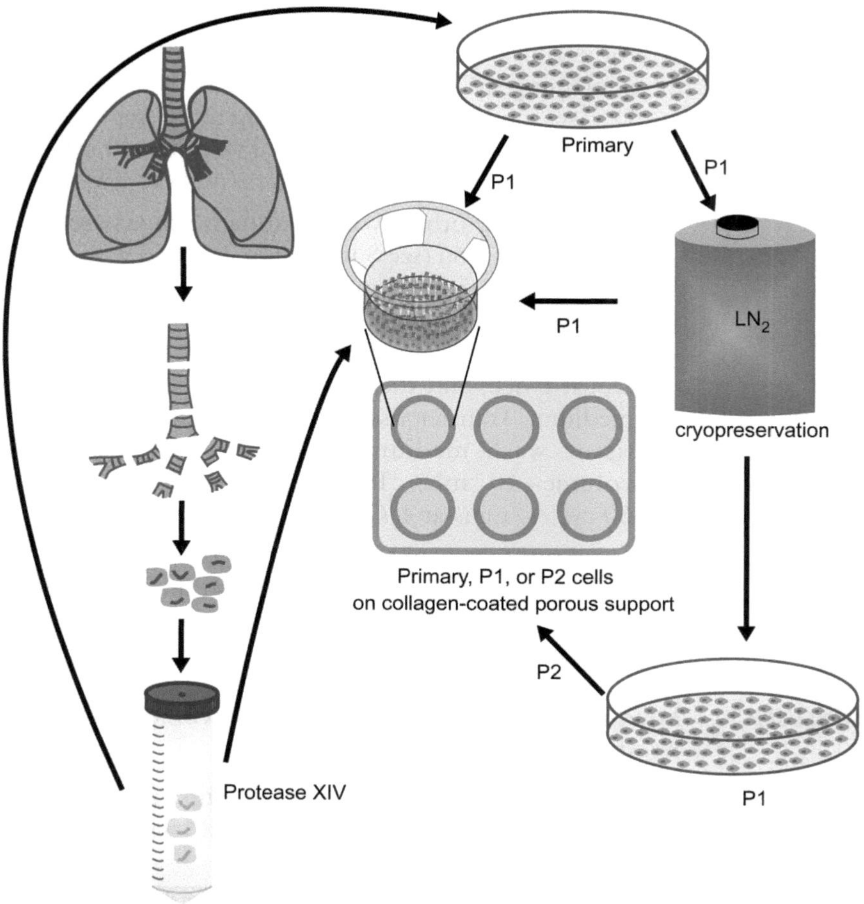

Fig. 1. Overview of the process for creating well-differentiated air–liquid interface (ALI) cultures of primary human airway epithelial (hAE) cells, P = passage.

(c) Ice bucket containing sterile specimen cups filled with LR solution.

(d) See a list of suggested tools in Subheading 2.2, step 6.

2. Dissect airways by removing all excess connective tissue and cutting into 5–10 cm segments. Clean tissue segments, removing any additional connective tissue and lymph nodes, and rinse by "dipping" in LR solution. Slit segments longitudinally and cut into 1 × 2 cm portions. Transfer to specimen cup containing chilled LR solution.

3. Since human tissue samples are likely to contain yeasts, bacteria, or fungi, begin antibiotic exposure as soon as possible. Prepare 300 mL of tissue wash medium (see Subheading 2.2, step 8).

Aspirate LR solution, add tissue wash medium, swirl, and repeat three times. Transfer washed tissue to 50 mL conical tubes containing 30 mL tissue wash medium plus 4 mL protease/DNase solution (see Subheading 2.7, step 2) (approximate tissue-to-fluid ratio of 1:10, final volume = 40 mL). Place tubes on platform rocker (50–60 cycles/min) at 4°C for 24 h.

4. CF tissues or others with abundant mucus are soaked in tissue soak solution (see Subheading 2.2, step 9). To prepare tissue soak solution, add 63 mg DTT and 1.25 mg DNase to 125 mL of tissue wash medium and filter sterilize. Aspirate LR solution, add 40 mL tissue soak solution, swirl, soak for 5 min on ice, repeat twice, and then rinse tissue three times in tissue wash medium. Transfer tissue to 50 mL tubes containing 30 mL tissue wash medium plus 4 mL protease/DNase (final volume = 40 mL). Place tubes on platform rocker (50–60 cycles/min) at 4°C for 24 h.
5. Nasal turbinates, polyps, and small bronchial specimens can be dissociated in 4–24 h, depending on the size of the tissue, in a 15 mL tube containing 9 mL tissue wash medium plus 1 mL protease solution.

3.3. Harvesting Cells

Follow standard sterile tissue culture techniques in a laminar flow hood.

1. End dissociation by pouring contents of 50 mL tubes into a 150 mm tissue culture dish; add FBS to a final concentration of 10% (v/v).
2. Gently scrape epithelial surface with a #10 scalpel blade. Rinse tissue and plate surface with PBS, collect and pool solutions, and distribute into 50 mL conical tubes. Centrifuge at 500 × *g* for 5 min at 4°C. Aspirate supernatant and add 12 mL of declumping solution (see Subheading 2.3, step 6). Incubate for 15 min to 1 h at 37°C, visually monitoring clump dissociation. Add FBS to 10% (v/v), centrifuge at 500 × *g* for 5 min, remove supernatant, and resuspend pellet in F12 for counting using a hemocytometer.

3.4. Type I/III Collagen Coating of Plastic Dishes

Passage 0 and freshly thawed, cryopreserved cells are plated on collagen-coated plastic dishes whereas cells passaged without freezing do not require coated dishes. Add 3.0 mL of a 1:75 dilution of Purecol® (see Subheading 2.7, step 6) in dH_2O per 100 mm dish. Incubate for 2–24 h at 37°C. Aspirate the remaining liquid and expose open dishes to UV in a laminar flow hood for 30 min. Plates can be stored for up to 8 weeks at 4°C.

3.5. Type IV Collagen Coating of Porous Supports

There are multiple porous support options for ALI cultures. See Subheading 2.4 for a description of the various porous supports for collagen coating.

Thaw frozen stock of collagen (see Subheading 2.7, step 7) and dilute 1:10 with dH_2O. Add 100, 400, or 500 μL per 12, 24, or 30 mm insert, respectively, in a Petri dish and dry in a laminar flow hood overnight. Expose to UV, uncovered, in a laminar flow hood for 30 min, cover dishes and wrap edges with parafilm, and store at 4°C for up to 1 month.

3.6. Plating Cells

Culture dissociated P0 hAE cells directly on porous supports in ALI medium with additional antibiotics (see Subheading 2.6) at a density of $0.1–0.25 \times 10^6$ cells/cm^2 ($0.8–2.0 \times 10^5$ cells per 12 mm support or $0.7–1.75 \times 10^6$ cells per 24–30 mm support, see Note 2). To generate P1 or P2 cells, plate cells in antibiotic-supplemented BEGM on type I/III collagen-coated plastic dishes at $2–6 \times 10^6$ cells per 100 mm dish. Change media at 24 h and every 2–3 days as needed to prevent acidification.

3.7. Cell Culture Maintenance

1. *P0 cells on plastic*: Assess attachment 24 h after plating P0 cells; if few clumps of floating cells, wash with PBS and feed with BEGM plus antibiotics (see Subheading 2.6). Floating clumps of cells can be "rescued" by washing dishes with PBS, harvesting into 50 mL conical tubes, pelleting at $500 \times g$ for 5 min, and repeating the "declumping" procedure (see Subheading 3.3, step 2).
2. *Passaging primary cells on plastic*: Passage primary cultures at 70–90% confluence. Harvest hard to detach cells, while minimizing trypsin exposure of cells that release quickly, using "double trypsinization." Rinse cells with PBS, add 3 mL of trypsin/EDTA (see Subheading 2.7, step 4) per 100 mm dish, and incubate for 5–10 min at 37°C. Gently tap dish to detach cells, rinse with PBS, and harvest into a 50 mL conical tube containing 3 mL STI solution (see Subheading 2.7, step 3) on ice. Add another 3 mL of trypsin/EDTA to the dish and repeat, visually monitoring detachment. Pool harvested cells and centrifuge at $500 \times g$ for 5 min at 4°C. Aspirate supernatant and resuspend cells in media for counting.
3. *Media change in ALI cultures*: For P0, P1, or P2 hAE cells grown on collagen-coated porous supports, remove apical media and rinse the apical surface with PBS. Prior to confluence, replace apical and basolateral media volumes as specified for the porous support, but after confluence do not add media apically. During periods of rapid cell growth, cells on Transwell® inserts in the standard configuration will acidify the media rapidly and require daily changes (see Note 3).

3.8. Cryopreservation of Cells

1. P0 hAE cells are trypsinized from plastic dishes (now P1 cells) and cryopreserved for long-term storage in liquid nitrogen. Cells are re-suspended in Ham's F-12 at a concentration of $2–6 \times 10^6$ cells/mL.

2. Keep cells on ice and slowly add an equal amount of freezing solution (see Subheading 2.8, step 1) to the cell suspension.
3. Place cryovials in freezing container and place in −80°C freezer for 4–24 h.
4. Transfer vial(s) from the −80°C freezer to liquid N_2 (−196°C) for long-term storage.

3.9. Thawing Cells

1. Thaw the cryovial at 37°C and wipe outside with 70% ethanol. Transfer cells to a 15 mL conical tube.
2. Dilute the cell suspension by slowly filling the tube with Ham's F-12. Centrifuge at 500 × *g* for 5 min at 4°C.
3. Gently resuspend cells in media, count, and assess viability.

4. Notes

1. In our experience, autopsy specimens must be procured within approximately 8 h after the time of death, but surgical pathology specimens can be stored for up to 3 days at 4°C. To protect personnel, do not accept specimens posing a known infection risk for HIV, Hep B and C, or tuberculosis. Samples from individuals on long-term immunosuppressive therapy may pose increased risk. All human tissue samples must be treated as a potential biohazard and handled using standard precautions. Steps for specimen procurement may breach sterility and/or tissues (especially CF) are likely infected with bacteria or fungi; thus appropriate antibiotics are necessary (see Subheading 2.6). A range of clinical data (laboratory values including blood gases, X-ray, or bronchoscopy findings) can guide acceptability of sub-transplant quality donor lungs. There are no hard-and-fast rules, but an arterial PO_2 of greater than 100 mm Hg on 100% inspired oxygen is a reasonable lower limit for acceptance.
2. Seeding densities: Primary hAE cells are mortal and require sufficient seeding density. Attachment and growth of cells from different individuals and preparations may vary. Generous seeding densities of passage 0 cells on porous supports ($>1.5 \times 10^5$ cells/cm^2) are required to obtain consistent, confluent, well-differentiated ALI cultures. Although it is tempting to expand primary cells on plastic, "overexpansion" should be avoided. Passage 0 cells first grown on plastic dishes should be seeded at no less than 1×10^6, and preferably $2–6 \times 10^6$, cells per 100 mm collagen-coated dish (or as calculated mathematically for other dish sizes). Under these conditions the cells should grow to >70% confluence within 5–7 days;

if a longer period is required, subsequent growth may be impaired. Cells at >70% confluence, but not >95% confluence, should be trypsinized for cryopreservation or subpassage to a porous support, or expanded one more round to passage 2 by seeding $>1 \times 10^6$ cells per 100 mm tissue culture dish. Passage 1 and 2 cells seeded on porous supports at $\sim 1.5 \times 10^5$ cells/cm^2 (~170,000 and $\sim 0.7 \times 10^6$ cells per 12 and 24 mm Transwell® membranes, respectively) should result in confluence within 3–5 days after seeding, at which point an ALI should be established. Lower seeding densities may be fully successful with some specimens, which can be determined empirically with aliquots of frozen cells, but is not possible when plating passage 0 cells, and greater variability is anticipated between different preparations.

3. We have devised teflon adapters to enable 12 mm Transwell® inserts to be kept in 6-well plates with a 2.5 mL basolateral reservoir, which decreases media change frequency. The 24 mm Transwell® insert may be kept in "Deep Well Plates" (BD Bioscience) with 12.5 mL medium.

References

1. Lechner JF, Haugen A, McLendon IA, Pettis EW (1982) Clonal growth of normal adult human bronchial epithelial cells in a serum-free medium. In Vitro 18:633–642
2. Fulcher ML, Gabriel S, Burns KA, Yankaskas JR, Randell SH (2005) Well-differentiated human airway epithelial cell cultures. Methods Mol Med 107:183–206
3. Lechner JF, LaVeck MA (1985) A serum-free method for culturing normal human bronchial epithelial cells at clonal density. J Tissue Culture Meth 9:43–48
4. Gray TE, Guzman K, Davis CW, Abdullah LH, Nettesheim P (1996) Mucociliary differentiation of serially passaged normal human tracheobronchial epithelial cells. Am J Respir Cell Mol Biol 14:104–112
5. Randell SH, Walstad L, Schwab UE, Grubb BR, Yankaskas JR (2001) Isolation and culture of airway epithelial cells from chronically infected human lungs. In Vitro Cell Dev Biol Anim 37:480–489

Chapter 9

Culture and Differentiation of Mouse Tracheal Epithelial Cells

Yingjian You and Steven L. Brody

Abstract

Airway epithelial cell biology has been greatly advanced by studies of genetically defined and modified mice; however it is often difficult to isolate, manipulate, and assay epithelial cell-specific responses in vivo. In vitro proliferation and differentiation of mouse airway epithelial cells are made possible by a high-fidelity system for primary culture of mouse tracheal epithelial cells described in this chapter. Using this method, epithelial cells purified from mouse tracheas proliferate in growth factor-enriched medium. Subsequent culture in defined medium and the use of the air–liquid interface condition result in the development of well-differentiated epithelia composed of ciliated and non-ciliated cells with characteristics of native airways. Methods are also provided for manipulation of differentiation and analysis of differentiation and gene expression. These approaches allow the assessment of global responses and those of specific cell subpopulations within the airway epithelium.

Key words: Mouse, Trachea, Culture, Airway epithelium, Cilia, Mucous

1. Introduction

1.1. Overview of mTEC Culture

The culture of the primary mouse tracheal epithelial cells (mTEC) provides a powerful approach for the investigation of specific functions of airway epithelial cells in respiratory tract development and disease. The method enables the culture and differentiation of mouse tracheal epithelial cells isolated from wild-type and genetically manipulated mice to study the function of a specific gene product in isolation. This culture system can be further manipulated to alter gene expression and cell differentiation. Compared to human and other species, there have been few approaches for primary culture of differentiated mouse airway epithelial cells (1–3). The protocol in this chapter for the culture of mTEC provides a high-fidelity model for the culture, differentiation, and assay of lung airway epithelial cells (4). The method relies on an efficient

Scott H. Randell and M. Leslie Fulcher (eds.), *Epithelial Cell Culture Protocols: Second Edition*, Methods in Molecular Biology, vol. 945, DOI 10.1007/978-1-62703-125-7_9, © Springer Science+Business Media, LLC 2012

approach for cell isolation and the use of media containing a combination of growth factors leading to cell proliferation and differentiation using the air–liquid interface (ALI) condition. Since the introduction in 2002, the protocol has been widely used and adapted for other experimental purposes (e.g., refs. 5–7).

1.2. Proliferation of mTEC

The general approach and conditions used for mTEC proliferation and differentiation are similar to those used to culture airway epithelial cells from human and other species (8–10). Those studies showed that growth factors, especially insulin, epidermal growth factor, and retinoic acid, are critical for cell proliferation, ciliogenesis, and mucous cell differentiation. Our studies indicate that the protocol described here is dependent on the isolation of an adequate number of cells and subsequent proliferation of an amplifying progenitor cell population (4). Pluripotent epithelial cells are present within the mouse tracheal epithelium and the small number of paratracheal glands located at the most proximal region of the trachea (11, 12). Experimental evidence shows that although well-differentiated cells (ciliated and Clara cells) and basal cells are initially present in the cells harvested, after 3 days only a fraction of seeded cells adhere to the membrane (4, 10). It is likely that most of these progenitor cells are basal cells and that the highly differentiated cells either die or dedifferentiate during culture. Proliferation of the progenitor population results in a confluent layer of cells. Further evidence of the presence of a progenitor population is supported by in vitro wound-repair and cell passage studies that demonstrate re-initiation of the proliferation and differentiation programs (4).

1.3. Differentiation of mTEC

In this system, cell differentiation is dependent on the establishment of confluent cell layers and an ALI condition, but is suppressed if cells remain submerged (13, 14). The protocol favors ciliogenesis and the generation of a layer of ciliated cells, rather than the secretory (Clara) or mucous cells. Cells that express Clara cell marker CCSP are transiently present during differentiation though abundance of these cells varies between preparations (4). Few goblet cells are present when identified by the expression of MUC5AC. The number of goblet cells can be markedly increased by the addition of IL-13 to the media (4, 15). Cells expressing basal cell marker protein cytokeratin 14 are abundant in the confluent layer at the initiation of differentiation.

1.4. Organization of the mTEC Culture Protocol

An overview and timeline of the mTEC preparation protocol are in Fig. 1. Materials (see Subheading 2) and Methods (see Subheading 3) are organized to match this sequence. Prior to the harvest of epithelial cells from mouse trachea, media and reagents required for cell isolation and proliferation must be prepared. Isolation of tracheal epithelial cells is accomplished over 2 days using pronase to release cells, and then selective adherence of non-epithelial cells for

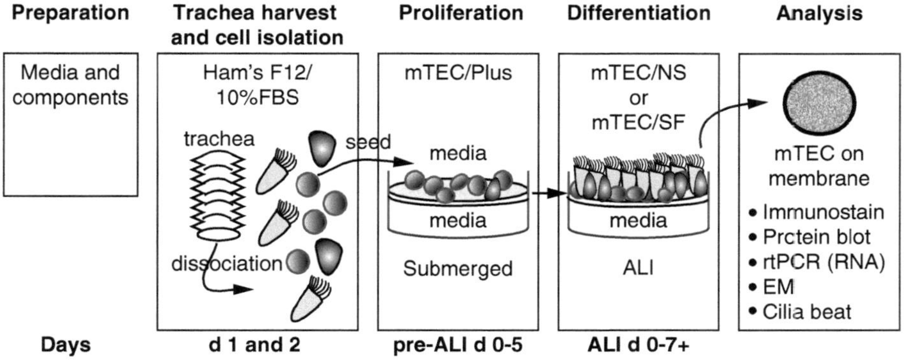

Fig. 1. Timeline for culture of mTEC. Preparation of components and media for each phase of the procedure is done prior to trachea harvest and cell isolation. Tracheal epithelial cells are released by enzyme dissociation, purified over 2 days, and then seeded on supported membranes. Cells undergo proliferation while submerged in mTEC/Plus medium for approximately 5 days until confluent. Cells are then differentiated by air–liquid interface (ALI) conditions using either mTEC/NS or mTEC/SF medium. Cells differentiate after approximately 7 days and may be analyzed as indicated (*EM* electron microscopy).

recovery of mTEC. mTEC are seeded on supported, semipermeable membranes and cultured in an enriched medium, called mTEC/Plus to favor proliferation. Medium is provided in the apical and basal chamber of the membranes during the proliferation phase. After approximately 5 days, confluent layers of cells permit the establishment of the ALI condition for differentiation. The differentiation phase is initiated by aspirating medium from the apical chamber and replacing the basal chamber medium with one of the two lower concentration growth factor media: mTEC/NS containing serum with proprietary additives (NuSerum™), or mTEC/SF, a serum-free, defined medium. During the differentiation phase, ciliogenesis occurs in the first week and after 7–14 days, cells are well differentiated. Cell differentiation may be altered by changing the cell culture environment, introducing bioactive agents or using recombinant viruses for gene transfer (4, 13, 16). Cells on supported membranes are amenable to analysis using multiple approaches to characterize the status of differentiation, gene expression, and ultrastructural features.

2. Materials

2.1. Stock Components for mTEC Medium

Prepare components prior to cell isolation (see Note 1).

1. HBSS/BSA: Bovine serum albumin (BSA, Fraction V, Fisher Scientific) is dissolved in sterile Hanks' Balanced Salt Solution (HBSS) at 1 mg/mL. Filter and store at 4°C.
2. BSA: Dissolved in HBSS at 100 mg/mL. Filter and store in aliquots at −20°C.

3. Retinoic acid stock solution A (RA stock A) [5×10^{-3} M]: Dissolve one vial of 50 mg retinoic acid (Sigma-Aldrich®) in 33.3 mL 100% ethanol, under dim lighting. Filter, protect from light, store in 1 mL aliquots at –80°C or liquid nitrogen up to 12 months (see Note 1).
4. Retinoic acid stock solution B (RA stock B) [10,000×, 5×10^{-4} M]: 1:10 dilute RA stock A in 100% ethanol. Protect from light, store in 1 mL aliquots at –20°C up to 1 month while using to supplement media (see Note 1).
5. Insulin (Sigma-Aldrich®): 2 mg/mL. Reconstitute in 4 mM HCl. Filter and store in 1,250 μL and 625 μL aliquots at –20°C.
6. Transferrin (Sigma-Aldrich®): 5 mg/mL. Reconstitute in HBSS/BSA. Filter and store in 250 μL aliquots at –20°C.
7. Epidermal growth factor (mouse EGF; BD Biosciences): 5 μg/mL. Reconstitute in HBSS/BSA. Filter and store in 1,250 μL and 250 μL aliquots at –20°C.
8. Cholera toxin (CT, Sigma-Aldrich®): 100 μg/mL. Reconstitute in HBSS/BSA. Filter and store in 250 μL and 62.5 μL aliquots at –20°C.
9. NuSerum™ (BD Biosciences): Store in 5 mL aliquots at –20°C.
10. Bovine pituitary extract (BPE): Prepare from frozen bovine pituitaries (Pel-Freeze) (8), or use Pel-Freeze BPE. Store in aliquots containing 7.5 mg total protein at –80°C.

2.2. mTEC Medium

mTEC/Basic is the core medium used to make the proliferation medium, called mTEC/Plus, and the differentiation media are referred to as mTEC/NS and mTEC/SF. RA Stock B (10,000×) must be freshly added to the proliferation and differentiation medium prior to use (see Note 1).

1. mTEC Basic medium (mTEC/Basic): Dulbecco's modified Eagle's medium/Ham's F-12 (DMEM/F-12) supplemented with 15 mM HEPES, 4 mM L-Glutamine, 3.6 mM NaHCO3, 100 U/mL penicillin, and 100 μg/mL streptomycin. Filter sterilize, store at 4°C.
2. mTEC Plus medium (mTEC/Plus): mTEC/Basic medium supplemented with 10 μg/mL insulin, 5 μg/mL transferrin, 0.1 μg/mL CT, 25 ng/mL EGF, 30 μg/mL BPE, 5% FBS (v/v). For 250 mL mTEC/Plus, use stock solutions from Subheading 2.1 (1,250 μL insulin, 250 μL transferrin, 250 μL CT, 1,250 μL EGF, 7.5 mg BPE) and 12.5 mL FBS. Add mTEC/Basic medium to a final volume of 250 mL. Filter sterilize, store at 4°C. Add RA stock B immediately before use (see Note 1).
3. mTEC NuSerum™ medium (mTEC/NS, serum-supplemented medium): mTEC/Basic medium supplemented with

2% (v/v) NuSerum™. Filter sterilize, store at 4°C. Add RA stock B immediately before use (see Note 1).

4. mTEC Serum-Free medium (mTEC/SF): mTEC/Basic medium supplemented with 5 μg/mL insulin, 5 μg/mL transferrin, 0.025 μg/mL CT, 5 ng/mL EGF, 30 μg/mL BPE, 1 mg/mL BSA. For 250 mL mTEC/SF, use stock solutions from Subheading 2.1 (625 μL insulin, 250 μL transferrin, 62.5 μL CT, 250 μL EGF, 7.5 mg BPE, 2.5 mL BSA). Add mTEC/Basic medium to a final volume of 250 mL. Filter sterilize, store at 4°C. Add RA stock B immediately before use (see Note 1).

2.3. Reagents for mTEC Isolation and Initiation of Culture

1. Ham's F-12/Pen–Strep: Ham's F-12 medium is supplemented with penicillin (100 U/mL) and streptomycin (100 μg/mL).
2. Pronase (Roche Applied Science): Dissolved in Ham's F-12/Pen–Strep at 0.15% (w/v). Make fresh pronase solution (about 2–5 mL) in a 15 mL tube, mix by gently rocking the tube, then filter sterilize.
3. Rat tail collagen (type I, BD Biosciences): Diluted at 50 μg/mL in 0.02 N acetic acid (in tissue culture grade water), filter. Store at 4°C up to 8 weeks.
4. DNase solution: Ham's F-12/Pen–Strep, 0.5 mg/mL crude pancreatic DNase I (Sigma-Aldrich®), 1 mg/mL BSA, filter. Aliquot and store in 5 mL aliquots at −20°C.

2.4. Trachea Harvest

1. Adult mice 8 weeks old (weighing over 20 g): Cells may be cultured from newborn mice to those aged 2 years using this protocol. Genetic backgrounds for culture include C57Bl/6, SV129/J, C57Bl/6-SV129/J hybrid, Balb/c, FVB, and Swiss Webster strains (see Note 2).
2. Surgical instruments: Harvesting and handling of tracheas are facilitated by the use of high-quality surgical tools (e.g., from Roboz). Use scissors with 4–5 cm blades to cut skin. To dissect and handle tracheas, use forceps 10 cm or less in length: one slightly curved and one straight, each with serrated, fine, 0.8 mm tips (e.g., Roboz 5135, 5130). To cut tracheas, use small straight scissors (blades < 3 cm). Tools are autoclaved for use.
3. Drugs for mouse euthanasia as recommended by the Institutional Animal Care and Use Committee.
4. Ethanol, 70%, 250 mL in a beaker for cleansing and wetting euthanized mice prior to dissection.
5. Ham's F12/Pen–Strep (see Subheading 2.3, item 1).
6. Freshly prepared 0.15% pronase (see Subheading 2.3, item 2).
7. Disposable gloves (non-sterile).
8. Sterile polypropylene plastic screw top conical tubes, 15 and 50 mL.

2.5. Tracheal Epithelial Cell Isolation

1. Plastic sterile Petri dishes, 100 mm, to hold tracheas. Tissue culture dishes are not required for this purpose.
2. Ham's F12/Pen–Strep, on ice.
3. Fetal bovine serum (FBS, 3–5 mL), warmed to 37°C.
4. Glass Pasteur pipettes, 9″, autoclaved, for handling tracheas.
5. High cell binding Primaria™ (BD Bioscience) plastic tissue culture plates for differential adherence of fibroblasts and tracheal epithelial cells. Use 35 mm or 100 mm plates depending on the number of tracheas used.
6. DNase solution (see Subheading 2.3, item 4).

2.6. Tracheal Epithelial Cell Seeding and Proliferation

1. Supported permeable membranes ("inserts") and culture plates: Polycarbonate (Transwell®, Corning) or polyester (polyethylene terephthalate, Transwell®-Clear) membranes fixed on plastic supports permit cell polarization by providing apical (upper) and basal (lower) chambers. The 6.5 mm size inserts (0.33 cm^2 surface area, pore size 0.4 μm, fit 24-well plates) allow multiple samples from a preparation. Polycarbonate membranes are translucent, so cell visibility is poor, but have a slight proliferation and differentiation advantage over the transparent polyester membranes. If using polycarbonate membranes, include at least one polyester membrane insert for each plate to enable inspection of cells by microscopy (see Table 1 and Note 3).
2. Rat tail collagen (type I collagen), 50 μg/mL (see Subheading 2.3, item 3).
3. A high-quality tissue culture type inverted microscope with phase optics is necessary to examine cells in tissue culture dishes. Hoffman modulation contrast is an additional filter that creates phase gradients to enhance cell imaging and visualization of beating cilia.

Table 1
Media volumes and cell numbers for initiation of mTEC cultures on supported membranes (Transwell®, Corning)

Plate size	Insert size (diameter) (mm)	Membrane area (cm^2)	Basal compartment medium volume (mL)	Apical compartment medium volume (mL)	Seeding: Cell number/insert (for $1.0 \times 10^5/cm^2$)
6 well	24	4.67	3.0–4.0	1.50	4.67×10^5 cells
12 well	12	1.12	1.0	0.50	1.12×10^5 cells
24 well	6.5	0.33	0.5	0.15	0.33×10^5 cells

4. Hemocytometer and Trypan blue to count live cells by microscopy.
5. Cell proliferation medium mTEC/Plus (see Subheading 2.2, item 2 and Note 1). mTEC/Plus contains high concentrations of growth factors and serum for cell proliferation prior to ALI.

2.7. Tracheal Epithelial Cell Differentiation Using the ALI Condition

1. A voltohmmeter (EVOM, World Precision Instruments) may be used for measuring transmembrane resistance (R_t) to determine when cells on membranes are confluent and have tight junctions, and to guide creation of the ALI conditions (see Note 4).
2. Cell differentiation media and components (see Subheadings 2.1 and 2.2): One of the two differentiation media may be used: mTEC/NS or mTEC/SF. mTEC/SF can be used when serum must be avoided during special treatment protocols; it results in a greater number of cell layers than mTEC/NS.

2.8. Manipulation of Differentiation

1. mTEC can be maintained in an undifferentiated state on rat tail collagen-coated (see Subheading 2.3, item 3) plastic tissue culture plates or membranes when continuously submersed. Proliferation in mTEC/Plus medium is followed by maintenance of cell populations in mTEC/NS or mTEC/SF.
2. Mucous cell differentiation can be induced by culture of mTEC in recombinant mouse IL-13 (Peprotech), 10 ng/mL.
3. Genetic modification using recombinant adenovirus or lentivirus, each with a titer of at least 10^7 infectious units/mL, may be used in mTEC. Virus should be generated using standard protocols and handled using biosafety guidelines recommended by the user's institution (see Note 5).

2.9. mTEC Passage and Freezing

1. Cells are removed from membranes for passage using 0.25% trypsin with 0.1% EDTA in Cell Dissociation Solution (Sigma-Aldrich®).
2. Cells are frozen using 10% DMSO in mTEC/Plus.

2.10. Analysis of mTEC Differentiation by Immunofluorescence

Immunofluorescent detection of cilia marker expression on fixed membranes.

1. Fixative 4% paraformaldehyde (see Note 5) in PBS. Prepared fresh or freshly defrost and use PBS to wash cells on the inserts.
2. A scalpel blade (#22) on a handle to cut the membrane from the plastic supports and watchmaker or embryo forceps to hold membranes on the edges without scraping off cells.
3. A blocking solution of 5% donkey serum (Sigma-Aldrich®), 3% BSA, and 0.1% Triton™ X-100 (Sigma-Aldrich®) in PBS to block nonspecific binding.

4. Antibodies to cilia proteins: Mouse monoclonal anti-β-tubulin-IV (BioGenex) or mouse monoclonal acetylated α-tubulin (clone 6-11B-1, Sigma-Aldrich®) and donkey anti-mouse, fluorescent-labeled secondary antibodies.
5. Mounting medium with anti-fade activity and containing nuclear DNA stain Hoechst or 4′, 6 diamidino-2-phenylindole (DAPI) such as Vectashield® (Vector Laboratories). DNA-binding chemicals are potentially carcinogenic (see Note 5).
6. Glass microscope slides and large coverslips (24 × 50 mm) to cover several pieces of mTEC membranes on a single slide.

2.11. Additional Methods to Assess Gene Expression and Differentiation of mTEC

1. mTEC flow cytometry: Use 0.1% EDTA or 0.25% trypsin with 0.1% EDTA in Cell Dissociation Solution for releasing cells from membrane (see Subheading 2.9 item 1). Use 2% FCS in PBS (2% FCS/PBS) for resuspending the cells and staining with antibodies. Use standard protocols for flow cytometry.
2. Protein blot analysis of mTEC.
3. RNA isolation from mTEC: Guanidinium thiocyanate–phenol extraction solution (e.g., TRIzol®, Life Technologies™; TRI Reagent®, MRC) for total RNA isolation. These chemicals are toxic (see Note 5).
4. Electron microscopy (scanning and transmission) of mTEC: High-quality glutaraldehyde, 2.5% in filtered sodium cacodylate buffer (0.1 M, pH 7.4). Glutaraldehyde and sodium cacodylate buffer are toxic (see Note 5). Use additional materials for EM processing as recommended by the microscopy facility.
5. Cilia beat frequency (CBF) can be detected in differentiated mTEC preparations. Cells on clear membranes are imaged using an inverted microscope with phase contrast filters and objectives (20×), high-speed video camera, and software appropriate for analysis of CBF (e.g., Sisson-Ammons Video Analysis, Ammons Engineering) (17).

3. Methods

Prepare media required for mTEC proliferation (mTEC/Plus) as in Subheading 2.2, item 2. Media for differentiation, mTEC/NS or mTEC/SF, should be prepared prior to cell differentiation phase.

3.1. mTEC Isolation and Initiation of Culture

1. Prepare Ham's F12/Pen–Strep as in Subheading 2.3, item 1.
2. Obtain inserts and multi-well plates. Include at least one transparent insert for each plate to monitor cell growth by microscopy.

3. Coat the apical surface of the insert membrane with rat tail type I collagen solution in the hood. Prepare approximately three 0.33 cm^2 inserts for each trachea to be harvested. Add coating solution to the apical compartment of the inserts using apical chamber volumes in Table 1. Cover and incubate at room temperature for 18–24 h or for a minimum of 4 h at 37°C. Aspirate the collagen coating solution and allow the membrane to air-dry for at least 5 min. Rinse both apical and basal surfaces with sterile PBS three times to remove free collagen. Allow the membrane to dry for 5 min before use. Collagen-coated inserts are stable for several months stored at room temperature in sealed plates.
4. Sterilize surgical instruments.
5. Prepare fresh pronase solution (about 2–5 mL) as in Subheading 2.3, item 2 (see Note 6).

3.2. Trachea Harvest (Day 1)

1. Wear clean, disposable gloves. Sterile gloves are not required. Wipe gloves with 70% ethanol during the procedure and change often, such as after handling animals, to avoid contamination.
2. In the tissue culture hood, prepare two 100 mm plastic dishes (non-etched, Petri type) with 10 mL cold sterile Ham's F-12/Pen–Strep to hold resected tracheas. Cover and place on ice at the bench to use as tracheas are resected.
3. Euthanize mice using the method recommended by the Institutional Animal Care and Use Committee. If using cervical dislocation, be cautious not to disrupt the neck and tracheal tissues.
4. Grasp each euthanized mouse by the incisors and immerse the entire body in a 250 mL beaker containing 70% ethanol. Spare the nose/mouth so as not to allow entrance of ethanol into the respiratory tract. The wet fur aids with dissection and enhances the sterility. Lay mice supine on a clean surface.
5. Using large scissors and forceps, grasp and incise the abdominal skin circumferentially, and then invert the entire layer of skin toward the head to reveal the neck. Avoid touching the neck area at this time.
6. Use small scissors to make a longitudinal, midline incision from the clavicles to the submandibular region to expose the trachea. Separate the muscles that overlie the trachea, and make a midline incision in the sternum to access the trachea within the thoracic cavity as it divides into the mainstem bronchi.
7. Lift the trachea and bluntly dissect away the esophagus from the posterior surface of the trachea with curved forceps. Cut the trachea with small scissors at the bifurcation of the mainstem bronchi and lift gently. Resect the trachea by cutting the

proximal trachea with scissors, just distal to the larynx, leaving the larynx within the mouse. Alternatively, the trachea may be resected first at the most proximal end (to include the tracheal submucosal glands) and the dissection carried down to the tracheal bifurcation. Place the resected trachea in the dish of cold, sterile Ham's F-12/Pen–Strep on ice. Repeat for each mouse.

8. In a tissue culture hood or under a dissecting microscope remove large pieces of the remaining adherent tissues from trachea. Perform this by grasping the proximal end of the trachea with a small forceps and using a second forceps, strip off adherent tissues. This will require stretching the trachea. Place each cleaned trachea in second dish of cold sterile Ham's F-12/Pen–Strep on ice.
9. Transfer the tracheas, one at a time, using a Pasteur pipette passed into the tracheal lumen, to a new dish containing cold Ham's F-12/Pen–Strep.
10. Cut each trachea open lengthwise with a small scissors and place into the tube of freshly made pronase. The pronase solution should completely cover all tissue, typically 3 mL to cover 10–30 tracheas. Gently mix the tube to assure that the pronase solution displaces air inside tracheas.
11. Incubate at 4°C overnight (18–24 h) in the tube, kept upright.

3.3. Tracheal Epithelial Cell Isolation (Day 2)

1. Thaw DNase solution on ice. Thaw and warm FBS to 37°C in a water bath.
2. Remove the tube with the tracheas in pronase from 4°C. Gently invert the tube containing the tracheas five times. Warm the tube to room temperature for 10 min and mix gently.
3. Add warmed FBS to a final concentration of 10% (v/v) and invert again gently 15–20 times (down and up is a single cycle) to dislodge epithelial cells that are attached to the trachea.
4. Take each trachea out of the tube with a sterile Pasteur pipette and place it in a new 15 mL tube containing 3 mL of Ham's F-12/10% FBS. Invert the tube 10–15 times again.
5. Again, remove tracheas from the tube using a fresh Pasteur pipette and place in a second fresh 15 mL tube containing 3 mL of Ham's F-12/10% FBS. Then, for a third iteration, invert the tube 10–15 times.
6. Using a fresh Pasteur pipette, remove the tracheas from the medium and place into a tube for discard.
7. Combine the contents of all tubes containing the enzyme-released cells in a 15 mL tube. Centrifuge at $500 \times g$, 4°C, for 10 min.
8. Carefully aspirate the supernatant, and resuspend the cells in DNase solution (~200 μL per trachea).

9. Put the tube on ice for 5 min, and then collect the cells by centrifugation at 500 × *g*, 4°C, for 5 min.
10. Resuspend the cells in mTEC/Basic medium containing 10% FBS, using 2–3 mL per ten tracheas. Plate the cells in Primaria™ tissue culture dish (35 mm or 100 mm depending on the volume). Incubate at 37°C, 5% CO_2 for 3–4 h, allowing fibroblasts to attach, while the epithelial cells are non-adherent or remain suspended in the medium.
11. Gently swirl the medium in the tissue culture dish a few times. Carefully collect the supernatant containing nonattached epithelial cells and place in a sterile tube. These floating cells are the mTEC that will be cultured.
12. Gently rinse the dish one or two times with warm mTEC Basic medium/10% FBS to recover additional airway epithelial cells, and collect the wash and add to sterile cell collection tube (from step 11). Do not wash the plate with excessive force or volume of medium to avoid detaching fibroblasts (see Note 7).
13. Centrifuge at 500 × *g*, 4°C, for 5 min.
14. Add RA Stock B (10,000×) to mTEC/Plus medium for use in subsequent steps (see Note 1).
15. Keep the cells in the hood at room temperature and work quickly. Aspirate the supernatant and resuspend the cell pellet in a small, measured volume (e.g., 100–200 μL/trachea) of mTEC/Plus medium with RA. Do not try to pipette the cell clumps vigorously to form a single cell suspension.
16. Mix 10 μL of the cell suspension with 10 μL 0.4% Trypan blue. Count the cells with a hemocytometer. Cell viability should be greater than 90%. Calculate the number of viable cells in suspension. Single cells and clumps will be present. Do not overestimate cell number within clumps and exclude red blood cells. Yields are $1–2 \times 10^5$ cells per trachea from a 20 g mouse. There are several causes of failure to isolate sufficient number of cells (see Note 6).

3.4. Tracheal Epithelial Cell Seeding and Proliferation

Proliferation requires adequate number of cells with progenitor potential and is density dependent. Seeding density is critical for ultimately generating differentiated mTEC. It may vary from 0.4 to 2.5×10^5 cells/cm^2. Cell proliferation (based on the length of time required to achieve cell confluence) varies with medium component activity. In most cases, a seeding density of 1.0×10^5 cells/cm^2 is recommended but this may vary with mouse strain and the individual user's technique (see Note 8).

1. To calculate the number of inserts that will be seeded at the recommended seeding density (1.0×10^5 cells/cm^2): Divide the total number of cells recovered by the number of cells per

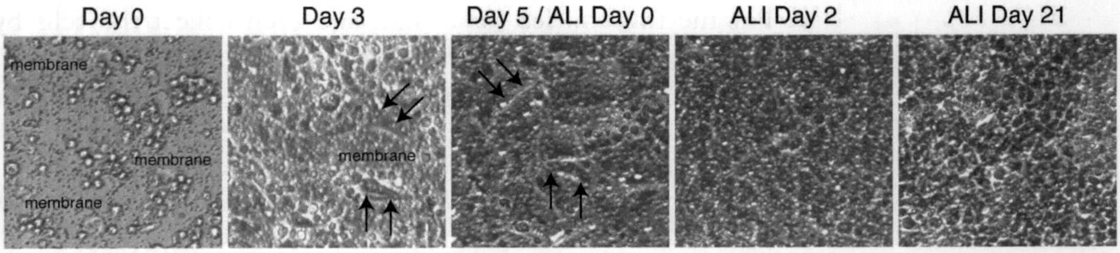

Fig. 2. Appearance of mTEC on membranes by light microscopy during culture. When initially seeded onto membranes, cells are rounded and in small clumps. After 3 days, cells adhere, elongate (*arrows*), and proliferate to form islands. After approximately 5 days, cells are confluent and are switched to ALI culture conditions (ALI day 0). At ALI day 0 some large, elongated cells are present. At ALI day 2, cells are smaller and cuboidal. By ALI days 7–21, cells are well differentiated and have a cobblestone appearance. Cilia beating is observed using appropriate optics. Cells were imaged by phase microscopy using a 10× objective.

insert according to Table 1. Then multiply the number of inserts to be seeded by the apical compartment medium volume (see Table 1). Adjust the cell volume to the total volume used for seeding the inserts. For example, if 5.0×10^5 cells are recovered and a seeding density of 1.0×10^5 cells/cm^2 is used, then 15 of the 0.33 cm^2 inserts will be seeded with an apical compartment volume of 150 μL each, requiring a total of 2.25 mL medium (15 × 150 μL).

2. Add mTEC/Plus with RA to the basal compartments.
3. Add the cell suspension into the apical chamber at the recommended seeding density as determined in Subheading 3.4, step 1. Inspection of the cells on the membrane by microscopy should show rounded cells and cell clumps (see Fig. 2).
4. Gently move the covered plate back and forth to spread the cells uniformly over the membranes. Place at 37°C with 5% CO_2 (day 0 of culture). Do not disturb the plate during the first 1–2 days to favor cell attachment.
5. On Day 3 of culture, change the medium. The medium may have an orange color due to cell proliferation. Yellow color media may indicate bacterial contamination. Inspect the cells to see that the medium is clear (i.e., not clouded by microorganisms). By microscopy confirm that the cells appear as islands adherent and elongated on the membrane (see Fig. 2, Day 3). Some floating cells may be present. Aspirate the old medium and replace with fresh medium using the volumes shown in Table 1.
6. On Day 5 of culture, the cells are typically confluent if seeded at a density of 1.0×10^5 cells/cm^2. The medium may be orange colored. Inspect the cells by microscopy (see Fig. 2, Day 5). Contaminating fibroblasts may be observed at this time (see Note 7). Confirm that cells are confluent or if cells are cultured on translucent membranes (polycarbonate) and cannot be

visualized, measure the R_t. When the cells initially become confluent, the R_t should be greater than 1,000 Ωcm^2 (typical range 1,500–2,500 Ωcm^2; R_t decreases as cells differentiate (4), see Note 4). If the R_t is greater than 1,000 Ωcm^2 or the cell layer is confluent, create ALI as described in Subheading 3.5, step 1. If cells are near confluent (R_t less than 1,000 Ωcm^2) at days 5–8, replace fresh mTEC/Plus with RA in both chambers and continue to observe. If cells are not confluent after Day 8, it is likely that the technique or culture conditions used will need to be adjusted (see Notes 7 and 8).

3.5. Tracheal Epithelial Cell Differentiation Using the ALI Condition

1. Cells that are confluent or with an R_t that is greater than 1,000 Ωcm^2 are ready for differentiation using ALI conditions with either mTEC/NS or mTEC/SF. This time point is ALI day 0. Medium is aspirated from the apical chamber and fresh medium is supplied only in the basal chamber. The cell layer should remain appearing dry. Medium leaking into the apical chamber indicates that either the cells are not confluent or that integrity of the cell layer is not maintained. There are several possible causes and solutions (see Notes 6–11).
2. Change the medium in the basal chamber every other day, consistently using the mTEC/NS or mTEC/SF with freshly added RA. Wash the apical surface gently with warm PBS. Medium can be changed less often if cells are maintained in culture longer than 14 days of ALI.
3. Continue to follow cells by inspection and microscopy. The apical surface should appear dry. Cells become smaller and cuboidal, and develop a cobblestone appearance by ALI day 7 (see Fig. 2). Ciliated cells appear at ALI days 3–4 and gradually increase in number (see Fig. 3). Beating cilia may be seen by microscope as early as ALI day 5. Washing the cell surface with PBS enhances visualization of cilia. Typically, the apical surface has at least 30% ciliated cells by ALI day 14. During ALI culture, a small ring of mucous may be present at the edge of the membrane that should be washed off. The cells can be maintained at ALI for over 2 months without loss of differentiation. There are several potential causes and solutions for failure of cells to differentiate (see Notes 7–11).

3.6. Manipulation of Differentiation

1. To prevent differentiation of mTEC, seed the cells on collagen-coated tissue culture plates using the same density as for generating ALI-differentiated cells. Proliferate the cells using mTEC/Plus with RA. Once cells are near confluent, change the medium to mTEC/NS or mTEC/SF with RA. Cells cultured on membranes may similarly be maintained submerged after confluence. At ALI day 0, fill the apical and basal chambers with mTEC/NS or mTEC/SF with RA. This condition

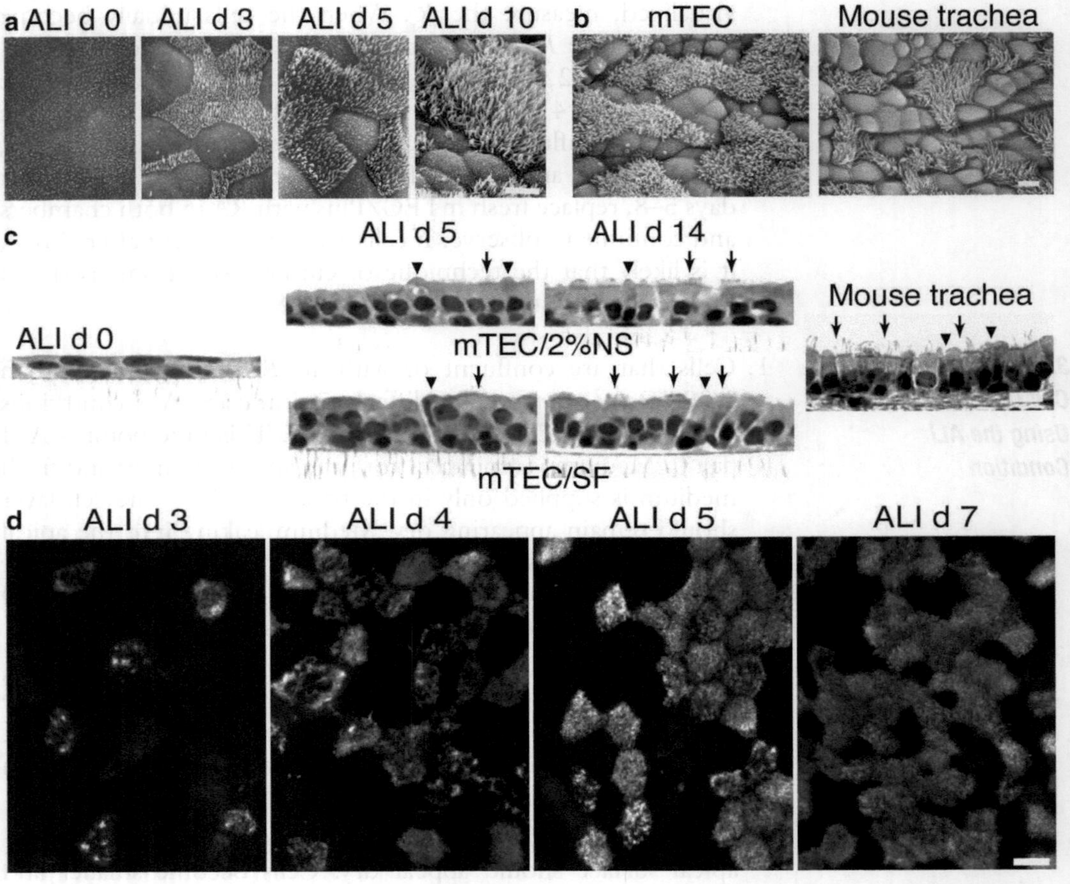

Fig. 3. mTEC differentiation. Differentiation of mTEC, initiated using ALI conditions. Preparations assessed by (**a**) scanning EM at indicated days; (**b**) at ALI day 10 mTEC appear similar to native trachea. (**c**) Hematoxylin and eosin stained paraffin sections of mTEC at indicated times demonstrate histologic appearance of ciliated (*arrow*) and Clara-like (*arrowhead*) cells and the differences in the number of cell layers produced when cultured in mTEC/NS compared to mTEC/SF medium. (**d**) Cilia biogenesis in mTEC preparations immunostained for cilia protein acetylated α-tubulin photographed on membranes *en face*. Ciliated cells are found at ALI day 3 and the number of ciliated cells increases through day 7. Bar (**a**) = 5 μm; (**b–d**) = 10 μm. Panels (**a–c**) are adapted from ref. 4 with permission from the American Physiological Society.

will markedly block ciliogenesis and proliferation. Reversion to the ALI condition can induce differentiation.

2. Mucous cell differentiation can be induced by adding mouse IL-13 to the basal chamber medium during the ALI phase. At ALI day 0, add IL-13, 10 ng/mL to mTEC/NS or mTEC/SF with RA. Maintain the ALI state and change the medium every 2 days. Mucous may be observed to accumulate on the apical surface of the cells. Quantitative real-time PCR or immunostaining for MUC5AC can be used to characterize the IL-13 response. Increasing the initial seeding density will enhance the percentage of mucous cells generated.

3. Genetic modification by gene transfer: Recombinant adenovirus or lentivirus vectors can be used for gene transfer as described (13, 16, 18). Adenovirus is added to the apical and basal chambers (the adenovirus receptor is basolateral) within 1 or 4 days after seeding cells using a multiplicity of infection of 25–1,000. Lentivirus collected in mTEC/Plus medium using standard protocols is used to resuspend the harvested tracheal epithelial cells for seeding and maintained in both chambers for the first 24 h. Successful proliferation leading to differentiation following lentivirus infection may require increasing the seeding density up to $2 \times 10^5/cm^2$.

3.7. mTEC Passage and Freezing

1. Removing cells from the membrane: mTEC preparations may be passaged two or three times with increasing loss of capacity to proliferate and differentiate (4). Wash the cells twice with PBS. Incubate mTEC with Trypsin/EDTA/Cell Dissociation Solution (see Subheading 2.9, item 1) for 10–15 min at 37°C. Mix the apical solution with a pipette to remove cells from the membrane. Add PBS with 10% FBS to the membrane to recover cells, repeating three times. Pool cells in a tube on ice. Pellet the cells by centrifugation at $500 \times g$, 5 min, 4°C.
2. Cell passage: Resuspend the cells in mTEC/Plus with RA. Count and seed cells as in step 3, Subheading 3.1, and in Subheading 3.4 using a seeding density greater than $1 \times 10^5/cm^2$.
3. Cell freezing and thawing: Resuspend cells in mTEC/Plus with 10% DMSO (see Subheading 2.9, item 2), slowly freeze in an isopropanol bath at –80°C overnight, and then transfer to liquid nitrogen vapor. To thaw cells, defrost rapidly in a 37°C water bath and then slowly add warm FBS to an equal volume. Centrifuge to pellet the cells and then resuspend in mTEC/Plus with RA. Seed the cells at a density $>1 \times 10^5/cm^2$.

3.8. Analysis of mTEC Differentiation by Immunofluorescence

The extent of ciliation of the cells on the membrane correlates with differentiation status (see Fig. 3d). For additional methods, see Subheading 3.9.

1. Fix the cells. Wash the inserts gently with PBS, aspirate the PBS, and then fill the chambers with 4% paraformaldehyde (this chemical is toxic, see Note 5). Incubate at room temperature for 10 min. Remove and properly discard the paraformaldehyde. Wash the cells on the membrane three times for 5 min each, by adding PBS to both chambers and slowly rocking the plate.
2. Cut the membrane from the plastic support. Do not allow the membrane to dry. Invert the insert and with a scalpel cut along the bottom edge between the membrane and plastic support. Prepare a wet surface for delivery of the membrane by adding 1 mL of PBS to a tissue culture dish. Stand the insert, with the

basal surface on the plate, and use fine forceps to completely release the membrane while firmly holding the plastic rim. To produce multiple samples from one membrane, cut into quarters by rocking the scalpel blade across the membrane. Notch the outer bottom edge of each quarter to mark the orientation to identify the cell surface. Handle the membrane at the edge with fine forceps. Fixed cells on membranes can be stored in sterile PBS at 4°C in a Parafilm®-sealed plate for several months.

3. Block nonspecific antibody binding on the cells (see Subheading 2.10, item 3). Transfer the membrane pieces to a 24- or a 96-well plate. Cover the membrane with blocking solution and slowly rock at room temperature for 30 min.
4. Immunostain cells with an anti-cilia antibody. Dilute anti-cilia antibody (see Subheading 2.10, item 4) in the blocking solution. Use an isotype-matched antibody as a control. Incubate antibodies with cells on membrane for 1 h at room temperature or overnight at 4°C. Wash membrane with PBS three times for 5 min each. Add a fluorescent-labeled secondary antibody for 30 min, and then wash again three times (see Note 12).
5. Mount immunostained membranes on slides. Transfer the membrane to a glass microscope slide and apply 10–20 μL of mounting medium containing DAPI. Use the notched edge as a guide for orienting the cell-containing side of the membrane upward. Inspect the membrane under the fluorescence microscope if the membrane orientation is not certain. Apply the coverslip and seal the edges with nail polish. Examine by fluorescent microscopy.

3.9. Additional Methods to Assess Gene Expression and Differentiation of mTEC

1. Flow cytometry of mTEC: To obtain a cell suspension for detecting proteins on the cell surface, use 0.1% EDTA in Cell Dissociation Solution. To detect an intracellular protein, use 0.25% trypsin with 0.1% EDTA in Cell Dissociation Solution. Put the appropriate cell dissociation solution in both chambers and then place in the tissue culture incubator for 5 min or longer up to 45 min. Aspirate the basal solution and release the cells from apical by mixing with a pipette. Transfer the cells to a tube. Add 2% FCS/PBS, 200 μL (0.33 cm^2 membrane) to the apical chamber, and pipette to recover additional cells. Repeat this twice more and pool all washes. Pass cell suspension through a 70 μM cell strainer to obtain a single cell suspension. Centrifuge the cells at 500×*g*, 4°C, for 5 min. Resuspend the cell pellet in 2% FCS/PBS. The typical yield is 2×10^5 cells per 0.33 cm^2 insert. Proceed using relevant flow cytometry protocols.
2. Protein blot analysis of mTEC: Put the plate with the inserts on ice, wash the cells twice with ice-cold PBS. Add 25–30 μL

of the appropriate lysis buffer on the apical surface of a 0.33 cm^2 of membrane. Incubate at 4°C, rocking slowly. After 20 min, pipette and gently scrape the surface of the membrane with the pipette tip to release cells. Transfer the lysate solution to a microcentrifuge tube on ice and process using standard protocols. Typical yield of protein is 25–40 μg per 0.33 cm^2 membrane.

3. RNA isolation from mTEC: Aspirate the medium, but do not wash the cells. For a 0.33 cm^2 insert, add 200 μL of guanidinium thiocyanate–phenol extraction solution (the maximum that can be added to the apical surface), pipette up and down several times, and transfer to an RNase-free microcentrifuge tube. The typical yield of total RNA is 1–2 μg per 0.33 cm^2 membrane. Purify and analyze RNA expression by standard protocols.
4. Electron microscopy of mTEC for SEM or TEM: Wash cells with cold PBS. Fix the cells in 2.5% glutaraldehyde in sodium cacodylate buffer at 4°C overnight. Consultation with the EM facility should guide sample preparation.
5. Cilia beat frequency determination of mTEC: Cells must be cultured on transparent membranes. Wash the cells with warm medium and then return to incubator for at least 30 min prior to analysis as washing may alter CBF.

4. Notes

1. Reagents and components for cell isolation should be filter sterilized using a 0.22 μM syringe filter with low protein binding (e.g., Pall PN 4602). Aliquots of stock components for mTEC media should be in volumes appropriate for preparing 250 mL of media. Different aliquot volumes are indicated for preparing mTEC/Plus and mTEC/SF media. Treat mTEC with Amphotericin B at a concentration of 0.25 μg/mL (added to media without filtering) until cells are changed to ALI condition if prior fungal contamination has occurred. Amphotericin B may be toxic to cells. Sustained use over several weeks should be avoided. RA should be freshly added to media to a final concentration of 5×10^{-8} M. RA-supplemented media should be used within 48 h. RA stock B at a concentration of 5×10^{-4} M (10,000×) is used throughout the protocol. However, pipette transfer of less than 1 μL of 10,000× RA stock B for small amounts of medium (<10 mL) may be inaccurate. In this case, Retinoic acid Stock B can be made at the lower concentration of [5×10^{-5} M] (1,000×) in HBSS/BSA. Store in 1 mL aliquots at −80°C up to 1 month.

2. We have not noted difference in growth and differentiation in mTEC from mice aged 4 weeks to over 18 months. Isolation of cells from younger and smaller mice is technically more challenging and total cell number recovered is diminished. However, growth and differentiation are not affected. Some strain-dependent differences (e.g., SV129, FVB) in proliferation can be overcome by increasing the seeding density.
3. An alternative to Transwell® inserts are Millicell® (Millipore) inserts. Both companies have similar materials and sizes with similar performance. Polyester (transparent) membranes must be used if cells are to be visualized by light microscopy and if cilia beating will be assessed. Handling polyester membranes is more difficult due to transparency. If sections are cut from paraffin-embedded preparations, the property of the polyester makes it difficult to maintain the cell layer on the membrane.
4. An electrode probe pair (STX2, World Precision Instruments) for transmitting and detecting current designed to fit into the apical and basal compartments of the supported membrane system is available. Medium is placed in both chambers for measurement of R_t. A coated membrane without cells is used to obtain baseline R_t. The baseline value is subtracted from the observed R_t to obtain the R_t from cells. Before use, probes are cleaned with 70% ethanol, air dried, and then rinsed with sterile water or PBS. The device is helpful for confirming the status of cell confluence or assessing cells that cannot be visualized by microscopy when cultured on translucent membranes (e.g., Transwell®). Measuring R_t is also useful for troubleshooting cell culture problems, but is not essential for generating the mTEC cultures.
5. Cautions regarding potentially toxic materials: Use protection and handling according to institutional guidelines. Fixatives, paraformaldehyde and glutaraldehyde, and organic chemicals, guanidinium thiocyanate and phenol, are potentially toxic. DNA-binding compounds Hoechst and DAPI and sodium cacodylate are potentially carcinogenic. Recombinant viruses used for gene transfer should be handled according to biosafety guidelines at the user's institution.
6. Poor yield of epithelial cells from tracheas: Causes may include: not resecting the entire trachea, insufficient removal of cells from trachea, and inadequate pronase activity. Corrective actions may include: proper dissection of the complete length of the trachea, increasing the number and force of "shakes" of tracheas in pronase, assuring that the pronase is fresh, and increasing pronase concentration (e.g., to 0.20%) or time of digestion. Pronase activity varies with supplier and lot.
7. Fibroblast contamination: Causes include: shaking tracheas in pronase too vigorously resulting in an increased number of

fibroblasts released, processing the larynx, large pieces of smooth muscle and esophagus with the trachea, allowing inadequate time for adherence of fibroblast in the culture dish, and over-washing the culture dish after fibroblast adherence. Corrective actions may include: review of anatomy and resection method, increasing the incubation time of dissociated cells to 3–4 h, and decreasing washing culture dish post fibroblast adherence.

8. Failure of Cells to Proliferate to Confluence (R_t < 1,000 Ωcm^2): Causes include: failure to isolate sufficient number of cells capable of proliferation, using seeding density that is too low, overestimating cell count after isolation, errors in media preparation or culture conditions, fibroblast contamination, and infection. Corrective actions may include: increasing the seeding density by 50%, re-estimation of the number of cells in clusters when counting cells, reviewing the status and activity of EGF and RA, and directly measuring CO_2 concentration and temperature of the incubator.
9. Failure of Cells to Differentiate: Causes may include: low recovery of tracheal cells and low seeding density, error in media preparation, and error in temperature, humidity, or CO_2 concentration in incubator. Corrective actions may include: use of higher seeding density and change to alternative differentiation media mTEC/NS or mTEC/SF. See Note 8 for causes of failure of cells to become confluent. NuSerum™ lots may vary in the ability to support differentiation.
10. Infection: In mTEC cultures, bacteria contamination is typically the result of errors in technique. Fungal contamination may come from the mouse fur or the lab environment. Sources include: technique-related handling and environmental contamination (e.g., incubator, water bath). Corrective actions may include: improving hand washing or use of gloves and cleaning the incubator; if Amphotericin B is not effective (see Note 1), add antifungal/antibiotic Primocin™ 50 mg/mL (InvivoGen) to medium.
11. Medium Leaking into the Apical Compartment after Establishment of ALI: Causes may include: Infection, fibroblast contamination and poor proliferation, and error in media preparation. Corrective actions include those discussed in Notes 6–10.
12. Immunostaining: The conditions provided are not appropriate for all antibodies. Alternative fixatives or detergents (e.g., 0.1% Triton™ X-100) in the block and wash solutions may be required. Altering cellular tubulin with Triton X-100, 0.5% in PHEM buffer (45 mM PIPES, 45 mM HEPES, 10 mM EGTA, 5 mM $MgCl_2$, and 1 mM phenylmethylsulfonyl fluoride) for 1 min before fixing will enhance the improve

immunostaining (13). To prepare paraffin sections of mTEC on membrane, after fixing and washing the membrane, create a bed of warm 2% agarose, apply the membrane piece, and then cover with the agar. After cooling, the block can be trimmed, processed for embedding, and then sectioned as for tissue (see Fig. 3c).

Acknowledgement

The authors thank Brody laboratory members for photomicrographs and review. This work was supported by awards from the National Institute of Health and the Children's Discovery Institute of Saint Louis Children's Hospital and Washington University.

References

1. Clarke LL, Burns KA, Bayle JY, Boucher RC, Van Scott MR (1992) Sodium- and chloride-conductive pathways in cultured mouse tracheal epithelium. Am J Physiol 263:L519–525
2. Kumar RK, Maronese SE, O'Grady R (1997) Serum-free culture of mouse tracheal epithelial cells. Exp Lung Res 23:427–440
3. Davidson DJ, Kilanowski FM, Randell SH, Sheppard DN, Dorin JR (2000) A primary culture model of differentiated murine tracheal epithelium. Am J Physiol Lung Cell Mol Physiol 279:L766–778
4. You Y, Richer EJ, Huang T, Brody SL (2002) Growth and differentiation of mouse tracheal epithelial cells: selection of a proliferative population. Am J Physiol Lung Cell Mol Physiol 283:L1315–1321
5. Shah AS, Farmen SL, Moninger TO, Businga TR, Andrews MP, Bugge K, Searby CC, Nishimura D, Brogden KA, Kline JN, Sheffield VC, Welsh MJ (2008) Loss of Bardet-Biedl syndrome proteins alters the morphology and function of motile cilia in airway epithelia. Proc Natl Acad Sci U S A 105:3380–3385
6. Aujla SJ, Chan YR, Zheng M, Fei M, Askew DJ, Pociask DA, Reinhart TA, McAllister F, Edeal J, Gaus K, Husain S, Kreindler JL, Dubin PJ, Pilewski JM, Myerburg MM, Mason CA, Iwakura Y, Kolls JK (2008) IL-22 mediates mucosal host defense against Gram-negative bacterial pneumonia. Nat Med 14:275–281
7. Wong AP, Keating A, Lu WY, Duchesneau P, Wang X, Sacher A, Hu J, Waddell TK (2009) Identification of a bone marrow-derived epithelial-like population capable of repopulating injured mouse airway epithelium. J Clin Invest 119:336–348
8. Lechner J, LaVeck M (1985) A serum-free method for culturing normal human bronchial epithelial cells at clonal density. J Tissue Cult Meth 9:43–48
9. Whitcutt MJ, Adler KB, Wu R (1988) A biphasic chamber system for maintaining polarity of differentiation of cultured respiratory tract epithelial cells. In Vitro Cell Dev Biol 24:420–428
10. Wu R (1997) Growth and differentiation of tracheobronchial epithelial cells. In: MacDonald J (ed) Growth and development of the lung. Deckker, New York, NY, pp 211–241
11. Borthwick DW, Shahbazian M, Krantz QT, Dorin JR, Randell SH (2001) Evidence for stem-cell niches in the tracheal epithelium. Am J Respir Cell Mol Biol 24:662–670
12. Hong KU, Reynolds SD, Watkins S, Fuchs E, Stripp BR (2004) In vivo differentiation potential of tracheal basal cells: evidence for multipotent and unipotent subpopulations. Am J Physiol Lung Cell Mol Physiol 286:L643–649
13. You Y, Huang T, Richer EJ, Schmidt JE, Zabner J, Borok Z, Brody SL (2004) Role of f-box factor foxj1 in differentiation of ciliated airway epithelial cells. Am J Physiol Lung Cell Mol Physiol 286:L650–657
14. Ostrowski LE, Nettesheim P (1995) Inhibition of ciliated cell differentiation by fluid submersion. Exp Lung Res 21:957–970
15. Tyner JW, Kim EY, Ide K, Pelletier MR, Roswit WT, Morton JD, Battaile JT, Patel AC, Patterson GA, Castro M, Spoor MS, You Y, Brody SL, Holtzman MJ (2006) Blocking airway mucous cell metaplasia by inhibiting EGFR

antiapoptosis and IL-13 transdifferentiation signals. J Clin Invest 116:309–321

16. Pan J, You Y, Huang T, Brody SL (2007) RhoA-mediated apical actin enrichment is required for ciliogenesis and promoted by Foxj1. J Cell Sci 120:1868–1876
17. Sisson JH, Stoner JA, Ammons BA, Wyatt TA (2003) All-digital image capture and whole-field analysis of ciliary beat frequency. J Microsc 211:103–111
18. Schmid A, Bai G, Schmid N, Zaccolo M, Ostrowski LE, Conner GE, Fregien N, Salathe M (2006) Real-time analysis of cAMP-mediated regulation of ciliary motility in single primary human airway epithelial cells. J Cell Sci 119:4176–4186

Chapter 10

Isolation and Culture of Alveolar Epithelial Type I and Type II Cells from Rat Lungs

Robert F. Gonzalez and Leland G. Dobbs

Abstract

The pulmonary alveolar epithelium, comprised of alveolar Type I (TI) and Type II (TII) cells, covers more than 99% of the internal surface area of the lungs. The study of isolated and cultured alveolar epithelial TI and TII cells has provided a large amount of information about the functions of both cell types. This chapter provides information about methods for isolating and culturing both of these cell types from rat lungs.

Key words: Lung, Alveolar epithelium, Type I cells, Type II cells

1. Introduction

More than 99% of the internal surface area of the lung is contained within the alveolar compartment (1) which is lined by two types of cells, Type I (TI) and Type II (TII) cells. Type I cells comprise ~8% of the cells in the lung, but cover ~95–98% of the surface area; TII cells comprise ~14% of all lung cells and cover the remaining ~2–5% of the internal surface area (2). TI cells are very large squamous cells (calculated diam. ~50–100 μm) whose thin (50–100 nm) cytoplasmic extensions form the epithelial portion of the air–blood barrier essential for normal gas exchange; TI cells participate in ion and fluid transport (3) and are likely to have other important functions within the lung (4). In vitro, TI cells have the capacity to proliferate (5). TII cells are smaller (diam. ~10 μm), cuboidal cells that synthesize, secrete, and recycle surfactant components, transport ions, participate in lung immune responses, and in vivo function as progenitor cells in response to injury.

There have been many attempts to establish passaged cell lines as models of both cell types. Unfortunately, these cell lines have been disappointing because none display the full range of morphologic,

Scott H. Randell and M. Leslie Fulcher (eds.), *Epithelial Cell Culture Protocols: Second Edition*, Methods in Molecular Biology, vol. 945, DOI 10.1007/978-1-62703-125-7_10,

biochemical, and molecular characteristics of freshly isolated TI or TII cells. Although the primary culture of epithelial cells has disadvantages in that it involves repeated cell isolations and in that cells undergo different extents of phenotypic drift in culture, much has been learned from the study of primary cultures of both TI and TII cells. We have previously published a detailed method for the isolation of TII cells from rat lungs using the "panning method" to purify cells using IgG-coated plates (6–8). In this communication, we focus on methods using TI or TII cell-specific antibodies to isolate these cell types from rat lungs.

2. Materials

2.1. Isolation of TI and TII Cells

1. Specific-pathogen-free Sprague-Dawley rats, 100–150g (Charles River™).
2. Heparin (1,000 U/mL) and sodium pentobarbital 50 mg/mL.
3. Suture: Size 3-0, silk.
4. Scissors, Mayo-Noble (6.5 in.) and microdissecting (4 in.).
5. Luer Stub Adaptor, 15 gauge (Intramedic®, Becton Dickinson) (for tracheal cannulation).
6. Angiocath, 20 gauge plastic catheter, connected to the perfusion device containing Dulbecco's phosphate-buffered saline (DPBS).
7. DPBS.
8. Perfusion device constructed from a 50 mL disposable syringe with the plunger removed and 24 in. of plastic tubing attached to the end.
9. Disposable plastic syringes (3, 10, 20, and 50 mL).
10. Ca–Mg-free DPBS containing 5 mM EGTA and 5 mM EDTA.
11. "Solution A": RPMI 1640-HEPES medium (Gibco/BRL™).
12. "Solution B": For TI cell isolations use 40 mL "Solution A" containing 1.75 mg elastase/mL (Worthington®) and 10% dextran (35–45,000 MW, Sigma-Aldrich®).
13. "Solution C": For TII cell isolations use 40 mL "Solution A" containing 1 mg elastase/mL (Worthington).
14. Tissue culture centrifuge tubes (15 and 50 mL sizes), disposable plastic beakers (500 and 50 mL sizes), and 250 mL disposable plastic Erlenmeyer flasks.
15. Fetal Bovine Serum (Hyclone™, Thermo Fisher Scientific).
16. DNAse (Sigma-Aldrich®) 2 mg/mL "Solution A."
17. Rat IgG and mouse IgG (Sigma-Aldrich®).

18. Filtering tubes made from plastic syringes (10 mL and 20 mL capacity syringes with the end of the bore and the tips cut off so that the bore is of uniform diameter) with 20, 40, or 100 μm nylon mesh filters (Tetko Inc.®) secured at the cutoff ends with rubber bands. Make at least two with the 100 μm filter, and 3–5 each with the 40 μm and 20 μm nylon mesh filters.
19. Fluorescein diacetate (FDA) (Sigma-Aldrich®) 2 mg/mL acetone.
20. Percoll™ (Life Technologies™).
21. Monoclonal antibody against RTI-40 (available from: http://www.lungantibodyinfo.com).
22. Monoclonal antibody against RTII-70 (available from: http://www.lungantibodyinfo.com).
23. Rat anti-mouse IgG1-, goat anti-mouse IgG-, and goat anti-rat IgG-ferromagnetic beads (Miltenyi Biotec).
24. Two types of capture columns were prepared from 3 mL disposable plastic syringes, one with 0.8 g of loosely packed "coarse grade" steel wool (#3 grade) and one with 0.5 g of loosely packed "fine grade" steel wool (#00 grade) (Homax® Products Inc.). Columns are gas sterilized.
25. Three-way Luer stopcocks.
26. Twenty three gauge hypodermic needles.
27. Capture magnet for cell isolation (VarioMACS™, Miltenyi Biotec).
28. Nine inch glass Pasteur Pipets.

2.2. FACS Sorting and Assessing Purities of TI and TII Cell Populations After Isolation

1. Monoclonal antibodies against RTI-40 and RTII-70 (see Subheading 2.1, items 21 and 22).
2. Zenon anti-mouse IgG1 Alexa Fluor® 610-RPE labeling kit (Life Technologies™/Molecular Probes®).
3. Antibodies against aquaporin-5 (AQP-5, Alpha Diagnostic) and pro-surfactant protein SP-C (pro-SP-C, Chemicon International®).
4. Secondary antibodies specific for either mouse IgG3 or IgG1 labeled with Alexa Fluor® 488 or Alexa Fluor® 594 (Life Technologies™).
5. BD FACS Aria Flow Cytometer (BD Biosciences™) or equivalent Flow Cytometer.
6. Cytocentrifuge.
7. Superfrost/Plus Microscope Slides, pre-cleaned.
8. 4% Paraformaldehyde (Sigma-Aldrich®) made fresh in DPBS.
9. "Prolong" anti-fade reagent (Life Technologies™/Molecular Probes®).
10. Trypan Blue.

2.3. Culture of TI Cells

1. TI cell culture medium: Dulbecco's modified Eagle's medium H-16 (DME-H16) (Gibco/BRL™) containing 20% FBS and 50 μg gentamicin/mL.
2. Tissue culture 6-well Transwell® plates (Corning) previously coated with bovine fibronectin (see Subheading 3.3, step 2).
3. Syringe filter (0.2 μm syringe filter).

2.4. Culture of TII Cells

1. TII cell culture medium: Dulbecco's modified Eagle's medium (DMEM) (Gibco/BRL™) containing 1% rat serum, 10 ng/mL keratinocyte growth factor (KGF, FGF-7, R&D Systems®), 10^{-4} M 8-bromocyclic AMP (Sigma-Aldrich®), and 50 μg gentamicin/mL.
2. Engelbreth-Holmes-Swarm Matrix (EHS Matrix, Matrigel™, BD™).
3. Tissue culture 6-well Transwell® plates (Corning) previously coated with EHS Matrix (see Subheading 3.4, step 2).

3. Methods

Alveolar epithelial cells are fragile and are easily injured (see Fig. 1). The keys to successful isolation and culture are rapid isolation (the entire procedure should take <4.0 h) while handling tissues and cells gently. Two indices of cell health are viability indices (>95%) and plating efficiency (>70% at 24 h). Although both TI and TII cell types comprise relatively small percentages of all lung cells [TI, 9%; TII, 14% (from ref. 2)], enzymatic digestion of the lung with carefully timed intratracheal instillation of elastase provides an initial cell enrichment because elastase effectively liberates epithelial cells from the underlying basement membrane, leaving most of the interstitial and vascular compartments intact (9).

3.1. Perfusion of the Lungs and Digestion of Lungs with Elastase

1. Anesthetize the rat with an i.p. injection of pentobarbital (50 mg/kg/body weight) and heparin (400 U/kg/body weight). After the rat is well anesthetized, it is important to perform the following surgical procedures rapidly. The entire surgical procedure, from the time of the abdominal incision to removal of the perfused lungs, should take no more than 4 min.
2. Make an abdominal incision, cut the descending aorta, and make a small incision in the diaphragm to deflate the lungs.
3. Incise the neck, remove the submandibular tissue and the thyroid gland, pass a suture under the trachea, nick the trachea, and cannulate it with a 15 gauge Intramedic® Luer Stub Adaptor, securing it with the suture.

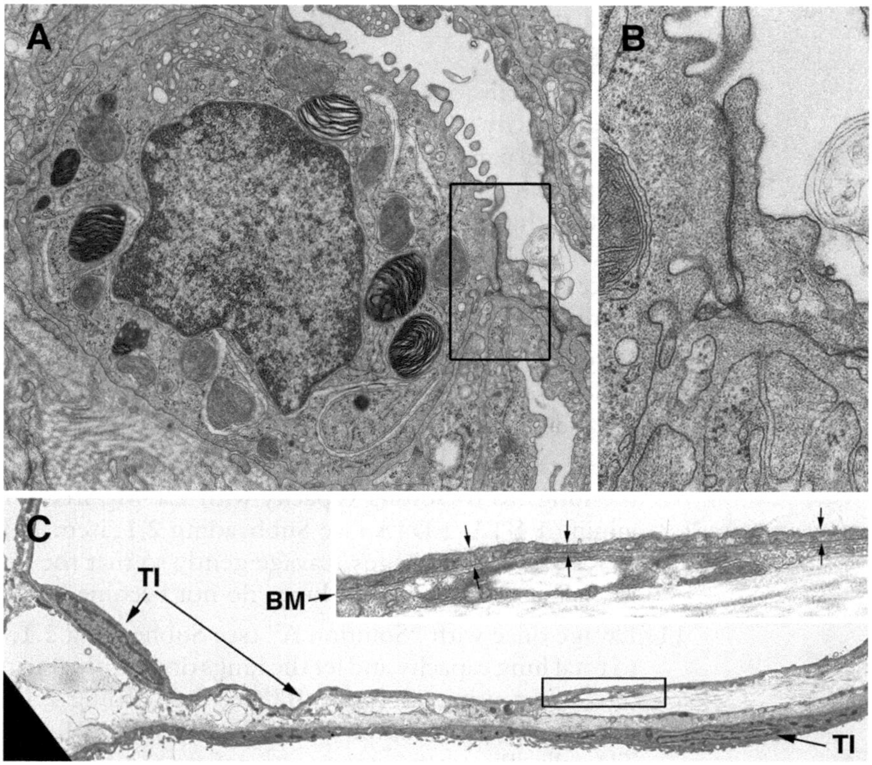

Fig. 1. Electron micrograph showing TI and TII cells. (**a**) Rat TII cells with electron-dense lamellar bodies. The *box* contains a junction between a TI cell and an adjacent TII cell. (**b**) Higher magnification of the image within the *box* in panel (**a**), illustrating the tight junction between the two cell types. (**c**) A portion of a rat TI cell showing the nucleus and a portion of the thin cytoplasmic extension (TI, *large arrows*) overlying the interstitial. Extensions of a different TI cell can be seen on the other side of the interstitium. A portion of the nucleus is visible (TI, *arrow*, *lower right*). The *box* and higher magnification (*inset*) view show the very thin cytoplasmic extensions of the TI cell on the basement membrane (BM) that separates the TI cell and the interstitial compartment (electron micrographs courtesy of Lennell Allen, C.V.R.I., U.C.S.F.).

4. Incise the chest wall along the entire length of the sternum and the anterior portions of the diaphragm so that the chest is open.
5. Pull the rib cage laterally and horizontally to expose the heart and the lungs. The sharp cut edges of the ribs should not be close to the lungs, once they are inflated in subsequent steps.
6. Identify the pulmonary artery, pass a suture underneath it, nick the inferior portion of the right ventricle, and cannulate the pulmonary artery with a 20 gauge plastic catheter (occluding the puncture site) connected to a perfusion bottle containing DPBS at 20 cm H_2O pressure. The perfusion buffer and length of tubing should be free of any bubbles prior to starting the perfusion, because bubbles will lodge in the pulmonary capillary bed, obstructing perfusion.

7. Nick the left atrium to allow the perfusion fluid to exit the lungs.
8. During perfusion, gently inflate the lungs several times to total lung capacity (7–10 mL) using a 10 mL syringe several times; this will expand the lung and allow perfusion to clear the lungs completely of blood. The heart should still be spontaneously beating during this step; if the heart is not spontaneously beating, it will be difficult to clear blood from the lungs. Following successful perfusion, the lungs should be completely white; there should not be red or pink areas.
9. Carefully remove the lungs, trachea (leaving the trachea still attached to the cannula), and heart from the chest cavity. Trim off and discard the heart.
10. Using the tracheal cannula to instil liquid, sequentially lavage the lungs to total lung capacity with Ca–Mg-free DPBS containing EGTA/EDTA (see Subheading 2.1, item 10) at 37°C to remove macrophages. Lavage gently so that the lung integrity is maintained and the lungs do not become leaky.
11. Lavage once with "Solution A" (see Subheading 2.1, item 11) to total lung capacity and let the lungs drain with gravity. Lavage once more using "Solution B" (see Subheading 2.1, item 12) for TI cells or "Solution C" (see Subheading 2.1, item 13) for TII cells. Be careful not to overinflate the lungs, which may create a leak.
12. Inflate the lungs to total lung capacity with "Solution B" (TI cells) or "Solution C" (TII cells) while the lungs are suspended in a beaker of saline in a 37°C water bath; then suspend the enzyme-filled lungs so that they float in the warm saline while additional enzyme is gradually instilled. We use a 3 mL plastic syringe taped to the side of the beaker and gradually add the remaining elastase-containing solution by gravity over a 15-min period, for a total enzymatic digestion period of 20 min for TII cells. For TI cell isolation, the time of enzymatic digestion is 40 min.
13. Dissect the lung parenchyma away from the major airways (do not spend the time to do this meticulously) and add the lungs to a 50 mL disposable plastic beaker containing 20 mL of fetal bovine serum (FBS) at 4°C and 0.5 mL of DNAse solution (see Subheading 2.1, item 16).
14. Quickly mince the lung pieces with very sharp Mayo-Noble scissors to a final size of ~1 mm^3 (~150–200 scissor strokes, duration <1 min).
15. Add additional "Solution A" to bring the final liquid volume to 30 mL and transfer the lung minces and liquid to a 250 mL plastic Erlenmeyer flask. Add rat IgG to a final concentration of 50 μg/mL.

16. Shake the flask vigorously side to side in a reciprocating water bath at 130 cycles/min for 2 min. The goal of this step is to apply shear force to the lung minces, freeing the epithelial cells. The liquid in the flask should move in a back-and-forth fashion, rather than swirling. If liquid starts to swirl in the flask, stop shaking until the swirling stops before resuming shaking.
17. Filter the lung minces sequentially through filters pre-wetted with "Solution A" (see Subheading 2.1, item 18):
    (a) Nylon mesh, 100 μm.
    (b) Nylon mesh, 40 μm.
    (c) Nylon mesh, 20 μm.

    Collect the filtered cells into 50 mL disposable beakers.
18. Let the filtering occur by gravity, using additional filters as needed when flow decreases due to filter clogging. We routinely use several separate filters for each of the 40 μm and 20 μm filter sizes in order to prevent the filters from clogging. When filters are clogged, the flow rates decrease. It should take less than 5 min to complete the pass through all three filter sizes.
19. From the cell suspension, remove 1 mL for measurement of viability, total cell count, and differential cell counts. At this stage of the cell isolation, representative yields and purities for each respective cell type should be the following: TI cells: $\sim 1\times 10^8$ cells, ~5% of which are TI cells; TII cells: $\sim 4\times 10^7$ cells, of which 30–50% are TII cells. Viabilities should be >95% (see Note 1). Cell viability is assessed by the ability of cells to internalize and hydrolyze FDA. To a 100 μL cell aliquot containing less than 1×10^5 cells add 0.2 μL of FDA (see Subheading 2.1, item 19). Incubate, shielded from light, for 20 min at room temperature and then wash the cells twice by centrifuging and resuspending the pellet in medium. Live cells will exhibit fluorochromasia.
20. Transfer the cell suspension to a 50 mL tube. For "magnetic sorting," add both rat and mouse IgGs to final concentrations of 50 μg/mL. For "FACS sorting," add only rat IgG to 50 μg/mL. From this point in the cell isolation protocol, use sterile-technique.

3.2. Isolation of TI and TII Cells by Magnetic Sorting or FACS

Two different separation techniques can be used to isolate either TI or TII cells from the cell suspension. Each method has its advantages and disadvantages. Magnetic sorting produces greater yields of cells, while FACS sorting produces cells of greater purity (see Table 1). These two techniques are described in the following three Subheadings.

3.2.1. Magnetic Separation of TI Cells from the Cell Suspension

Alveolar TI cells are isolated by depleting (negative selection) the crude cell suspension of TII cells and macrophages on a "coarse grade" separation column. This is followed by positive selection of

Table 1
Comparison of yields and purities of TI and TII cells isolated by either ferromagnetic beads or FACS (mean SD, $n = 10$)

Isolation method	Cell yield	Cell purity
Magnetic sorting of TI cells	$8.9 \pm 2.6 \times 10^6$	80–91%
Magnetic sorting of TII cells	$21.0 \pm 3.8 \times 10^6$	84–92%
FACS isolation of TI cells	$2.5 \pm 0.5 \times 10^5$	>98%
FACS isolation of TII cells	$1.7 \pm 0.7 \times 10^6$	>98%

TI and TII cells isolated by FACS have a greater purity, but lower yields

TI cells on a "fine grade" separation column (see Subheading 2.1, item 24). During the column selection steps it is important not to allow the columns to run dry, which traps air in the column and diminishes the final cell yield.

1. Add 1 mL of monoclonal antibody supernatant against RTII-70 (specific to TII cells) to the cell suspension from Subheading 3.1, step 20, mix gently, and incubate on ice for 5 min. Dilute to 35 mL with "Solution A" containing 20% FBS, and place a Percoll™ cushion (150 μL) at the bottom of the centrifuge tube using a 9 in. glass Pasteur Pipet after adding the cell suspension to the tube (see Note 2). Centrifuge the cell suspension at 350×*g* for 15 min at 4°C.
2. After centrifugation, remove and discard the supernatant above the pellet, add 200 μL of DNAse (2 mg/mL "Solution A"), and gently resuspend the pellet. Add 200 μL of goat anti-mouse IgG ferromagnetic beads (see Subheading 2.1, item 23) to the suspended cells and incubate for 10 min at 4°C.
3. Pre-equilibrate a "coarse grade" syringe capture column with "Solution A" containing 20% FBS attached to a three-way stopcock. Dilute the cell suspension to 2 mL with "Solution A" containing 20% FBS. Insert a 23 gauge needle at the tip of the column and place the column in the capture magnet's magnetic field, pass the cell suspension through the column five times, being careful not to let the column dry out (see Note 3). After the final pass, collect the unbound cells in a 15 mL centrifuge tube, add 20 μL of purified IgG monoclonal antibody against RTI-40 (~2 μg/mL), mix gently, and incubate on ice for 5 min. Dilute to 15 mL with "Solution A" and add ~150 μL of Percoll™ to the bottom of the tube to act as a cushion. Centrifuge at 350×*g* for 15 min at 4°C.
4. Remove the supernatant and add 200 μL of DNAse (2 mg/mL "Solution A"), gently resuspend the pellet, and add 100 μL of

rat anti-mouse IgG1 ferromagnetic beads to the cell suspension. Incubate for 10 min on ice. Fit a "fine grade" syringe capture column with a 3-way stopcock and a 23 gauge needle, and pre-equilibrate the column with "Solution A" containing 20% FBS. Place the column in the capture magnet's magnetic field. Dilute the cell suspension to 2 mL with "Solution A" containing 20% FBS, and pass the cell suspension through the column five times. After the final pass, wash the column five to eight times with "Solution A" containing 20% FBS (see Note 4). Remove the column from the magnetic field and collect the bound cells by back-washing 5 mL of buffer through the 3-way stopcock so that bound cells are collected from the top of the column bed. Collected TI cells are then centrifuged onto a Percoll™ cushion at 350 × *g* for 15 min at 4°C, and resuspended in sterile culture medium.

3.2.2. Magnetic Separation of TII Cells from the Cell Suspension

Alveolar TII cells are isolated by depleting the cell suspension of TI cells and macrophages by negative selection on a "coarse grade" separation column, followed by positive selection of TII cells on a "fine grade" separation column.

1. Add 20 μL of monoclonal antibody supernatant against RTI-40 (TI cell specific) to the cell suspension (see Subheading 3.1, step 20), mix gently, and incubate on ice for 10 min. Dilute to 35 mL with "Solution A" containing 10% FBS and centrifuge the cell suspension at 350 × *g* for 10 min at 4°C. After centrifugation, remove and discard the supernatant above the pellet, add 200 μL DNAse (2 mg/mL "Solution A"), and gently resuspend the pellet. Add 200 μL of rat anti-mouse IgG1 and goat anti-rat IgG ferromagnetic beads and incubate on ice for 10 min.
2. Fit a "coarse grade" syringe capture column with a 3-way stopcock and a 23 gauge needle and pre-equilibrate the column with "Solution A" containing 10% FBS. Place the column in the capture magnet's magnetic field. Dilute the cell suspension to 2 mL with "Solution A" containing 20% FBS, and pass it through the column five times, being careful not to let the column dry out. After the final pass, collect the unbound cells in a 15 mL conical plastic centrifuge tube, add 1 mL of monoclonal antibody supernatant against RTII-70 (TII cell specific) to the cell, mix gently, and centrifuge at 350 × *g* for 10 min at 4°C.
3. Remove the supernatant and add 200 μL of DNAse (2 mg/mL "Solution A") to the cell pellet, gently resuspend it, and add 100 μL of goat anti-mouse IgG ferromagnetic beads to the cell suspension. Incubate for 10 min on ice. Fit a "fine grade" syringe capture column with a 3-way stopcock and a 23 gauge needle and pre-equilibrate the column with "Solution A" containing 10% FBS. Place the column in the capture magnet's magnetic field. Dilute the cell suspension to 2 mL with "Solution A"

containing 20% FBS and pass the cell suspension through the column five times. After the final pass, wash the column five times with "Solution A" containing 10% FBS. Collect the bound cells and centrifuge them at 350 × *g* for 10 min at 4°C.

3.2.3. Isolation of TI and TII Cells by Flow Cytometry

To isolate both cell types by flow cytometry, we differentially label both cell types with fluorescently labeled antibodies and sort the labeled cells on a flow cytometer.

1. Directly conjugate anti-RTI-40 to Alexa Fluor® 610-RPE using the Zenon technology (see Note 5). Approximately 3 μL of anti-RTI-40 IgG1 (1 mg/mL) is labeled with 8 μL of Zenon labeling reagent (see Subheading 2.2, item 2). The anti-RTII-70 antibody is of the IgG3 subclass. Because the Zenon technology is not available for antibodies of the IgG3 subclass and because other methods of direct conjugation we have tried destroy the antigen binding capacity of anti-RTII-70, we use Alexa Fluor® 488 anti-IgG3 as a secondary antibody for FACS sorting.
2. To the cell suspension from Subheading 3.1, step 20, add 1 mL of hybridoma supernatant containing anti-RTII-70 (see Subheading 2.1, item 22), incubate on ice for 10 min, and centrifuge at 350 × *g* for 12 min at 4°C onto a 150 μL Percoll™ cushion.
3. The supernatant liquid is removed and the centrifuge tube is tapped gently to dislodge the pellet. 50 μL of DNAse (2 mg/mL "Solution A") is added directly to the pellet and the cells are gently resuspended in 1 mL of "Solution A" containing 20% FBS. The volume should be approximately 1.3 mL.
4. Add a 100 μL aliquot of suspended cells to three 15 mL disposable tissue culture tubes and label them "A," "B," and "C." Add 1 μL of Alexa Fluor® 610-RPE-labeled anti-RTI-40 to Tube "A," and add 0.5 μL of goat anti-mouse IgG3-Alexa Fluor® 488 to Tube "B;" tube "C" receives no labeled antibodies. Tube "C" serves as the unlabeled sorting control. The cells in these three tubes will be used to adjust the Alexa Fluor® 488 and Alexa Fluor® 610-RPE FACS sorting gates. Transfer the remaining 1 mL of the cell suspension to a fourth 15 mL tube labeled "D," and add 10 μL of Alexa Fluor® 610-RPE-labeled anti-RTI-40 and 5 μL of goat anti-mouse IgG3-Alexa Fluor® 488. The TI cells will be labeled with Alexa Fluor® 610-RPE and the TII cells will be labeled with Alexa Fluor® 488. Incubate at room temperature for 10 min.
5. Dilute each tube to 15 mL with "Solution A" containing 20% FBS, add a 150 μL Percoll™ cushion to the bottom of each tube, and centrifuge at 350 × *g* for 15 min at 4°C.
6. Remove the supernatant from each of the pellets and add 50 μL of DNAse (2 mg/mL "Solution A") to each of the

pellets. Gently tap each tube to dislodge the pellet and resuspend each pellet in 0.5 mL of "Solution A" containing 20% FBS. The cells are now labeled and ready for sorting. An aliquot of differentially labeled TI and TII cells in tube "D" should be checked for differential cell staining prior to sorting using a fluorescence microscope.

7. Dilute the cell suspensions in tubes "A," "B," and "C" to 0.5 mL with "Solution A" containing 20% FBS and filter them through a 20 μm filter. Run the cells from tube "C" containing "unlabeled control cells" on the FACS to adjust the voltage. Then run labeled cells from tubes "A" and "B" and set the gates for Alexa Fluor® 610-RPE- and Alexa Fluor® 488-labeled cells. Dilute the cells in tube "D" to a volume of 3 mL, filter through a 20 μm nylon mesh filter, and sort.

8. Sort the cells on a FACS Aria using a 100 μm orifice and a slow flow rate (Aria flow rate setting "3"); this setting maximizes cell purity. (If an investigator desires to obtain greater number of cells at a cost of lower purities, higher flow rates can be selected.) Cells are sorted for 120 min; collect sorted cells into a tube containing 0.5 mL FBS at 4°C.

9. Cell sorting results are analyzed using BD™ Diva software. Alveolar TI cell purities are typically >98% by immunofluorescence using anti-RTI-40 or anti-aquaporin 5 to identify TI cells; yields are ~2.5×10^5 TI cells per rat. Because TI cells are fragile, viability by FDA is lower, ~85–90% (see Subheading 3.1, step 19), than that of TII cells. Type II cell purities are >98% by immunofluorescence using anti-RTII-70 or anti-pro-SP-C as primary antibodies. Yields average 1.7×10^6 TII cells per rat. TII cells have a viability of >90% by Trypan blue exclusion. Substantially higher yields can be achieved with longer sorting times, but cell viability decreases with longer sorting periods.

3.2.4. Criteria for Identification of Alveolar Epithelial Cells

Cells are identified by assessing purities of cytocentrifuged cells with cell-specific antibodies, such as anti-RTI-40 and anti-aquaporin 5 for TI cells and anti-RTII-70 and anti-pro-SP-C for TII cells. For representative images of cytocentrifuged, stained cells at different stages of the cell isolation procedure, see Fig. 2a–f.

1. Preparation of cytocentrifuged cells: Cytocentrifuged preparations should be made at each step of the isolation. Dilute the cell suspension to 0.3×10^6 cells/mL. Make several (~6) slides, using ~0.3 mL of the cell suspension in each cytocentrifuge well. Centrifuge at $350 \times g$ for 10 min at room temperature onto Superfrost Slides. The slides are then placed in a freshly made (daily) solution of 4% paraformaldehye/DPBS. Cell density should be adjusted as needed to achieve even distribution of individual cells on the slide; clumps of cells are difficult to count accurately.

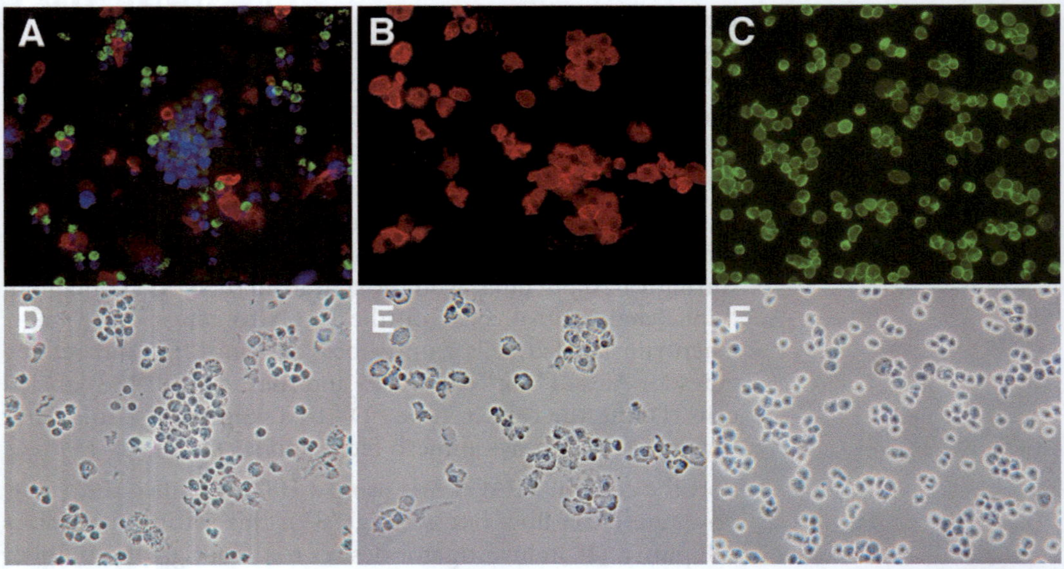

Fig. 2. Cytocentrifuged preparations of cells at different stages of the cell isolation procedure. The images shown are paired immunofluorescence (**a**, **b**, **c**) and phase contrast (**d**, **e**, **f**) views of cytocentrifuged preparations. Cells are stained for CC10 (*blue*, Clara cells+), RTI-40 (*red*, TI cells+), and RTII-70 (*green*, TII cells+) (**a**, **d**): Crude cell digest (Subheading 3.1, step 19). (**b**, **e**): FACS sorted TI cells. (**c**, **f**): FACS sorted TII cells.

2. To label cells with antibodies, remove the slides from fixative and rinse them five times with DPBS. Apply a ring of rubber cement with a 2 cm diameter around the sample. Pipet solutions onto the slide, but remove the solution by aspirating with a syringe and needle, taking care that the deposited cells do not dry out or are damaged. The procedure is as follows. Add blocking solution consisting of 10% serum (use the same serum type as the secondary antibody, which is usually goat serum) in DPBS to the slide and incubate the slide at room temperature for 1 h. Rinse slides once with DPBS, add primary antibody diluted in blocking buffer, and incubate for 20 min. Slides are then rinsed five times with DPBS, add secondary antibody diluted in blocking buffer, and incubate for 20 min. Carefully remove the rubber cement ring by slowly "peeling" the ring from the slide using small forceps, rinse the slides ten times with DPBS once with water, and briefly air-dry. Add Prolong fixative, apply a coverslip, and seal the edges of the coverslip with nail polish. Allow the fixative to dry overnight in a light-protected environment (see Prolong protocol insert). Slides can be examined immediately or can be stored in a freezer.

3.3. Culturing TI Cells

TI cells are cultured on fibronectin-coated Transwells® under the following conditions:

1. Alveolar TI cells are cultured in DME-H16 containing 20% FBS and 50 μg gentamicin/mL in 6-well tissue culture

Transwell® plates previously coated with fibronectin, and maintained in a 10% CO_2/air incubator at 37°C.

2. To coat the inserts, use fibronectin isolated from bovine blood according to Engvall et al. (10). Dilute fibronectin to 100 μg/mL DPBS and filter the solution through a 0.2 μm syringe filter; add 0.5 mL to each insert. After 3 h at room temperature, rinse each insert six times with DPBS at room temperature.
3. After isolating TI cells by magnetic sorting or FACS, centrifuge cells onto a Percoll™ cushion, 350 × *g* for 15 min at 4°C. Add 50 μL of DNAse (2 mg/mL "Solution A") to the cell pellet and carefully resuspend the cells in 0.5 mL of culture medium. Remove a small aliquot for measurement of cell yield, viability, and purity. Dilute the cell suspension to the desired density and plate onto the inserts (we seed cells at 1×10^4 cells/cm^2). Change the growth medium every 4 days.

3.4. Culturing TII Cells

TII cells are cultured on EHS-coated Transwells® under the following conditions:

1. Alveolar TII cells are cultured in DMEM containing 1% rat serum, 10 ng/mL KGF (see Note 6), 10^{-4} M 8-bromocyclic AMP, and 50 μg gentamicin/mL on EHS Matrix in a 10%CO_2/air incubator at 37°C.
2. Coat Transwell® inserts with 0.5 mL of EHS Matrix diluted 2:1 with DMEM. Slowly thaw the EHS Matrix on ice and DMEM at 4°C; immediately use the solution to coat each insert. Polymerization of the matrix is initiated by warming the gels to 37°C in an incubator.
3. Centrifuge the isolated TII cells at 350 × *g* for 10 min at 4°C.
4. Add 50 μL of DNase (2 mg/mL "Solution A") to the cell pellet and carefully resuspend the cells in 1.0 mL of culture medium. Remove a small aliquot for measurement of cell purity, viability, and yield.
5. Dilute TII cells to the desired density (we seed cells at 2×10^5 cells/cm^2) and plate on the EHS Matrix-coated Transwell® plates; change the medium every 4 days.

4. Notes

1. As with all primary cell isolations, viabilities and purities of TI and TII cells can vary substantially among isolations. Therefore, it is important to assess the quality of the cell preparations used for the intended experiments and not to use cells with lower viabilities and purities than the threshold each investigator wishes to accept. Furthermore, if there are difficulties in the

cell isolations, analysis of the cell yield and viability at successive steps is very useful in troubleshooting problems. We assess cell number, viability, and purity after filtering the mixed cell preparation (see Subheading 3.1, step 19) and also after cells have been purified prior to culture. Prolongation of the time of isolation and rough handling of tissues and cells are major causes of low viability. Surgical techniques should be practiced so that the lung perfusion takes less than 4 min. Scissors should be new and/or recently sharpened to avoid shear stress on tissue and cells. Low purities after the initial enzymatic digests may be caused by inactive enzyme, omission of the initial step to lavage the lungs with elastase before instilling it, inadequate digestion times, or temperatures of instillate and/or water bath lower than 37°C. Low purities after magnetic bead sorting should be further investigated after negative cell sorting to assess whether the negative sort is working properly. It is difficult to obtain high purities at the end of isolation if the purities are too low after the initial digestion.

2. To place a Percoll™ cushion at the bottom of the centrifuge tube, resuspend any cells that may have settled during the incubation period. With a 9 in. disposable sterile glass Pasteur Pipet, slowly place ~150 μL of Percoll™ at the bottom of the tube beneath the cell suspension.
3. Columns are washed with ten bed volumes of "Solution A" prior to use. Care is taken not to introduce air or allow the column to dry during the entire procedure.
4. We carefully monitor the wash buffer to make certain that unbound cells are not present in the wash buffer.
5. Conjugate anti-RTI-40 to Alexa Fluor® 610-RPE 30 min before using.
6. Do not filter the KGF through a 0.2 μm filter, as most activity will be lost.

Acknowledgments

We wish to thank Lennell Allen for the electron micrograph and preparation of the figures. This work was supported by NIH Grants 5RC1HL099820 and HL-24075.

References

1. Mercer RR, Russell ML, Roggli VL, Crapo JD (1994) Cell number and distribution in human and rat airways. Am J Respir Cell Mol Biol 10:613–624
2. Stone KC, Mercer RR, Gehr P, Stockstill B, Crapo JD (1992) Allometric relationships of cell numbers and size in the mammalian lung. Am J Respir Cell Mol Biol 6:235–243

3. Dobbs LG, Johnson MD (2007) Alveolar epithelial transport in the adult lung. Respir Physiol Neurobiol 159:283–300
4. Dahlin K, Mager EM, Allen L, Tigue Z, Goodglick L, Wadehra M, Dobbs L (2004) Identification of genes differentially expressed in rat alveolar type I cells. Am J Respir Cell Mol Biol 31:309–316
5. Gonzalez RF, Allen L, Dobbs LG (2009) Rat alveolar type I cells proliferate, express OCT-4, and exhibit phenotypic plasticity in vitro. Am J Physiol Lung Cell Mol Physiol 297: L1045–L1055
6. Dobbs LG (1990) Isolation and culture of alveolar type II cells. Am J Physiol 258: L134–L147
7. Dobbs LG, Gonzalez R, Williams MC (1986) An improved method for isolating type II cells in high yield and purity. Am Rev Respir Dis 134:141–145
8. Dobbs LG, Gonzalez RF (2002) Isolation and culture of pulmonary alveolar epithelial type II cells. In: Freshney RI, Freshney MG (eds) Culture of epithelial cells, 2nd edn. Wiley-Liss, New York, NY, pp 277–303
9. Dobbs LG, Mason RJ, Williams MC, Benson BJ, Sueishi K (1982) Secretion of surfactant by primary cultures of aleveolar type II cells isolated from rats. Biochim Biophys Acta 713:118–127
10. Engvall E, Ruoslahti E (1977) Binding of soluble form of fibroblast surface protein, fibronectin, to collagen. Int J Cancer 20:1–5

Chapter 11

Classical Human Epidermal Keratinocyte Cell Culture

Cathy Rasmussen, Christina Thomas-Virnig, and B. Lynn Allen-Hoffmann

Abstract

It has been more than 30 years since the serial cultivation of human keratinocytes in monolayer culture was first described by Rheinwald and Green. Initially, isolation of primary keratinocytes from disaggregated human skin tissue and subsequent propagation was promoted through use of replication-inactivated murine fibroblast feeder layers. Since then numerous advances have been made to the cultivation of human keratinocytes in both two-dimensional monolayer and three-dimensional organotypic culture. Monolayer culture facilitates keratinocyte proliferation, whereas organotypic culturing techniques promote keratinocyte differentiation using conditions permissive for stratification. The protocols presented here describe traditional culturing methods, providing guidance for isolation and serial cultivation of primary human keratinocytes and dermal fibroblasts, as well as the use of these cells types for generation of stratified skin tissue.

Key words: Primary human keratinocytes, Dermal fibroblasts, Murine 3T3 fibroblasts, Cell isolation and cultivation, Organotypic culture, Skin equivalent

1. Introduction

Serial cultivation of epidermal keratinocytes was first successfully established in 1975 by Rheinwald and Green (1). Propagation of keratinocytes is complicated by the tendency of this cell type to become committed to terminal differentiation, thereby irreversibly losing division potential. *In vitro* keratinocytes possess three distinct proliferative capacities, discernable by the frequency of abortive colony formation when grown at clonal density (2). Clonogenic keratinocytes that almost exclusively generate proliferative colonies are termed holoclones and typically are able to undergo 120–160 divisions. Cells which only generate abortive colonies are termed paraclones, whereas meroclones represent a transitional stage. These seminal works established the basis of *in vitro* human keratinocyte cultivation enabling further studies

Scott H. Randell and M. Leslie Fulcher (eds.), *Epithelial Cell Culture Protocols: Second Edition*, Methods in Molecular Biology, vol. 945, DOI 10.1007/978-1-62703-125-7_11,

into the proliferation (3), differentiation (4), and adhesion (5) properties of this cell type.

The use of replication-inactivated murine 3T3 embryonic fibroblasts was integral to the initial isolation and propagation of human keratinocytes. Cells of the feeder layer not only secrete soluble factors into the culture medium, but deposit ECM molecules on the culturing surface which facilitate attachment and growth of cocultured cells (6). As described by Todaro and Green, embryonic fibroblasts isolated from 17–19 day old mouse embryos generated an immortal cell line when routinely plated at a cell density of 3×10^5 cells/20 cm^2 and passaged every 3 days (7). Although the original 3T3 line was developed from cells isolated from random-bred Swiss mice, 3T3 lines from Balb-c (8) and NIH Swiss mice (9) have also been successfully generated. Subclones of the 3T3 line generated from random-bred Swiss mice, particularly clone 3T3J2, have been used extensively for primary epidermal keratinocyte propagation. Methods employing feeder layers have also been used by our laboratory to isolate and serially cultivate NIKS cells, a source of long-lived keratinocyte progenitors (10). In addition to propagating keratinocytes directly on feeder layers, culturing systems which are now commercially available have been devised to maintain cells in defined medium without using feeder layers such as Keratinocyte Growth Medium® (Lonza, Walkersville, MD) and Epilife® (Invitrogen, Carlsbad, CA). The protocol presented here focuses on the traditional culturing methods of Rheinwald and Green and provides a guideline for isolation and propagation of epidermal keratinocyte and dermal fibroblast cells.

Full stratification and histological differentiation of epidermal keratinocytes can be achieved through use of organotypic culturing techniques which promote keratinocyte growth and differentiation under *in vivo*-like conditions (11, 12). A cellularized matrix consisting of type I collagen and dermal fibroblasts, termed a dermal equivalent, provides structure and critical signaling elements which support terminal differentiation of epidermal keratinocytes. Although keratinocytes seeded onto the dermal equivalent initially proliferate and epithelialize the culture surface, stratification is induced by increasing the calcium concentration and by maintaining the culture at the air–medium interface. Under these conditions proliferating basal cells remain in close proximity to the gradient of nutrients provided by diffusion through the dermal equivalent. The resulting three-dimensional tissue recapitulates the architecture of stratified squamous epithelia and expresses terminal differentiation-specific proteins that closely mimic those of the human interfollicular epidermis (13–15). These advanced three-dimensional culturing techniques have enabled the study of cutaneous cancer progression and wound repair (16), facilitated examinations into the response of human skin to chemical and environmental insult (17, 18), and fostered the development of engineered tissues for clinical use (19–21). The protocol for organotypic culturing describes the steps required for generation of a full-thickness skin equivalent containing both epidermal and dermal components. Numerous reports exist that employ alternative organotypic culturing conditions (22–25), often optimized for specific conditions (26) or for inclusion of other cell types such as melanocytes, endothelial cells, and immune cells (27–30).

2. Materials

2.1. General Tissue Culture

1. Dulbecco's modified Eagle's medium (DMEM), High glucose with L-glutamine (Life Technologies™).
2. Ham's F-12 with L-glutamine.
3. Fetal bovine serum (FBS, Hyclone).
4. FetalClone II (FCII, Hyclone).
5. Sterile solution of 0.02% ethylenediamine tetraacetate (EDTA) (Sigma-Aldrich®) in phosphate buffered saline (PBS).
6. Sterile solution of 0.1% trypsin (Sigma-Aldrich®) in PBS.
7. HEPES-Buffered Earle's Salts solution: 10× Earle's salts (Sigma-Aldrich®) diluted to 1× in tissue culture grade water, buffered with 1 M HEPES for a final HEPES concentration of 25 mM, and filter-sterilized.
8. Sterile cell-culture grade glycerol.
9. Tissue culture plates.
10. Sterile polystyrene or glass pipets.
11. Polypropylene cryogenic vials.
12. Hemocytometer.
13. Freezing container or a controlled rate freezer.

2.2. 3T3 Cultivation and Feeder Layer Preparation

1. 3T3 Medium: DMEM supplemented with 10% bovine calf serum (CS, Hyclone) (see Note 1).
2. Solution of 200 μg/mL mitomycin-C in HEPES-Buffered Earle's Salts, filter-sterilized, aliquoted, and frozen at −20°C (see Note 2).

2.3. Isolation and Cultivation of Epidermal Keratinocytes and Human Dermal Fibroblasts

1. Hydrocortisone (HC, EMD Chemicals Inc.) is prepared to 40 μg/mL in 5% FBS and HEPES-Buffered Earle's Salts, filter-sterilized, and stored at −20°C. This stock solution is 100× the final concentration in medium.
2. Cholera Enterotoxin (CT, MP Biomedicals) is prepared to 840 ng/mL in 0.1% bovine serum albumin (BSA) and

HEPES-Buffered Earle's Salts, filter-sterilized, and stored at 4°C. This stock solution is 100× the final concentration in medium.

3. Insulin (Ins, Sigma-Aldrich®) is dissolved at 500 μg/mL in 0.005 N HCl, filter-sterilized, and stored at 4°C for up to 1 week. This stock solution is 100× the final concentration in medium.
4. Adenine (Ade, Sigma-Aldrich®) is prepared to 2.4 mg/mL in 0.05 N HCl, filter-sterilized, and stored at -20°C. This stock solution is 100× the final concentration in medium.
5. Epidermal growth factor (EGF, R&D Systems) is dissolved at 1 μg/mL in 0.1% BSA and HEPES-Buffered Earle's Salts, filter-sterilized, and stored at −20°C. This stock solution is 100× the final concentration in medium.
6. Keratinocyte medium: 3 parts Ham's F-12 plus 1 part DMEM supplemented with 2.5% FBS or FCII, 0.4 μg/mL HC, 8.4 ng/mL CT, 5 μg/mL Ins, 24 μg/mL Ade, and 10 ng/mL EGF. During initial isolation of keratinocytes from skin tissue, Keratinocyte medium lacking the EGF supplement should be used.
7. Fibroblast medium: Ham's F-12 supplemented with 10% FBS.
8. 125 mL Spinner flask for tissue disaggregation.
9. Sterile solution of 0.1% trypsin (Sigma-Aldrich®) in PBS for disaggregation of neonatal tissue or sterile solution of 0.25% trypsin (Sigma-Aldrich®) and 0.2% collagenase type I (Life Technologies™) in PBS for disaggregation of adult tissue.

2.4. Organotypic Cultivation

1. Culture inserts are available from Corning, Millipore, or Thermo Scientific or Nunc in a variety of sizes.
2. 10× F-12: 10.63 g Ham's F-12 powder (Life Technologies™) and 117.6 mg sodium bicarbonate dissolved in 100 mL tissue culture grade water. Adjust pH to 7.2, filter-sterilize, and store at 4°C for up to 6 months.
3. Collagen Type I solution suitable for gelation is supplied in acetic acid and should be at least 1 mg/mL.
4. Plating medium: 3 parts Ham's F-12 plus 1 part DMEM supplemented with 0.2% FBS or FCII, 0.4 μg/mL HC, 8.4 ng/mL CT, 5 μg/mL Ins, and 24 μg/mL Ade. Final calcium concentration is adjusted to 1.88 mM by supplementing with 1.3% 100 mM $CaCl_2$.
5. Cornification medium I: 3 parts Ham's F-12 plus 1 part DMEM supplemented with 2% FBS or FCII, 0.4 μg/mL HC, 8.4 ng/mL CT, 5 μg/mL Ins, and 24 μg/mL Ade. Final calcium concentration is adjusted to 1.88 mM by supplementing with 1.3% 100 mM $CaCl_2$.

6. Cornification medium II: 3 parts Ham's F-12 plus 1 part DMEM supplemented with 1% FBS or FCII, 0.4 μg/mL HC, 8.4 ng/mL CT, 5 μg/mL Ins, and 24 μg/mL Ade. Final calcium concentration is adjusted to 1.88 mM by supplementing with 1.3% 100 mM $CaCl_2$.

3. Methods

Several procedures are common to serial cultivation, including replenishment of medium (feeding), harvesting and replating cells for expansion (passaging), and cryopreservation of cell stocks (freezing and subsequent thawing). The following procedures describe common cultivation methods with cell-type specific handling requirements noted as appropriate. To ensure successful cultivation, several specific culturing practices should always be adhered to. Proper sterile technique must be utilized at all times. To maintain a proliferative cell population, cells should be passaged or frozen prior to reaching confluence (30–60% confluent is ideal). Figure 1a–f depicts the morphology of each cell type prior to and at confluence. Similarly, 3T3 cells should be used for feeder layer preparation prior to, or at confluence (see Note 3).

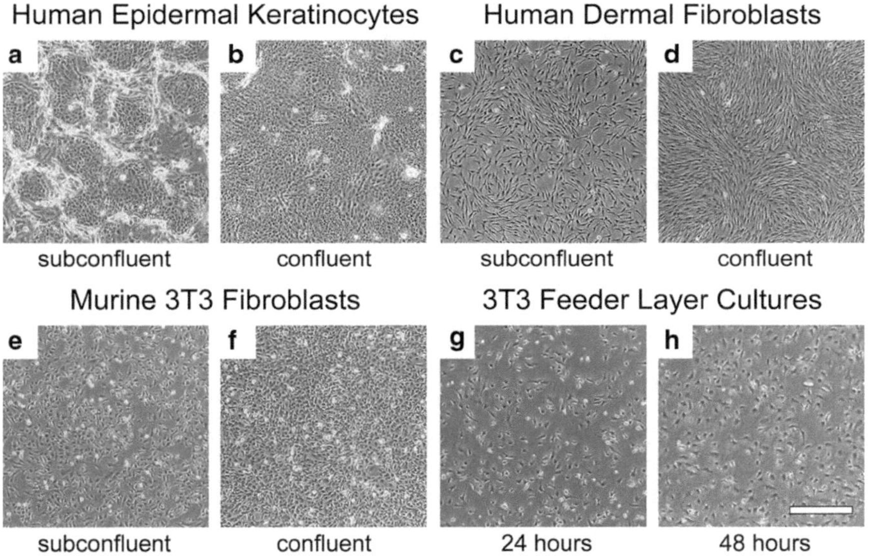

Fig. 1. Morphological appearance of proliferating keratinocytes, fibroblasts, and replication-inactivated 3T3 feeder layer cultures. Subconfluent human keratinocytes display a uniform, cobblestone morphology and grow in colonies with clearly defined edges that push aside the cells of the feeder layer (**a**). At confluence, keratinocytes pack together closely as remaining feeder layer cells are displaced (**b**). Human dermal fibroblasts have a fusiform morphology (**c**) and adopt a whorl appearance as cultures become confluent (**d**). Murine 3T3 fibroblasts prior to (**e**) and at (**f**) confluence have a cobblestone appearance, but do not grow in compact colonies. After mitomycin-C-treatment, 3T3 cells attach and flatten over the culture surface within 24 h (**g**), a process which continues as demonstrated at 48 h (**h**). Scale bar = 500 μm.

It is not recommended to use post-confluent 3T3 cells because the resulting feeder layer cultures do not support optimal keratinocyte growth. Also, feeder layer cultures must be prepared in sufficient time for use in keratinocyte cultivation, typically the day prior to subculture.

A wide selection of culture inserts are commercially available, therefore organotypic culturing techniques have been described broadly to accommodate variations in culture insert size. Several instructions are written in relation to culture insert surface area and final volumes will need to be calculated based on the insert style utilized.

3.1. Cultivation of Cells

3.1.1. Thawing Cells from a Frozen Stock

1. Warm medium to 37°C.
2. Transfer vial of frozen cells from storage to 37°C water bath and thaw until no ice crystals remain (<5 min).
3. Transfer vial and medium to a biosafety cabinet.
4. Carefully remove vial cap and slowly pipet contents up and down to mix.
5. Transfer cell suspension to sterile tube that is larger than the final volume of the diluted cell suspension. For example, a frozen suspension of 1×10^6 cells should be diluted to 20 mL that will be divided equally to two 100 mm plates.
6. Add 1 mL medium dropwise to tube containing 1 mL cell suspension and gently swirl to mix.
7. Add an additional 2 mL medium dropwise, again swirling gently to mix.
8. Add remaining medium and mix by slowly pipetting up and down.
9. Transfer suspensions of 3T3 or human dermal fibroblasts to culture plates, or keratinocytes to previously prepared feeder layer cultures, randomizing delivery pattern to ensure an even distribution of the cells to the culture plate surface.
10. Place culture plates in a humidified, 37°C, 5% CO_2 atmosphere making sure that the plates are level to ensure an even cell distribution.
11. For optimal cell health, medium should be refreshed on the day following the thaw.

3.1.2. Refreshing Growth Medium for Cell Maintenance

1. Warm medium to 37°C.
2. Transfer warmed medium and cell culture plates to be fed to biosafety cabinet.
3. Using an aspirating pipet, remove spent medium from cell culture plates.

4. Using a pipet, gently deliver an appropriate volume of fresh medium to cell cultures. Typically 10 mL of medium is used for a 100 mm plate. See Note 4 for recommendations on volumes to use for various culturing plasticware.
5. Return cultures to humidified, 37°C, 5% CO_2 incubator.
6. For optimal cell health, medium should be refreshed every 2 days. Additional medium may be used to occasionally support 3 days of cultivation (see Note 5).

3.1.3. Passaging Cells for Serial Cultivation

1. Warm medium and 0.1% trypsin solution to 37°C.
2. Transfer 0.02% EDTA solution, medium, trypsin, and cells to be passed to biosafety cabinet.
3. Using an aspirating pipet, remove medium from cell culture plates.
4. Add 5 mL EDTA solution to each 100 mm culture.
   (a) For 3T3 and human dermal fibroblast cultures, proceed to step 8.
   (b) For human keratinocyte culture, continue with steps 5–7 for removal of feeder layer cells.
5. Remove feeder layer cells by repeatedly pipetting up and down with a 5 mL or 10 mL pipet. Pipetting should be vigorous and in a defined pattern to ensure removal of feeder layer cells from the entire surface area of the culture. Care should be taken to not dislodge keratinocyte colonies, which can occur after lengthy exposure to EDTA.
6. Aspirate cell-containing EDTA and rinse culture with additional EDTA solution (see Note 6).
7. Using a microscope, confirm that feeder layer cells have been dislodged and removed. If a large number of feeder layer cells remain, repeat steps 5–7.
8. Aspirate EDTA and add 1 mL trypsin to each 100 mm culture.
9. Tilt plate to evenly distribute the trypsin and place directly onto a metal shelf in a 37°C incubator.
10. Incubate for 4 min. Remove and examine for cellular detachment. If trypsinization is not complete, incubate for an additional 2 min and check again. Repeat as necessary, but avoid trypsinization times longer than 12 min to maintain cell health.
11. Return cultures to biosafety cabinet and neutralize trypsin with 9 mL medium appropriate for the cell type.
12. Thoroughly resuspend cells by pipetting up and down and determine cell density using a hemocytometer.

13. Centrifuge cell suspension for 5 min at $200 \times g$.
14. Aspirate cleared supernatant from cell pellet and resuspend cells to a density convenient for subsequent dilution, such as 1×10^6 cells/mL.
15. Transfer a volume of cell suspension to the volume of medium required to achieve the desired cell dilution. For example, plating 3×10^5 cells in 10 mL to a 100 mm culture plate would require transferring 300 μL of a 1×10^6 cell/mL suspension into a total volume of 9.7 mL medium. See Note 7 for recommendations on plating densities for each cell type.
16. Transfer diluted cell suspensions of 3T3 or human dermal fibroblasts to new culture plates. Transfer suspensions of keratinocytes to previously prepared feeder layer cultures, randomizing the delivery pattern to ensure an even distribution of the cells to the culture plate surface.
17. Place culture plates in a humidified, 37°C, 5% CO_2 atmosphere making sure that the plates are level to ensure an even cell distribution.

3.1.4. Freezing Cell Stocks

1. Cells should be frozen prior to reaching confluence.
2. Label vials suitable for cryopreservation prior to initiating cell harvest. In addition to the cell type, label should contain information pertaining to cell density and passage number.
3. Prepare a solution of 20% glycerol in the medium appropriate for the cell type and warm to 37°C.
4. Warm medium and 0.1% trypsin solution to 37°C.
5. Harvest cells following steps 2–13 of the protocol describing passaging (see Subheading 3.1.3).
6. Aspirate supernatant from cell pellet and resuspend cells to 2×10^6 cells/mL in medium.
7. Slowly dilute cell suspension to 1×10^6 cells/mL with an equal volume of the 20% glycerol solution.
8. Carefully transfer 1 mL of the cell suspension to each labeled vial. Care should be taken to fully tighten the vial cap prior to freezing.
9. Transfer vials to a freezing container or a controlled rate freezer. Follow manufacturer's instruction for use.
10. Frozen cell stocks are ideally maintained in liquid nitrogen vapor phase storage.

3.2. Preparation of the Feeder Layer Cultures

1. Warm 3T3 medium (see Subheading 2.2, item 1) to 37°C.
2. Transfer 200 μg/mL mitomycin-C solution, warmed medium, and cells to be inactivated to biosafety cabinet (see Note 8).

3. Prepare 4 μg/mL mitomycin-C solution by adding 100 μL of the stock mitomycin solution to 5 mL 3T3 medium for each 100 mm 3T3 culture to be inactivated.
4. Using an aspirating pipet, remove medium from cell culture plates.
5. Add 5 mL of the 4 μg/mL mitomycin-C solution to each 100 mm 3T3 culture.
6. Return cultures to humidified, 37°C, 5% CO_2 incubator for 2 h (see Note 9).
7. Warm 3T3 medium and 0.1% trypsin at 37°C.
8. Transfer 0.02% EDTA, warmed medium and trypsin, and mitomycin-C-treated cells to biosafety cabinet.
9. Using an aspirating pipet, remove mitomycin-C solution from cell culture plates.
10. Rinse treated 3T3 cultures by adding 4 mL of 3T3 medium to each 100 mm culture.
11. Aspirate medium and repeat rinse step two more times to thoroughly remove mitomycin-C.
12. Although cultures of treated cells can be refed with fresh medium and used directly as feeder layers, it is more common to trypsinize and replate treated cells as follows.
13. Rinse each 100 mm culture with 5 mL EDTA solution.
14. Harvest cells following steps 8–13 of the protocol describing passaging (see Subheading 3.1.3).
15. Aspirate supernatant from cell pellet and resuspend cells to 1×10^6 cells/mL in 3T3 medium.
16. Perform a second dilution step to generate a cell suspension at 1×10^5 cells/mL in a volume sufficient for the desired number of feeder layer cultures. For example, add 8 mL of the 1×10^6 cells/mL suspension to 72 mL 3T3 medium for production for eight 100 mm feeder layer cultures.
17. Transfer 10 mL of diluted cell suspension to each 100 mm culture for a final density of 1×10^6 replication-inactivated 3T3 cells per 100 mm plate. When transferring cell suspension, randomize the delivery pattern to ensure an even cell distribution.
18. Place feeder layer cultures in a humidified, 37°C, 5% CO_2 incubator, making sure that the plates are level to ensure an even cell distribution.
19. Cultures may be used as feeder layers for keratinocyte cultivation after a minimum of 2 h. Since replication-inactivated 3T3 cells secrete matrix proteins which aid in epithelial colony initiation, feeder layer cultures are ideally prepared the day prior

to use in keratinocyte cultivation. Cultures may be used as feeder layers for up to 2 days after preparation (Fig. 1g–h).

3.3. Isolation of Keratinocytes and Dermal Fibroblasts from Human Skin Tissue

1. The day prior to cell isolation, prepare eight 100 mm feeder layer cultures (see Subheading 3.2).
2. Warm F-12, Keratinocyte (see Subheading 2.3, item 6), and Fibroblast media (see Subheading 2.3, item 7) and trypsin solution to 37°C. See Note 10 regarding use of antibiotics. Use 0.1% trypsin for disaggregation of neonatal tissue or a solution of 0.25% trypsin and 0.2% collagenase for adult tissue.
3. Transfer media, trypsin solution, and skin tissue sample to a biosafety cabinet.
4. Add 10 mL of F-12 to a 100 mm culture plate.
5. Transfer skin tissue to this plate and drag tissue through medium using a sterile forceps to rinse away red blood cells.
6. Place skin sample in a new culture plate and remove any fatty tissue using a sterile scissors (see Note 11).
7. Mince skin tissue very finely with sterile scissors.
8. Transfer minced tissue to a sterile 125 mL spinner flask using sterile forceps.
9. Add 10 mL of the trypsin solution to the spinner flask.
10. Trypsinize tissue in spinner flask on a magnetic stirrer for 45 min at room temperature.
11. Remove cell suspension/ trypsin solution with a Pasteur pipet and transfer to a 15 mL conical tube. Be careful to avoid large pieces of tissue.
12. Centrifuge cell suspension for 5 min at $200 \times g$.
13. Prepare culture plates. (a) For keratinocyte isolation, aspirate medium from two previously prepared 3T3 feeder layer cultures and replace with 10 mL Keratinocyte medium that does not contain EGF (see Subheading 2.3, item 6). (b) For dermal fibroblast isolation, add 10 mL Fibroblast medium to two 100 mm culture plates.
14. Aspirate trypsin from cell pellet. Add 4 mL Fibroblast medium, resuspend cells by pipetting.
15. To establish epidermal keratinocyte cultures, transfer 1 mL of the cell suspension to each of the feeder layer cultures prepared in step 13a. For dermal fibroblast cultures, transfer 1 mL of cell suspension to each 100 mm plate containing Fibroblast medium from step 13b. Add cell suspensions dropwise using a randomized pattern to ensure even cell distribution.
16. Repeat steps 9–15, performing a total of four rounds of trypsinization to yield eight 100 mm keratinocyte cultures and eight 100 mm dermal fibroblast cultures.

17. Approximately 24 h after initiating cells into culture, gently change the medium to remove cellular debris (see Subheading 3.1.2). For keratinocyte cultures, medium containing EGF should be used for this and all subsequent feedings to aid in cell growth.
18. Prior to confluence, cells should be frozen (see Subheading 3.1.4) to establish passage one stocks and passaged (see Subheading 3.1.3) for cultivation of passage two.
19. Cells can also be frozen from passage two for cell stocks (see Note 12).

3.4. Organotypic Culture to Generate a Stratified Skin Equivalent

3.4.1. Preparation of Dermal Equivalent

1. Assemble tissue culture inserts in final culturing plates.
2. Warm Fibroblast medium and 0.1% trypsin solution to 37°C.
3. Harvest dermal fibroblasts following steps 2–13 of the protocol describing passaging (see Subheading 3.1.3).
4. Aspirate cleared supernatant from cell pellet and resuspend cells in Fibroblast medium to a density of 2.5×10^6 cells/mL.
5. Prepare a solution consisting of 45% FCII, 40% 10× F-12, and 15% sterile tissue culture grade water in a 50 mL conical tube.
6. Mix by gentle pipetting, taking care not to introduce air bubbles into the mixture.
7. Transfer tube containing concentrated medium solution into a container of ice in the biosafety cabinet.
8. For every 1 mL of concentrated medium, add 3.25 mL ice cold collagen type I solution and mix by slowly pipetting up and down once. Collagen solution will be extremely viscous and may be retained in the pipet. If necessary, allow excess collagen solution to drain from pipet.
9. At this point, the color of the collagen solution should range from light pink to light yellow. A pronounced yellow coloration indicates that the solution is too acidic to support gelation. To adjust pH, titrate in 1 N NaOH dropwise until the appropriate color is observed (see Note 13).
10. Relative to every 1 mL of concentrated medium used, deliver 375 μL of the fibroblast suspension into the collagen mixture.
11. Mix by slowly pipetting up and down twice to evenly disperse cells.
12. Add 500 μL/cm^2 of cellularized collagen solution to each insert (see Fig. 2a and Note 14). Move gently but quickly. Tilt culture plates as needed to ensure solution covers entire surface area of insert.
13. Place culture plates in a humidified, 37°C, 5% CO_2 incubator for 30 min to gel collagen.

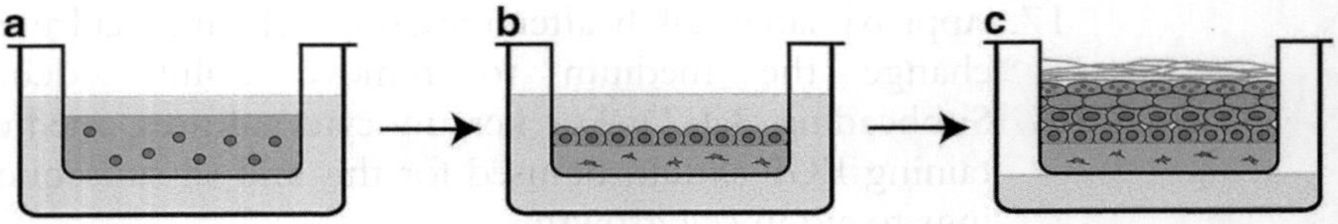

Fig. 2. Organotypic culturing techniques generate a stratified skin equivalent. A collagen gel containing dermal fibroblasts is prepared inside specialized culture inserts that contain the cellular component but permit a free flow of medium with the outer well (**a**). The cellularized collagen gel contracts after several days, becoming a dermal equivalent for subsequent plating of keratinocytes (**b**). After a brief period of submerged culture, the developing skin tissue is cultivated at the air–medium interface to promote full stratification (**c**).

14. Transfer cultures to biosafety cabinet and carefully add Fibroblast medium to each outer well so as to completely cover the level of the gelled collagen in the inserts.
15. Return cultures to the incubator to enable contraction.
16. Gels will begin to contract within 48 h and will contract completely within 3–5 days, at which point keratinocytes can be added (see Note 15). During this time, the plates are left in the incubator and do not need to be fed.

3.4.2. Plating Keratinocytes on Dermal Equivalent

1. Warm Keratinocyte medium, Plating medium (see Subheading 2.4, item 4), and 0.1% trypsin solution to 37°C.
2. Harvest keratinocytes following steps 2–13 of the protocol describing passaging (see Subheading 3.1.3).
3. Aspirate cleared supernatant from cell pellet and resuspend cells in plating medium to a density of 2.5×10^6 cells/mL.
4. Transfer dermal equivalents to the biosafety cabinet.
5. Aspirate medium from both outer wells and inserts, carefully avoiding the contracted collagen.
6. Add 200 μL/cm² of keratinocyte cell suspension to the surface of the dermal equivalent.
7. Add plating medium to the outer well in a volume sufficient to cover the keratinocyte-seeded dermal equivalent upon equilibration (see Fig. 2b).
8. Place culture plates in a humidified, 37°C, 5% CO_2 incubator.
9. After 2 days, remove spent medium from outer wells.
10. Similar to step 7, add Cornification medium I (see Subheading 2.4, item 5) to the outer well in a volume sufficient to cover the keratinocyte-seeded dermal equivalent.
11. After an additional 2 days, remove spent medium from outer wells.
12. Allow medium from culture insert to drain into outer well and aspirate this additional medium. Alternatively, carefully remove

medium from interior of the culture inserts without disturbing the keratinocyte layer.

13. Lift inserts and/or reduce medium volume to achieve culturing at the air–medium interface. Add Cornification medium II (see Subheading 2.4, item 6) to each outer well in a volume sufficient to reach the bottom of the culture insert (Fig. 2c). Specifics for air–medium interface culturing varies depending on the type of insert and outer well assembly used.
14. Refresh Cornification medium II every 2–3 days. Maintain air–medium interface culture for at least 10 days to enable full stratification.

4. Notes

1. 3T3 cells grow well from high or low density plating's in DMEM + CS, but do not grow well with FBS.
2. Mitomycin-C solution is light-sensitive, so aliquots should be covered with foil. Once thawed for use, solution is only stable for 2 weeks at 4°C. Mitomycin-C is toxic, so appropriate precautions must be taken.
3. If during serial cultivation the morphology of 3T3 cells begins to appear spindly or if a number of unusually large cells are apparent, the resulting replication-inactivated cells may not adhere or spread well after mitomycin-C treatment and therefore work poorly as feeder layer cultures. A lower passage number 3T3 cell stock should be initiated into culture for use. Typically 3T3 cells may be serially cultivated for up to 15 passages before morphological changes are evident.
4. The following volumes are recommended for use during cultivation: 4 mL per 60 mm plate, 10 mL per 100 mm plate, 30 mL per 150 mm plate, 60 mL per T225 flask.
5. Cells will be healthier for subsequent cultivation if fed within 2 days prior to passage or freezing.
6. During the initial passage of keratinocytes after isolation, rinsing with EDTA removes feeder layer cells and any human dermal fibroblasts present in the epithelial cultures.
7. A plating density of $5.5 \times 10^3/\text{cm}^2$ is recommended for keratinocyte and fibroblast cultivation. For 3T3 cells, typically several cell densities ($2 \times 10^3/\text{cm}^2$, $3.5 \times 10^3/\text{cm}^2$, $7 \times 10^3/\text{cm}^2$, $1.4 \times 10^4/\text{cm}^2$) are used to generate a series of cultures suitable for either feeder layer preparation or serial cultivation.
8. One confluent 100 mm culture of 3T3 cells typically yields four 100 mm or eight 60 mm feeder layer cultures.

9. The exposure time and concentration of mitomycin-C used for replication inactivation was determined empirically. These parameters enable retention of cell viability, but block replication potential. If the effective mitomycin-C concentration is insufficient, some 3T3 cells will survive treatment and may proliferate in the keratinocyte cultures.
10. For initial isolation of cells from skin tissue, antibiotics (100 units penicillin and 100 μg streptomycin per mL) may be added to the media to prevent bacterial contamination.
11. Approximately 3×10^6 keratinocytes are released per square centimeter of neonatal foreskin.
12. Cell passage should always be noted especially when culturing primary cells. Early passage cells with proven high colony forming efficiency should be used whenever possible. Due to senescence, it is not recommended to use primary keratinocytes for experimentation after passage six.
13. Collagen gelation is dependent on temperature and pH. It is recommended that each new lot of collagen is evaluated for gelation properties prior to use in culturing.
14. Although a volume of 500 $\mu L/cm^2$ of cellularized collagen solution is typically added to each culture insert for generation of the dermal equivalent, the geometry and composition of an insert from a particular vendor may yield different results from an insert procured from an alternative source. Minor modifications may be required.
15. Contraction of the dermal equivalent should be evident after 24 h. If no contraction is observed after 48 h, new dermal equivalents should be made.

Acknowledgments

The authors would like to thank Sandy Schlosser and Kelly Van Winkle for comments and suggestions on the manuscript.

References

1. Rheinwald JG, Green H (1975) Serial cultivation of strains of human epidermal keratinocytes: the formation of keratinizing colonies from single cells. Cell 6:331–343
2. Barrandon Y, Green H (1987) Three clonal types of keratinocyte with different capacities for multiplication. Proc Natl Acad Sci USA 84:2302–2306
3. Rochat A, Kobayashi K, Barrandon Y (1994) Location of stem cells of human hair follicles by clonal analysis. Cell 76:1063–1073
4. Green H (1977) Terminal differentiation of cultured human epidermal cells. Cell 11:405–416
5. Muller EJ, Williamson L, Kolly C, Suter MM (2008) Outside-in signaling through integrins

and cadherins: a central mechanism to control epidermal growth and differentiation? J Invest Dermatol 128:501–516

6. Alitalo K, Kuismanen E, Myllyla R, Kiistala U, Asko-Seljavaara S, Vaheri A (1982) Extracellular matrix proteins of human epidermal keratinocytes and feeder 3T3 cells. J Cell Biol 94:497–505
7. Todaro GJ, Green H (1963) Quantitative studies of the growth of mouse embryo cells in culture and their development into established lines. J Cell Biol 17:299–313
8. Aaronson SA, Todaro GJ (1968) Development of 3T3-like lines from Balb-c mouse embryo cultures: transformation susceptibility to SV40. J Cell Physiol 72:141–148
9. Jainchill JL, Aaronson SA, Todaro GJ (1969) Murine sarcoma and leukemia viruses: assay using clonal lines of contact-inhibited mouse cells. J Virol 4:549–553
10. Allen-Hoffmann BL, Schlosser SJ, Ivarie CA, Sattler CA, Meisner LF, O'Connor SL (2000) Normal growth and differentiation in a spontaneously immortalized near-diploid human keratinocyte cell line, NIKS. J Invest Dermatol 114:444–455
11. Bell E, Ehrlich HP, Buttle DJ, Nakatsuji T (1981) Living tissue formed in vitro and accepted as skin-equivalent tissue of full thickness. Science 211:1052–1054
12. Prunieras M, Regnier M, Woodley D (1983) Methods for cultivation of keratinocytes with an air-liquid interface. J Invest Dermatol 81:28s–33s
13. Andriani F, Margulis A, Lin N, Griffey S, Garlick JA (2003) Analysis of microenvironmental factors contributing to basement membrane assembly and normalized epidermal phenotype. J Invest Dermatol 120:923–931
14. El Ghalbzouri A, Jonkman MF, Dijkman R, Ponec M (2005) Basement membrane reconstruction in human skin equivalents is regulated by fibroblasts and/or exogenously activated keratinocytes. J Invest Dermatol 124:79–86
15. Lee DY, Cho KH (2005) The effects of epidermal keratinocytes and dermal fibroblasts on the formation of cutaneous basement membrane in three-dimensional culture systems. Arch Dermatol Res 296:296–302
16. Garlick JA (2007) Engineering skin to study human disease–tissue models for cancer biology and wound repair. Adv Biochem Eng Biotechnol 103:207–239
17. Gibbs S (2009) In vitro irritation models and immune reactions. Skin Pharmacol Physiol 22:103–113
18. Gibbs S, van de Sandt JJ, Merk HF, Lockley DJ, Pendlington RU, Pease CK (2007) Xenobiotic metabolism in human skin and 3D human skin reconstructs: a review. Curr Drug Metab 8:758–772
19. Pham C, Greenwood J, Cleland H, Woodruff P, Maddern G (2007) Bioengineered skin substitutes for the management of burns: a systematic review. Burns 33:946–957
20. MacNeil S (2007) Progress and opportunities for tissue-engineered skin. Nature 445: 874–880
21. Mansbridge JN (2009) Tissue-engineered skin substitutes in regenerative medicine. Curr Opin Biotechnol 20:563–567
22. Parenteau N (1994) Skin equivalents. In: Leigh IM, Watt FM (eds) Keratinocyte methods. Cambridge University Press, Cambridge, Great Britain, pp 45–54
23. Margulis A, Zhang W, Garlick JA (2005) In vitro fabrication of engineered human skin. Methods Mol Biol 289:61–70
24. Black AF, Bouez C, Perrier E, Schlotmann K, Chapuis F, Damour O (2005) Optimization and characterization of an engineered human skin equivalent. Tissue Eng 11:723–733
25. Carlson MW, Alt-Holland A, Egles C, Garlick JA (2008) Three-dimensional tissue models of normal and diseased skin. Curr Protoc Cell Biol Chapter 19, Unit 19 9
26. Vicanova J, Boelsma E, Mommaas AM, Kempenaar JA, Forslind B, Pallon J, Egelrud T, Koerten HK, Ponec M (1998) Normalization of epidermal calcium distribution profile in reconstructed human epidermis is related to improvement of terminal differentiation and stratum corneum barrier formation. J Invest Dermatol 111:97–106
27. Todd C, Hewitt SD, Kempenaar J, Noz K, Thody AJ, Ponec M (1993) Co-culture of human melanocytes and keratinocytes in a skin equivalent model: effect of ultraviolet radiation. Arch Dermatol Res 285:455–459
28. Boyce ST, Medrano EE, Abdel-Malek Z, Supp AP, Dodick JM, Nordlund JJ, Warden GD (1993) Pigmentation and inhibition of wound contraction by cultured skin substitutes with adult melanocytes after transplantation to athymic mice. J Invest Dermatol 100:360–365
29. Black AF, Berthod F, L'Heureux N, Germain L, Auger FA (1998) In vitro reconstruction of a human capillary-like network in a tissue-engineered skin equivalent. FASEB J 12:1331–1340
30. Regnier M, Patwardhan A, Scheynius A, Schmidt R (1998) Reconstructed human epidermis composed of keratinocytes, melanocytes and Langerhans cells. Med Biol Eng Comput 36:821–824

Chapter 12

Mouse Epidermal Keratinocyte Culture

Luowei Li

Abstract

Cultured mouse epidermal keratinocytes provide a powerful model for studying epidermal proliferation, differentiation, pathogenesis, and oncogenic transformation. Primary mouse keratinocytes can be isolated from newborn mice, and their growth and differentiation can be manipulated by changing calcium concentrations in culture medium. Primary mouse keratinocytes proliferate in medium containing 0.05 mM calcium but cease to proliferate and start to terminally differentiate, when the calcium concentration in the culture medium is increased to 0.10 mM or greater.

Key words: Keratinocytes, Epidermis, Skin, Mouse, Differentiation, Keratin, Loricrin, Filaggrin, Immunofluorescence

1. Introduction

The epidermal compartment of skin is composed of four layers of keratinocytes: basal, spinous, granular, and cornified (1). The basal layer of the epidermis consists of a single layer of proliferative keratinocytes that expresses keratin 5 (K5) and keratin 14 (K14). Above the basal layer is the spinous layer which is made of several layers of keratinocytes that have lost their proliferative potential and have stopped expressing K5 and K14 transcriptionally. Instead, the keratinocytes in the spinous layer start to express keratin 1 (K1) and 10 (K10) (2). Further out is the granular layer, in which the keratinocytes contain keratohyalin and lamellar granules and express loricrin, filaggrin, and involucrin (3, 4). Activation of transglutaminase in the granular layer catalyzes the formation of cornified envelopes (5), insoluble structures that play a pivotal role in skin barrier function. The outermost layer of the epidermis is the stratum corneum, where most of the keratinocytes, which have cornified envelopes, are dead and lack nuclei. Mutations in keratins and other skin specific proteins cause disruption in skin structure and are linked

Scott H. Randell and M. Leslie Fulcher (eds.), *Epithelial Cell Culture Protocols: Second Edition*, Methods in Molecular Biology, vol. 945, DOI 10.1007/978-1-62703-125-7_12,

to several human skin diseases (6, 7). Many of the epidermal differentiation programs can be recapitulated in cultured keratinocytes (8–10). Primary mouse keratinocytes from normal and genetically modified mice provide an unlimited source of cells to study the molecular mechanisms that maintain skin homeostasis.

Primary mouse keratinocytes isolated from newborn mice (1–5 days) can be cultured in vitro for up to 10 days (11, 12). The proliferation and differentiation of cultured mouse keratinocytes are manipulated by changing the extracellular calcium concentration (13). With low levels of calcium (0.02–0.05 mM) in the culture medium, mouse keratinocytes grow as a cobblestone-like monolayer with a high proliferative rate and express basal cell protein makers K5 and K14. Elevation of calcium concentrations in culture medium from 0.05 to 0.1 mM or higher inhibits keratinocyte proliferation and induces keratinocyte differentiation as detected by morphological changes such as desmosome formation and cell stratification (14). The expression of spinous layer keratins, Kl and K10, and granular layer proteins, filaggrin, loricrin, and involucrin, are induced by elevating the calcium concentration in culture medium (15, 16). Enhanced calcium concentration in culture medium also activates transglutaminase that cross-links proteins to form cornified envelopes (17, 18). The existence of a calcium gradient across the epidermis of the living skin (19, 20) strongly supports the notion that calcium is a major determinant of the differentiation status of the epidermis in vivo and in vitro.

Genetically modified mouse models, especially with skin target-gene expression and deletion, allow for the study of specific proteins and pathways that are involved in skin homeostasis and pathogenesis (21–23). Primary mouse keratinocytes isolated from the genetically modified mice have been widely used as in vitro models to elucidate pathways and proteins that regulate many cellular functions such as cell proliferation, apoptosis, and differentiation. In addition, protein function in primary mouse keratinocytes can be studied by manipulating their levels via plasmid- and viral-vector mediated gene overexpression or siRNA/shRNA mediated gene downregulation.

2. Materials

2.1. Tissue Culture Media, Reagents and Chemicals

1. Calcium and magnesium (Ca/Mg)-free Eagle's Minimal Essential Medium (EMEM, Life Technologies™).
2. Fetal bovine serum (FBS).
3. FBS with reduced calcium (see Note 1).
4. 0.25% Trypsin (1×) (Life Technologies™).

5. 0.25% Trypsin–EDTA (1×) (Life Technologies™).
6. 100× Solution of penicillin–streptomycin (10,000 U penicillin; 10,000 μg streptomycin).
7. 100× Amphotericin B (250 μg/mL).
8. LipoFectamine™ transfection reagent (Life Technologies™).
9. Calcium chloride 3 M solution: Dissolved in H_2O, sterile filter with a 0.2 μm filter.
10. Polybrene: Dissolved in Ca/Mg-free EMEM medium at 400 μg/mL, and sterilized through 0.2 μm filter.
11. Chelex® 100 (Bio-Rad Laboratories).
12. Fibronectin/collagen solution: For 100 mL of DMEM medium, add 1 mg fibronectin (Life Technologies™); 1 mL bovine collagen type I (3 mg/mL, BD Bioscience); 0.5 mL of 2% bovine serum albumin (BSA) in PBS; 2 mL of 1 M HEPES, pH 7.0. The mixture is sterilized through a 0.2 μm filter and stored at 4°C.
13. RIPA buffer: 50 mM Tris, pH 7.4; 150 mM NaCl; 0.5% deoxycolate acid; 0.1% SDS; 1% NP-40; and 1 mM PMSF, freshly added.
14. Ca/Mg-free phosphate buffered saline (PBS).
15. PBS with 3% BSA (Fraction V) (PBS–BSA).
16. Tris buffered saline (TBS) and TBS with 0.1% Tween 20 (TTBS).
17. TTBS with 3% BSA (TTBS-BSA).
18. β-Mercaptoethanol (100%) and SDS (20%, w/v, in H_2O) solution.
19. Acetone–methanol: 1:1 (v/v).
20. Rabbit anti-mouse specific K1, K10, filaggrin, and loricrin antibodies (Convance): 1:1,000 diluted in TTBS-BSA.
21. HRP-conjugated goat anti-rabbit antibody (Cell Signaling): 1:2,000 diluted in TTBS-BSA.
22. Rhodamine-conjugated goat anti-rabbit antibody (Jackson ImmunoResearch): Dissolved in 50% glycerol in H_2O at 1 mg/mL and diluted 1:300 in PBS–BSA.
23. VECTASHIELD® Mounting Medium with DAPI (Vector Laboratory).
24. FITC BrdU Flow Kit (BD).
25. Click-iT® EdU Cell Proliferation Assays (Life Technologies™).
26. 5-Bromo-2′-deoxy-uridine Labeling and Detection Kit I (Roche).
27. HiPerFect transfection reagent (Qiagen).

28. Super PAP Pen HT™ (RPI Corp.).
29. Whatman filter paper (No. 1).
30. Beckman Coulter Counter® Cell Counter.
31. Cell scraper.
32. Nitrocellulose membrane (0.45 μm).
33. SuperSignal West Substrate (Pierce).
34. X-ray films (Kodak).
35. 4% Paraformaldehyde: Freshly diluted from 16% paraformaldehyde solution (Electron Microscopy Sciences) in PBS.
36. Ponceau S solution (Sigma-Aldrich®).
37. Cytofix/Cytoperm Buffer (BD).
38. EdU (5-ethynyl-2′-deoxyuridine) 10 mM stock: Dissolved in H_2O (Life Technologies™).
39. Blocking buffer: TTBS containing 5% nonfat dry milk.
40. 60 mm tissue culture treated dish.
41. 150 mm Petri dishes.
42. BCA Protein assay (Pierce).
43. 5× SDS Sample Buffer (Laemmli Buffer): 195 mM Tris HCl, pH 6.8, 50% glycerol, 10% SDS, 0.0125% bromophenol blue.

2.2. Mouse Epidermal Keratinocyte Isolation

1. Surgical instruments, including 4.5″ operating straight scissors sharp/sharp and 5″ rounded tip and serrated tissue forceps.
2. Newborn mice (1–5 days after birth).
3. Betadine.
4. 70% Ethanol.
5. 100 μm Cell Strainer.
6. Cell freezing container, Mr. Frosty (Nalgene®).
7. PE Aanalyst 200 (Perkin Elmer).
8. Keratinocyte freezing medium: HiCa medium containing 10% final concentration of DMSO and HEPES buffer (final concentration at 100 mM), pH 7.0. Store at −20°C.
9. Cryovials.

3. Methods

3.1. Preparation of Cell Culture Medium with Reduced Calcium

1. Add 5 mL of penicillin–streptomycin (100×) solution and 44 mL of chelex-treated FBS (see Note 1) to one bottle (500 mL) of Ca/Mg-free EMEM.

2. Examine calcium concentration in the prepared medium using atomic absorption spectroscopy. Adjust the final calcium concentration in the medium to 0.05 mM by adding concentrated calcium chloride solution (see Subheading 2.1, step 9); this medium is designated as LoCa medium.
3. Add 5 mL of amphotericin B and concentrated calcium chloride solution to LoCa medium to adjust calcium concentration to 1.4 mM; this medium is designated as HiCa medium.
4. Adjust calcium concentration in Ca/Mg-free EMEM to 0.05 mM by adding calcium chloride solution; this serum- and antibiotics-free LoCa medium will be used for plasmid transfection and viral infection.

3.2. Removal of Whole Skin from Newborn Mice

1. Sacrifice newborn mice (1–5 days after birth) using carbon dioxide (CO_2) asphyxiation until the mice stop breathing and moving and their skin turns from pink to blue (at least 30 min), and then place the mice on ice for an additional 15 min.
2. Immerse the mice in Betadine twice, rinse them with deionized water twice, and then rinse them with 70% ethanol twice to remove excesses Betadine. Place the sacrificed and sterilized mice on ice in a sterile petri dish.
3. Use one pair of sterile forceps to hold a mouse in one hand and use a pair of scissors to remove tail (close to body and slanted to the back leaving a 1–2 mm hole) and limbs (cut directly above the up joints). Put the mouse down.
4. Pick up the mouse and firmly hold the body of the mouse with a pair of forceps along the body side with the tail toward outside, insert the tip of a pair of sharp scissors into the hole at the tail, gently expand the scissors under the skin, and then reposition the scissors at the hole and start to carefully cut the dorsal skin along the midline to the tip of the nose. Place the mouse back on the petri dish.
5. Use one pair of forceps to hold the body and the other pair of forceps to pull the cut dorsal skin away from body until most of the dorsal skin is separated from body. Now use one pair of forceps to hold the middle part of the body and another pair of forceps to grab the loose skin, and gently pull the skin over the leg stumps and the nose until it completely separated from the body.
6. Place the separated skin on a new sterile petri dish, dermis-side down. Stretch and flatten the skin, making sure that the edge of the skin is not curled in. The flattened skin can be stored at 4°C for up to 2 h.
7. Add an appropriate volume, e.g., 50 mL for a 150 mm petri dish, of freshly thawed 0.25% trypsin to a new plastic culture dish.

8. Loosen the attached skin at the edge from the petri dish using a pair of forceps. Carefully lift the skin with two pairs of forceps and then transfer it to the dish containing trypsin with the dermis side down. Use forceps to go under the skin to lift any part that is curled downward. The skin should float on the surface of trypsin.
9. Incubate the skin floated on trypsin at 4°C for 18 h.

3.3. Isolation of Primary Mouse Keratinocytes

1. Transfer floated skin, epidermis-side down, onto the dry surface of a new petri dish. Starting from the edge, using a pair of forceps, gently separate the dermis from the epidermis, which sticks to the dry surface of the petri dish. Peel back the dermis until it is completely separated from the epidermis, and place it in a test tube. The epidermal sheet remaining on the petri dish should appear very thin and semitransparent. Lift up the epidermis from the petri dish and place it in a new test tube containing HiCa medium (see Subheading 3.1, step 3). Repeat the procedure for multiple skin pieces and combine all the epidermises.
2. Pool the epidermal sheet(s) into a deep dish filled with HiCa medium, about 0.5 mL per epidermis. Mince the epidermal sheet(s) with a pair of scissors until all the pieces are small enough to pass through a 5 mL pipette. Further shred the epidermis by pipetting at least ten times with a 5 mL pipette to release the keratinocytes.
3. Transfer the suspension containing the released keratinocytes and epidermal debris to a 50 mL conical tube, rinse the dish with additional HiCa medium and transfer it to the same tube.
4. Centrifuge the cell suspension at 150 × *g* for 5 min, remove the supernatant, and resuspend the pellet in HiCa medium, 1 mL per epidermis.
5. Filter the cell suspension through a 100 μm cell strainer.
6. Pellet the isolated keratinocytes by centrifugation at 150 × *g* for 5 min. Resuspend the keratinocyte pellet in HiCa medium, one mouse equivalent per mL, and then store at 4°C. The cell suspension is safe to keep at 4°C for up to 3 days without losing the plating efficiency.
7. Determine cell number per mL in the cell suspension using a cell counter. The cell suspension contains small hair follicles, single basal and differentiated keratinocytes.

3.4. Short Term Storage of Freshly Isolated Primary Mouse Keratinocytes

Isolation and culture of skin keratinocytes from transgenic or knockout mice have been widely used. Sometimes, collecting enough primary keratinocytes for an experiment at one time becomes a problem due to limited breeding or survival of the genetically modified newborn mice. It is possible to freeze freshly isolated primary keratinocytes in liquid nitrogen and then combine several batches together for later use.

1. Centrifuge the prepared primary mouse keratinocytes (see Subheading 3.3, step 7) at 150 × *g* for 5 min and discard the supernatant. Resuspend the cell pellet in keratinocyte freezing medium (see Subheading 2.2, step 8) at one mouse equivalent per mL and then aliquot the cells into a cryotube.
2. Place the cryotube into a cell freezing container.
3. Place the cell freezing container into −70°C freezer for 24 h, and then transfer the cryotube to liquid nitrogen for storage. The cells can be stored in liquid nitrogen for up to 6 months.
4. Before use, thaw the cells from liquid nitrogen to room temperature, centrifuge the tube at 150 × *g* for 5 min, remove the freezing medium, rinse the cells with HiCa medium once, and then resuspend the cells in HiCa medium.

3.5. Culture of Primary Mouse Keratinocytes

1. Plate either the freshly isolated or previously frozen primary mouse keratinocytes into desired tissue culture ware. To obtain consistent cell-plating density, plate the same number of mouse equivalent (ME) rather than the same number of cells (see Note 2). For example, 0.5 ME in 0.5 mL is mixed with 3.5 mL of LoCa medium and then seeded into one 60 mm tissue culture dish. Alternatively, prepared keratinocytes are resuspended in medium containing 0.5 mM calcium before plating to enhance the plating efficiency. Since seeding efficiency varies among different mouse strains, one needs to find the best plating density by testing their isolated keratinocytes. We found that, for keratinocytes isolated from most of the commonly used laboratory mouse strains, including FVB/n, C57BL/6 and BALB/c mice, the best plating density can be obtained by plating 0.2–0.8 MEs per 60 mm dish. To plate mouse keratinocytes to tissue culture plates other than 60 mm dishes, adjust the number of cells to be plated according to the surface area of the new culture dish.
2. Enhance cell attachment by plating cells to fibronectin/collagen (see Subheading 2.1, step 12) coated culture dishes. Cover the bottom of a 60 mm culture dish with 2 μL of fibronectin/collagen solution. Incubate the culture dish at room temperature for 30 min and then aspirate the collagen solution. Plate the prepared keratinocytes into the culture dish. The fibronectin/collagen solution is stored in 4°C and can be reused if kept sterile.
3. Incubate the keratinocyte cultures in a 36°C incubator supplemented with 7% CO_2 (see Note 3).

4. On the next day, remove medium from the culture dish, rinse the cells with Ca/Mg-free PBS twice, and then feed the cells with fresh LoCa medium. At this point, many cells, including differentiated and undifferentiated basal keratinocytes remain on the culture dish.
5. After incubating the keratinocytes in LoCa medium overnight, only undifferentiated basal keratinocytes remain on the dish; the cells appear cobblestone-like (see Fig. 1).
6. Maintain the culture by changing the medium every other day. To obtain repeatable results from repeated experiments, only use the primary mouse keratinocytes that have been in culture for 4–8 days after initial plating. The cultured primary mouse keratinocytes start to senesce, detach, and eventually die after 10–14 days in culture.

3.6. Evaluation of Mouse Keratinocyte Differentiation Marker Expression

1. To evaluate the expression of epidermal differentiation markers by western blot, primary mouse keratinocytes cultured in 60 mm dishes were treated with medium containing 0.05 or 0.12 mM calcium (see Note 4). After 24–48 h, the cultured cells are rinsed with ice-cold PBS three times and then frozen immediately on dry ice for storage at −20°C or immediate lysis.
2. Add 100 μL RIPA buffer (see Subheading 2.1, step 13) to each 60 mm dish, scrap cells using a cell scraper and transfer the lysate into an eppendorf tube. Vortex the cell lysate for 20 s. Let the lysate sit on ice for 15 min. Do not centrifuge. Take 1 μL of the lysate for a protein assay immediately after vortexing the lysate. Determine protein concentrations in the lysates (see Note 5) using a protein assay (see Subheading 2.1, step 42).

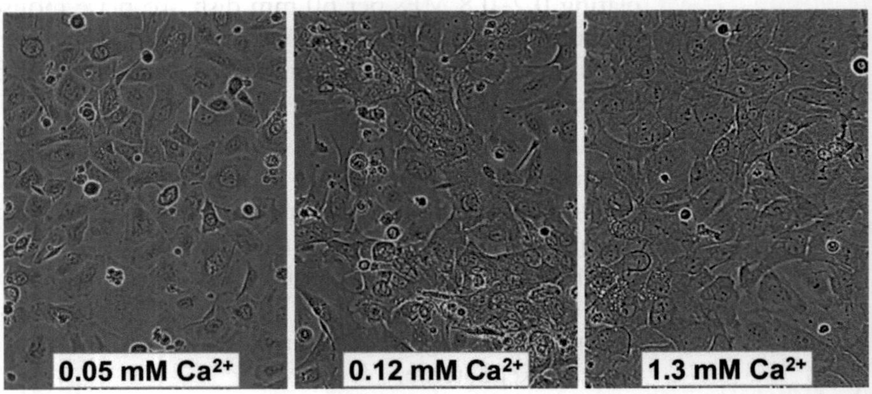

Fig. 1. Morphology of cultured primary mouse keratinocytes in media containing various calcium concentrations. Five-day-old primary mouse keratinocytes were maintained in medium containing 0.05 mM calcium or switched to media containing 0.12 mM or 1.4 mM calcium for 24 h. Pictures were taken with an inverted phase contrast microscope.

3. Transfer equal amount of protein from each sample into new 1.5 mL eppendorf tubes and adjust all the samples to the same volume with RIPA buffer. To dissolve keratins in the lysate, add β-mercaptoethanol and SDS (see Subheading 2.1, step 18) to each sample so the final concentrations of β-mercaptoethanol and SDS in each sample are 20 and 5%, respectively (see Note 5).
4. Heat all the samples at 95°C for 15 min and then cool them to room temperature.
5. Add 5× SDS sample buffer (see Subheading 2.1, step 43) to the lysate and mix well.
6. Load 10–30 μg of the cell lysate to an SDS-PAGE per lane and run electrophoresis.
7. Transfer protein from the SDS-PAGE onto a nitrocellulose membrane (see Subheading 2.1, step 32).
8. Use Ponceau S solution (see Subheading 2.1, step 36) to stain the nitrocellulose membrane. Record the protein staining on the membrane and then rinse with TTBS to remove the Ponceau S.
9. Block transferred nitrocellulose membrane with TTBS containing 5% nonfat dry milk for 30 min at room temperature and rinse the membrane in TTBS for 10 min. Repeat the wash step three times.
10. Incubate the membranes with anti-Kl, K10, filaggrin, or loricrin antibodies (see Subheading 2.1, step 20) in 4°C overnight or room temperature for 3 h. Rinse the membranes with TTBS at room temperature for 10 min. Repeat the wash step three times.
11. Incubate the membrane with HRP-conjugated anti-rabbit antibody (see Subheading 2.1, step 21) at room temperature for 60 min, and then rinse the membrane three times.
12. Incubate the membranes with 1–2 mL chemiluminescent reagents (see Subheading 2.1, step 33) for 5 min at room temperature; avoid direct light.
13. Detect chemiluminescent signal with either X-ray film in a darkroom or an imaging system designed for chemiluminescence.
14. Calcium and time dependent keratinocyte differentiation marker expressions are shown in Fig. 2.
15. Detect keratinocyte differentiation marker expression by indirect immunofluorescence. Primary mouse keratinocytes incubated in 0.05 or 0.12 mM calcium medium for 36 h are fixed with acetone–methanol (see Subheading 2.1, step 19) at room temperature for 15 min. If the cells were cultured in 60 mm

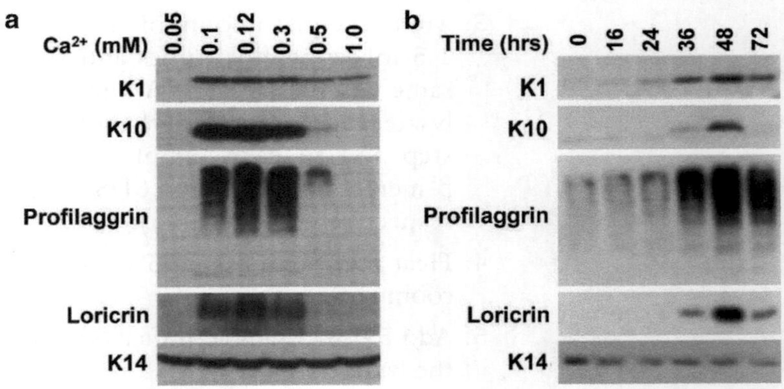

Fig. 2. Expression of keratinocyte differentiation markers, K1, K10, filaggrin, and loricrin. (**a**) Cultured primary mouse keratinocytes in 0.05 mM were switched to medium containing various concentrations of calcium for 36 h as indicated in the figure. (**b**) Culture medium for primary mouse keratinocytes were switched from 0.05 to 0.12 mM calcium and the cultures were terminated at the time indicated in the figure. Cell lysates were collected and immunoblotted with anti-mouse K1, K10, filaggrin, and loricrin antibodies.

dishes, air-dry the culture dishes after removal of the fixative and draw a hydrophobic barrier with a PAP PEN (see Subheading 2.1, step 28) to create small areas for staining.

16. Rehydrate culture dish with PBS.
17. Block fixed cells with PBS–BSA (see Subheading 2.1, step 15) for 30 min at room temperature. Rinse the cells with PBS three times.
18. Add anti-mouse K1 or K10 antibodies to the fixed cells and incubate at 4°C overnight. Rinse the fixed cells with PBS three times.
19. Incubate fixed cells in Rhodamine-conjugated secondary antibody (see Subheading 2.1, step 22) for 1 h at room temperature.
20. Wash the fixed cells with PBS three times. Use water to rinse the dish and air-dry.
21. Mount coverslips with mounting medium containing DAPI (see Subheading 2.1, step 23).
22. Visualize immunofluorescence under a fluorescent microscope (see Fig. 3).

3.7. Analysis of Keratinocyte Proliferation and Cell Cycle

1. Add BrdU (10 μM) to cultured primary mouse keratinocytes 2 h before the termination of the experiment.
2. Fix cells with 4% paraformaldehyde at room temperature for 10 min. Rinse the culture with PBS thoroughly.

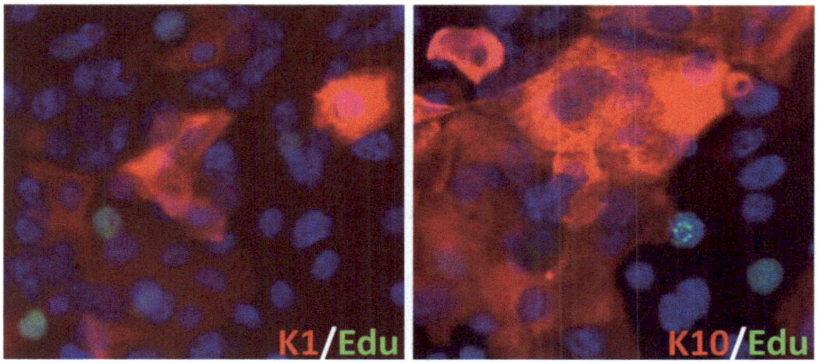

Fig. 3. Immunofluorescent detection of mouse K1 and K10 and proliferative cells. Primary mouse keratinocytes cultured in LoCa medium were switched to media containing 0.12 mM calcium for 30 h. Two hours before harvesting, EdU (10 μM) was added to the culture media directly to label cells in S-phase. The cultures were fixed with acetone–methanol for 15 min. Click-iT® reaction was carried out to detect EdU labeled nuclei, and then the indirect immunofluorescence for mouse K1 and K10 was performed using anti-mouse K1 and K10 antibodies, followed by Rhodamine-conjugated goat anti-rabbit antibodies. DAPI containing mount medium was used to stain nuclei. *Red*: K1 or K10; *Green*: EdU; and *Blue*: DNA.

3. Stain the fixed cells with the 5-bromo-2′-deoxy-uridine Labeling and Detection Kit I (see subheading 2.1, step 26) according to the instruction provided by the manufacture.
4. BrdU positive cells are observed using fluorescent microscopy.
5. To detect the percentage of cultured cell in S-phase of cell cycle, trysinize BrdU labeled cells with trypsin–EDTA (see Subheading 2.1, step 5) and then fix the cells with Cytofix/Cytoperm Buffer (see Subheading 2.1, step 37). BrdU incorporation is detected by FITC-conjugated anti-BrdU antibody using the FITC BrdU Flow Kit (see Subheading 2.1, step 24). Cell cycle analysis is carried out using flow cytometry.
6. Alternatively, cultured keratinocytes are labeled with EdU (see Subheading 2.1, step 38), a nucleoside analog to thymidine, at a final concentration of 10 μM. After 2 h, the cells are fixed with 4% paraformaldehyde for 15 min, and then stained for EdU using Click-iT® EdU detection kit (see Subheading 2.1, step 25) following the instruction provided by the manufacturer.
7. To perform double staining for cells in S-phase and keratins, the keratinocytes grown in 0.05 or 0.12 mM calcium media are labeled with EdU for 2 h, fixed with acetone–methanol and then subjected to the EdU reaction first. Indirect immunofluorescence for keratins is performed after the EdU reaction by following the procedures mentioned in Subheading 2.1, step 25.
8. EdU and keratin positive cells are visualized using fluorescent microscopy.

3.8. Transfection of Plasmid and SiRNA

Primary mouse keratinocytes cultured in LoCa medium for 3–5 days after initial plating are used for transfection.

1. Plasmid transfection is carried out using LipoFectamine™ reagent (see Subheading 2.1, step 8). The lipid and DNA are diluted in serum- and antibiotics-free LoCa medium according to the instruction provided by the manufacturer. Cells are rinsed with the same medium before the addition of the medium containing lipid–DNA mixture. After 5 h, fresh LoCa medium is added to cells. The transfection efficiency is <10% for primary mouse keratinocytes.
2. SiRNA transfection of mouse keratinocytes is achieved by using HiPerFect (see Subheading 2.1, step 27). Serum- and antibiotics-free LoCa medium is used to prepare lipid and siRNA (5–40 mM) according to the instruction provided by the manufacturer. Cultured primary mouse keratinocytes in LoCa medium are rinsed with the serum- and antibiotics-free EMEM first, then treated with medium containing the lipid–siRNA mixture. After 5 h, fresh LoCa medium is added to cultures. Downregulation of the targeted gene can be detected at both the protein and RNA level after 24 h. Depending on the gene, maximum reduction can be archived between 24 and 72 h.

3.9. Infection of Adenovirus, Retrovirus, and Lentivirus

1. For adenovirus infection, purified virus is diluted in serum-free and antibiotics-free EMEM containing 0.05 mM calcium and polyberene (2.5 μg/mL) (see Subheading 2.1, step 10) at 5–30 PFU (plague formation unit) per cell. A small volume of diluted virus (1 mL for a 60 mm dish) is added to cells, and then the cells are incubated for 1 h in the tissue culture incubator. The culture dishes are hand-rotated every 15 min. At the end of the 1 h of incubation period, 3 mL of LoCa medium per dish is added to the cells directly. Adenovirus mediated protein expression is detected 6–12 h after the initial infection.
2. To prepare retro or lentivirus for infection of mouse primary keratinocytes, the package cell lines need to be fed with LoCa medium 24 h before collection of the viral supernatant. Polybrene is added to the viral supernatant directly to a final concentration of 2.5 μg/mL. Viral supernatant may be diluted if the titer is too high. Viral supernatants are added to keratinocyte cultures at 1 mL/60 mm dish. The culture dishes are hand-rotated every 15 min. After 1 h of incubation, 3 mL of LoCa medium is added. The protein expression is detected 2–4 days after the initial infection.

4. Notes

1. Calcium in the serum is removed by treating FBS with a chelating resin, Chelex® 100 (24). The ratio of serum to the packed resin should be 30:1 (v/v). Column preparation and serum chelating should be carried out in a 4°C cold room. Rehydrate the Chelex® 100 resin in water, stir it for 30 min, and remove water by pouring the resin over a Whatman filter paper lined funnel. Resuspend the resin with 0.2 M sodium phosphate buffer, pH 7; stir it for an additional 30 min, and then remove the liquid. Resuspend the resin with 70% ethanol for 30 min and remove the liquid. Resuspend the resin in normal saline. Transfer the washed resin into a suitable size of column for the amount of serum to be chelated, and wash the packed resin with 2 volumes of normal saline. Start the flow of serum into the column. Monitor the eluent and start to collect it once the serum, which appears yellow, starts to come out. Flow normal saline into the column after the serum in the container is empty. Avoid air bubbles. Collect all the serum passed through the column and mix it well. Determine calcium concentration in the chelated and unchelated serum by an atomic absorption spectrometer (see Subheading 2.2, step 7). Calcium concentration in untreated serum and the efficiently chelated serum are around 3 mM and <0.01 mM, respectively. Mix the chelated and unchelated serum to reach a final calcium concentration of 0.620 mM. Sterilize the mixed serum with a 0.4 μm filter followed by a 0.2 μm filter. Store the sterile serum at −70°C for future use.
2. Cell yield from each preparation should be relatively stable if the protocol is followed correctly; usually $6–9 \times 10^6$ cells per ME are isolated from 1 to 3 day old pups of BALB/c mice. Although older pups may have a higher cell yield than the younger ones, the basal cell population in the skin is similar; therefore, the plating density is standardized with ME instead of cell number.
3. The temperature of the incubator for culturing primary mouse keratinocytes is set at 36°C; 1° less than the optimal temperature, 37°C, for most mammalian cell culture. This setting is to limit the growth of fibroblasts that are contaminated in the keratinocytes preparation (25).
4. Keratinocyte differentiation marker expression is regulated transcriptionally by extracellular calcium concentration (15). Although morphological changes and formation of cell–cell junctions in cultured mouse keratinocytes are positively correlated to calcium concentrations in the culture medium, the expression of spinous and granular layer markers, K1, K10,

filaggrin, and loricrin are optimal in medium containing 0.12 mM calcium (15).

5. Keratins are a major component of the intermediate filament in epidermal keratinocytes and insoluble in aqueous salt solution. Nonionic detergents, such as Triton X-100 and NP-40, found in most cell lysis buffer, fail to completely solubilize keratins. A lysis buffer containing 5% SDS and 20% β-mercaptoethanol is used to prepare a whole cell lysate that includes solubilized keratin (15). However, high concentrations of β-mercaptoethanol and SDS in lysis buffer interfere with commonly used protein assay methods, including the Bradford method, the protein DC assay and Piece's BCA* protein assay. The cell lysing method described here allows protein concentration determination and complete keratin solubilization.

Acknowledgements

The author would like to thank Dr. Ulrike Lichti for her advice and helpful discussion. This work was supported by the intramural program of the NIH, Center for Cancer Research, National Cancer Institute.

References

1. Yuspa SH (1997) Multistage carcinogenesis of the skin. In: Wells SA Jr, Sharp PA (eds) Accomplishments in cancer research 1997. Lippincott Co., Philadelphia, PA, pp 90–98
2. Roop DR, Krieg TM, Mehrel T, Cheng CK, Yuspa SH (1988) Cancer Res 48:3245–3252
3. Yuspa SH, Kilkenny A, Roop DR (1988) Aberrant differentiation in mouse skin carcinogenesis. In: Yamasaki H, Tomatis L, Sugimura T, Kakunaga T (eds) Cell differentiation and carcinogenesis: current concepts on critical gene expression in carcinogenesis. IARC, Lyon, pp 3–10
4. Fuchs E (1990) J Cell Biol 111:2807–2814
5. Goldsmith LA (1991) Physiology, biochemistry, and molecular biology of the skin. Oxford University Press, New York, NY
6. Vassar R, Coulombe PA, Degenstein L, Albers K, Fuchs E (1991) Cell 64:365–380
7. Fuchs E (1991) Biochem Soc Trans 19: 1112–1115
8. Green H (1977) Cell 11:405–416
9. Yuspa SH (1978) Mouse epidermal cell cultures as an in vitro model for the study of chemical carcinogenesis. In: Saffiotti U, Autrup H (eds) In vitro carcinogenesis, guide to the literature, recent advances and laboratory procedures. U.S. Government Printing Office, Washington, DC, pp 47–56
10. Yuspa SH, Hawley-Nelson P, Stanley JR, Hennings H (1980) Transplant Proc 12:114–122
11. Hennings H, Holbrook KA, Steinert P, Ben T, Yuspa SH (1980) Growth and differentiation of mouse epidermal cells in culture: effects of extracellular calcium. In: Bernstein IA, Seiji M (eds) Biochemistry of normal and abnormal epidermal differentiation. Tokyo University Press, Tokyo, pp 3–22
12. Yuspa SH, Lichti U, Ben T, Patterson E, Hennings H, Slaga TJ, Colburn N, Kelsey W (1976) Nature 262:402–404
13. Hennings H, Michael D, Cheng C, Steinert P, Holbrook K, Yuspa SH (1980) Cell 19:245–254
14. Hennings H, Holbrook KA (1983) Exp Cell Res 143:127–142
15. Yuspa SH, Kilkenny AE, Steinert PM, Roop DR (1989) J Cell Biol 109:1207–1217
16. Li L, Tucker RW, Hennings H, Yuspa SH (1995) J Cell Physiol 163:105–114

17. Lichti U, Yuspa SH (1988) Cancer Res 48:74–81
18. Dlugosz AA, Yuspa SH (1994) J Invest Dermatol 102:409–414
19. Forslind B (1986) Scan Electron Microsc III:1007–1014
20. Elias PM, Nau P, Hanley K, Cullander C, Crumrine D, Bench G, Sideras-Haddad E, Mauro T, Williams ML, Feingold KR (1998) J Invest Dermatol 110:399–404
21. Hanahan D (1988) Annu Rev Genet 22:479–519
22. Wang XJ, Greenhalgh DA, Eckhardt JN, Rothnagel JA, Roop DR (1994) Mol Carcinog 10:15–22
23. Hansen LA, Woodson Ii RL, Holbus S, Strain K, Lo YC, Yuspa SH (2000) Cancer Res 60:3328–3332
24. Lichti U, Anders J, Yuspa SH (2008) Nat Protoc 3:799–810
25. Jensen PK, Therkelsen AJ (1981) J Invest Dermatol 77:210–212

Chapter 13

Three-Dimensional Culture of Human Breast Epithelial Cells: The How and the Why

Pierre-Alexandre Vidi, Mina J. Bissell, and Sophie A. Lelièvre

Abstract

Organs are made of the organized assembly of different cell types that contribute to the architecture necessary for functional differentiation. In those with exocrine function, such as the breast, cell–cell and cell–extracellular matrix (ECM) interactions establish mechanistic constraints and a complex biochemical signaling network essential for differentiation and homeostasis of the glandular epithelium. Such knowledge has been elegantly acquired for the mammary gland by placing epithelial cells under three-dimensional (3D) culture conditions.

Three-dimensional cell culture aims at recapitulating normal and pathological tissue architectures, hence providing physiologically relevant models to study normal development and disease. The specific architecture of the breast epithelium consists of glandular structures (acini) connected to a branched ductal system. A single layer of basoapically polarized luminal cells delineates ductal or acinar lumena at the apical pole. Luminal cells make contact with myoepithelial cells and, in certain areas at the basal pole, also with basement membrane (BM) components. In this chapter, we describe how this exquisite organization as well as stages of disorganization pertaining to cancer progression can be reproduced in 3D cultures. Advantages and limitations of different culture settings are discussed. Technical designs for induction of phenotypic modulations, biochemical analyses, and state-of-the-art imaging are presented. We also explain how signaling is regulated differently in 3D cultures compared to traditional two-dimensional (2D) cultures. We believe that using 3D cultures is an indispensable method to unravel the intricacies of human mammary functions and would best serve the fight against breast cancer.

Key words: 3D culture, Breast epithelium, Basoapical polarity, Cancer progression, Tissue architecture

1. Introduction

1.1. Structure and Function of the Breast Epithelium

The mammary gland is composed of a series of branched ducts that connect the functional glandular units (acini) to the nipple (see Fig. 1). As early as 1840, Sir Astley Paston Cooper published his observation of a branched organization of the mammary gland with distinct lobes or ductal systems, which each opens at the

Scott H. Randell and M. Leslie Fulcher (eds.), *Epithelial Cell Culture Protocols: Second Edition*, Methods in Molecular Biology, vol. 945, DOI 10.1007/978-1-62703-125-7_13, © Springer Science+Business Media, LLC 2012

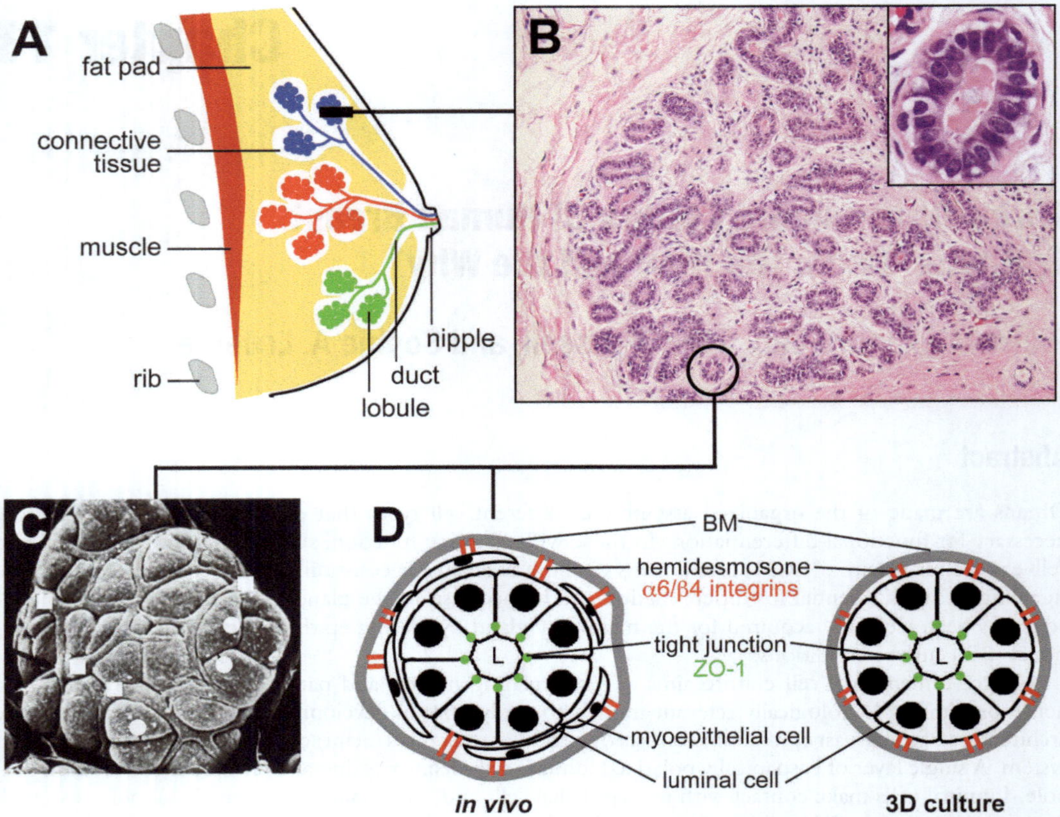

Fig. 1. Anatomy of the mammary gland. (**A**) Schematic representation of the breast. Distinct ductal systems are represented with different colors. (**B**) Cross section of a breast lobule stained with hematoxylin and eosin (nuclei appear in *blue*). A higher magnification view of the smallest structural and functional mammary gland unit (acinus) is shown in the inset. (**C**) Scanning electron micrograph from an acinus. Luminal cells (*white circles*) and myoepithelial cells (*white squares*) are visible. Reproduced from Rosen's Breast Pathology (Third Edition, 2009) with permission from Wolters Kluwer. (**D**) Schematic of cross sections through an acinus in vivo (*left*) and in 3D culture (*right*). Basoapical polarity is evidenced by cell–cell tight junctions (*green dots*) at the apical pole and cell–BM junctions (hemidesmosomes, *red*) at the basal side. Polarity markers are indicated. In 3D monoculture lacking myoepithelial cells, cell–BM contacts are exclusively mediated by luminal cells. Nuclei are shown in *black*. *L* lumen.

nipple (1). The existence of multiple ductal systems, the number of which varies between individuals, is now well established (2, 3).

The branching ducts, as well as acini located at the extremities of each ductal system, are composed of two cell layers: an inner layer of secretory luminal epithelial cells, with apical microvilli, surrounded by contractile myoepithelial cells. The luminal cells of acini are arranged radially with tight junctions between cells located at the narrow width of the cells, near the central lumen (see Fig. 1). The primary function of the acinus is milk secretion. Milk flow into the ducts is powered by the contractions of the myoepithelial cells.

The breast epithelium has a unique behavior compared to other tissues in the human body since it continues to develop after birth. It undergoes extended remodeling with cycles of branching, acini formation, and dissolution of epithelial structures during puberty, pregnancy, lactation, and involution. Moreover, there are less extensive and repeated modifications during menstrual cycles (4). The BM, a specialized form of ECM linking epithelial and connective tissues, and the adjacent stroma that traps an abundance of soluble factors constitute the microenvironment of the epithelium. Myoepithelial cells and luminal cells are in contact with the BM (5, 6), the composition of which also undergoes modifications according to the physiological status of the mammary gland (7, 8). The BM is constituted of a polymeric network of collagen IV and laminins, notably laminins-111 and -332. The laminins and collagen IV are interconnected by nidogen and perlecan (9). Many of the BM components are involved in crucial signaling events that regulate tissue-specificity and function. The framework for such signaling was proposed already in the early 1980s (10), and the first proof in the mammary gland was provided for the expression of the milk protein β-casein which is controlled by laminin-111/β1 integrin signaling (11). The BM is also a repository for growth factors and cytokines that upon binding to their receptors trigger specific intracellular signals (12, 13).

A key feature of all luminal epithelia is the basoapical polarity axis. Transmembrane integrins at the basal side of cells serve as anchorage points and receptors for BM components. They trigger intracellular signaling and participate in the perception of the cells' microenvironment. They cooperate with growth factor receptors to control essential cellular processes such as survival, proliferation, and differentiation (14–16). Among the cell–BM contacts, basal polarity is specifically determined by the interaction between laminin-332 and α6/β4 integrin dimers that form hemidesmosomes (15). Lateral cell–cell contacts are mediated by apical tight junctions, adherens junctions, and in some instances desmosomes (17). The location of tight junctions, the uppermost apical cell–cell adhesion complex, is paramount as it permits to separate cell membrane components and receptors between the apical and basolateral cell membranes and thus, strictly defines apical polarity. The tight seal generated by tight junctions prevents milk leakage in-between cells during lactation. The apical junctional complex formed by tight and adherens junctions also organizes the cytoskeleton and associated signaling pathways, which ultimately impinges on nuclear functions. Thus, the basoapical polarity axis permits unidirectional secretion of milk components in the lumen, as well as structured integration of hormonal and mechanical signals exerted by the microenvironment.

Characterizing the mechanisms underlying normal cell behavior in the context of an organized ductal system is critical to understanding which alterations are necessary for breast cancer to progress. This is particularly important for prevention research related to breast cancer that aims at reducing the burden of this important public health concern.

1.2. Architectural Alterations During Breast Cancer Development and Progression

Tissue architecture (i.e., the organized arrangement of cells into specific multicellular structures) has been shown to be critical for the maintenance of functional differentiation and cell survival (18). It comes as no surprise that alterations in tissue architecture are needed to permit tumor formation and that tissue and cellular organization is commonly used by pathologists to precisely diagnose breast cancer. It has been proposed that the loss of apical polarity is a critical event necessary for tumor development (19). Indeed, 3D culture models of breast acini show that only cells with disrupted apical polarity can be pushed into the cell cycle (19). This hypothesis is supported by the observations reported by several laboratories that tight junction proteins can influence cell proliferation and act as tumor suppressors (17). The impact of basal polarity on the maintenance of apical polarity is not clearly determined. Our recent results suggest that collagen IV and hemidesmosomes both influence the integrity of tight junction organization (20), indicating that microenvironmental alterations might be sufficient to perturb apical polarity. The steps that follow initial apical polarity alterations and ultimately lead to the multilayering of epithelial cells characteristic of preinvasive neoplastic stages (hyperplasia and carcinoma in situ) remain to be uncovered. Among the internal cellular changes associated with apical polarity loss is the relocation of the cell nucleus away from the basal side, as observed in cells of preinvasive neoplastic stages (21).

Breast cancer progression toward invasive stages is accompanied by the breakage of the BM (22) which allows cells to invade the underlying ECM and move into the surrounding tissue. Invasive breast tumors in vivo have been characterized by altered expression and localization of BM proteins (23) and receptors, indicating that both the organization of the microenvironment and receptor–ligand interactions are profoundly altered (24, 25). Changes in intracellular organization have also been observed during tumor progression. Notably the nucleus of invasive tumor cells displays striking changes in the distribution of splicing factor speckle components, heterochromatin and euchromatin domains, and the nuclear mitotic apparatus (NuMA) protein compared to phenotypically normal acinar cells in culture (refs. 19, 26, 27 and unpublished results from the Lelièvre laboratory). Interestingly, the distribution patterns of the chromatin-associated NuMA protein observed in 3D cultures that mimic phenotypically normal and cancerous tissues have been observed also in vivo using archival biopsy samples (Lelièvre and Knowles, unpublished data), suggesting that architectural changes observed in 3D cultures are good predictors of what could be seen in real tissues.

The pioneering work of Mintz and Illmesee (28) with teratocarcinomas brought the paradigm shifting concept that tissue architecture can override genetic and genomic changes (29). The subsequent demonstration that it is possible to induce acinus-like structures from cancer cells with profoundly altered genomes by simply modulating signaling pathways (see Subheading 3.3) has unambiguously revealed the critical role played by epithelial architecture in controlling cell fate (30).

1.3. The Third Dimension Factor in Cell Culture

Monolayer cultures (2D cultures) have been very useful models for gene discovery and early work on viral transformation but are a far cry from physiologically relevant models. Flattened cell morphologies and the spatial plane of cell–cell contacts obtained in 2D cultures are strikingly different from those observed in tissues. Cell shape is known to influence cell behavior including growth and nutrient uptake (31, 32) and gene expression (29) (see Subheading 1.7), which may partially explain the decreased expression of tissue-specific genes often observed in 2D cultures.

Organ cultures are physiologically relevant model systems but are often technically challenging and typically short-lived due to necrosis in tissue explants. Culturing cells in 3D is a more flexible alternative to organ culture. One important goal is to enable the formation of tissue structures that have precise geometrical and functional signatures. This can be achieved by providing cells with proper mechanical and chemical signals from both specific types of architectural components of the ECM and soluble molecules. Serum varies in composition and contains high levels of growth factors and hormones in unpredictable concentrations. It is known to disrupt the ability of cells to express their tissue-specific functions (33) and its use should be avoided if at all possible.

Although most cells in culture synthesize ECM components, the establishment of 3D structures from single cells usually requires the use of hydrogels (viscoelastic meshworks consisting of two or more components, one of which is water) containing exogenous ECM components that provide the structural and biochemical signaling necessary for the formation of the correct architecture and differentiation status. Originally, 3D cultures were performed in floating collagen gels, as demonstrated with murine cells (34). Using this method, morphological characteristics of differentiation, including basoapical polarity, was maintained in culture. Cells at the surface of the gel formed monolayers and some cells below the gel surface rearranged themselves to form acinus-like structures. Nowadays, Engelbreth-Holm-Swarm (EHS) extracts are commonly used as hydrogels (see Subheading 1.5). While a large number of human breast cancer cell lines can be cultured in 3D to mimic tumor development (35), the recapitulation of phenotypically normal acinar phenotypes is more challenging and usually requires the presence of BM components in the exogenous hydrogel (20, 36). A few non-neoplastic breast epithelial cell lines have been used in 3D cultures including MCF-10A (37) and HMT-3522S1 (38) cells (for a comparison of the MCF-10A and S1 models please refer to ref. 20). This chapter mainly focuses on the 3D culture of the HMT-3522 series derived from S1 cells that provides a well-studied model of breast cancer progression. The use of primary cells as complements to immortalized cell lines is also discussed.

1.4. The Example of the HMT-3522 Series: A Progression of Breast Neoplastic Phenotypes in 3D Culture

The HMT-3522 progression series is derived from a benign mammary fibrocystic lesion (38). HMT-3522 cells became spontaneously immortalized in culture, giving rise to the non-neoplatic S1 cell line. In the presence of the appropriate substratum (i.e., BM components and a specific mechanical environment) (20), S1 cells differentiate into basoapically polarized acinus-like structures of approximately 30 μm in diameter containing 25–35 cells (36) (see Fig. 2). Basoapical polarity is a critical feature of normal breast epithelia. The polarity axis is evidenced by the presence of basal BM components (laminin-332 and collagen IV) deposited by the cells, basally localized α6/β4 integrins—the bona fide laminin receptor dimer in breast epithelia—lateral cell–cell adherens junctions (with Ecadherin and β-catenin used as markers), and lateroapical tight junctions (with ZO-1 and ZO-2 core plaque proteins used as markers). The presence of a tiny lumen (often less than the size of a single cell) in S1 acini further indicates close resemblance to acini in the resting mammary gland (20). S1 cells express the luminal marker cytokeratin 18 (30). They have a number of genetic alterations (39) probably linked to their immortalized cell line status. A sub-population of S1 cells carries a mutation in p53 (His to Asp at codon 179). This mutation may confer a slight growth advantage, since its frequency in the cell population progressively increases with passage numbers (40). This phenomenon requires that the use of the S1 cells be restricted to passages below 60, to avoid the drift of the cell population. Acini formed by HMT-3522S1 cells are typically composed of one layer of luminal epithelial cells. In contrast, acini in the mammary gland are constituted by a layer of luminal cells surrounded by contractile myoepithelial cells (see Subheading 1.1 and Fig. 1). Therefore, in 3D monoculture models, luminal cell–ECM contacts are prevalent, whereas in mammary glands in vivo, luminal cells are mostly in contact with myoepithelial cells and make only punctual contacts with the BM (5, 6). Myoepithelial cells are largely responsible for making BM components (41). Surprisingly, although S1 cells are luminal in their behavior and main characteristics, the basal portion of these cells displays myoepithelial characteristics (e.g., presence of vinculin). We believe that these myoepithelial characteristics permit the formation of the appropriate endogenous BM necessary for acinar differentiation.

Malignant transformation of the S1 cells was achieved in vitro by altering the composition of the culture medium (39, 42). The

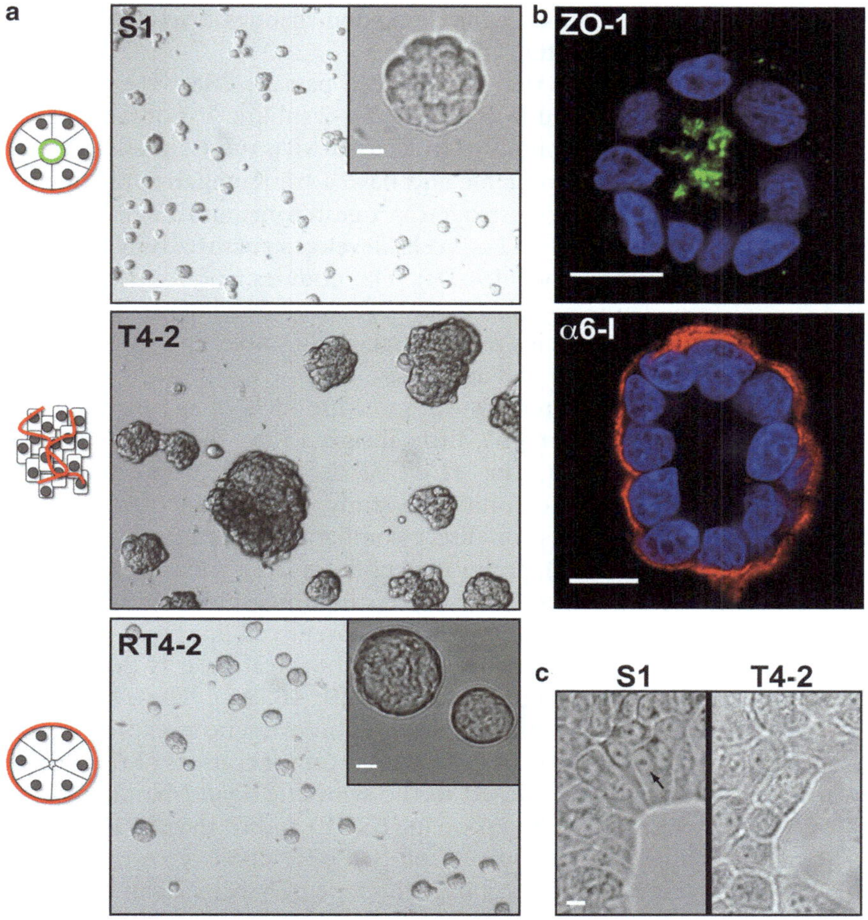

Fig. 2. Examples of phenotypes of the HMT-3522 progression series. (**a**) Bright field images of HMT-3522 cells in 3D drip cultures. Non-neoplastic HMT-3522S1 cells form structures reminiscent of mammary acini, whereas HMT-3522T4-2 malignant cells develop into disorganized and heterogeneous nodules. T4-2 cell phenotype can be reverted (RT4-2) into spheroid structures similar to S1 acini but lacking apical polarity by blocking specific signaling pathways (see Subheading 3.3). Schematic representation of the tissue structures are shown on the *left*. The distribution of basal and apical polarity markers is highlighted in *red* and *green*, respectively. Size bar, 200 μm (10 μm for insets). (**b**) Immunostaining for ZO-1 (*green*) apical and α6-integrin (*red*) basal polarity markers in S1 acini differentiated in 3D culture. Images were captured by confocal microscopy. Nuclei are stained with DAPI (*blue*). (**c**) Bright field images of 2D cultures on plastic. After an initial proliferation stage, most of the S1 cells exit the cell cycle and organize into "islands." Cells at the periphery of these islands have a trapezoidal morphology and nuclei are located toward the inside of the island (*arrow*). Size bar, 10 μm (for **b** and **c**).

HMT-3522S2 subline, growing independently of EGF and bearing characteristics of preinvasive carcinoma (43–45), was isolated. These cells are actually sensitized to EGF and would not thrive in its presence. They also require coating the flasks with collagen I for 2D culture. S2 structures in 3D culture are heterogeneous in size. Homogeneous S3 sublines (HMT-3522S3-A, S3-B, and S3-C) have been derived from S2 3D cultures by selecting for colony size (45). S3 cell lines display progressive loss of polarity, genomic anomalies, and gene expression changes characteristic of preinvasive to invasive transition (45).

S2 cells that had reached passage 238 were found to produce tumors in nude mice. The T4-2 subline was obtained from a tumor after two rounds of in vitro–in vivo mouse passage. T4-2 cells are highly tumorigenic and have a triple-negative phenotype (i.e., no expression of estrogen, progesterone, and ErbB2 receptors) (45). In 3D culture, T4-2 cells develop structures reminiscent of invasive breast tumors (19, 30). The nodules reach approximately 200 μm in diameter after 10 days in culture (see Fig. 2). Thereafter, the size of the nodules remains relatively stable due to a balance between cell division and apoptosis.

Compared to widespread models of aggressive and metastatic cancers that give little insights into the early events of cancer progression, the HMT-3522 cancer progression series offers unique opportunities to study early phases of tumorigenesis. The series allows for direct comparisons between nonmalignant acini models (S1), preinvasive nodules (S2), and invasive tumors (T4-2). There exists other interesting breast cancer progression series. Non-neoplastic MCF-10A parent cell lines were reported to form acinus-like structures devoid of apical polarity as their acini usually lack apical tight junctions (20, 46, 47). These cells can undergo malignant transformation upon exogenous expression of chimeric ErbB2 receptors (48). This receptor consists of the ErbB2 intracellular domain fused to the synthetic ligand-binding domain of the FK506-binding protein (FKBP) and to the extracellular and transmembrane domains of the p75 nerve growth factor receptor. Homodimerization of chimeric p75-ErbB2 can be induced with a FKBP ligand, leading to the activation of ErbB2 signaling without interfering with endogenous ErbB2 receptors. Activation of p75-ErbB2 in MCF-10A acini produced in the presence of EHS extracts induces cell proliferation and multilayering. The resulting structures retain epithelial properties and are not invasive; thus, they may represent a model of early stage mammary cancer in vitro (48). MCF-10A cells have also been transformed using T24 Ha-ras, giving rise to the MCF10-AneoT cell line (49). The MCF10-AT cell lines, derived from xenograft-passaged MCF10-AneoT, represent premalignant stages. However, a subset of MCF10-AT xenographs developed into preinvasive and invasive carcinomas in immunodeficient mice. One such tumor was isolated and cultured in vitro, leading to the malignant MCF10CA1 cell line (50, 51).

1.5. Challenges and Limitations of 3D Cultures

One of the major challenges of 3D culture is to reproduce a microenvironment close enough to the in vivo situation to permit proper differentiation into phenotypically normal tissues, and to accurately mimic different tumor stages. During the development of 3D models of specific tissues, it is important to test for compliance with the architecture and physiology of the in vivo counterparts.

An ECM-like hydrogel substratum and an appropriate medium need to be carefully chosen, in order to solve the problem under study using conditions relevant to the organ in vivo. Ideally, well-characterized hydrogels with defined ECM components should be used. However, with the exception of collagen I, pure ECM components are very costly and difficult to formulate as hydrogels. As an alternative, an ECM mixture isolated from Engelbreth-Holm-Swarm mouse sarcoma cells (52) provides an acceptable BM approximation in terms of components and organization, especially when studying noninvasive breast tissue structures and tumors. Commercially available EHS extracts contain laminin-111, type IV collagen, proteoglycans, and entactin. The composition of commercially available EHS extracts is not fully elucidated and varies between lots; this implies that lots need to be tested for a particular application (see Note 1). Engineered ECM-like hydrogel substrata are being developed as alternatives to EHS extracts, but, to our knowledge and to this date, there is no report that they allow non-neoplastic breast epithelial cells to recapitulate acinar differentiation.

In the mammary glands, acini are connected to ducts (see Fig. 1); whereas human acinar models in 3D cultures are isolated sphere-like entities that rarely branch out. The 3D culture protocols aim at keeping the growth conditions as constant as possible using chemically defined cell culture media and carefully tested lots of EHS-derived hydrogels. Despite these efforts, variations inherent to the biological origin of the additives used in the cell culture media and components of the hydrogel are difficult to control, as is the heterogeneity of primary cultures or even established cell lines. Moreover, the physiological changes found in vivo (e.g., estrous cycles and functional changes associated with puberty, pregnancy and menopause) are yet to be readily explored using 3D culture models.

Ultimately, results obtained using 3D models need to be validated in vivo. This is a challenge when working with human cells. Available in vivo alternatives are murine models bearing human cells (e.g., cells cultured in cleared fat pad) and human tissues from biopsies and reduction mammoplasties. Experiments with xenograft models can be influenced by the animals' physiology and the context into which the cells are injected, and human biological samples provide only frozen moments in time. Nevertheless, there have been several compelling examples showing that what was discovered in 3D cultures indeed illuminated important phenomena in vivo (see Subheading 1.7).

1.6. Primary Cells Versus Cell Lines

Primary cells are nonimmortalized cells obtained directly from tissues. Breast tumor cells can be obtained from breast cancer patients undergoing surgical treatment whereas non-neoplastic mammary cells are typically derived from reduction mammoplasty or from milk. Primary cells have proven to be useful to study breast phenotypes (53). They are heterogeneous and hence truly represent their tissue of origin. Primary cells do not carry genetic (polyploidy, mutations) and phenotypic (e.g., rapid growth) alterations linked to immortalization. However, their use in culture is limited due to the small number of divisions achievable in vitro, which hampers long-term studies. Variability within primary cell populations also represents a challenge since it may reduce experimental reproducibility. Finally, compared to immortalized cell lines, the access to primary cells is limited.

1.7. Examples of Applications of 3D Lumen Breast Epithelial Cell Cultures

This section is a nonexhaustive description of the use of 3D human breast epithelial cultures in a number of research areas (54). Due to space limitations, we have focused only on a few studies presenting compelling and original demonstrations of the usefulness of 3D cultures to study different aspects of breast tissue biology.

Three-dimensional tissue models have provided unique information pertaining to cell signaling, notably that there exist crosstalks between signaling pathways that are only established under certain architectural conditions (i.e., when an acinus is formed). The crosstalk between integrin and growth factor receptors signaling in the glandular epithelium is now well established (16). Originally, it was shown that epidermal growth factor receptor (EGFR) and β1 integrin signal transduction pathways were coupled in breast acini obtained in 3D culture, but not in 2D cultures of the same breast epithelial cell line (55). The predominance of tissue architecture in establishing specific signaling networks was confirmed by the restoration of the crosstalk in cancer cells induced to form basally polarized acinus-like structures. This example illustrates the large impact of tissue architecture on signaling from the microenvironment (29, 56).

Three-dimensional cultures are being used extensively also to study how basoapical polarity is established, maintained, and compromised in mammary epithelia. In particular, laminin-111 was shown to be essential for proper polarization. Luminal epithelial cells cultured on collagen I substratum adopted an inverted polarity; but proper polarity could be rescued by laminin-111 (23). Blocking ECM signaling through β4 integrins prevented the phenotypic reversion of the malignant T4-2 HMT3522 cells (30, 57), and blocking the same signaling pathway in S1 acini compromised apical polarity (20), suggesting that the establishment of basal polarity is needed to maintain apical tight junctions. The underlying mechanisms remain to be discovered.

Intracellular organization in 3D cultures often closely resembles the organization in vivo. In particular, the nuclear structure of acini produced in 3D culture and that observed on sections of acini in resting mammary glands are strikingly similar. This observation is illustrated by the distributions of the nuclear mitotic apparatus

(NuMA) protein (ref. 58, Knowles and Lelièvre, unpublished), and certain markers of higher order chromatin organization (59). A tremendous advantage of 3D culture models over fixed human tissues is that they can be used for functional experiments with physiological relevance. Studies conducted with 3D acini and 3D tumor models have revealed that nuclear organization actively modulates cell and tissue phenotypes (60). A clear demonstration of the impact of nuclear organization on cell behavior in epithelial cells came from studies performed with mammary acini in 3D culture (26). Alterations induced in the distribution of the nuclear protein NuMA using function blocking antibodies, expression of dominant negative truncated forms of the protein, and siRNAs were shown to impair acinar differentiation, alter BM integrity and lead to proliferation and cell death (19, 26, 59). This "dynamic reciprocity" between NuMA and the cell phenotype was not observed in 2D cultures. More particularly, the work performed in 3D culture demonstrated that the protein NuMA was influencing the higher order organization of chromatin (59) normally achieved upon acinar differentiation (19). Recurring findings that changes in cell shape during differentiation are accompanied by the remodeling of nuclear organization and alterations in gene expression profiles (60) strengthen the importance of studying biological processes in tissue contexts.

Specific recognition of the influence of tissue architecture on tumor development has come from the demonstration that cancer cell phenotype can be reverted by selectively modulating cell communication with the microenvironment (18). Relatively simple experiments had already indicated that the arrangement of tumor cells could impact their behavior. Indeed tumor cells were found to change their sensitivity to cancer chemotherapeutic drugs when placed on soft agarose (61). The behavior displayed by cells organized into a tumor was in fact similar to that of xenografts, indicating that the formation of a tumor nodule (instead of a flat monolayer of cells) mimicked the in vivo situation concerning the response to treatments. Later on, it was demonstrated that in breast cancer cells placed in 3D culture, drug sensitivity was influenced by basal polarity, and notably that hemidesmosome-directed signaling was conferring resistance to treatments aimed at killing cancer cells (57). These findings have important implications for research on DNA repair (Rizki, Jasin, and Bissell, unpublished) and apoptosis, and for the development of chemotherapeutics. They clearly indicate that 2D cultures, although easily amenable to high-content assays, are poor predictors of the effect of drugs in vivo.

Finally, 3D cultures in hydrogels have revealed the importance of mechanical stimuli for tissue differentiation. Manipulation of the mammary gland's microenvironment stiffness (using glutaraldehyde fixation of collagen I gels or by mixing collagen I to laminin-111 gels) was found to affect β-casein expression (24). Interestingly, increased ECM stiffness characteristic of malignant tumors resulted in increased cell growth and altered organization of non-neoplastic mammary epithelial cells (62).

In summary, virtually all cellular processes studied so far appear to be influenced by the architectural and microenvironmental contexts. Hence, 3D culture models provide invaluable tools to elucidate fundamental biological questions under physiologically relevant conditions.

2. Materials

2.1. Cell Culture Medium for HMT-3522 Cells

Unless indicated otherwise, additives are dissolved in Milli-Q water, filter-sterilized (0.22 μm), aliquoted, fast-frozen in liquid nitrogen and stored at −80°C.

1. Dulbecco's Modified Eagle's Medium (DMEM/F12, Life Technologies™), stored at 4°C.
2. Prolactin (Sigma-Aldrich®): Diluted to 1 mg/mL (30.3 IU/mL) in 26 mM (2.22 mg/mL) sodium bicarbonate. Stable for 1 year at −80°C and 1 month at 4°C.
3. Insulin (Sigma-Aldrich®): Dissolved as a 2 mg/mL concentrated stock in 5 mM HCl. The stock solution is filter-sterilized and diluted to 100 μg/mL in sterile Milli-Q water. Stable for 6 months at −80°C and 1 month at 4°C.
4. Hydrocortisone (Sigma-Aldrich®): Dissolved as a 5 mg/mL concentrated stock solution in 95% (v/v) ethanol. The stock is further diluted to 0.5 mg/mL (1.4×10^{-3} M) in ethanol and kept at −80°C. Stable for 1 year at −80°C and 1 month at 4°C.
5. β-Estradiol (Sigma-Aldrich®): Dissolved as an 8 mg/mL concentrated stock solution in 95% (v/v) ethanol. The stock solution is then serial-diluted in ethanol to 2.67×10^{-5} mg/mL (10^{-7} M). β-Estradiol should be protected from light. Aliquots can be stored for 6 months at −80°C and 1 month at 4°C.
6. Sodium selenite (BD Biosciences): Dissolved as a 20 mg/mL concentrated stock and serial-diluted to 2.6 μg/mL. Stable for 1 year at −80°C and 1 week at 4°C.
7. Transferrin (Sigma-Aldrich®): Diluted to 20 mg/mL. Stable for 3 months at −80°C and 1 month at 4°C.
8. Epidermal growth factor (EGF, BD Biosciences): Dissolved to 20 μg/mL. Stable for 3 months at −80°C and 1 week at 4°C.

2.2. Propagation of HMT-3522 Cells in Monolayer (2D) Cultures

1. Cell culture flasks (T-75).
2. Solution of trypsin (0.25%) and ethylenediamine tetraacetic acid (EDTA, 1 mM) (Life Technologies™).

3. Soybean trypsin inhibitor (SBTI, Sigma-Aldrich®): Prepared as a 10 mg/mL dilution in sterile Milli-Q water, filter-sterilized, aliquoted, fast-frozen, and stored at −80°C. SBTI solution is stable for 6 months at −80°C and 2 weeks at 4°C.
4. Type I collagen solution (PureCol®, Advanced Biomatrix) and sterile PBS, used for coating plastic surfaces for the culture of HMT-3522S2, S3, and T4-2 cells: 9 mL of a (1:44) PureCol®:PBS dilution is added to a 75 cm^2 flask for a minimum of 24 h at 4°C. Before use, the PureCol® dilution is aspirated and the flask is rinsed with 5 mL DMEM/F12.

2.3. Culture of HMT-3522 Cells in 3D

1. Four-well chamber slides (Lab-Tek™, Nunc) are used for immunostaining experiments and are amenable for direct imaging using the drip or HTP 3D culture methods (see Subheading 3.2).
2. Four-well plates (Nunc) are used for immunostaining experiments on frozen sections of 3D cultures obtained from embedded 3D culture methods (see Subheadings 3.2 and 3.6).
3. Tissue culture dishes (Falcon, 35 mm in diameter) are used for biochemical assays (protein, RNA, and DNA extracts).
4. Basement membrane matrix from EHS extracts (Matrigel™, BD Biosciences; other sources may be appropriate). Matrigel™ is kept at −80°C and thawed on ice 24 h before use. If rigorously kept at 0°C, Matrigel™ is stable for 1 month after thawing; it can only be thawed twice. Thus, usually a 10 mL bottle is thawed to prepare 1–2 mL aliquots that will be frozen and kept at −80°C until use. For the optimal culture of human breast cell lines currently available, it was observed that Matrigel™ should have low levels of endotoxins (<4 U/mL) and a protein concentration below 13 mg/mL. In addition, certain research teams prefer to use growth-factor depleted Matrigel™. Due to the variability between lots, Matrigel™ lots need to be tested before use (see Note 1).
5. Collagen I gel (PureCol®), used to induce the formation of nonpolarized nodules from S1 cells of sizes similar to that of acini obtained with Matrigel™ (see Subheading 3.3).

2.4. Retrieval of Cells and Tissue Structures from the Extracellular Matrix

1. Dispase (50 U/mL, BD Biosciences).
2. Collagenase type IV (250 U/mg, Life Technologies™).

2.5. Introduction of Small Peptides and Antibodies into Live Cells in 3D Culture

1. Permeabilization buffer: 25 mM HEPES buffer (pH 7.2), 78 mM potassium acetate, 3 mM magnesium acetate, 1 mM ethylene glycol-bis(2-aminoethylether)-*N*,*N*,*N*′,*N*′-tetraacetic acid (EGTA), 300 mM sucrose, 1.0% (w/v) bovine serum

albumin. Filter-sterilize and store aliquots at −20°C. Permeabilization buffer can be stored at 4°C for up to 1 month.

2. Digitonin (caution, toxic material): Dissolved as a 0.5% solution in Milli-Q water, filter-sterilized and kept at 4°C. Heating (90°C, up to 4 h) is necessary for solubilization when making the stock solution.
3. Peptides and function-blocking antibodies (this is application-specific and normally used to modify the action of an endogenous protein of interest).

2.6. Cryosections from Embedded 3D Cultures

1. 18% (w/v) sucrose (ACS grade) in PBS.
2. 30% (w/v) sucrose in PBS.
3. Tissue-Tek® optimum cutting temperature (O.C.T.).
4. Dry ice.
5. Cryostat.
6. DAPI (Life Technologies™).
7. Superfrost glass slides.

3. Methods

3.1. Propagation of Non-Neoplastic and Neoplastic Breast Epithelial Cells

Culture cell lines from the HMT-3522 breast cancer progression series (39) in the absence of serum in chemically defined H14 medium (38, 63) (see Note 2 and Subheading 3.1, step 1). The same culture medium is used for nonmalignant S1, preinvasive S2/S3 cells, and malignant T4-2 cells, except that EGF is omitted from S2, S3, and T4-2 cultures. HMT-3522 cell lines are propagated in 75 cm^2 cell culture flasks. For S2, S3, and T4-2 cultures, PureCol®-coated flasks are used (see Subheading 2.2, step 4). Rigorous attention should be paid to seeding density to avoid changes in phenotypes (see Note 3).

Splitting cultures of HMT-3522S1, S2, and T4-2 cells:

1. Prepare fresh H14 medium (volumes of additive stocks, see Subheading 2.1, per 10 mL medium are indicated in brackets): DMEM/F12, 5 μg/mL (0.15 IU/mL) prolactin (50 μL), 250 ng/mL insulin (25 μL), 1.4 μM hydrocortisone (10 μL), 0.1 nM β-estradiol (10 μL), 2.6 ng/mL sodium selenite (10 μL), 10 μg/mL transferrin (5 μL), 5 ng/mL EGF (2.5 μL).
2. Remove the medium, rinse briefly with 1 mL trypsin/EDTA, aspirate the solution, and incubate for 7–12 min in 1 mL trypsin/EDTA solution at 37°C.

3. Suspend the dislodged cells in 10 mL DMEM/F12 supplemented with 180 μL of 10 mg/mL SBTI for trypsin inactivation and count cells using a hemocytometer.
4. Centrifuge (115 ×*g*, 5 min, room temperature) volumes of cell suspension corresponding to 1.75×10^6 cells (S1, 2.3×10^4 cells/cm^2) or 875,000 cells (S2 and T4-2, 1.15×10^4 cells/cm^2). Resuspend cell pellets in 10 mL H14 medium and seed in 75 cm^2 flasks.
5. Change the culture medium every 2–3 days. A medium change before 2 days would deplete cells of important ECM components, whereas a medium change after 3 days might be damaging as nutrients will become too scarce. Use strictly 5% CO_2 and 37°C for the cell incubator environment as deviating from these settings would lead to profound changes in cell behavior, and possibly death, in 3D culture.

3.2. 3D Culture Systems: Variations on a Theme

Cells are cultured in 3D in the presence of the hydrogel. Several techniques have been developed (refs. 20, 64, see Fig. 3) for this type of culture. Irrespective of the culture method used, normal and malignant phenotypes obtained in 3D culture need to be validated (see Note 4). The embedded method was initially used

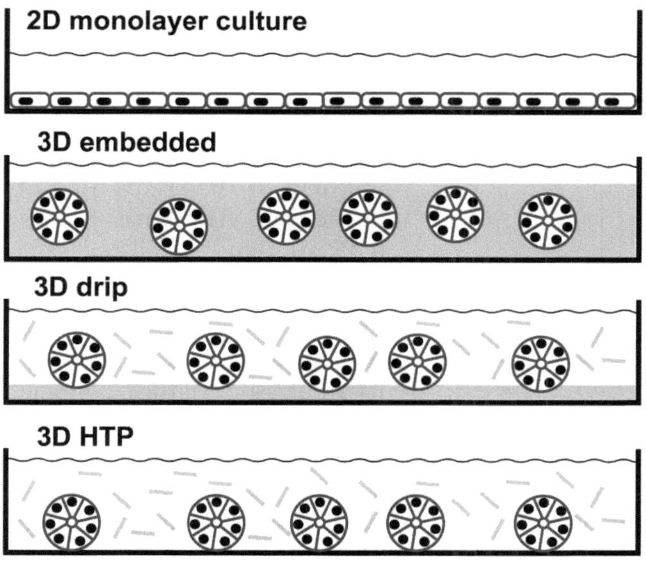

Fig. 3. Drawings of different cell culture systems. Cells are propagated as monolayer cultures on plastic (2D culture). For 3D assays, cells can be embedded in EHS-derived gel (*gray*, 3D embedded), cultured on top of a thin EHS-derived gel coat with a 5% final gel drip (3D drip), and (for S1 cells) grown directly on glass or tissue culture-treated plastic with a 5% final gel drip (3D HTP). For the 3D drip and HTP methods, the EHS-derived gel is diluted in the culture medium (*gray bars*) and dripped onto the cells at the time of seeding. Drawings represent acini. Cell nuclei are shown in *black*.

for HMT-3522 cultures (36). In this setting, individual cells are included within EHS-derived gels covered by culture medium. The technique offers the advantage of providing a homogeneous microenvironment to the cells. It is useful to study the invasive phenotype of tumors rather than the invasive capabilities of individual cells measured in Boyden chambers (65). However, the thick gel layer renders a number of applications challenging, in particular direct fluorescence imaging. Embedded cultures can be cryofixed and sectioned for immunostaining experiments (see Subheading 3.6).

3.2.1. Embedded Culture

1. Prepare H14 medium (see Subheading 3.1).
2. Apply a thin coat of EHS-derived gel to the tissue culture surface (10 μL/cm²). Dispense the EHS-derived gel quickly by spreading the drop deposited in the center of the culture vessel using a pipette tip and let solidify by incubating 5 min at 37°C in the cell culture incubator. The EHS-derived gel aliquot used to prepare 3D cultures should be kept on ice at all times to avoid solidification and degradation.
3. Detach cells with trypsin and count cells (see Subheading 3.1).
4. Centrifuge a cell suspension volume corresponding to the desired amount of cells per individual culture device (if several wells or culture dishes are used, cell amounts cannot be pooled due to the nature of the embedding technique as described below). For S1 cells, use 90,000 cells/cm² of culture device. For S2, S3, or T4-2 cultures, use 45,000 cells/cm² of culture device.
5. Resuspend the cell pellet in a volume of DMEM/F12 that corresponds to 10% of the total EHS-derived gel volume that will be needed. The total EHS-derived gel volume is typically 0.15 mL/cm².
6. Mix EHS-derived gel (0.15 mL/cm² of tissue culture surface) to the cell suspension by pipetting up and down briefly (usually once or twice). Avoid forming bubbles. Transfer immediately to the tissue culture container, adding the cell suspension evenly across the surface precoated with EHS-derived gel.
7. Place at 37°C in the cell culture incubator for 30 min to let the EHS-derived gel solidify.
8. Add H14 medium (volume = volume of EHS-derived gel × 1.5) slowly to the side of the culture vessel to avoid disturbing the gel.
9. Place 3D cultures back in the cell culture incubator. Replace H14 medium every 2–3 days. For S1 cultures, it might be necessary to use H14 without EGF starting day 7 of culture to optimize acinar differentiation. For S2, S3, and T4-2 cultures, always use H14 without EGF.

3.2.2. Drip Culture

The drip method is an alternative to the embedded method which offers greater flexibility in the experimental design, especially for high-resolution imaging applications.

1. Coat the tissue culture surface with EHS-derived gel (see Subheading 3.2.1, step 2), but use 50 μL of gel per cm². Let solidify by incubating 30 min at 37°C in the cell culture incubator.
2. Prepare H14 medium (see Subheading 3.1).
3. Detach cells with trypsin and count cells (see Subheading 3.1).
4. Centrifuge a cell suspension volume corresponding to the desired amount of cells. For S1 cells, use 42,000 cells/cm² (e.g., 400,000 cells per 35 mm dish or 50,000 cells per well in 4-well chamber slides). For S2, S3, or T4-2 cultures, use 21,000 cells/cm².
5. Resuspend the cell pellet in H14 medium in half of the final volume of H14 culture medium (e.g., 650 μL for a 35 mm dish or 200 μL per well for 4-well chamber slides) and add the cell solution drop by drop evenly over the cell culture surface.
6. Let the cells settle down for 5 min.
7. Drip a volume equivalent to that used for the cell suspension of 10% (v/v) Matrigel™ in H14 onto the cells (evenly over the entire culture surface) and return the vessels to the incubator. The resulting final EHS-derived gel concentration of 5% in the culture medium appears to be optimal to obtain apical polarity with non-neoplastic HMT-3522S1 cells (20).
8. Change the culture medium (H14) every 2–3 days (see Subheading 3.2.1). Tissue structures are typically analyzed between day 10 and day 12 of culture, unless specific treatments of tissue structures need to be tested (see Subheading 3.3). After the day of plating, the gel drip does not need to be repeated when changing the medium.

   The drip method has also been used for 3D culture of other non-neoplastic (e.g., MCF-10A) and malignant breast cell lines including MCF-7, MDA-MB-231, BT20, and BT-474 (35).

3.2.3. "High Throughput" Culture

A new technique has been developed for HMT-3522S1 acini culture (ref. 20, see Fig. 3). This technique, referred to as "high throughput" (HTP), can be used to obtain S1 acini with basoapical polarity and morphological characteristics indistinguishable from acini in "drip" or "embedded" cultures. The HTP technique circumvents the need to coat cell culture surfaces with EHS-derived gel, hence reducing handling time and permitting the production of large quantities of acini for high-content screening methods and biochemical analyses. The absence of a gel coat on the culture surface greatly reduces background signals in live cell and immunostaining experiments.

1. Follow steps 2–8 from Subheading 3.2.2. Deposit S1 cell suspensions directly on the glass or plastic substratum.
2. Change the culture medium every 2–3 days, removing EGF on day 6. Cells are used after 8 days for imaging or biochemical experiments. Cells in 3D culture are in general more sensitive to alterations in their environment compared to 2D culture (see Note 5), irrespective of the method used for 3D culture.

3.3. Reversion of Tumor Cells and Manipulation of Nonmalignant Phenotypes

The phenotype of malignant cells can be "reverted" to mimic phenotypically normal acinus-like spheroids or partial acinar differentiation (see Fig. 2). This approach is particularly useful to determine if a difference in phenotype or behavior between non-neoplastic and neoplastic cells results from differences in tissue architecture, cell cycle status, or the genetic background. Moreover, models of non-neoplastic multicellular structures like the acini can be manipulated, notably to affect the basoapical polarity status. Established protocols for tumor cell reversion and for the manipulation of nonmalignant phenotypes in 3D culture are referenced in Table 1. A powerful method is to add function-blocking antibodies targeting integrins to the culture medium starting the day of cell plating. 1 mL of cell suspension is first incubated for 30 min with 15 μg/mL antibodies at 37°C; then cells are plated in 3D culture in the presence of 15 μg/mL antibodies. This technique has been used to obtain reverted T4-2 structures resembling S1 acini but lacking apical polarity.

The same approach was applied to block the establishment of basoapical polarity (57) in S1 cultures. It is also possible to treat with antibodies or small molecules after the establishment of nonmalignant acinar structures and tumor-like colonies, hence mimicking clinically relevant situations to assess specifically the effect of potential anticancer drugs on their tumor targets and/or the side effects on normal tissue (66). An example of the application of this method is illustrated by the suppression of β4-integrin signaling in S1 acini (20). Preformed acini are released from the drip culture with dispase (see Subheading 3.5) and incubated with function blocking antibodies (30 min at 37°C), before replating in the presence of antibodies using the drip method (see Subheading 3.2.2).

3.4. Design of Functional Experiments in a 3D Culture Setting: Mechanical and Biochemical Aspects

Cells in 3D culture usually have a morphology that does not seem amenable to the use of classical transfection reagents. Moreover, non-neoplastic breast epithelial cells usually become quiescent after a few days in the presence of EHS-derived or collagen 1 gels. To circumvent these limitations, cells can be transiently transfected in 2D culture, just before plating them under 3D culture conditions. This approach can be applied to silence the expression of genes

Table 1
Tumor cell reversion and manipulation of non-neoplastic phenotypes

Cell line	Treatment	Phenotype	Reference
T4-2	(a) EGFR inhibition using function-blocking antibodies or Tyrphostin (AG1478) (b) β1 integrin blocking (AIIB2 function-blocking antibodies) (c) PI3K inhibition (LY294002)	Spheroid[a]	(30, 55, 69, 70)
MDA-MB-231; MCF-7	(a) β1 integrin blocking + PI3K inhibition (b) β1 integrin blocking + MAPK inhibition (PD98059) (c) E-cadherin expression + β1 integrin blocking (d) E-cadherin expression + PI3K inhibition (e) E-cadherin expression + MAPK inhibition	Spheroid[a]	(69)
MCF-7	CEACAM1 (carcinoembryonic antigen-related cell adhesion molecule 1) expression	Acinar morphology, "lumen" formation[b]	(71)
S1	Type I collagen	"Reversely polarized" spheroid[c]	(19, 72)
S1	(a) β4 integrin blocking (CD104) (b) α6 integrin blocking (CD49F)	Acini with compro mised cell–BM interactions[d]	(20, 57)

[a]Basally polarized spheroid lacking apical polarity
[b]Polarity status not determined
[c]Spheroid with inverted polarity (i.e., apical polarity markers located at the basal side)
[d]Disruption of BM signaling resulting in loss of basal and apical polarity; *EGFR* epidermal growth factor receptor

coding for proteins with relatively slow turnover using RNA interference (59). Stable transfection of cells in 2D culture prior to use in 3D culture is also possible (57, 59), but clonal selection should be avoided since some of the clonal populations might lack/lose differentiation capability. Other possibilities include viral infection followed by selection of stably transduced cells in 2D culture prior to placing them in 3D cultures (67). If the transgene should only be expressed following differentiation in 3D culture, conditional expression should be used (68).

Function-blocking antibodies or peptides can be introduced inside cells in 2D and 3D cultures to disrupt protein functions or to manipulate signaling pathways. This method requires plasma membrane permeabilization (58, 59):

1. Remove the cell culture medium and cover the cells with 0.01% (w/v) digitonin in permeabilization buffer (see Subheadings 2.5, step 1 and 2.5, step 2).
2. Incubate for exactly 1 min (30 s in 2D culture) at room temperature.
3. Aspirate digitonin solution promptly and wash twice 5 min at 37°C with permeabilization buffer.
4. Remove the permeabilization buffer and add interfering peptides or function-blocking antibodies (generally at 15 μg/mL) in H14 medium. Proper controls should also be used (e.g., corresponding nonspecific immunoglobulin for function-blocking antibody experiments).
5. Incubate at 37°C. Incubation times need to be empirically determined, but they typically range between 30 min and 5 h. Longer incubations are not useful as the cell membrane pores created by digitonin will close. The cells can be kept in culture for several days. Normally, cells should get rid of the antibodies after a couple of days, but if the antibodies were trapped in a specific cellular compartment, they might stay until the protein that they target undergoes turnover.
6. Proceed with the analysis. Antibodies introduced upon cell permeabilization can be detected in immunostaining experiments by adding only the fluorescently labeled secondary antibodies.

Although technically challenging, microinjection can be applied to 3D tissue-like structures (20) and could be used for nucleic acid delivery.

3.5. Dissolution of Hydrogels for Biochemical Assays

Classical biochemical analyses of RNAs and proteins are possible using 3D cultures. Although some of these experiments can be performed by extracting the cellular material directly from 3D cultures containing the hydrogel, it is often better to first isolate the tissue structures from the exogenous substratum, typically by solubilization of the ECM using proteases (collagenase for collagen I gels or dispase for EHS-derived gel). Cell suspensions are then processed using standard protocols. Below is a protocol used for the isolation of HMT-3522S1 acini and T4-2 tumor nodules:

1. Remove the cell culture medium and add dispase solution (0.75 mL/mL of EHS-derived gel).
2. Incubate for 30 min at 37°C.

3. Wash cells three times in DMEM/F12 medium and once in a buffer compatible with downstream applications (e.g., PBS supplemented with protease and phosphatase inhibitors). Each wash is followed by centrifugation ($450 \times g$, 5 min, room temperature). *Note*: A similar method is used to release tissue structures to be treated with function-blocking antibodies against BM receptors before replating the structures in the presence of EHS-derived gel. In the latter case, all four washes are in DMEM/F12 medium. Do not reduce the number of washes; otherwise, cells will be injured by leftover proteases.
4. Cell pellets are further processed for RNA or protein analysis, or subcellular fractionation using standard protocols.

In contrast to trypsin/EDTA treatments, dispase (or collagenase) treatment does not affect cell–cell contacts. This is important since disruption of cell–cell junctions with EDTA would result in profound changes in cellular organization and gene expression that may influence the outcome of the experiments. The recently developed HTP culture for HMT-3522S1 acini (ref. 20, see Subheading 3.2.3) is an alternative for larger-scale production of acinus-like structures. With the HTP culture, dispase treatments with shorter incubation time (10 min) can be used to release the acini.

3.6. High-Resolution Imaging

Immunostaining can be performed on tissue structures obtained with different 3D culture techniques (see Subheading 3.2). Embedded cultures are too thick for direct immunolabeling and imaging. Therefore, cryosections have to be prepared (20, 30):

1. Remove the cell culture medium from 3D embedded cultures in four-well plates.
2. Incubate for 15 min in 18% sucrose at room temperature, rocking slowly.
3. Remove the solution and incubate for 15 min in 30% sucrose at room temperature, rocking slowly.
4. Remove the sucrose solution as much as possible and add Tissue-Tek® O.C.T. freezing solution to the cells.
5. Freeze 3D cultures by placing the 4-well plate on dry ice for 30 min.
6. Gently remove 3D culture blocks using a tweezer and store at −80°C. When ready to use, make frozen sections with a cryostat using a thickness setting between 4 and 20 μM. Check that sections contain cells by staining with DAPI and viewing under a fluorescence microscope every few slices.
7. Deposit the cryosection on superfrost glass slides and maintain slides at −20°C in the cryostat chamber until use or storage.
8. Store sections at −80°C until immunofluorescence labeling.

Immunostaining can also be performed directly on 3D cultures ("drip" and "HTP" methods). These types of 3D cultures offer the advantage of avoiding artifacts caused by sectioning and permit volume imaging. Direct immunostaining often results in high fluorescence background due to the EHS-derived gel, in particular when using the drip method. To control background issues, blocking and washing steps are doubled compared to standard immunostaining protocols. S1 acini range between 25 and 35 μm in diameter, whereas T4-2 nodules are typically 50–200 μm in diameter after 10 days of 3D culture (see Fig. 2). These intact structures can be easily imaged using confocal laser scanning microscopy.

4. Notes

1. Matrigel™ is only good for 2 years when stored at −80°C. Due to variability inherent to biological materials, new gel lots need to be tested in order to ensure reproducibility between experiments using different lots. A typical test procedure consists of culturing side-by-side S1 cells using new as well as currently used gel lots. After 10 days in culture, morphological characteristics (diameter and smoothness of the multicellular structures) are evaluated and basoapical polarity and cell proliferation are scored after immunostaining for specific markers (see Note 4). T4-2 tumor phenotypes also need to be validated when changing the gel lot.
2. A major reason for not using serum for the culture of HMT-3522 cells is to have a defined medium to monitor tissue phenotypes and signaling. It is also desirable to prevent exposure to concentrations of growth factors and cytokines that the cells do not generally experience in vivo.
3. In order to avoid phenotype drifts (i.e., selection of cells with more or less aggressive behavior or less differentiation capabilities), cells are only used within defined passage numbers (passages 52–60 for S1 cells; T4-2 cells can be split usually ten times without seeing a shift in aggressiveness compared to the first passage used in 3D culture). In addition, S1 cell cultures are split no sooner than day 8 and no later than day 12. These cells need time to deposit BM components on the cell culture surface. Passages performed before day 8 would select for rapidly adhering cells that do not differentiate well. Waiting longer than 12 days may lead to difficulties in detaching cells with trypsin. T4-2 cell cultures are split when 70–80% confluence is reached (typically after 4–5 days in culture) as waiting longer would select for cells with less aggressive phenotype in 3D culture. Finally, the culture medium is replaced no earlier than

48 h after plating the cells in order to avoid selection for fast-adhering (more aggressive) S1 cells and to allow cells to benefit from the production of paracrine and autocrine factors during the initial phase of the culture. In 2D culture, HMT-3522S1 cells organize into patches (islands) delineated by a layer of cells that display a morphology typical of epithelial cells, with nuclei basally located against the inside of the island and cytoplasm extending toward the outside of the island (see Fig. 2). The proliferation rate of S1 and T4-2 cells in seed cultures needs to be monitored since aberrant growth rates often reflect changes in cellular metabolism that have profound consequences on the phenotypes obtained in 3D cultures.

4. Quality control is essential to confirm that tissue structures obtained in 3D culture indeed mimic physiologically relevant tissue structures in vivo. The first step in the validation process is the morphological comparison between structures obtained in 3D culture and their counterpart in vivo. Under optimal conditions, S1 cells in 3D culture should differentiate into smooth spherical structures of ~30 μm in diameter, whereas T4-2 cells should produce irregular structures with a broader size distribution (50–200 μm range). Differentiated mammary acini are quiescent (i.e., cells have exited the cell cycle) and basoapically polarized (see Figs. 1 and 2). In a typical S1 acini culture, close to 100% of the acini display correctly distributed basal markers and approximately 75% of the acini display correctly distributed apical markers under optimal culture conditions. The percentage of apically polarized acini may drop to 50% or 60% from one experiment to another since apical polarity seems very sensitive to microenvironmental conditions (20). In contrast, T4-2 tumor-like structures contain proliferating cells and, although certain basal and apical polarity markers are expressed, these makers typically show highly disorganized distributions. The percentage of quiescent cells can be monitored by Ki67 immunostaining. Basal polarity markers include type IV collagen, laminin-332, and α6 and β4 integrins. Tight junction protein ZO-1 is used as a robust apical polarity marker.

5. In our experience, cells are much more resilient to alterations in their environment when cultured in 2D compared to 3D. For example, subtle variations in the cell culture medium—such as altered water quality or variations in additive concentrations—may have no visible effect on cells cultured in 2D while profoundly altering cell behavior (differentiation, proliferation, survival) in 3D culture. Moreover, treatments with certain drugs or dyes that may be well tolerated by cells in 2D cultures might drastically change phenotypes in 3D culture and even lead to cell death.

Acknowledgments

We thank Dr. Kurt Hodges for providing micrographs from tissue sections used in Fig. 1. Support was provided by the National Institutes of Health (R01CA112017 and R03CA112613) and the Susan G. Komen Breast Cancer Foundation (BCTR-0707641) to SAL; the US Department of Energy, Office of Biological and Environmental Research, via a Distinguished Fellow Award and Low Dose Radiation Program (DE-AC02-05CH1123), the National Institutes of Health (R37CA064786, U54CA126552, R01CA057621, U54CA112970, U01CA143233, U54CA143836—Bay Area Physical Sciences–Oncology Center, University of California, Berkeley, California), and the US Department of Defense (W81XWH0810736) to MJB; postdoctoral fellowships from the Novartis Foundation and the Swiss National Science Foundation (PBNEA–116967) to PAV.

References

1. Cooper AP (1840) Anatomy of the breast. Longman Orme, Greene Brown and Longman, London, UK
2. Love SM, Barsky SH (2004) Anatomy of the nipple and breast ducts revisited. Cancer 101:1947–1957
3. Ramsay DT, Kent JC, Hartmann RA, Hartmann PE (2005) Anatomy of the lactating human breast redefined with ultrasound imaging. J Anat 206:525–534
4. Ferguson JE, Schor AM, Howell A, Ferguson MW (1992) Changes in the extracellular matrix of the normal human breast during the menstrual cycle. Cell Tissue Res 268:167–177
5. Gusterson BA, Warburton MJ, Mitchell D, Ellison M, Neville AM, Rudland PS (1982) Distribution of myoepithelial cells and basement membrane proteins in the normal breast and in benign and malignant breast diseases. Cancer Res 42:4763–4770
6. Bergstraesser LM, Srinivasan G, Jones JC, Stahl S, Weitzman SA (1995) Expression of hemidesmosomes and component proteins is lost by invasive breast cancer cells. Am J Pathol 147:1823–1839
7. Haslam SZ, Woodward TL (2001) Reciprocal regulation of extracellular matrix proteins and ovarian steroid activity in the mammary gland. Breast Cancer Res 3:365–372
8. Schedin P, Mitrenga T, McDaniel S, Kaeck M (2004) Mammary ECM composition and function are altered by reproductive state. Mol Carcinog 41:207–220
9. Tzu J, Marinkovich MP (2008) Bridging structure with function: structural, regulatory, and developmental role of laminins. Int J Biochem Cell Biol 40:199–214
10. Bissell MJ, Hall HG, Parry G (1982) How does the extracellular matrix direct gene expression? J Theor Biol 99:31–68
11. Streuli CH, Bailey N, Bissell MJ (1991) Control of mammary epithelial differentiation: basement membrane induces tissue-specific gene expression in the absence of cell–cell interaction and morphological polarity. J Cell Biol 115:1383–1395
12. Folkman J, Klagsbrun M, Sasse J, Wadzinski M, Ingber D, Vlodavsky I (1988) A heparin-binding angiogenic protein—basic fibroblast growth factor—is stored within basement membrane. Am J Pathol 130:393–400
13. Bashkin P, Doctrow S, Klagsbrun M, Svahn CM, Folkman J, Vlodavsky I (1989) Basic fibroblast growth factor binds to subendothelial extracellular matrix and is released by heparitinase and heparin-like molecules. Biochemistry 28:1737–1743
14. Boudreau N, Sympson CJ, Werb Z, Bissell MJ (1995) Suppression of ICE and apoptosis in mammary epithelial cells by extracellular matrix. Science 267:891–893
15. Taddei I, Faraldo MM, Teuliere J, Deugnier MA, Thiery JP, Glukhova MA (2003) Integrins in mammary gland development and differentiation of mammary epithelium. J Mammary Gland Biol Neoplasia 8:383–394

16. Alam N, Goel HL, Zarif MJ, Butterfield JE, Perkins HM, Sansoucy BG, Sawyer TK, Languino LR (2007) The integrin-growth factor receptor duet. J Cell Physiol 213:649–653
17. Lelièvre SA (2010) Tissue polarity-dependent control of mammary epithelial homeostasis and cancer development: an epigenetic perspective. J Mammary Gland Biol Neoplasia 15:49–63
18. Bissell MJ, Radisky DC, Rizki A, Weaver VM, Petersen OW (2002) The organizing principle: microenvironmental influences in the normal and malignant breast. Differentiation 70:537–546
19. Chandramouly G, Abad PC, Knowles DW, Lelièvre SA (2007) The control of tissue architecture over nuclear organization is crucial for epithelial cell fate. J Cell Sci 120:1596–1606
20. Plachot C, Chaboub LS, Adissu HA, Wang L, Urazaev A, Sturgis J, Asem EK, Lelièvre SA (2009) Factors necessary to produce basoapical polarity in human glandular epithelium formed in conventional and high-throughput three-dimensional culture: example of the breast epithelium. BMC Biol 7:77
21. Konska G, Guillot J, De Latour M, Fonck Y (1998) Expression of Tn antigen and *N*-acetyllactosamine residues in malignant and benign human breast tumors detected by lectins and monoclonal antibody 83D4. Int J Oncol 12:361–367
22. Sternlicht MD, Lochter A, Sympson CJ, Huey B, Rougier JP, Gray JW, Pinkel D, Bissell MJ, Werb Z (1999) The stromal proteinase MMP3/stromelysin-1 promotes mammary carcinogenesis. Cell 98:137–146
23. Gudjonsson T, Ronnov-Jessen L, Villadsen R, Rank F, Bissell MJ, Petersen OW (2002) Normal and tumor-derived myoepithelial cells differ in their ability to interact with luminal breast epithelial cells for polarity and basement membrane deposition. J Cell Sci 115:39–50
24. Kass L, Erler JT, Dembo M, Weaver VM (2007) Mammary epithelial cell: influence of extracellular matrix composition and organization during development and tumorigenesis. Int J Biochem Cell Biol 39:1987–1994
25. Levental KR, Yu H, Kass L, Lakins JN, Egeblad M, Erler JT, Fong SF, Csiszar K, Giaccia A, Weninger W, Yamauchi M, Gasser DL, Weaver VM (2009) Matrix crosslinking forces tumor progression by enhancing integrin signaling. Cell 139:891–906
26. Lelièvre SA, Weaver VM, Nickerson JA, Larabell CA, Bhaumik A, Petersen OW, Bissell MJ (1998) Tissue phenotype depends on reciprocal interactions between the extracellular matrix and the structural organization of the nucleus. Proc Natl Acad Sci U S A 95: 14711–14716
27. Knowles DW, Sudar D, Bator-Kelly C, Bissell MJ, Lelièvre SA (2006) Automated local bright feature image analysis of nuclear protein distribution identifies changes in tissue phenotype. Proc Natl Acad Sci U S A 103:4445–4450
28. Mintz B, Illmensee K (1975) Normal genetically mosaic mice produced from malignant teratocarcinoma cells. Proc Natl Acad Sci U S A 72:3585–3589
29. Bissell MJ, Weaver VM, Lelièvre SA, Wang F, Petersen OW, Schmeichel KL (1999) Tissue structure, nuclear organization, and gene expression in normal and malignant breast. Cancer Res 59:1757s–1763s (discussion 1763s–1764s)
30. Weaver VM, Petersen OW, Wang F, Larabell CA, Briand P, Damsky C, Bissell MJ (1997) Reversion of the malignant phenotype of human breast cells in three-dimensional culture and in vivo by integrin blocking antibodies. J Cell Biol 137:231–245
31. Bissell MJ, Farson D, Tung AS (1977) Cell shape and hexose transport in normal and virus-transformed cells in culture. J Supramol Struct 6:1–12
32. Folkman J, Moscona A (1978) Role of cell shape in growth control. Nature 273:345–349
33. Bissell MJ (1981) The differentiated state of normal and malignant cells or how to define a "normal" cell in culture. Int Rev Cytol 70:27–100
34. Emerman JT, Pitelka DR (1977) Maintenance and induction of morphological differentiation in dissociated mammary epithelium on floating collagen membranes. In Vitro 13:316–328
35. Kenny PA, Lee GY, Myers CA, Neve RM, Semeiks JR, Spellman PT, Lorenz K, Lee EH, Barcellos-Hoff MH, Petersen OW, Gray JW, Bissell MJ (2007) The morphologies of breast cancer cell lines in three-dimensional assays correlate with their profiles of gene expression. Mol Oncol 1:84–96
36. Petersen OW, Ronnov-Jessen L, Howlett AR, Bissell MJ (1992) Interaction with basement membrane serves to rapidly distinguish growth and differentiation pattern of normal and malignant human breast epithelial cells. Proc Natl Acad Sci U S A 89:9064–9068
37. Soule HD, Maloney TM, Wolman SR, Peterson WD Jr, Brenz R, McGrath CM, Russo J, Pauley RJ, Jones RF, Brooks SC (1990) Isolation and characterization of a spontaneously immortalized human breast epithelial cell line, MCF-10. Cancer Res 50:6075–6086
38. Briand P, Petersen OW, Van Deurs B (1987) A new diploid nontumorigenic human breast epithelial cell line isolated and propagated in chemically defined medium. In Vitro Cell Dev Biol 23:181–188

39. Briand P, Lykkesfeldt AE (2001) An in vitro model of human breast carcinogenesis: epigenetic aspects. Breast Cancer Res Treat 65:179–187
40. Moyret C, Madsen MW, Cooke J, Briand P, Theillet C (1994) Gradual selection of a cellular clone presenting a mutation at codon 179 of the p53 gene during establishment of the immortalized human breast epithelial cell line HMT-3522. Exp Cell Res 215:380–385
41. Gudjonsson T, Adriance MC, Sternlicht MD, Petersen OW, Bissell MJ (2005) Myoepithelial cells: their origin and function in breast morphogenesis and neoplasia. J Mammary Gland Biol Neoplasia 10:261–272
42. Schmeichel KL, Weaver VM, Bissell MJ (1998) Structural cues from the tissue microenvironment are essential determinants of the human mammary epithelial cell phenotype. J Mammary Gland Biol Neoplasia 3:201–213
43. Madsen MW, Lykkesfeldt AE, Laursen I, Nielsen KV, Briand P (1992) Altered gene expression of c-myc, epidermal growth factor receptor, transforming growth factor-alpha, and c-erb-B2 in an immortalized human breast epithelial cell line, HMT-3522, is associated with decreased growth factor requirements. Cancer Res 52:1210–1217
44. Briand P, Nielsen KV, Madsen MW, Petersen OW (1996) Trisomy 7p and malignant transformation of human breast epithelial cells following epidermal growth factor withdrawal. Cancer Res 56:2039–2044
45. Rizki A, Weaver VM, Lee SY, Rozenberg GI, Chin K, Myers CA, Bascom JL, Mott JD, Semeiks JR, Grate LR, Mian IS, Borowsky AD, Jensen RA, Idowu MO, Chen F, Chen DJ, Petersen OW, Gray JW, Bissell MJ (2008) A human breast cell model of preinvasive to invasive transition. Cancer Res 68:1378–1387
46. Fogg VC, Liu CJ, Margolis B (2005) Multiple regions of Crumbs3 are required for tight junction formation in MCF10A cells. J Cell Sci 118:2859–2869
47. Underwood JM, Imbalzano KM, Weaver VM, Fischer AH, Imbalzano AN, Nickerson JA (2006) The ultrastructure of MCF-10A acini. J Cell Physiol 208:141–148
48. Muthuswamy SK, Li D, Lelièvre SA, Bissell MJ, Brugge JS (2001) ErbB2, but not ErbB1, reinitiates proliferation and induces luminal repopulation in epithelial acini. Nat Cell Biol 3:785–792
49. Basolo F, Elliott J, Tait L, Chen XQ, Maloney T, Russo IH, Pauley R, Momiki S, Caamano J, Klein-Szanto AJ et al (1991) Transformation of human breast epithelial cells by c-Ha-ras oncogene. Mol Carcinog 4:25–35
50. Heppner GH, Miller FR, Shekhar PM (2000) Nontransgenic models of breast cancer. Breast Cancer Res 2:331–334
51. Santner SJ, Dawson PJ, Tait L, Soule HD, Eliason J, Mohamed AN, Wolman SR, Heppner GH, Miller FR (2001) Malignant MCF10CA1 cell lines derived from premalignant human breast epithelial MCF10AT cells. Breast Cancer Res Treat 65:101–110
52. Kleinman HK, Martin GR (2005) Matrigel: basement membrane matrix with biological activity. Semin Cancer Biol 15:378–386
53. Gudjonsson T, Villadsen R, Ronnov-Jessen L, Petersen OW (2004) Immortalization protocols used in cell culture models of human breast morphogenesis. Cell Mol Life Sci 61:2523–2534
54. Schmeichel KL, Bissell MJ (2003) Modeling tissue-specific signaling and organ function in three dimensions. J Cell Sci 116:2377–2388
55. Wang F, Weaver VM, Petersen OW, Larabell CA, Dedhar S, Briand P, Lupu R, Bissell MJ (1998) Reciprocal interactions between beta1-integrin and epidermal growth factor receptor in three-dimensional basement membrane breast cultures: a different perspective in epithelial biology. Proc Natl Acad Sci U S A 95:14821–14826
56. Bissell MJ, Labarge MA (2005) Context, tissue plasticity, and cancer: are tumor stem cells also regulated by the microenvironment? Cancer Cell 7:17–23
57. Weaver VM, Lelièvre SA, Lakins JN, Chrenek MA, Jones JC, Giancotti F, Werb Z, Bissell MJ (2002) beta4 integrin-dependent formation of polarized three-dimensional architecture confers resistance to apoptosis in normal and malignant mammary epithelium. Cancer Cell 2:205–216
58. Lelièvre SA, Bissell MJ (1998) Communication between the cell membrane and the nucleus: role of protein compartmentalization. J Cell Biochem Suppl 30–31:250–263
59. Abad PC, Lewis J, Mian IS, Knowles DW, Sturgis J, Badve S, Xie J, Lelièvre SA (2007) NuMA influences higher order chromatin organization in human mammary epithelium. Mol Biol Cell 18:348–361
60. Lelièvre SA (2009) Contributions of extracellular matrix signaling and tissue architecture to nuclear mechanisms and spatial organization of gene expression control. Biochim Biophys Acta 1790:925–935
61. Kerbel RS (1994) Impact of multicellular resistance on the survival of solid tumors, including micrometastases. Invasion Metastasis 14:50–60
62. Paszek MJ, Zahir N, Johnson KR, Lakins JN, Rozenberg GI, Gefen A, Reinhart-King CA,

Margulies SS, Dembo M, Boettiger D, Hammer DA, Weaver VM (2005) Tensional homeostasis and the malignant phenotype. Cancer Cell 8:241–254

63. Blaschke RJ, Howlett AR, Desprez PY, Petersen OW, Bissell MJ (1994) Cell differentiation by extracellular matrix components. Methods Enzymol 245:535–556
64. Lee GY, Kenny PA, Lee EH, Bissell MJ (2007) Three-dimensional culture models of normal and malignant breast epithelial cells. Nat Methods 4:359–365
65. Lelièvre SA, Bissell MJ (2005) Three dimensional cell culture: the importance of microenvironment in regulation of function. In: Meyers RA (ed) Encyclopedia of molecular cell biology and molecular medicine, 2nd edn. Wiley-VCH Verlag GmbH, Weinheim, pp 383–420
66. Park CC, Zhang H, Pallavicini M, Gray JW, Baehner F, Park CJ, Bissell MJ (2006) Beta1 integrin inhibitory antibody induces apoptosis of breast cancer cells, inhibits growth, and distinguishes malignant from normal phenotype in three dimensional cultures and in vivo. Cancer Res 66:1526–1535
67. Kaminker P, Plachot C, Kim SH, Chung P, Crippen D, Petersen OW, Bissell MJ, Campisi J, Lelièvre SA (2005) Higher-order nuclear organization in growth arrest of human mammary epithelial cells: a novel role for telomere-associated protein TIN2. J Cell Sci 118: 1321–1330
68. Zhan L, Rosenberg A, Bergami KC, Yu M, Xuan Z, Jaffe AB, Allred C, Muthuswamy SK (2008) Deregulation of scribble promotes mammary tumorigenesis and reveals a role for cell polarity in carcinoma. Cell 135:865–878
69. Wang F, Hansen RK, Radisky D, Yoneda T, Barcellos-Hoff MH, Petersen OW, Turley EA, Bissell MJ (2002) Phenotypic reversion or death of cancer cells by altering signaling pathways in three-dimensional contexts. J Natl Cancer Inst 94:1494–1503
70. Liu H, Radisky DC, Wang F, Bissell MJ (2004) Polarity and proliferation are controlled by distinct signaling pathways downstream of PI3-kinase in breast epithelial tumor cells. J Cell Biol 164:603–612
71. Kirshner J, Chen CJ, Liu P, Huang J, Shively JE (2003) CEACAM1-4S, a cell-cell adhesion molecule, mediates apoptosis and reverts mammary carcinoma cells to a normal morphogenic phenotype in a 3D culture. Proc Natl Acad Sci U S A 100:521–526
72. Gudjonsson T, Ronnov-Jessen L, Villadsen R, Bissell MJ, Petersen OW (2003) To create the correct microenvironment: three-dimensional heterotypic collagen assays for human breast epithelial morphogenesis and neoplasia. Methods 30:247–255

Chapter 14

Three-Dimensional Cultures of Mouse Mammary Epithelial Cells

Rana Mroue and Mina J. Bissell

Abstract

The mammary gland is an ideal "model organism" for studying tissue specificity and gene expression in mammals: it is one of the few organs that develop after birth and it undergoes multiple cycles of growth, differentiation and regression during the animal's lifetime in preparation for the important function of lactation. The basic "functional differentiation" unit in the gland is the mammary acinus made up of a layer of polarized epithelial cells specialized for milk production surrounded by myoepithelial contractile cells, and the two-layered structure is surrounded by basement membrane. Much knowledge about the regulation of mammary gland development has been acquired from studying the physiology of the gland and of lactation in rodents. Culture studies, however, were hampered by the inability to maintain functional differentiation on conventional tissue culture plastic. We now know that the microenvironment, including the extracellular matrix and tissue architecture, plays a crucial role in directing functional differentiation of organs. Thus, in order for culture systems to be effective experimental models, they need to recapitulate the basic unit of differentiated function in the tissue or organ and to maintain its three-dimensional (3D) structure. Mouse mammary culture models evolved from basic monolayers of cells to an array of complex 3D systems that observe the importance of the microenvironment in dictating proper tissue function and structure. In this chapter, we focus on how 3D mouse mammary epithelial cultures have enabled investigators to gain a better understanding of the organization, development and function of the acinus, and to identify key molecular, structural, and mechanical cues important for maintaining mammary function and architecture. The accompanying chapter of Vidi et al. describes 3D models developed for human cells. Here, we describe how mouse primary epithelial cells and cell lines—essentially those we use in our laboratory—are cultured in relevant 3D microenvironments. We focus on the design of functional assays that enable us to understand the intricate signaling events underlying mammary gland biology, and address the advantages and limitations of the different culture settings. Finally we also discuss how advances in bioengineering tools may help towards the ultimate goal of building tissues and organs in culture for basic research and clinical studies.

Key words: Mouse mammary epithelium, 3D culture, Tissue architecture, Polarity, Tissue-specific signaling

Scott H. Randell and M. Leslie Fulcher (eds.), *Epithelial Cell Culture Protocols: Second Edition*, Methods in Molecular Biology, vol. 945, DOI 10.1007/978-1-62703-125-7_14, © Springer Science+Business Media, LLC 2012

1. Introduction

1.1. Rationale for Using Rodent Culture Models to Understand Human Disease

Biological research has been hampered by the inability to sustain functional differentiation in organs and tissues ex vivo for long enough to allow for adequate experimentation. In addition, tissue samples of human origin are scarce, and the ethical constraints guiding their use necessitate that we find alternative model organisms for study. Rodents and particularly mice have emerged as the species of choice in biological research: they are sufficiently similar to humans anatomically and physiologically, although the differences need to be kept in mind at all times. In addition, they are small and easy to handle and breed, and their short life span and rapid reproductive rate make it possible to study development and disease processes and to generate statistically significant data sets. Standardization of mouse research was possible by the generation of inbred lines of laboratory mice. This was essential initially to circumvent issues of genetic variability among animals, and as such made it possible to replicate experiments more rigorously. Genetic engineering in mice was another milestone for biological research. The versatility of engineered mice allows investigators to probe for the role of specific genes in growth, development, and disease, and thus, these animals have become important model systems for human diseases and an essential component for clinical and pharmaceutical research (1).

Nevertheless, research with rodents is not ideal, not only because it is constrained by ethical guidelines as it should be, but also because of issues of systemic and physiological differences between humans and rodents and variability among different strains of animals, as well as among litter mates. Also, the generation of pure laboratory strains, while useful for genetic analysis, does not encapsulate the heterogeneity of human population. Both concerns are especially relevant when large numbers of samples are required. Furthermore, whole animal experiments make it difficult to dissect molecular events at the cellular level. Thus, the use of cultured cells remains a necessity.

1.2. Structure and Function of the Rodent Mammary Gland

The basic function of the mammary gland is to make milk for the infant. As such the structural unit of mammary function is the acinus: a bilayered tube of two epithelial cell types that surround a hollow lumen where milk is secreted vectorially en route to the nipple via a branched ductal system. Luminal epithelial cells (Leps) are surrounded by contractile myoepithelial cells (Meps) and both cell layers are encapsulated by a basement membrane (BM) and surrounded by stroma including the immune cells (see Fig. 1a). Despite differences between the mouse mammary gland and the human breast, the acinar unit of function is largely similar in the two species, making it a suitable surrogate for the human

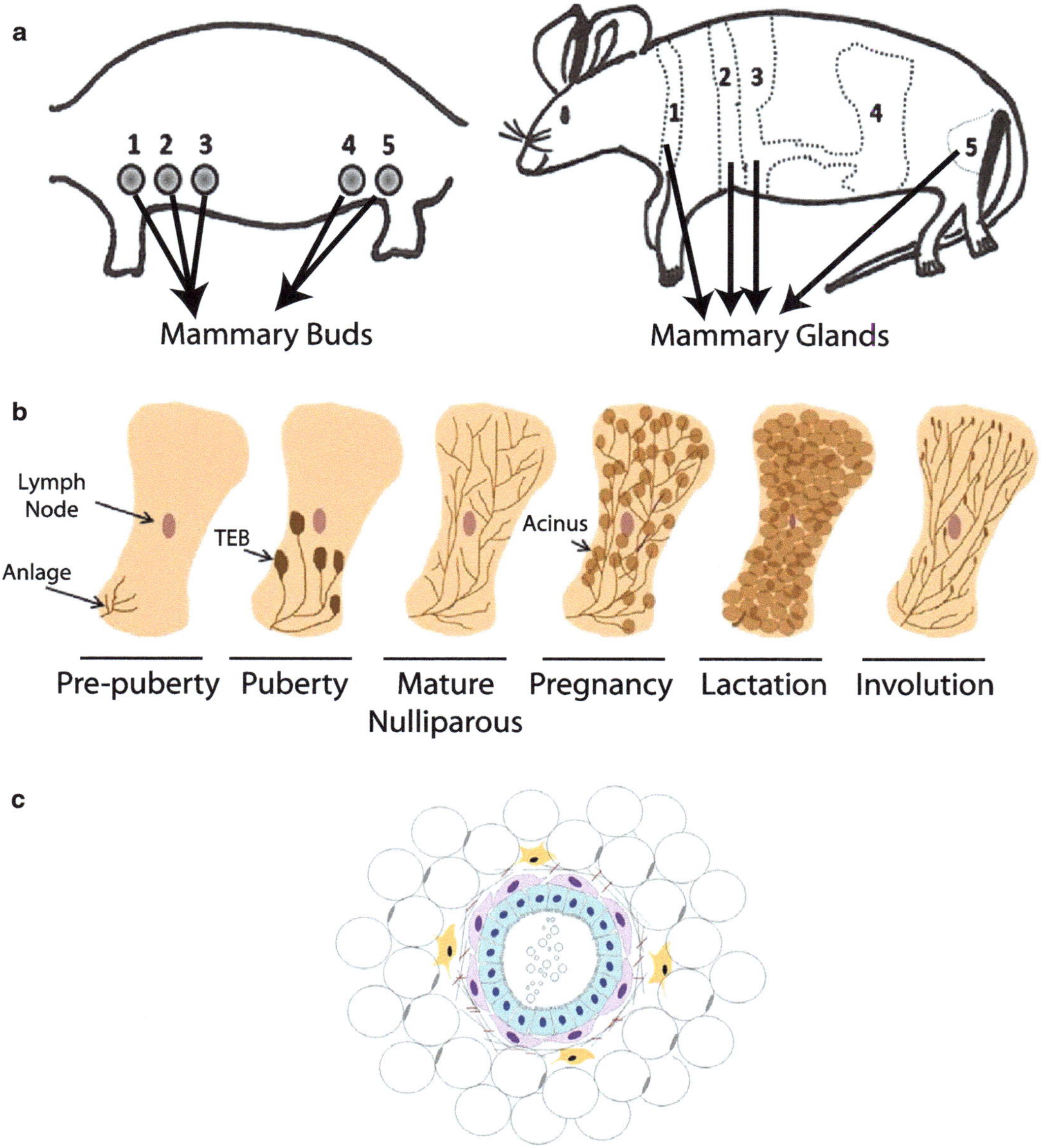

Fig. 1. Representative schematics of the different stages of mammary gland development and of the mammary acinus. (**a**) Schematic showing the five mammary buds in the embryo along the milk line which will later develop into mammary glands. (**b**) Representative drawings of the different stages of mammary gland development, starting from birth when the mammary anlage is present, to puberty where TEB's extend through the fat pad and the mature nulliparous gland filled with epithelial ducts. Pregnancy and lactation are modeled where the gland fills with alveoli which regress in involution such that the gland remodels to a pre-pregnant like state. (**c**) Representative drawing of the mammary acinus: Cuboidal luminal epithelial cells (*light blue*) are surrounded by contractile spindle shaped myoepithelial cells (*pink*) and facing a lumen on their apical side where milk droplets are deposited during lactation on their way to the ducts that terminate in the nipple. The entire structure is surrounded by a basement membrane with stromal cells including fibroblasts (*yellow*) and adipose cells (*clear round cells*).

mammary unit of function. Observations from the mouse mammary gland can be transferred to the human breast, and this exchange of information has been invaluable for the gradual development of mammary culture systems from monotypic 3D cultures, to multicellular cocultures, to in vivo xenograft models.

Each of the 3D models may be used independently and represents a physiologically relevant assay capable of answering a specific question. However, when used as a series and engineered with common components, these models become invaluable tools for identifying and testing disease-related mechanisms, and enable the design or the use of effective drug therapies. Finally, the mouse mammary fat pad may be "humanized" by repopulating it with human breast cells, and may serve then as a more faithful model to the human microenvironment to circumvent the problems pertaining to species differences (2–4).

To use the mammary acinus as an experimental model in culture, it is important to appreciate how the mammary gland develops and attains its function in vivo. In the mouse, there are five pairs of mammary glands, which reside in the fat pads located just below the skin. They extend from the thoracic (three pairs) to the inguinal (two pairs) regions of the animal along the milk line (see Fig. 1b). Each fat pad has an exterior nipple connected to the primary epithelial duct where milk is released after it is pumped by myoepithelial cells from the luminal space of the acini (5, 6). Mammary gland development occurs in two phases: ductal growth and early alveolar development during puberty. In addition, alveolar growth, expansion and differentiation occur with each pregnancy and lactation, followed by a period of epithelial remodeling during post-lactational involution. Before birth, a rudimentary ductal structure that originates from the parenchyma extends in the fat pad (5). This epithelial structure then remains dormant until approximately 3 weeks of age, when it becomes stimulated by the ovarian hormones (7). At this time, a cluster of epithelial cells termed the terminal end buds (TEBs) appears at the end of the mammary ducts and the process of ductal elongation begins (8–10), and continues until approximately 10–12 weeks of age at which time the TEBs reach the limits of the fat pad and regress. With the onset of the estrous cycle at puberty, the gland begins to branch and alveolar buds are formed. During pregnancy, the changes in the hormonal and local environment allow alveolar development to progress and to establish a gland that is filled densely with alveolar bodies (11). Concomitantly, mammary epithelial cells within the gland begin to attain their differential ability to synthesize specific milk constituents, such that by parturition, functional lactogenesis is realized. At parturition, the alveoli begin abundant milk secretion, which continues for about 3 weeks (12–16).

At weaning, the gland begins a process of tissue remodeling (involution) which involves orchestrated apoptosis of mammary epithelial cells and the remodeling back to a state somewhat similar, but not identical, to the pre-pregnant gland architecture (17–20) (see Fig. 1c). The process of involution takes approximately 2 weeks to complete in rodents, after which the gland is ready to initiate another cycle of pregnancy, lactation, and involution.

1.3. Modeling Mammary Gland Functional Differentiation and Morphogenesis in Culture

The mammary acinus exists in the context of a rich stromal and extracellular matrix (ECM) milieu that is constantly changing and is subject to signaling cues from a variety of hormones and growth factors. It has been known for decades that whereas cells in monolayers lose functional differentiation even in the presence of lactogenic hormones, the same cells transplanted into the gland-free fat pads of mice form tubular structures and are able to respond to the correct hormonal cues (21, 22). It is obvious that either the cellular microenvironment surrounding mammary epithelial cells exerts an essential role in driving their functional differentiation, or systemic factors not present in culture are operating in vivo. The work of Michalopoulos and Pitot (1976), who used floating collagen gels for hepatocytes, was followed by studies from Emerman and Pitelka in the 70s and subsequently in our laboratory where both floating collagen-I (Col-I) gels and laminin-rich ECM gels (lrECM) were used to grow mouse primary mammary epithelial cells (23–26). In contrast to cells cultured on attached Col-I gels, mammary epithelial cells grown on floated collagen gels reorganized and formed secretory structures capable of de novo expression of milk proteins (14, 26–28). On floating collagen (but not flat) gels, primary mammary epithelial cells (MECs) clustered with basoapical polarity and junctions, expressed the milk protein β-casein but not whey acidic protein (WAP) and did not form luminal alveolar structures. When cultured in lrECM, primary mouse Leps assumed an in vivo-like structural organization, secreted milk proteins into the lumen and under these conditions, they were capable of expressing WAP (13), not expressed even when Col-I gels were floated (29). Expression of WAP was possible because tight junctions were formed and were necessary to keep TGFα, an inhibitor of WAP, away from its receptor (EGFR). In later studies, laminin-111 (LN1) was shown to be the main component of the ECM that drives milk protein expression (30). Cultures of single mouse cells within the collagen gels required LN1 to express β-casein. It was later shown that LN1 modulates β-casein expression partially by allowing cells to polarize, thus exposing the basolaterally localized prolactin receptor (PrlR) to its ligand (Prl) which signals for β-casein synthesis (30, 31). Interestingly, a study done in collaboration with the Petersen laboratory, using primary human Leps revealed that Leps cultured *inside* collagen gels form clusters of similar size and quiescence to those formed inside lrECM but were "inside out," i.e., had reverse polarity. Correct polarity could be established if LN1 was added to collagen gels (32) confirming the role of LN1 in inducing acinar polarity. Whether or not purified mouse Leps also form inside out clusters in collagen has not been reported. In the experiments performed by Streuli et al. (30) with unfractionated epithelial cells (i.e., containing both Leps and Meps) in collagen gels, mouse cells were able to make endogenous LN1 providing the needed ligand for β1 integrin and signaling for milk production.

The structural scaffold that surrounds the cells in their native 3D environment is the basement membrane (BM), a specialized and heterogeneous entity within the ECM (33, 34). In fact, the ECM is not a static standalone entity: its composition is under the control of physiological effectors such as growth factors, cytokines, and hormones and thus is continuously changing during developmental stages, aging, tissue repair, as well as during tumor progression (33, 35). In turn, the ECM structure and its constituents regulate growth, differentiation and survival of cells within tissues. For example in the mammary gland, the specialized BM containing collagen IV and laminin-1 yields better expression of genes encoding milk proteins (36–39), but a Col-I rich ECM would favor tubular growth under the right hormonal stimulation (40–42) For further description of these assays please refer to the upcoming Subheadings 1.3.1 and 1.3.2.

Signals from the ECM are relayed to the cells via surface receptors that translate the biochemical and mechanical stimuli into a cellular and nuclear response (30, 43). ECM receptors are formed largely by the integrin family of proteins although for milk proteins, dystroglycan has been shown to be involved as well (44). As mentioned earlier, the prolactin receptor is also required for milk proteins to be expressed, and in order for these receptors to receive the signal from their ligands, they need to be present in the correct mechanical and structural platform (31, 45).

In addition to ECM and hormone signaling, the role of neighboring cells and intercellular interactions within the tissue are important for functional differentiation (46–48). Currently, mammary culture systems range from monotypic cell culture models grown in the appropriate 3D microenvironment to cocultures of Leps and Meps (32, 49) and cocultures of epithelial and endothelial cells as well as Leps and stromal fibroblasts (50–53), in addition to culturing pieces of the epithelial tissue in different gels referred to as organoid cultures (54, 55). More models still need to be developed to include the entire spectrum of cell–cell interactions found within the mammary gland. Equipped with this growing array of culture systems, it would be easier for investigators to probe for a variety of molecular signaling events that otherwise would be difficult in the whole organism, and artifactual on monolayer cultures (56) (see Subheadings 1.3.1 and 1.3.2).

1.3.1. Differentiation

The mammary acinus in culture provides a reliable functional readout, namely, the expression of milk proteins and vectorial secretion of milk into the lumen (13). The concept was extended to the human cells and used as an assay for distinguishing normal and malignant phenotypes rapidly and reproducibly (57) (for further details please refer to the accompanying chapter by Vidi et al.). Despite our inability to recapitulate the whole organ structure in culture as yet, 3D models of the mammary acinus have provided ample information on the molecular events that contribute to differentiation. These assays have allowed investigators to dissect the hierarchy of the processes that establish tissue function (58). Functional differentiation in the mammary acinus comprises two components: architectural reorganization that allows the formation of a polarized structure, which ultimately enables synthesis and vectorial secretion of milk.

Using the floating Col-I gel methodology developed by Michalopoulos and Pitot (25), Emerman and Pitelka first demonstrated that mammary epithelial cells (MECs) cultured on top of collagen gels which subsequently were allowed to float assumed a cuboidal morphology and expressed some milk proteins (26). In a follow up study, we showed that MECs cultured in floating Col-I gels, were able to become polar and functional by depositing Collagen-IV and LN1 vectorially to the basal surface of the floating gels, which once assembled into a BM would allow the de novo expression of some of the milk proteins the most prominent being β-casein (59). To determine the minimum requirement for expression of β-casein and whether or not single mammary cells could produce this milk protein without cell–cell interactions, we cultured primary mouse Leps inside collagen gels. We then added different BM molecules and showed that LN-1—but not other BM proteins—could induce expression of β-casein (30). Cell–cell interactions however, are essential to allow β-casein expression in the *absence* of exogenously provided LN1-rich BM (30). This conclusion correlated with the previous findings from floating Col-I gel where cells were able to deposit a BM and make some of the milk proteins (26, 27). When grown on top or inside lrECM, however, primary mouse cells (see Fig. 2a) can express abundant milk proteins including WAP, indicating the importance of the acinar morphology and tight junctions in formation of WAP and presumably other milk-specific molecules not measured in those studies.

Signals from the BM and specifically from LN1 were found to be relayed mainly through the integrin family of heterodimeric proteins and specifically through β1- and β4-integrins (60–62). Dystroglycan (DG), another LN1 binding partner, was found to be required along with integrin to induce polarization and β-casein expression associated with LN1 signaling, and functions by helping laminin anchoring at the cell surface (63). Whereas cultures on floating Col-I gels and lrECM gels enabled copious expression of

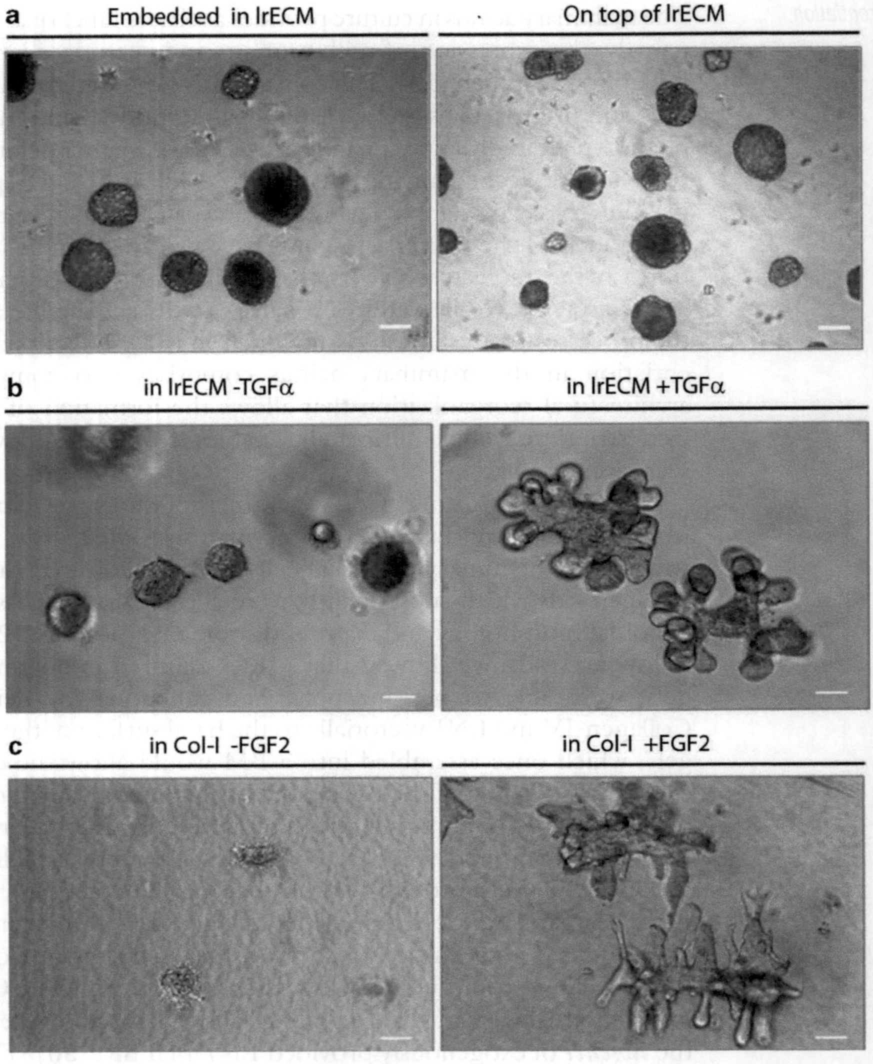

Fig. 2. Primary mammary organoids in 3D. (**a**) primary mammary organoids isolated from the mammary glands of 14-week-old nulliparous mice are grown *inside/embedded* in lrECM (*left*) and *on top* of lrECM (*right*) and imaged by bright-field light microscopy on day 4 of culture, size bar, 50 μm. (**b**) Primary organoids isolated from the mammary glands of 8–12-week-old mice are embedded in lrECM for 4 days in additive-free medium as control (*left*), and stimulated with TGFα to induce alveolar growth (*right*). (**c**) Primary organoids isolated from the mammary glands of 8–12-week-old mice are embedded in collagen I (1 mg/mL) for 4 days as control (*left*), and stimulated with FGF2 to induce branch formation (*right*), size bar, 50 μm.

β-casein, WAP, another major milk protein, was poorly expressed even in primary cultures. This is because unlike β-casein which can be expressed even in single cells as long as LN1 and prolactin are present, WAP is not expressed until cells form complete acini with tight junctions, allowing separation of WAP's inhibitor, TGF-α from its receptor, EGFR (29, 64). This temporally regulated event whereby tight junction sealing separates the distinct apical and basal sides of MEC's is an important step in the completion of functional differentiation in the tissue. In fact, polarization of cells and tissues is paramount for modulating the reception and transmission of cues within and between cells and is a major contributor to tissue-specific functions. In the mammary acinus, correct polarization of Leps ensures that the basal side of the cells receives the appropriate signaling cues from the surrounding BM, ECM, Meps, and from circulating hormones, and that the apical side of the cell, delivers milk vectorially into the lumen for further transport into the ductal system towards the nipple. Tissue polarity is important for proper presentation of surface molecules and receptors to their ligands. Indeed even in the case of β-casein, we showed this principle again recently: mammary epithelial cells on 2D culture plastic, under differentiation-permissive conditions were unable to sustain activation of transcription factor STAT5 by its upstream regulator, prolactin (Prl). The study revealed that the Prl receptor was localized basally under these conditions and thus it is not available to the ligand, Prl, which is usually placed in the culture medium. However, when cells were grown either on a nonadhesive substratum with LN1 or on top of 3D lrECM gels, Prl was able to bind to the exposed receptor and could activate STAT5 transiently. Nevertheless, continuous exposure to LN1 was further essential to allow reorganization of the chromatin and to sustain activation of STAT5 permitting continued milk protein expression (31).

Using β-casein as the functional readout, it was possible to develop assays to probe the relationship between the extracellular environment and the nucleus and how the cross-talk between the two is essential for tissue-specific functions (31, 65–67). We discovered the first ECM-response element in the promoter of the β-casein gene and later established that the chromatin context and histone acetylation/deacetylation played essential roles in ECM-mediated nuclear changes that in turn lead to the establishment of tissue-specific gene expression supporting the postulated model of "Dynamic Reciprocity" (65, 68–70).

1.3.2. Branching Morphogenesis and Pattern Formation

Branching morphogenesis in the nulliparous mouse results from an intricate interplay between the cells and the extrinsic factors. Earlier studies from our laboratory and others established the importance of ECM remodeling enzymes (metalloproteinases: MMPs) in driving invasion into the fat pad and branching of the mammary epithelium (71–73). The process could be simulated in culture by embedding mammary epithelial cells in collagen-I gels and stimulating them with growth factors that favor branch formation (42, 55, 74).

In the assay described above, it is not easy to investigate the role of positional control in the process of branching. To separate positional control from simple growth and invasion, we developed a micropatterned assay where cells were placed between a collagen "sandwich" and where the results of 50 or more wells could be imaged and binarized, allowing a rapid read out and meaningful statistics to study how positional cues dictate the initiation of branching points (75). Using different geometric contexts to place the cells into, the assay allowed us to discover that geometry influenced the secretion and activation of endogenous TGF-β, which then would inhibit branching depending on the shape of the micropatterns (75).

Morphogenesis in the gland of pregnant mice takes the form of alveologenesis, a process by which lobules of the milk-producing acini are formed. This process can be modeled by replacing the composition of the ECM gel from Col-I to lrECM (see Fig. 2b, c). This is yet another example of how the nature of the ECM molecules allows unique aspects of growth and functional differentiation to be studied. In this latter assay, MECs (primary or established lines that are clustered as described later, or organoids from the mammary gland) are embedded in an lrECM gel and stimulated with growth factors such as TGF-α (76). Under these conditions, mammary epithelial cells undergo a growth and remodeling process that resembles the structure of the "alveoli" in the mammary gland (see Fig. 2c) during pregnancy and lactation, rather than branching ducts (see Fig. 2b). The use of the alveologenesis assay enabled our laboratory to uncover a critical role for MAPK signaling and its downstream molecules in promoting alveolar growth (76).

These studies clearly demonstrate that one can create "designer microenvironments" (77) in 3D to answer questions that are otherwise too difficult to address in the organism. This is made possible by modulating the type and organization of the culture substrata to model different aspects of mammary growth and development. Nevertheless, many challenges remain in making physiologically relevant culture systems and we refer to these in the section below.

1.4. Current Limitations and Future Perspectives

3D mammary epithelial culture systems have greatly evolved since their early inception and are still undergoing refinements aimed ultimately at recreating the entire organ in culture. This will however require concerted efforts from many fields of science. The advent of tools such as bio-engineered materials and nanotechnology has been a driving force in improving available culture systems by allowing investigators to manipulate and control the cellular microenvironment at multiple levels and scales. That is, they have provided means to control the chemistry, geometry, and mechanics of culture settings with increasing precision. These tools will ultimately allow us to replicate the gland in culture, and with rigorous controls it will be possible to answer questions pertaining to particular developmental stages and environmental constraints in vivo.

There are still, however, many challenges ahead and many questions to answer. We still lack a profound knowledge of the complex array of signals generated from the ECM. Signaling from laminin-1 through both integrin and non-integrin (e.g., dystroglycan) receptors is required for BM and acinus formation and differentiation (62, 63). With a few exceptions, currently we have limited knowledge of the functional roles of other BM components. Moreover, it is not yet clear why it is that cells on 2D culture plastic are able to synthesize BM molecules, but why they do not deposit a structurally intact BM that would eventually drive reorganization of cells into their correct conformation in the tissue. However, it is now possible to start to tackle these questions with the use of synthetic inert gels where purified components of the BM are added. Ultimately, different components can be recombined together to study the hierarchy of signaling events from the BM to the epithelium (78).

Unfortunately, simply adding different ECM and BM membrane components is not sufficient to model the complexity of the microenvironment in vivo, especially given the temporal regulation of signaling within a tissue. Our inability to model such complexity confounds our interpretation of data derived from culture studies. One particular example of this complexity is observed with the alveologenesis assay. Even though we are able to model a process that phenotypically resembles alveolar growth in culture, we still do not know if our alveologenesis assay is a true surrogate for alveolar growth and expansion that happens during pregnancy, ultimately leading to lactation. However, it is evident that in the advances we have made in building mammary epithelial culture systems, we have answered a number of questions but also opened a whole array of new questions. It is important to recognize that in order to obtain pertinent knowledge from culture systems, we should ask the relevant questions and admit to the limitations of the models available. In the end the organ itself remains the unit of function, and the integration of signals within the organ is what ultimately provides tissue specificity (79).

2. Materials

The sources listed for the reagents are the current materials used in the Bissell laboratory. We are not advocating their use. Other sources can be substituted, and we do that from time to time depending on the availability, price, and purity.

2.1. Culture Medium for Primary Cells, Organoids, and EpH4 Cells

1. Dulbecco's modified Eagle's medium/ Ham's F-12 1:1 Mix Medium (DMEM/F12): 3.15 g/L Glucose, 0.365 g/L L-Glutamine, 3.754 g/L HEPES, 1.2 g/L NaHCO3 pH 7.3 (DMEM/F12, Life Technologies™).
2. Fetal Bovine Serum (FBS).

3. Prolactin (Sigma-Aldrich®): 1 mg/mL (30.3 IU/mL) in 26 mM (2.22 mg/mL) sodium bicarbonate, filter-sterilized (0.22 μm filter), aliquoted, fast-frozen on dry ice, and stored at −80°C. Prolactin is stable for 1 year at −80°C and 1 month at 4°C.
4. Insulin (Sigma-Aldrich®): dissolved as a 2 mg/mL concentrated stock in 5 mM HCl. The stock solution is filter-sterilized, diluted to 100 μg/mL in filter-sterilized MilliQ (ddH_2O) water, aliquoted, fast-frozen on dry ice, and stored at −80°C. Insulin solution is stable for 6 months at −80°C and 1 month at 4°C.
5. Hydrocortisone (Sigma-Aldrich®): Prepare a 5 mg/mL concentrated stock solution in ethanol. The stock is further diluted to 0.5 mg/mL ($1.4\ 10^{-3}$ M) in ethanol and kept at −80°C. Hydrocortisone solution is stable for 1 year at −80°C and 1 month at 4°C.
6. Insulin-Transferrin-Sodium Selenite media supplement (Sigma-Aldrich®): lyophilized powder, diluted in MilliQ (ddH_2O) water and filter-sterilized to a stock of 1,000×. Stock solutions are aliquoted and stored for 6 months at −20°C or 1 month at 4°C.
7. Penicillin-Streptomycin Solution (Penicillin: 6,370 mg/mL; Streptomycin: 10,000 mg/mL): 100×.
8. Gentamicin solution (50 mg/mL): 100×.
9. Poly (2-hydroxyethyl methacrylate), polyHEMA (Sigma-Aldrich®): dissolved in 95% ethanol at 6 mg/mL.
10. Growth medium for organoids and primary cells (100 mL): 94 mL of DMEM/F12, 100 μL of ITS (1,000× dilution), 5 mL of FBS, 100 μL of gentamycin for a final concentration of 50 μg/mL, 1 mL of Pen/Strep (100× dilution) (see Note 1).
11. Growth medium for cell lines (100 mL): 95 mL of DMEM/F12, 100 μL insulin for a final concentration of 5 μg/mL, 2 mL of FBS, 100 μL of gentamycin for a final concentration of 50 μg/mL.
12. Differentiation medium for organoids and primary cells (100 mL): 99 mL of DMEM/F12, 100 μL of ITS (1,000× dilution), 100 μL of gentamycin for a final concentration of 50 μg/mL, 1 mL of Pen/Strep (100× dilution), 100 μL of prolactin for a final concentration of 3 μg/mL, 100 μL of hydrocortisone for a final concentration of 1 μg/mL (see Note 1).
13. Differentiation medium for cell lines (100 mL): 100 mL DMEM/F12, 100 μL of gentamycin for a final concentration of 50 μg/mL, 100 μL of insulin for a final concentration of 5 μg/mL, 100 μL of prolactin for a final concentration of 3 μg/mL, 100 μL of hydrocortisone for a final concentration of 1 μg/mL.

2.2. Propagation of Primary Organoids, Cells, or EpH4 Cell Lines in 2D

1. 10 cm Cell culture plates.
2. Solution of trypsin (0.25%) and ethylenediamine tetraacetic acid (EDTA, 1 mM) (Life Technologies™).
3. Purecol® (Advanced BioMatrix) diluted to a concentration of 1 mL in 44 mL of phosphate buffered saline (PBS).
4. Cell Freeze solution: 10% Dimethyl Sulfoxide (DMSO) in 90% FBS.

2.3. Monoculture of Primary Cells, Organoids, or EpH4 Cells in 3D

1. 4-Well chamber slides (LabTek, Nunc) are used for immunostaining experiments.
2. 35-mm Tissue culture dishes are used for biochemical assays (protein, RNA, and DNA extracts).
3. Multiwell plates (6, 12, 24, 48, and 96) are used depending on the type of assay and the amount of cells needed.
4. DMEM (powder, Sigma-Aldrich®): to prepare 10× solution, dissolve DMEM powder into 0.1 volume of H_2O. Filter-sterilize and store at –20°C in working aliquots.
5. Basement membrane matrix from EHS extracts (Matrigel™, BD Biosciences). Matrigel™ is kept at –80°C and thawed on ice 24 h before use. If rigorously kept at 0°C, Matrigel™ is stable for 1 month after thawing and can only be thawed twice. Thus, usually a 10 mL bottle is thawed to prepare 1–2 mL aliquots that will be frozen and kept at –80°C until use. For the optimal culture of human breast cell lines currently available, it was observed that Matrigel™ should have low levels of endotoxins (less than 4 U/mL) and a protein concentration below 13 mg/mL. In addition, certain research teams prefer to use growth-factor depleted lrECM.
6. Collagen-I/acid-soluble collagen: 7.5 volume of a 0.5% solution mixed gently on ice with 1 volume of 10× DMEM/F12, followed by 1 volume of 0.1 N NaOH. (Cellagen™ AC-5, ICN). Collagen 1 can also be purchased from Cosmo Bio Co., as Native Collagen, Bovine Dermis, 5 mg/mL or 3 mg/mL or from BD Biosciences as acid-extracted rat tail tendon collagen.

2.4. Branching Assay and Alveologenesis Assay

1. TGFα: 9 nM (Sigma-Aldrich®). The TGFα working solution is prepared by diluting the supplied powder in a balanced salt solution such as PBS under sterile conditions. The working solution is then diluted to desired concentration of 9 nM in desired medium.
2. FGF2: 9 nM (Sigma-Aldrich®). The FGF2 working solution is prepared by reconstituting the powder in sterile water to 100 μg/mL.
3. FGF7: 9 nM (Sigma-Aldrich®). The FGF7 working solution is prepared by reconstituting the contents of the vial using sterile PBS.

The working solution is then diluted to desired concentration of 9 nM in desired medium.

4. FGF10: 9 nM (Life Technologies™). The FGF10 working solution is prepared by reconstituting the contents of the vial using sterile PBS. The working solution is then diluted to desired concentration of 9 nM in desired medium.
5. 2% Agarose in 2% DMEM solution: Add 2 g of Agarose to 100 mL of 2% DMEM/F12.
6. Branching morphogenesis medium for organoids and primary cells (100 mL): 99 mL of DMEM/F12, 100 μL of ITS (1,000× dilution), 100 μL of gentamycin for a final concentration of 50 μg/mL, 1 mL of Pen/Strep (100× dilution), choice of growth factor (see steps 1–4) at designated concentration.
7. Branching morphogenesis medium for cell lines (100 mL): 100 mL of DMEM/F12, 100 μL insulin for a final concentration of 5 μg/mL, 100 μL of gentamycin for a final concentration of 50 μg/mL, choice of growth factor (see steps 1–4) at designated concentration.
8. Alveologenesis medium for organoids and primary cells (100 mL): 99 mL of DMEM/F12, 100 μL of ITS (1,000× dilution), 100 μL of gentamycin for a final concentration of 50 μg/mL, 1 mL of Pen/Strep (100× dilution), choice of growth factor (see steps 1–4) at designated concentration.
9. Alveologenesis medium for cell lines (100 mL): 100 mL of DMEM/F12, 100 μL insulin for a final concentration of 5 μg/mL, 100 μL of gentamycin for a final concentration of 50 μg/mL, choice of growth factor (see steps 1–4) at designated concentration.

2.5. Isolation of Primary Organoids and Fractionation of Mammary Epithelial Cell Types

1. VWR® Razor Blades (VWR®).
2. Collagenase Type IV (GIBCO®): 0.2% (w/v), dilute 0.1 g of collagenase in 47.5 mL of DMEM/F12 mixed with 2.5 mL of FBS.
3. Trypsin 1:250 powder: 0.2% (w/v), dilute 0.1 g of trypsin in 47.5 mL of DMEM/F12 mixed with 2.5 mL of FBS.
4. DNase-I (Sigma-Aldrich®): Dilute DNase-I powder in DMEM/F12 for a 2 U/mL working solution.
5. 50 mL Conical tube.
6. Collagenase solution for organoid extraction (50 mL): 0.2% collagenase Type IV (see step 2), 0.2% trypsin (see step 3), 5% FBS, 5 μg/mL gentamycin, 47.5 mL DMEM/F12. Prewarm at 37°C (shake occasionally). Filter-sterilize after the collagenase has gone into solution through a 0.2-μm filter (see Note 2).
7. Joklik's Modification of Minimal Essential Medium (Sigma-Aldrich®).
8. Hanks' Balanced Salt Solution (HBSS).

9. L-15 medium (Sigma-Aldrich®).
10. Alexa Fluor® 647 anti-mouse CD326 (Ep-CAM) Antibody (Biolegend®).
11. FITC anti-mouse CD104 Antibody (Biolegend®).
12. Unconjugated anti-mouse Ep-CAM antibody (Abcam®).
13. Unconjugated anti-mouse CD104/β4 Integrin antibody (R&D systems).
14. MACS® micro-beads (Miltenyi Biotec).
15. MACS® Pre-separation filter (Miltenyi Biotec).
16. MACS® Separation Filter (Miltenyi Biotec).
17. MACS® Magnet (Miltenyi Biotec).
18. MACS® Columns (Miltenyi Biotec).
19. Bovine Serum Albumin (BSA, Sigma-Aldrich®).
20. Ca^{2+}/Mg^{2+}-free PBS 10×.
21. MACS Buffer: 0.5% (w/v), BSA in 1× Ca^{2+}/Mg^{2+}-free PBS.
22. FACS Buffer: 1% BSA in 1× Ca^{2+}/Mg^{2+}-free PBS.

2.6. Biochemical Assays

1. EDTA (GIBCO®): 0.5 M Stock.
2. Dispase 100 mL solution (BD Biosciences): recommended concentration is 10 U/cm^2 of BD Matrigel™ Matrix (for example: 100 U for a 35 mm culture dish).
3. Phosphatase inhibitor cocktail 1 (100×) (Sigma-Aldrich®).
4. Phosphatase inhibitor cocktail 2 (100×) (Sigma-Aldrich®).
5. 100× Protease Inhibitor Cocktail (EMD-Calbiochem Merck®).
6. PBS–EDTA buffer: 50 mL of cold 1× PBS buffer, 500 μL of 0.5 M EDTA for a final concentration at 5 mM, 500 μL of 100× phosphatase inhibitor cocktail 1 (1×), 500 μL of 100× phsophatase inhibitor cocktail 2 (1×), 500 μL of 100× protease inhibitor cocktail to 1×.
7. Polyethylene cell lifter.
8. RLT RNA lysis Buffer (Qiagen).
9. Ripa Buffer (Sigma-Aldrich®): supplement with 1× protease inhibitor and 1× phosphatase inhibitors (cocktail 1 and 2).
10. Microcentrifuge tube shaker.
11. 18–21 Gauge syringe.
12. Sonic Dismembrator Model 100.

2.7. High-Resolution Imaging

1. Triton™-X 100 (Sigma-Aldrich®): 0.5% (v/v). Add 500 L of Triton™-X 100 to 100 mL of 1× PBS.
2. 16% Paraformaldehyde (Electron microscopy sciences): Dilute 25 mL of 16% paraformaldehyde into 75 mL of PBS to make 4% paraformaldehyde solution.

3. Superfrost Plus Micro Slides.
4. Spot RT camera attached to an upright epifluorescence microscope (Carl Zeiss, Inc.).
5. Zeiss LSM 710 Meta confocal microscope (Carl Zeiss, Inc.).

3. Methods

3.1. Isolation of Primary Organoids and Epithelial Cells

To isolate primary epithelial cells or primary organoids from the mouse mammary gland we use either the third thoracic or the fourth inguinal mammary glands from the mouse. The age of the animal is critical since some assays cannot be performed with older mice (see Subheading 3.5.4). But in general, mammary glands could be removed at different ages, and the choice of age depends on the stage of mammary gland growth desired to answer a specific question (see Notes 3 and 4).

1. Mince glands with two razor blades on sterile surface (see Subheading 2.5, item 1). For optimal mincing, hold two razor blades and make 10–15 cuts (based on mammary gland size) in two perpendicular directions across the entire gland.
2. Shake minced tissue in collagenase solution (see Subheading 2.5, item 6) at 100 rpm/37°C for 30 min.
3. Spin at 405×*g* for 10 min at room temperature.
4. The top layer is going to be the fatty layer, but it will often contain uncollected organoids, so collect 10 mL of the top layer in a 15 mL conical tube. Pipette it up and down to break it up and free organoids prior to another round of spinning and add another 5 mL of fresh serum-free medium on top (see Note 5).
5. The pellet at the bottom will contain epithelial cells, fibroblasts, and debris. Remove all the rest of the medium. Resuspend the pellet at the bottom in 10 mL of serum-free DMEM/F12.
6. Take the two 15 mL conical tubes, the one that contains the pellet (see step 5) and the one that contains the fatty top layer (see step 4) and spin at 405×*g* for 10 min at room temperature.
7. Remove the medium from the two conical tubes and combine the two pellets by resuspending one then the other in the same 4 mL of 0% DMEM/F12. Now there is one 15 mL conical tube of organoids with 4 mL of 0% DMEM/F12. Add 40 μL of DNase, (see Subheading 2.5, item 4) and shake by hand for 2–5 min at room temperature (until clumps are broken up), then add 6 mL of 0% DMEM/F12 to obtain a final volume of 10 mL.
8. Spin at 405×*g* for 10 min at room temperature.

9. Discard supernatant and resuspend the pellet in 10 mL of 0% DMEM/F12.
10. Spin at 405×g for a brief pulse then cut the machine and hit the brakes as the centrifugation speed reaches 405×g. This step is done to separate the fibroblasts from the epithelial cells/organoids. Fibroblasts are lighter and will float in the supernatant while the epithelial organoids will pellet down (see Notes 6 and 7).
11. Resuspend the pellet at the desired concentration into Growth medium (see Subheading 2.1, items 10 and 11, see Note 8).

3.2. Fractionation of Cell Types

1. To fractionate the epithelial subtypes from each other (i.e., the Leps and Meps from each other) the primary step is the isolation of mammary epithelial organoids as described in Subheading 3.1. Then proceed as follows: Resuspend organoids in JokliK's Ca-free medium at 37°C for 15 min.
2. Spin down the cells at 405×g for 10 min at room temperature and resuspend them in 2 mL of Hank's Balanced Salt Solution at 37°C for 5 min.
3. Vigorously flush organoids up and down with Pasteur pipette and incubate for 2 min.
4. Add 5 mL of serum-free DMEM/F12 supplemented with 50 μL type-I DNase (see Subheading 2.5, item 4) to remove clumps. Incubate for 5 min.
5. To stop the reaction: add 10% FBS in 5 mL of L-15 medium or in calcium-free DMEM/F12 with DNase (50 μL, 2 U/mL), then spin down the cells at 405×g for 10 min at room temperature and wash cells 2× in medium to clean.
6. Remove large multicellular clumps with filtration through preseparation filter (see Subheading 2.5, item 15).
7. Resuspend cells in 500 μL MACS or 500 μL FACS buffer (see Subheading 2.5, items 21 and 22) depending on the method used (see Subheadings 3.2.1 and 3.2.2).

3.2.1. Fractionation Using MACS Protocol

1. Pre-wet separation filter (see Subheading 2.5, item 16) with 500 μL MACS buffer (see Subheading 2.5, item 21).
2. Apply cell suspension to filter.
3. Wash 3× with MACS buffer.
4. Count cells.
5. Label cells by adding 0.25 μg Ab/10,000,000 cells EpCAM (labels Leps) or β4 integrin (labels Meps).
6. Incubate cells with primary antibody for 30 min in the cold room.
7. Wash/spin 2× with 10× volume MACS buffer.
8. Incubate cells with MACS® micro-beads at 1:10 dilution 15 min in the cold room.

9. Wash/spin 2× with 10× volume MACS buffer.
10. Resuspend in 500 μL MACS buffer.
11. Pre-wet MACS columns.
12. Apply the cell suspension.
13. Wash columns 3× with 500 μL MACS buffer.
14. Freeze negative fraction in cell freeze solution (see Subheading 2.2, item 4).
15. Remove column from magnet (see Subheading 2.5, item 17) to elute positive fraction with 1 mL MACS buffer.
16. Spin down the positive fraction at 405 ×g for 10 min at room temperature and resuspend in 500 μL MACS buffer.
17. Repeat steps 11–16 with a new column for double purification.
18. Count the cells in the positive fraction.

3.2.2. Fractionation Using FACS Protocol

1. Count cells.
2. Label cells: 0.25 μg Ab /10,000,000 cells EpCAm (luminal epithelial) or β4 integrin (myoepithelial).
3. Incubate cells with primary antibody for 30 min in the cold room.
4. Wash/spin 2× with 10× volume FACS buffer (see Subheading 2.5, item 22).
5. Resuspend in 500 μL FACs buffer.
6. Run on FACS sorter.

3.3. Maintaining Primary Mouse Mammary Epithelial Cells

Primary mammary epithelial cells are not easy to maintain in culture. To do that, it is preferable first to change the medium the next day and every 2 days after that. Passage the cells at 1:1 or 1:2 split ratio for the first two passages, because the cells are not easy to grow and they have been stressed by the isolation procedure. With time it is possible to increase the split to a 1:5 ratio. They should survive 5–7 passages before reaching senescence.

1. To passage cells use dispase (see Subheading 2.6, item 2).
2. Incubate for 20–30 min at 37°C.
3. Tap culture or pipette up and down to release cells from plate.

3.4. Mouse Mammary Epithelial Cell Lines: EpH4 as a Cell Line Model and as a Surrogate for Primary Cultures

EpH4 cells were originally isolated from the mammary tissue of a mid-pregnant Balb/c mouse yielding a spontaneously immortalized, non-tumorigenic mouse mammary cell line, designated IM-2 (80, 81). They provide a powerful tool for the identification of factors that modulate epithelial architecture. EpH4 cells exhibit a polarized epithelial phenotype (as revealed by their high transepithelial electrical resistance and appropriate localization of apical and basolateral marker proteins), and form branching tubular structures when grown in collagen gels (see Notes 9 and 10).

1. EpH4 cells are thawed and initially maintained in growth medium (see Subheading 2.1, items 10 and 11).
2. Cells are plated at a density of 10,000 cells/cm^2, and initially allowed to attach for 16–24 h.
3. Cells are propagated by trypsinization (see Subheading 2.2, item 2) and splitting in 1:4 ratio at around 70% confluency.

3.5. Design of Functional Assays in 3D

The ultimate goal of cell culture models is to allow us to ask biological questions that are either not possible in the organism or that are too costly or inhumane to perform. Useful culture models should recreate at least some aspects of the tissue and organ microenvironment to ask relevant questions, and to provide a functional readout. Mammary epithelial cells (primary and cell lines) provide a variety of functional outputs we can recreate in culture such as recapitulation of formation of acini, milk synthesis, invasion, and branching, etc. We, and others, have taken advantage of this ability and in this next section we try to provide investigators with detailed protocols for culturing these cells under relevant conditions and describe assays that can be used to probe molecular signals.

3.5.1. 3D Laminin-Rich ECM Embedded Assay

In this culture set up, (Barcellos-Hoff et al. (13), mammary epithelial cells are cultured inside lrECM (Matrigel™) such that they are fully surrounded by an lrECM thick coat (see Note 11, Fig. 2a).

1. Coat tissue culture plates with lrECM and leave them at room temperature for 30 min (see Subheading 2.3, item 5). The desired amount of coating is 100 μL per 0.75 cm^2 (48- well plate). In this case, smaller size culture dishes are used because the amount of lrECM needed is larger than in the on top assay.
2. After obtaining a pellet of EpH4 cells, put cells on ice and add on top of them 200 μL of lrECM and keep on ice while you suspend the cells in lrECM by mixing with a p1000 pipette until the cells are evenly distributed in lrECM. Take the entire volume of cells/lrECM and add it on top of the coated plates and allow cells to attach, then add medium (see Subheading 2.1, items 10 and 11) on top of cells (see Note 12).
3. Add fresh growth medium (see Subheading 2.1, items 10 and 11) and incubate for 24 h at 37°C, 5% CO_2.
4. The next day, change growth medium into differentiation medium (see Subheading 2.1, items 12 and 13). Make sure to wash the cells from residual growth medium with 1× PBS.
5. Harvest cells for analysis after 24 h or 48 h from treatment with lactogenic hormones (see Subheading 3.6.).

3.5.2. 3D Laminin-Rich ECM On-Top Assay

In this culture assay, mammary epithelial cells are cultured on top of a thin layer of lrECM and dripped with lrECM (see Note 12, Fig. 2a).

1. Coat tissue culture plates with lrECM and leave them at room temperature for 30 min. The desired amount of coating is 500 μL/9.62 cm^2 (35 mm dish).
2. Add EpH4 cells on the coated plates at a density of 500×10^3/9.62 cm^2 (35 mm dish).
3. Add fresh growth medium (see Subheading 2.1, items 10 and 11) and drip 2% lrECM (vol/vol) (Matrigel™) on top of cells and incubated for 24 h at 37°C and 5% CO_2.
4. The next day, change growth medium into differentiation medium (with lactogenic hormones) (see Subheading 2.1, items 12 and 13). Make sure to wash the cells from residual growth medium with 1× PBS.
5. Harvest cells for analysis after 24 h or 48 h from treatment with lactogenic hormones.

3.5.3. Suspension Culture Assay

In this assay cells are allowed to cluster on top of a nonadhesive substratum, polyHEMA.

1. Dissolve polyHEMA in 95% ethanol at 6 mg/mL to make the working solution.
2. Coat plates with polyHEMA at 0.25 mg/cm^2 as previously described (60) by adding 4 mL of the working solution in a 10 cm plate and drying overnight in an incubator at 55°C.
3. Plate EpH4 cells on polyHEMA-coated plates at 30,000 cells/cm^2.
4. 24–48 h after plating, collect cells by centrifugation (either at 1,154 × *g* for primary organoids, or at 405 × *g* for cell lines) and resuspend them in DMEM/F12 either for growth conditions (see Subheading 2.1, items 10 and 11) or for differentiation conditions (see Subheading 2.1, items 12 and 13).

3.5.4. Branching Morphogenesis Assay

Branching assays are divided into two types in concordance with two distinct types of branching that occur in the mammary gland in vivo. The first assay is used to model morphogenesis in the early mammary gland or the phase of tubular growth. Accordingly, the cells in the assay are embedded in a Collagen-I rich matrix to simulate the early status of the epithelium in the mammary gland where cells that are invading the fat pad are constantly remodeling and invading through a layer of collagenous ECM. The second assay (see Subheading 3.5.5) is used to model the alveolar growth in the mammary gland and as such cells are embedded in an lrECM rich matrix similarly to the differentiation assay, but hormones that induce branching are used instead of the lactogenic hormones. The latter assay is referred to from here on as "alveologenesis" (see Subheading 3.5.5).

For the branching morphogenesis assay, EpH4 cells are pre-clustered before embedding in collagen gels. Pre-clustering can be done on top of polyHEMA (see Subheading 3.5.3) or on top of Agarose coated plates. In both cases a differential centrifugation step is required to remove the single cells and only keep the clustered EpH4 for embedding in Collagen-I (see Note 13 and Fig. 2b).

1. Day 1: Make 2% Agarose (see Subheading 2.4, item 6) and with a 10 mL pipette drop 1 mL into 4 center wells of a 24-well plate and put medium in surrounding wells.
2. Let it gel as needed and then add 1.5 mL of 2% medium in the well and incubate at 37°C for ~30 min to 1 h.
3. Trypsinize cells (90% confluent EpH4).
4. Spin for 5 min at 115 ×*g* and aspirate the medium and leave 100 μL of medium on top of cell pellet.
5. Add 5 μL of DNase I (2 U/μL) to the cells. Finger tap for 2 min.
6. Add 2 mL of 2% medium to the cells.
7. Aspirate the medium from the four wells. Leave half the amount of medium in the surrounding wells.
8. Add 500 μL of the 2 mL medium in each of the four wells and incubate in a shaker overnight at 37°C, 100 rpm.
9. Day 2: differential centrifugation, take 2 mL out of the four wells in the plate and transfer them to 15 mL conical tube, top off to 10 mL with DMEM/F12.
10. The tube will contain various sizes of clusters and you want to pull out the larger clusters via centrifugation. Centrifuge the sample at 405 ×*g*. When it reaches this speed, stop the centrifugation.
11. Discard supernatant and resuspend pellet each time in 5–10 mL of DMEM/F12. This step may have to be repeated three times until you get rid of the single cells.
12. Count and check the number of clusters using a hemocytometer. Calculate how much volume required to get the number of clusters needed (see Note 14). Make Collagen-I (see Subheading 2.3, item 6): 800 μL collagen (take 840 μL because it is viscous to account for loss of volume on the pipette edges) and add: 100 μL 10× DMEM (see Subheading 2.3, item 4) to 100 μL 0.1 N NaOH. Adjust pH to 7.4. Mix well and put 100 μL in each well of 48-well plate and incubate at 37°C to allow Collagen to gel.
13. Resuspend cells in Collagen-I at concentrations of either 1 mg/mL or 3 mg/mL depending on the experiment (see Note 15). Similarly allow to gel by incubation at 37°C.
14. Add the branching medium (see Subheading 2.4, items 7 and 8) on top of the cells.

3.5.5. Alveologenesis Assay

lrECM assay (basement membrane mix) is performed in 96-well plates, due to the need for large amounts of Matrigel™. Cells are pre-clustered as per the Collagen-I assay on polyHEMA (see Subheading 3.5.3) or on top of Agarose coated plates (see Subheading 3.5.4, steps 1–12), and then embedded in 50 μL of Matrigel™ in the wells (see Fig. 2c).

1. Overlay 50 μL of Matrigel™ on the bottom of the well in a 96-well plate.
2. Decant off the medium from clusters. It is desired that the cells are diluted in Matrigel™ to get 50 clusters per 50 μL of Matrigel™. Mix quickly and swirl but minimize the bubbles when mixing and put 50 μL in each well of the 96-well plate (that was pre-coated with Matrigel™).
3. Incubate the cells at 37°C for 30 min until the Matrigel™ solidifies.
4. Add 150–200 μL growth medium (see Subheading 2.1, items 10 and 11) to each 96-well plate.
5. Change growth medium the next day to alveologenesis medium (see Subheading 2.4, items 9 and 10) and then every other day, and probably after 2–3 days branching will start.

3.6. Biochemical Assays for Isolations of Cells from LrECM (Matrigel™)

Biochemical analysis, including RNA profiling and quantitation and protein analysis, is possible with cells in 3D cultures. However, it is better to isolate the 3D multicellular cell clusters from the surrounding matrix whether it is collagen or Matrigel™. This is usually done by solubilizing the gels using ECM-specific proteases (dispase or PBS–EDTA for Matrigel™). Gel-extracted cells are then subjected to the typical protocols used for RNA extraction and isolation or cell lysis for protein extraction.

3.6.1. Extraction Using Dispase

1. Remove the medium and add dispase solution (see Subheading 2.6, item 2) (0.75 mL per mL of Matrigel™).
2. Incubate 30 min at 37°C.
3. Remove proteases by washing three times in DMEM/F12 medium and once with RNA lysis buffer (see Subheading 2.6, item 8) or protein lysis buffer (see Subheading 2.6, item 9). Each wash is followed by centrifugation (1,154 ×*g*, 5 min at room temperature).
4. Cell pellets are further processed for RNA or protein analysis, or subcellular fractionation using standard protocols.

3.6.2. Extraction Using PBS–EDTA

1. Aspirate medium and rinse culture with 2 mL of cold 1× PBS, for a 35 mm dish, and leave the dish on ice.
2. Add 2–3 mL PBS–EDTA (see Subheading 2.6, item 6) to the culture and use a polyethylene cell lifter to gently scrape the cells and Matrigel™ off the bottom of the dish. Transfer the

cell/Matrigel™ mixture with a serological pipet to a 15 mL conical tube.

3. Repeat 2–3 times and use about 10 mL total PBS–EDTA (depends on amount of Matrigel™ in culture).
4. Shake the tubes gently in cold room or on ice for 30 min to 1 h to dissolve the Matrigel™ (also dependent on amount of Matrigel™) (see Note 15).
5. Centrifuge 5 min at 1,154 ×*g* at room temperature.
6. Transfer the cell pellet to a 1.5 mL Eppendorf tube on ice (see Note 16).
7. Wash with PBS and spin at full speed for 5 min and get rid of supernatant, and keep the cell pellet (see Note 17). Follow protocols for RNA or protein lysis in Subheading 3.6.2.

3.6.3. Cell Lysis for RNA or Protein Extraction

1. Add RNA lysis buffer (see Subheading 2.6, item 8) or RIPA buffer (see Subheading 2.6, item 9) (100–150 μL according to pellet size) and keep in mind that a concentrated protein solution is desired.
2. Sheer cells with an 18–21 gauge syringe (see Subheading 2.6, item 11) or sonicate with sonic dismembrator (see Subheading 2.6, item 12) on ice about 20 s at power setting 3 or 4.
3. Microcentrifuge for 10 min at full speed (17,000×*g*) at 4°C.
4. Transfer supernatant to a new Eppendorf for immediate use or store at −80°C for later use.

3.7. Biochemical Assays for Isolations of Cells from Collagen Gels

To extract RNA or protein from cells and organoids embedded in collagen gels, detach the collagen gel from the culture well and add RNA lysis buffer (see Subheading 2.6, item 8) or Ripa buffer for protein isolation (see Subheading 2.6, item 9).

For collagen gels in 96-well plates:

1. Add 300 μL of the desired buffer (either RNA or protein).
2. Shake for 6 h at room temperature in a microcentrifuge tube shaker (see Subheading 2.6, item 10).
3. Microcentrifuge for 10 min at full speed (17,000×*g*) at room temperature: collagen will pellet and the RNA or protein will be in the supernatant.
4. Discard pellet and keep the supernatant with the RNA or the protein either for immediate use or store at −80°C for later use.

3.8. High-Resolution Imaging

Imaging is essential for cell biological studies, and thus, it was important to find the best means to image cells in 3D. Confocal microscopy and live imaging are very important tools for studying cellular behavior in 3D.

1. Cells grown on tissue culture plastic are fixed with 4% paraformaldehyde (see Subheading 2.7, item 2) and permeabilized with 0.5% Triton™ X-100 (see Subheading 2.7, item 1).
2. Cells in lrECM gel are either smeared on slides, dried briefly, and fixed with 4% paraformaldehyde and permeabilized with 0.5% Triton™ X-100, or smeared after extraction using PBS–EDTA or dispase solutions (see Subheading 2.6, items 2 and 6). Choice of smearing depends on the strength of the antibody and background considerations with Matrigel™.
3. Cell clusters on polyHEMA-coated plates are processed for immunofluorescence analysis as follows: cells are pelleted by centrifugation at 500×*g* for 2 min and resuspended in PBS. The clusters are plated on slides, dried briefly, and then fixed with 4% paraformaldehyde.
4. Stained samples are imaged using a Spot RT camera attached to an upright epifluorescence microscope or viewed and acquired with a digital axiocam camera attached to an upright Zeiss LSM 710 Meta confocal microscope.
5. Time-lapse movies are collected using a Zeiss LSM 710 Meta confocal microscope. Images of dimensions, 512×512 pixels (lateral dimensions) with a maximum axial displacement of 75 μm (axial step size, 2 μm) are acquired using a 0.8 NA 20× air objective at digital zoom 0.6 corresponding to an area of 701×701 μm^2 at desired rate. Samples were maintained at 37°C and 5% CO_2. Immunofluorescence images are exported using Zen 2009 (Zeiss) software.

4. Notes

1. Fungizone (Amphotericin B) 100× can be used in primary organoid preparations to reduce the risk of contamination with mold but is not necessary. It is suggested because primary preparations are messy and are easily contaminated, especially if you dissect in the open air. If you do not want to use Fungizone in the medium, the mouse should originally be dissected in the hood and dipped in 70% ethanol including the tail before dissection. In addition, the tools should be sterilized before the preparation. It is necessary to avoid mouse hair and/or any other contamination in the preparation.
2. The collagenase solution must be made fresh right before the extraction procedure as it will inactivate at room temperature if kept for a long time.

3. The size and yield of epithelial structures/organoids from the mammary glands is affected by the strain of the mouse used for the preparation.
4. When performing an organoid/epithelial extraction from the mammary gland it is essential to pre-coat all pipettes and conical (Falcon™ or other) tubes with 5% BSA-PBS to prevent epithelial structures from adhering to the surface of the tubes or pipettes which would decrease the final yield.
5. After the first round of centrifugation, most organoids will settle in the bottom of the conical tube, but some will remain trapped in the fatty layer floating on top of the supernatant. To extract the organoids from this layer, gently remove about 10 mL of medium from the top layer of the supernatant (with the fatty layer) and transfer to another conical tube, then, add 5 mL of fresh, serum-free medium and flush up and down before spinning.
6. If you want a fibroblast culture, they are in the supernatant at this step, thus, you can put them into a dish and culture them.
7. You must clean up the epithelial organoids by repeating the pulse 4–10 times. Each time, monitor the composition of the pellet by examining 10 μL of the resuspended pellet under the microscope. Look for mainly organoids and a few single cells. Single cells are mainly the fibroblasts, so you want to get rid of them. Some epithelial cells are in the single cell form, so you do not have to completely remove all single cells. Ten pulses are generally more than sufficient to get a clean epithelial culture.
8. If primary cells are to be plated as a monolayer on tissue culture plastic, then it is necessary to plate the cells at a high density (20,000 cells/cm^2). Low-density cultures do not do well or maintain proper culture morphology. It is possible to coat the plate before plating with a thin layer of Collagen-I. Our lab uses Purecol® which is diluted as 1 mL in 44 mL of PBS. We add 3 mL of this mixture for every 25-cm^2 surface area of culture dish. It must gel over night in an airtight case at 4°C. Before use, it is necessary to aspirate off the excess collagen.
9. All protocols can be performed with primary organoids with slight modifications that relate to medium composition. Base medium for primary organoids is DMEM/F12 with ITS-PS with 5% FBS (see Subheading 2.1, item 10) while base medium for EpH4 is DMEM/F12 with GI (gentamycin-insulin) with 2% FBS (see Subheading 2.1, item 11).
10. Primary cells are more difficult to handle than cell lines, mostly due to the inherent variability across different isolations and possible variations in the amount of epithelial cells in each extraction. Thus, it is essential to perform the extraction techniques as rigorously as possible and it is paramount to seed

primary cells/organoids at similar numbers when starting the experiment.

11. Due to variability inherent to biological materials, Matrigel™ lots need to be tested in order to ensure reproducibility between experiments using different lots of Matrigel™. A typical test procedure consists of culturing EpH4 cells using new as well as currently used Matrigel™ lots. After 2 days in culture, cells are extracted from Matrigel™ and their ability to produce milk proteins such as β-casein is assessed by RT-PCR or western analysis. Assessment of morphological characteristics of cell clusters can also be used to infer if the Matrigel™ used is conducive to optimal differentiation or if the batch should be discarded.

12. LrECM available as Matrigel™ is difficult to handle at room temperature as it gels when removed from 4°C. As it gels, it might create bubbles at the bottom of the culture dish. To avoid that, make sure to keep Matrigel™ on ice inside the cell culture hood.

13. When using primary organoids for morphogenesis assay, it is crucial that cells be isolated from young animals (less than 14 weeks) to be able to recapitulate the phenotype in culture.

14. Counting cells: After centrifugation, dilute the cellular pellet in 1 mL of medium. Take 10 μL of the 1 mL cell suspension and add to an Eppendorf tube with 990 μL of medium such that your cell suspension is diluted at 1:100. Each square of the hemocytometer has a total volume of 0.1 mm^3 or 10^{-4} cm^3. 1 cm^3=1 mL. So to calculate the number of cells per milliliter of your suspension:

    Cells/mL = the average count per square (count 10 squares) × the dilution factor × 10^4

    Example: If the average count per square is 45 cells × 100 × 10^4 = 45,000,000 or 45 × 10^6 cells/mL.

15. Using 3 mg/mL versus 1 mg/mL of Collagen-I: Type I collagen is used typically at a concentration of 3 mg/mL. However, it has been observed that in some functional assays, a lower concentration of type-I collagen can be used to induce specific phenotypes. The optimal concentration of type I collagen is to be determined by the researcher when designing the experiment.

16. If Matrigel™ is not all dissolved, add additional PBS–EDTA, pipet gently, and invert the tube a few times before centrifuging (does not require prolonged shaking this time). If there is still a problem in dissolving the Matrigel™, check your EDTA concentration.

17. Cell pellet in 15 mL conical tube can also be lysed directly and then lysate can be transferred to Eppendorf.

Acknowledgments

We thank Dr. Joni Mott for critical reading of the manuscript and Dr. Hidetoshi Mori for providing images of cells in Fig. 2b, c. The work from MJB's laboratory is supported by grants from the US Department of Energy, Office of Biological and Environmental Research, a Distinguished Fellow Award to MJB and Low Dose Radiation Program (contract no. DE-AC02-05CH1123); by National Cancer Institute (awards R37CA064786, U54CA126552, R01CA057621, U54CA112970, U54CA143836, U01CA143233); and by US Department of Defense (W81XWH0810736). RM is supported by US Department of Defense, BCRP fellowship (W81XWH0810481).

References

1. Babinet C (2000) Transgenic mice: an irreplaceable tool for the study of mammalian development and biology. J Am Soc Nephrol 11(Suppl 16):S88–S94
2. Proia DA, Kuperwasser C (2006) Reconstruction of human mammary tissues in a mouse model. Nat Protoc 1:206–214
3. Ronnov-Jessen L, Petersen OW, Bissell MJ (1996) Cellular changes involved in conversion of normal to malignant breast: importance of the stromal reaction. Physiol Rev 76:69–125
4. Schmeichel KL, Weaver VM, Bissell MJ (1998) Structural cues from the tissue microenvironment are essential determinants of the human mammary epithelial cell phenotype. J Mammary Gland Biol Neoplasia 3:201–213
5. Richert MM, Schwertfeger KL, Ryder JW, Anderson SM (2000) An atlas of mouse mammary gland development. J Mammary Gland Biol Neoplasia 5:227–241
6. Bolander FF Jr (1990) Differential characteristics of the thoracic and abdominal mammary glands from mice. Exp Cell Res 189:142–144
7. Fendrick JL, Raafat AM, Haslam SZ (1998) Mammary gland growth and development from the postnatal period to postmenopause: ovarian steroid receptor ontogeny and regulation in the mouse. J Mammary Gland Biol Neoplasia 3:7–22
8. Wiseman BS, Sternlicht MD, Lund LR, Alexander CM, Mott J, Bissell MJ, Soloway P, Itohara S, Werb Z (2003) Site-specific inductive and inhibitory activities of MMP-2 and MMP-3 orchestrate mammary gland branching morphogenesis. J Cell Biol 162:1123–1133
9. Silberstein GB (2001) Postnatal mammary gland morphogenesis. Microsc Res Tech 52:155–162
10. Daniel CW, Robinson S, Silberstein GB (1996) The role of TGF-beta in patterning and growth of the mammary ductal tree. J Mammary Gland Biol Neoplasia 1:331–341
11. Nandi S (1958) Endocrine control of mammarygland development and function in the C3H/He Crgl mouse. J Natl Cancer Inst 21:1039–1063
12. Ramanathan P, Martin I, Thomson P, Taylor R, Moran C, Williamson P (2007) Genomewide analysis of secretory activation in mouse models. J Mammary Gland Biol Neoplasia 12: 305–314
13. Barcellos-Hoff MH, Aggeler J, Ram TG, Bissell MJ (1989) Functional differentiation and alveolar morphogenesis of primary mammary cultures on reconstituted basement membrane. Development 105:223–235
14. Lee EY, Lee WH, Kaetzel CS, Parry G, Bissell MJ (1985) Interaction of mouse mammary epithelial cells with collagen substrata: regulation of casein gene expression and secretion. Proc Natl Acad Sci U S A 82:1419–1423
15. Howlett AR, Bissell MJ (1993) The influence of tissue microenvironment (stroma and extracellular matrix) on the development and function of mammary epithelium. Epithelial Cell Biol 2:79–89
16. Rosen JM, Wyszomierski SL, Hadsell D (1999) Regulation of milk protein gene expression. Annu Rev Nutr 19:407–436
17. Li M, Liu X, Robinson G, Bar-Peled U, Wagner KU, Young WS, Hennighausen L, Furth PA (1997) Mammary-derived signals activate programmed cell death during the first stage of mammary gland involution. Proc Natl Acad Sci U S A 94:3425–3430

18. Furth PA, Bar-Peled U, Li M (1997) Apoptosis and mammary gland involution: reviewing the process. Apoptosis 2:19–24
19. Quarrie LH, Addey CV, Wilde CJ (1995) Apoptosis in lactating and involuting mouse mammary tissue demonstrated by nick-end DNA labelling. Cell Tissue Res 281:413–419
20. Li M, Hu J, Heermeier K, Hennighausen L, Furth PA (1996) Apoptosis and remodeling of mammary gland tissue during involution proceeds through p53-independent pathways. Cell Growth Differ 7:13–20
21. Daniel CW, Deome KB (1965) Growth of mouse mammary glands in vivo after monolayer culture. Science 149:634–636
22. Daniel CW, Young LJ, Medina D, DeOme KB (1971) The influence of mammogenic hormones on serially transplanted mouse mammary gland. Exp Gerontol 6:95–101
23. Emerman JT, Burwen SJ, Pitelka DR (1979) Substrate properties influencing ultrastructural differentiation of mammary epithelial cells in culture. Tissue Cell 11:109–119
24. Emerman JT, Pitelka DR (1977) Maintenance and induction of morphological differentiation in dissociated mammary epithelium on floating collagen membranes. In Vitro 13:316–328
25. Michalopoulos G, Pitot HC (1975) Primary culture of parenchymal liver cells on collagen membranes. Morphological and biochemical observations. Exp Cell Res 94:70–78
26. Emerman JT, Enami J, Pitelka DR, Nandi S (1977) Hormonal effects on intracellular and secreted casein in cultures of mouse mammary epithelial cells on floating collagen membranes. Proc Natl Acad Sci U S A 74:4466–4470
27. Lee EY, Parry G, Bissell MJ (1984) Modulation of secreted proteins of mouse mammary epithelial cells by the collagenous substrata. J Cell Biol 98:146–155
28. Aggeler J, Ward J, Blackie LM, Barcellos-Hoff MH, Streuli CH, Bissell MJ (1991) Cytodifferentiation of mouse mammary epithelial cells cultured on a reconstituted basement membrane reveals striking similarities to development in vivo. J Cell Sci 99(Pt 2): 407–417
29. Chen LH, Bissell MJ (1989) A novel regulatory mechanism for whey acidic protein gene expression. Cell Regul 1:45–54
30. Streuli CH, Bailey N, Bissell MJ (1991) Control of mammary epithelial differentiation: basement membrane induces tissue-specific gene expression in the absence of cell-cell interaction and morphological polarity. J Cell Biol 115:1383–1395
31. Xu R, Nelson CM, Muschler JL, Veiseh M, Vonderhaar BK, Bissell MJ (2009) Sustained activation of STAT5 is essential for chromatin remodeling and maintenance of mammary-specific function. J Cell Biol 184:57–66
32. Gudjonsson T, Ronnov-Jessen L, Villadsen R, Rank F, Bissell MJ, Petersen OW (2002) Normal and tumor-derived myoepithelial cells differ in their ability to interact with luminal breast epithelial cells for polarity and basement membrane deposition. J Cell Sci 115:39–50
33. Guo W, Giancotti FG (2004) Integrin signalling during tumour progression. Nat Rev Mol Cell Biol 5:816–826
34. Kalluri R (2003) Basement membranes: structure, assembly and role in tumour angiogenesis. Nat Rev Cancer 3:422–433
35. Mott JD, Werb Z (2004) Regulation of matrix biology by matrix metalloproteinases. Curr Opin Cell Biol 16:558–564
36. Streuli CH, Schmidhauser C, Bailey N, Yurchenco P, Skubitz AP, Roskelley C, Bissell MJ (1995) Laminin mediates tissue-specific gene expression in mammary epithelia. J Cell Biol 129:591–603
37. Li ML, Aggeler J, Farson DA, Hatier C, Hassell J, Bissell MJ (1987) Influence of a reconstituted basement membrane and its components on casein gene expression and secretion in mouse mammary epithelial cells. Proc Natl Acad Sci U S A 84:136–140
38. Wicha MS, Lowrie G, Kohn E, Bagavandoss P, Mahn T (1982) Extracellular matrix promotes mammary epithelial growth and differentiation in vitro. Proc Natl Acad Sci U S A 79:3213–3217
39. Medina D, Li ML, Oborn CJ, Bissell MJ (1987) Casein gene expression in mouse mammary epithelial cell lines: dependence upon extracellular matrix and cell type. Exp Cell Res 172:192–203
40. Daniel CW, Berger JJ, Strickland P, Garcia R (1984) Similar growth pattern of mouse mammary epithelium cultivated in collagen matrix in vivo and in vitro. Dev Biol 104:57–64
41. Montesano R, Soriano JV, Fialka I, Orci L (1998) Isolation of EpH4 mammary epithelial cell subpopulations which differ in their morphogenetic properties. In Vitro Cell Dev Biol Anim 34:468–477
42. Niemann C, Brinkmann V, Spitzer E, Hartmann G, Sachs M, Naundorf H, Birchmeier W (1998) Reconstitution of mammary gland development in vitro: requirement of c-met and c-erbB2 signaling for branching and alveolar morphogenesis. J Cell Biol 143:533–545
43. Ruoslahti E, Pierschbacher MD (1986) Arg-Gly-Asp: a versatile cell recognition signal. Cell 44: 517–518
44. Hynes RO (1987) Integrins: a family of cell surface receptors. Cell 48:549–554

45. Alcaraz J, Xu R, Mori H, Nelson CM, Mroue R, Spencer VA, Brownfield D, Radisky DC, Bustamante C, Bissell MJ (2008) Laminin and biomimetic extracellular elasticity enhance functional differentiation in mammary epithelia. EMBO J 27:2829–2838
46. Talhouk RS, Elble RC, Bassam R, Daher M, Sfeir A, Mosleh LA, El-Khoury H, Hamoui S, Pauli BU, El-Sabban ME (2005) Developmental expression patterns and regulation of connexins in the mouse mammary gland: expression of connexin30 in lactogenesis. Cell Tissue Res 319:49–59
47. El-Sabban ME, Sfeir AJ, Daher MH, Kalaany NY, Bassam RA, Talhouk RS (2003) ECM-induced gap junctional communication enhances mammary epithelial cell differentiation. J Cell Sci 116:3531–3541
48. Nguyen DA, Neville MC (1998) Tight junction regulation in the mammary gland. J Mammary Gland Biol Neoplasia 3:233–246
49. Talhouk RS, Mroue R, Mokalled M, Abi-Mosleh L, Nehme R, Ismail A, Khalil A, Zaatari M, El-Sabban ME (2008) Heterocellular interaction enhances recruitment of alpha and beta-catenins and ZO-2 into functional gap-junction complexes and induces gap junction-dependant differentiation of mammary epithelial cells. Exp Cell Res 314:3275–3291
50. Wang X, Sun L, Maffini MV, Soto A, Sonnenschein C, Kaplan DL (2010) A complex 3D human tissue culture system based on mammary stromal cells and silk scaffolds for modeling breast morphogenesis and function. Biomaterials 31:3920–3929
51. Ingthorsson S, Sigurdsson V, Fridriksdottir AJ, Jonasson JG, Kjartansson J, Magnusson MK, Gudjonsson T (2010) Endothelial cells stimulate growth of normal and cancerous breast epithelial cells in 3D culture. BMC Res Notes 3:184
52. Chen A, Cuevas I, Kenny PA, Miyake H, Mace K, Ghajar C, Boudreau A, Bissell M, Boudreau N (2009) Endothelial cell migration and vascular endothelial growth factor expression are the result of loss of breast tissue polarity. Cancer Res 69:6721–6729
53. Petersen OW, Nielsen HL, Gudjonsson T, Villadsen R, Rank F, Niebuhr E, Bissell MJ, Ronnov-Jessen L (2003) Epithelial to mesenchymal transition in human breast cancer can provide a nonmalignant stroma. Am J Pathol 162:391–402
54. Ip MM, Darcy KM (1996) Three-dimensional mammary primary culture model systems. J Mammary Gland Biol Neoplasia 1:91–110
55. Hamamoto S, Imagawa W, Yang J, Nandi S (1988) Morphogenesis of mouse mammary epithelial cells growing within collagen gels: ultrastructural and immunocytochemical characterization. Cell Differ 22:191–201
56. Bissell MJ (1981) The differentiated state of normal and malignant cells or how to define a "normal" cell in culture. Int Rev Cytol 70: 27–100
57. Petersen OW, Ronnov-Jessen L, Howlett AR, Bissell MJ (1992) Interaction with basement membrane serves to rapidly distinguish growth and differentiation pattern of normal and malignant human breast epithelial cells. Proc Natl Acad Sci U S A 89:9064–9068
58. Lin CQ, Bissell MJ (1993) Multi-faceted regulation of cell differentiation by extracellular matrix. FASEB J 7:737–743
59. Streuli CH, Bissell MJ (1990) Expression of extracellular matrix components is regulated by substratum. J Cell Biol 110:1405–1415
60. Roskelley CD, Desprez PY, Bissell MJ (1994) Extracellular matrix-dependent tissue-specific gene expression in mammary epithelial cells requires both physical and biochemical signal transduction. Proc Natl Acad Sci U S A 91:12378–12382
61. Faraldo MM, Deugnier MA, Lukashev M, Thiery JP, Glukhova MA (1998) Perturbation of beta1-integrin function alters the development of murine mammary gland. EMBO J 17:2139–2147
62. Muschler J, Lochter A, Roskelley CD, Yurchenco P, Bissell MJ (1999) Division of labor among the alpha6beta4 integrin, beta1 integrins, and an E3 laminin receptor to signal morphogenesis and beta-casein expression in mammary epithelial cells. Mol Biol Cell 10:2817–2828
63. Weir ML, Oppizzi ML, Henry MD, Onishi A, Campbell KP, Bissell MJ, Muschler JL (2006) Dystroglycan loss disrupts polarity and beta-casein induction in mammary epithelial cells by perturbing laminin anchoring. J Cell Sci 119:4047–4058
64. Lin CQ, Dempsey PJ, Coffey RJ, Bissell MJ (1995) Extracellular matrix regulates whey acidic protein gene expression by suppression of TGF-alpha in mouse mammary epithelial cells: studies in culture and in transgenic mice. J Cell Biol 129:1115–1126
65. Zhang Y, Sif S, DeWille J (2007) The mouse C/EBPdelta gene promoter is regulated by STAT3 and Sp1 transcriptional activators, chromatin remodeling and c-Myc repression. J Cell Biochem 102:1256–1270
66. Spencer VA, Xu R, Bissell MJ (2010) Gene expression in the third dimension: the ECM-nucleus connection. J Mammary Gland Biol Neoplasia 15:65–71

67. Lelievre S, Weaver VM, Bissell MJ (1996) Extracellular matrix signaling from the cellular membrane skeleton to the nuclear skeleton: a model of gene regulation. Recent Prog Horm Res 51:417–432
68. Xu R, Spencer VA, Bissell MJ (2007) Extracellular matrix-regulated gene expression requires cooperation of SWI/SNF and transcription factors. J Biol Chem 282:14992–14999
69. Schmidhauser C, Bissell MJ, Myers CA, Casperson GF (1990) Extracellular matrix and hormones transcriptionally regulate bovine beta-casein 5' sequences in stably transfected mouse mammary cells. Proc Natl Acad Sci U S A 87:9118–9122
70. Bissell MJ, Hall HG, Parry G (1982) How does the extracellular matrix direct gene expression? J Theor Biol 99:31–68
71. Talhouk RS, Chin JR, Unemori EN, Werb Z, Bissell MJ (1991) Proteinases of the mammary gland: developmental regulation in vivo and vectorial secretion in culture. Development 112:439–449
72. Talhouk RS, Bissell MJ, Werb Z (1992) Coordinated expression of extracellular matrix-degrading proteinases and their inhibitors regulates mammary epithelial function during involution. J Cell Biol 118:1271–1282
73. Sympson CJ, Talhouk RS, Alexander CM, Chin JR, Clift SM, Bissell MJ, Werb Z (1994) Targeted expression of stromelysin-1 in mammary gland provides evidence for a role of proteinases in branching morphogenesis and the requirement for an intact basement membrane for tissue-specific gene expression. J Cell Biol 125:681–693
74. Simian M, Hirai Y, Navre M, Werb Z, Lochter A, Bissell MJ (2001) The interplay of matrix metalloproteinases, morphogens and growth factors is necessary for branching of mammary epithelial cells. Development 128: 3117–3131
75. Nelson CM, Vanduijn MM, Inman JL, Fletcher DA, Bissell MJ (2006) Tissue geometry determines sites of mammary branching morphogenesis in organotypic cultures. Science 314:298–300
76. Fata JE, Mori H, Ewald AJ, Zhang H, Yao E, Werb Z, Bissell MJ (2007) The MAPK(ERK-1,2) pathway integrates distinct and antagonistic signals from TGFalpha and FGF7 in morphogenesis of mouse mammary epithelium. Dev Biol 306:193–207
77. Stoker AW, Streuli CH, Martins-Green M, Bissell MJ (1990) Designer microenvironments for the analysis of cell and tissue function. Curr Opin Cell Biol 2:864–874
78. Nelson CM, Bissell MJ (2006) Of extracellular matrix, scaffolds, and signaling: tissue architecture regulates development, homeostasis, and cancer. Annu Rev Cell Dev Biol 22:287–309
79. Bissell MJ, Barcellos-Hoff MH (1987) The influence of extracellular matrix on gene expression: is structure the message? J Cell Sci Suppl 8:327–343
80. Reichmann E, Ball R, Groner B, Friis RR (1989) New mammary epithelial and fibroblastic cell clones in coculture form structures competent to differentiate functionally. J Cell Biol 108:1127–1138
81. Reichmann E, Schwarz H, Deiner EM, Leitner I, Eilers M, Berger J, Busslinger M, Beug H (1992) Activation of an inducible c-FosER fusion protein causes loss of epithelial polarity and triggers epithelial-fibroblastoid cell conversion. Cell 71:1103–1116

Chapter 15

Isolation, Characterization, and Reaggregate Culture of Thymic Epithelial Cells

Natalie Seach, Maree Hammett, and Ann Chidgey

Abstract

The thymus organ is composed of a three-dimensional (3D) network of adjoining epithelium and stromal cells. Bone marrow-derived T cell precursors, upon entering the thymus, interact with and migrate through this cellular network as they differentiate and mature. An essential component of the stroma is the thymic epithelial cells (TEC), which play a vital role in T cell development and induction of self-tolerance for adaptive immunity. TEC can be isolated from the embryonic and adult thymus by a series of gentle enzymatic digestions and characterized into discrete subpopulations based on their expression of surface markers by flow cytometry. Enrichment of adult TEC can be achieved by depletion of hematopoietic cells, allowing sufficient numbers to be purified for subsequent functional and molecular analysis. Although monolayer cultures have been used to study TEC phenotype and T cell interaction, methods that mimic the 3D thymic microenvironment, such as fetal and reaggregate thymic organ cultures, are more accurate for the analysis of TEC function and support more complete T cell development. Herein, we describe methods for the efficient isolation and enrichment of TEC for downstream analyses as well as the reaggregation of embryonic progenitor epithelium to form a functional thymus graft under the kidney capsule.

Key words: Thymus, Epithelial cells, Reaggregate culture

1. Introduction

The complex derivation of T cell lineages from hematopoietic stem cell (HSC) precursors requires sequential stages of development within broad microenvironments of the thymus gland. Herein, defined processes of T cell maturation and selection are regulated, in part, via cell–cell interactions with resident cells of the thymus, collectively deemed "thymic stroma" (1). The importance of thymic stromal cells (TSC), in particular the thymic epithelium, is well documented in critical aspects of T cell development and tolerance (2–5). Understanding the role and function of these stromal subtypes is fundamental not only to our

Scott H. Randell and M. Leslie Fulcher (eds.), *Epithelial Cell Culture Protocols: Second Edition*, Methods in Molecular Biology, vol. 945, DOI 10.1007/978-1-62703-125-7_15,

understanding of T cell biology, but to gain insight into states of T cell dysfunction and disease.

TEC are a relatively rare thymic population comprising less than 1% of total thymic cellularity. Unlike other epithelial organs, TEC exist in a unique 3D meshwork-like structure to enable maximal thymocyte migration and interaction (6). To effectively analyze TSC populations, a method of gentle enzymatic digestions is required to breakdown the extracellular matrix and free stromal cells to form a single cell suspension. Flow cytometric analysis of TSC reveals distinct populations of both epithelial and non-epithelial nature (7, 8). Further, molecular and functional analysis of these stromal cell populations can be performed following depletion of $CD45^+$ hematopoietic cells and separation via fluorescence activated cell sorting (FACS) (9).

Although two-dimensional TEC culture has provided some insight into the morphology and nature of these cells, 3D fetal thymus organ culture (FTOC) systems, which spatially mimic the thymus microenvironment, are currently the only in vitro system able to support the full program of T cell development (10). FTOC and variations of this method, provide a powerful experimental system in which the study and/or manipulation of T cell precursor entry, development and lineage commitment can be assessed in vitro (11). The development of 3D reaggregate thymic organ culture (RTOC) techniques by the Owen/Jenkinson/Anderson group, provided further advances in the field by allowing the reaggregation of defined TSC populations and hematopoietic elements in vitro, in order to assess their specific roles in thymopoiesis (12, 13). Importantly, this technique has played a fundamental role in demonstrating the functional potential of putative thymic progenitor epithelial cell populations. The monoclonal antibody, MTS24 (recently demonstrated to bind the Plet-1 antigen (14)), was developed as part of a panel of antibodies raised against mouse thymic stromal determinants (15) and has been shown to label epithelial cells with increased progenitor potential in both the skin and embryonic thymus (16, 17). Reaggregates of $MTS24^+$ embryonic thymic epithelium were found to be capable of differentiating into multiple TEC lineages, attracting bone marrow derived HSC precursors and promoting full T cell development when engrafted under the kidney capsule of *nude* mice (16, 18). Although, subsequent studies have also demonstrated progenitor potential of the MTS24- embryonic epithelial compartment when reaggregated in larger numbers (19).

This chapter first describes the dissection and isolation of both embryonic and adult thymus tissue and their subsequent enzymatic digestion to form a single cell suspension. At this stage, immunostaining of individual adult thymus digests and analysis by flow cytometry reveals discreet stromal populations, including epithelial cells of both medullary and cortical phenotype. A cell depletion step, using CD45 antibody-bound magnetic microbeads, removes the major hematopoietic cell component of pooled adult thymus digests (thymocytes) and enriches for $CD45^-$ TSC populations to expediate their isolation through cell sorting. We provide examples of purified TSC subsets and demonstrate the expected cellular return as well as the quantity and quality of RNA for downstream molecular analysis. Finally, we describe the reaggregation of $MTS24^+$ progenitor epithelium isolated from gestation day 15 thymus and subsequent engraftment under the kidney capsule to form an ectopic thymus graft.

2. Materials

2.1. Dissection of Adult Thymus

1. Adult mice.
2. Dissection scissors and forceps.
3. 70% Ethanol and dissecting mat.
4. 15 mL centrifuge tubes containing 1–2 mL of RPMI-1640 medium to collect thymus tissue.

2.2. Dissection and Isolation of Fetal Thymus

1. Time-mated pregnant mice (see Note 1).
2. 70% Ethanol and dissecting mat.
3. Dissecting forceps and scissors. Nonsterile set for dissection of pregnant female and one sterile set for removing uteri.
4. 50 mL centrifuge tubes containing 20 mL of sterile RPMI-1640 to place embryonic sacs in.
5. Petri dishes: Sizes; 100, 60 and 35 mm for dissection of embryos and 1.5 mL eppendorf tubes for collection of thymic lobes.
6. Sterile carbon steel surgical blades and scalpel holder.
7. Autoclaved instrument dissection kit containing extra fine forceps for dissection of fetal thymus (e.g., Dumont number 5 straight forceps, Fine Science Tools®).
8. Dissecting microscope with zoom; wiped down with ethanol and placed in clean laminar flow hood.

2.3. Enzymatic Digestion of Thymus Tissue

2.3.1. Digestion of Adult Tissue

Thymus tissue from individual mice can be digested and analyzed separately when numbers are required for statistical interpretation. However, pooled thymic tissue digests will be required to obtain sufficient numbers of TSC for subsequent molecular and functional analyses.

1. 60 mm Petri dishes and forceps for cleaning thymus tissue.
2. 10 and 15 mL tubes for thymus digestion and collection of supernatant fractions.

3. A 100 mL schott bottle, magnetic flea and magnetic stirrer are required for the initial release of thymocytes when greater than five lobes are pooled for digestion. Performed in the cold room.
4. Glass pipettes of decreasing bore size. These are cut to size using a glass-cutter. A pipette bulb is attached to the end for pipetting.
5. 200 and 1,000 mL pipette tips.
6. RPMI-1640. Keep cold on ice.
7. FACS (fluorescence activated cell sorting)-buffer: 1× phosphate buffered saline (PBS) containing 0.1% bovine serum albumin (BSA) and 0.5 mM ethylenediamine tetraacetic acid (EDTA). Store at 4°C.
8. Enzymes: We have previously used a combination of DNAse I (Roche) and crude Collagenase/Dispase (Roche) for the digestion of thymus tissue (9). Recently we have found that the highly purified Liberase™ enzymes (Roche) dramatically improved TSC yield and viability (Seach et al., manuscript in press, J Immunol Methods). Liberase products containing medium to high levels of thermolysin (Liberase TM and TH respectively, see Note 2) are used at a concentration of between 0.5 and 1 standard Wünsch unit (U) per milliliter in RPMI-1640. DNAse I is added to a final concentration of 0.1% (w/v). Make up enzymes fresh on the day of digestion (see Note 3).
9. 100 μm filter mesh.
10. Red blood cell lysis buffer: 8.3 g/L Ammonium Chloride in 500 mL 0.01 M Tris–HCl, final pH = 7.5 ± 0.2.

2.3.2. Embryonic Thymic Digest

1. Digestion of embryonic lobes is performed in a 1.5 mL eppendorf.
2. 100 μm filter mesh.
3. 200 and 1,000 mL pipette tips.
4. FACS-buffer (see Subheading 2.3.1, item 7).
5. 1–2 mL of Liberase–DNAse enzyme mix (*see* Subheading 2.3.1, item 8).

2.4. Immuno-Staining of Thymic Stromal Cells for Flow Cytometry

1. Single cell thymus suspension prepared by enzymatic digestion (see Subheading 2.3).
2. FACS-buffer (see Subheading 2.3.1, item 7).
3. The following rat anti-mouse antibodies (Ab) were purchased from BD Biosciences unless otherwise stated: CD45 (clone 104), MHCII I-A/I-E (clone M5/114.15.2), Ly51 (clone 6C3), EpCAM (clone G8.8; eBioscience), MTS24 (Richard Boyd Laboratory), Ulex Europaeus Agglutinin 1 (UEA1;

Vector Laboratories). For detection of MTS24 we used anti-rat Ig Alexa Fluor 647 (Molecular Probes).

4. Normal rat serum: diluted 1/5 in PBS.
5. Propidium iodide (PI).
6. 96-well round-bottom plates for cell-staining and appropriate sized test tubes for running sample on flow-cytometer.
7. Cell samples in this chapter were run on a BD FACSCanto™ II flow cytometer (BD Biosciences) using up to eight separate color channels and data was analyzed using BD FACSDiva 6.0 software (BD Biosciences).

2.5. CD45 Bead Depletion for Sorting of Thymic Stromal Cell Subsets

Due to the high proportion of thymocytes within adult thymic tissue, hematopoietic depletion steps are required to enrich for TSC. In contrast, embryonic thymus does not require depletion due to enriched TSC to thymocyte ratio at this early age.

1. Single cell suspension of adult thymus prepared by enzymatic digestion (see Subheading 2.3.1).
2. 100 μm filter mesh.
3. CD45 MicroBeads and AutoMACS® Separator (Miltenyi Biotech).
4. FACS-buffer (see Subheading 2.3.1, item 7).
5. Rotating mixer in a 4°C cold room for incubation of samples.

2.6. Fluorescence Activated Cell Sorting

1. Single cell thymus suspension (see Subheading 2.3.1) depleted of $CD45^+$ cells (see Subheading 2.5) and stained with appropriate antibodies to detect TSC subsets (see Subheading 2.4).
2. 100 μm filter mesh.
3. 2 mL Eppendorfs for cell collection.
4. Collection medium: 30% fetal bovine serum (FBS) in RPMI-1640 and 100% FBS for coating of sides of collection Eppendorfs.
5. Appropriate sized samples tubes for the cell sorter.
6. All samples presented in this paper were sorted using a BD Influx (BD Biosciences).

2.7. RNA Isolation and cDNA Synthesis

1. Sorted TSC populations snap frozen in lysis buffer; stored at −80°C.
2. For samples greater than 500,000 cells, we used the RNEasy Mini kit (Qiagen). For samples less than 500,000 cells, we used the RNAqueous Micro kit (Ambion®).
3. RNAse-free DNAse set (Qiagen).
4. Nanodrop-1000 for assessment of RNA quantity and quality (Thermo Scientific).
5. Superscript™ III First-strand synthesis system for RT-PCR (see Note 4) (Life Technologies™).

6. Nuclease-free water.
7. Nuclease-free PBS.
8. Nuclease-free tips (Pathtech) and nuclease-free Eppendorf tubes.

2.8. Quantitative PCR

1. cDNA template prepared from sorted TSC populations (see Subheading 2.7).
2. Platinum® SYBR® Green qPCR SuperMix-UDG (12.5 mL) (Life Technologies™).
3. Nuclease-free water.
4. Nuclease-free tips and nuclease-free Eppendorf tubes.
5. Housekeeping gene (e.g., GAPDH) and pre-validated test primers (Qiagen).
6. Quantitative PCR machine: Rotor-Gene 3000 (Qiagen).
7. Strip Tubes and Caps, 0.1 mL for Rotor-Gene 3000 PCR machine (Qiagen).

2.9. Reaggregate Thymus Organ Culture

1. Purified E15 MTS24$^+$ TEC. Cell reaggregates of between 50 and 500×10^3 MTS24$^+$ E15 TEC are optimal for assessing graft formation under the kidney capsule (see Note 5).
2. 1.5 mL Eppendorf tubes.
3. 0.8-μm Isopore membrane filters.
4. 12-well plates.
5. Sterile Gelfoam® gelatin sponges (2×6 cm pieces, Pfizer).
6. RTOC medium: RPMI-1640 supplemented with 10% v/v FBS, 10 mM HEPES and 1 mM sodium pyruvate, 2 mM GlutaMAX™, 0.1 mM nonessential amino acids, 55 μM 2-Mercaptoethanol, 100 U/mL penicillin, and 100 μg/mL streptomycin.
7. Elongated glass Pasture pipettes. Place the end of a glass pipette over a medium-heat flame of a Bunsen burner. Move gently over flame until you can feel glass melting. Remove quickly from flame and pull apart to elongate and thin until the glass snaps cleanly (see Note 6).
8. Sealed 200 mL tip. Pass end of tip quickly through the yellow flame of a Bunsen burner to seal it. Hold tip vertically whilst it hardens.
9. Mouth-pipette set up: Flat mouthpiece (e.g., HPI Hospital Products) and conventional latex rubber laboratory tubing which will fit over the end of a glass pasture pipette (e.g., 1/8 in. inner diameter, 1/32 in. wall thickness, 3/16 in. outer diameter).

2.10. Kidney Capsule Grafting

1. 4–6 week old BALB/c Nude (CAnN.Cg-*Foxn1nu*/Crl) male mice only (the ovaries in female mice can be damaged when exposing the kidney).

2. Ketamine–Xylazine anesthetic. Dilute 1 mL ketamine (Ketalar 100 mg/mL) and 0.5 mL Xylazine (Ilium Xylazil 20 mg/mL), into 8.5 mL of sterile water. Dosage; intraperitoneal (IP) 0.2 mL/20 g of mouse body weight.
3. Eye gel (Lacrilube; Provet) to place on mouse cornea to prevent drying-out of the eye whilst under anesthetic.
4. Autoclaved instruments: Sharp scissors, two curved forceps, two fine straight forceps, needle holder.
5. Dissecting microscope for placing graft under kidney capsule.
6. Absorbable underlay mat.
7. 70% Ethanol spray bottle.
8. Sterile 15 cm cotton tips.
9. Betadine® antiseptic.
10. 4.0 Polysorb™ suture.
11. 18 and 30 g needles with 1 and 3 mL syringes for drug and fluid injection.
12. 9 mm autoclips for skin stapling with autoclip applier and autoclip remover (Braintree Scientific).
13. Antibiotic powder (Baytril 25 oral, Provet).
14. Sunflower seeds placed in cage with mice post surgery.
15. Heat pads for suturing and overnight recovery.
16. Carprofen (Rimadyl® sterile injectable solution, 50 mg/mL, Pfizer). Dilute 1:100 with sterile water. Dosage; subcutaneous (SC) injection of 0.2 mL diluted Carprofen/20 g of mouse body weight.
17. Antisedan® 5 mg/mL (Pfizer). Dilute 1:100 with sterile water. Dosage; IP 0.2 mL/20 g of body weight.
18. Fluids: compound sodium lactate (Hartmanns Solution). For surgery: SC injection of 0.2 mL/20 g of body weight. For therapy: SC or IP injection of 0.4 mL/20 g of mouse body weight.

2.11. Kidney-Thymus Graft Analysis

1. Tissue-Tek® O.C.T compound.
2. Liquid nitrogen and isopentane, mixed carefully to form a slurry.
3. Leica CM1850 Cryostat.
4. Glass slides and coverslips.
5. Fluorescent mounting medium (DakoCytomation).
6. Antibodies for detecting thymus graft: rabbit anti-bovine pan-cytokeratin (polyclonal; Dakocytomation), rat anti-mouse MTS10 (Richard Boyd Laboratory) and hamster anti-mouse CD3-FITC (145-2C11, BD Biosciences). Secondary antibodies:

anti-rabbit Ig-AlexaFluor647, anti-rat Ig-AlexaFluor568 (Life Technologies™).

7. Bio-Rad MRC 1024 confocal microscope for image capture and LaserSharp2000 software for analysis (Bio-Rad).

3. Methods

Unless otherwise stated, all mice used in this study are C57Bl/6 strain. For molecular analysis and in vitro RTOC assays, all techniques should be performed under aseptic conditions using sterile instruments and consumables.

3.1. Isolation of Adult Thymus

1. Cull mouse in a CO_2 chamber according to ethical and institutional guidelines.
2. Pin the animal on its back (exposing the ventral view) and spray liberally with ethanol.
3. Using forceps, pinch skin at the level of the pelvis and make an incision along the ventral midline from the groin to the chin. Make two more incisions from the bottom of midline cut, down each leg. Peel back skin and pin to expose peritoneal wall and rib cage.
4. Make an incision through the peritoneal cavity just below the diaphragm and cut down to groin region. Peel back peritoneal muscle. Push the liver down and out of the way to expose the diaphragm. Grasping the bottom of the sternum, cut horizontally across the diaphragm. Still holding the sternum, make an approximate 1.5 cm cut up both the right and left side of the rib cage (see Fig. 1a). Pull flap upwards to expose the thoracic cavity and organs.
5. The thymus is located above the heart (see Fig. 1b). Using one pair of curved serrated forceps, pull the pair of thymus lobes gently up towards you, freeing the connective tissue that holds the thymus in place. Place the second pair of forceps underneath both lobes and gently pull towards you to free thymus.
6. Place thymus lobes into a 10 mL tube containing RPMI-1640. Keep on ice.

3.2. Isolation of Fetal Thymus Lobes

1. Cull pregnant mice of desired gestational day (see Note 1) in a CO_2 chamber according to ethical and institutional guidelines.
2. Pin on back to expose ventral region and spray liberally with ethanol.
3. Pinch skin and make a small lateral incision across the abdomen of the mouse. Grab skin firmly above and below the incision

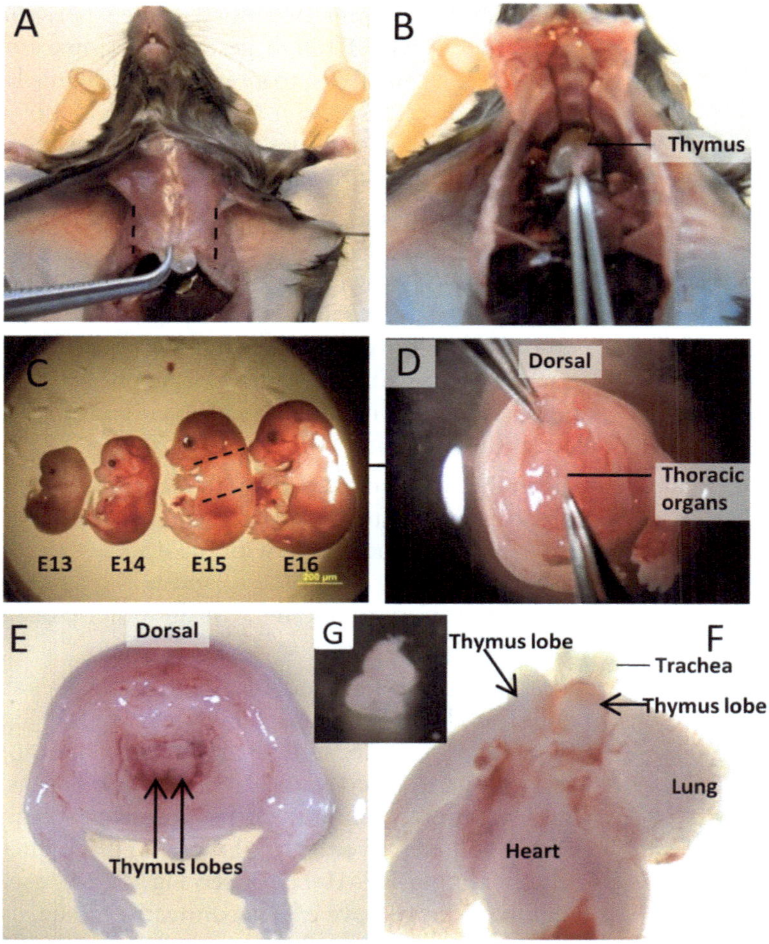

Fig. 1. Isolation of thymus lobes from adult and embryonic mice: (**a**) Thoracic region of an adult (6 week old) mouse demonstrating where the rib cage should be cut (*dashed lines*). (**b**) Thoracic cavity of adult mouse demonstrating thymus lobes. (**c**) Mouse embryos of defined ages; *dashed lines* indicate where embryo should be cut to isolate thoracic portion. Posterior view of the thoracic cavity of an embryonic day 15 (E15) embryo indicating the thoracic contents to be removed (**d**) and the thymic lobes which are left behind in the cavity (**e**). (**f**) Isolated thoracic organs removed from an E14 embryo with thymus lobes still attached. (**g**) Isolated E15 thymic lobes.

and pull apart to expose the abdomen. Change to another set of sterile instruments before continuing.

4. Taking care not to pierce the underlying embryos, cut though the peritoneal muscle to expose cavity.
5. Take out uteri and place in a 50 mL tube containing 20 mL RPMI-1640.
6. In a sterile hood, place embryos into a sterile 60 mm petri dish.
7. To separate embryo from the amniotic sac, gently grasp sac with forceps and make a small incision in the membrane. The embryo can be extracted by applying gentle pressure with the forceps. Place embryos into a new petri dish containing 10 mL RPMI-1640.

8. To dissect the thymus of embryos aged between embryonic day 13 to 18 (E13–E18) of gestation, we first isolate the thoracic section of the fetus, from which the thymic lobes can then be dissected (see Note 7). Other methods, however, employ equally effective techniques whereby thymus lobes are isolated directly from the chest cavity (20).
9. To isolate the thoracic portion, orientate embryos on their side in the Petri dish. Taking two sterile scalpels, cut through the neck region by drawing both scalpels across one another (see Fig. 1c). Make a similar cut through the midsection of the embryo, just above the liver (the liver portion should remain attached to the bottom section of the embryo). Discard head and bottom sections and place thoracic section into a new small Petri dish containing RPMI-1640.
10. Under the microscope, orientate the embryonic thoracic portion so that the anterior (neck-end) is facing down and the posterior view is facing up, with the dorsal region furthest away from you. Using fine forceps, grasp the spine and carefully remove the thoracic organs with the other pair of fine forceps (see Fig. 1d).
11. At embryonic days 15 and 16, the thymus lobes should remain behind in the anterior portion of cavity (see Fig. 1e). If the thymus cannot be located in the cavity, it has most likely been brought out attached to the other thoracic organs (see Fig. 1f). This is a more common occurrence with younger embryos (E13–E14). Remove thymus lobes and place into a small Petri dish containing RPMI-1640 (see Fig. 1g). Clean gently using fine forceps to remove excess connective tissue.
12. Transfer embryonic lobes from the Petri dish into a 2 mL Eppendorf using a pre-wet 1 mL pipette tip to prevent lobes from sticking to plastic and keep on ice until further use.

3.3. Enzymatic Digestion of Thymus Tissue

The following technique describes the digestion of individual adult thymuses in separate tubes for flow cytometric analysis.

3.3.1. Individual Adult Thymus Digestion

1. Place thymus lobes in Petri dish containing cold RPMI-1640 and clean gently using fine forceps to remove any surrounding fatty and connective tissue. Using scissors, make small incisions in the thymic lobes, which will be used to release initial thymocytes prior to digestion. Place thymus lobes in 10 mL tube containing 1–2 mL fresh RPMI-1640.
2. To release thymocytes, gently pipette lobes up and down using a wide-bore glass pipette (see Subheading 2.3.1, item 4). When solution is cloudy, remove supernatant containing thymocytes and store on ice. Replace with fresh RPMI-1640. Repeat one to two times (see Note 8).

3. For enzymatic digestion; add 1–2 mL of Liberase–DNAse enzyme mix (see Subheading 2.3.1, item 8) per 10 mL tube, mix gently with wide bore glass pipette. Place in 35–37°C waterbath for periods of 15–20 min incubations.
4. Gently agitate digesting thymus tissue every 5–10 min with glass pipette. Collect supernatant fractions every 15–20 min and replace with fresh enzyme mix. Collect supernatant fractions in 15 mL tubes containing 2 mL FACS-buffer (see Subheading 2.3.1, item 7) to neutralize enzymes. Keep on ice.
5. As the thymic fragments reduce in size, use progressively smaller-cut glass pipettes to agitate tissue. For the last fractions, tissue can be agitated with a 1,000 μL or 200 μL pipette tip. Thymus tissue should completely disperse in 3–5 fractions (see Note 9). After complete thymus digestion, pool all digest fractions into a 15 mL tube (see Note 10). If analysis of TSC percentage, as a proportion of total thymus cellularity is required, then the supernatant from the thymocyte wash (see Subheading 3.3.2) should be included to give an accurate representation of total thymus cellularity.
6. Filter sample using 100 μm mesh and centrifuge at $480 \times g$ for 5 min at 4°C to collect cells. If pellet is particularly contaminated with red blood cells, resuspend with 1 mL of pre-warmed red blood cell lysis buffer (see Subheading 2.3.1, item 10) for 1–2 min, wash with FACS-buffer and centrifuge to collect cells. Perform cell count.
7. For flow cytometric analysis, resuspend digested thymus cells in FACS-buffer to a final concentration of 5×10^6 cells per 100 μL for staining in 96-well plates.

3.3.2. Pooled Adult Thymus Digestion

The following method is used for digesting larger groups of thymuses ($n = 10$–20), when it is necessary to isolate greater numbers of TSC for molecular analysis.

1. After cleaning, place thymic lobes into approximately 50 mL of cold RPMI-1640 in a 100 mL schott bottle with magnetic flea. Stir gently using a magnetic stirrer for 20–30 min at 4°C. Collect supernatant and transfer thymuses into 10 mL tubes containing fresh RPMI-1640 (maximum 10 thymi/tube).
2. Remove any further thymocytes by gently mixing with a wide bore glass pipette until the supernatant is mostly clear of thymocytes. Discard the thymocyte washes.
3. Add 5 mL of Liberase/DNAse enzyme mix (see Subheading 2.3.1, item 8) per thymus digest tube, mix gently with wide bore glass pipette. Place in 37°C waterbath.
4. Continue digesting thymus as per Subheading 3.3.1, steps 4–6.
5. After complete digestion, pooled thymic digests are now ready for CD45-bead depletion (see Subheading 3.5).

3.3.3. Embryonic Thymus Digestion

1. Aspirate RPMI-1640 from Eppendorf tubes containing settled embryonic thymus lobes and replace with 0.5–1 mL Liberase/DNAse enzyme mix. Incubate for 20 min in a 37°C waterbath.
2. Using a pre-wet pipette tip, mix thymus lobes gently with a 1000 μL, then with a 200 μL pipette tip to disperse organs. Thymus lobes may require an extra incubation of 10–15 min to disaggregate lobes completely into a single cell suspension.
3. Filter cells using 100 μm filter mesh and collect via centrifugation at $480\times g$ for 5 min.
4. Resuspend in FACS-buffer and determine cell concentration.

3.4. Analysis of Adult TEC Subsets by Flow Cytometry

1. Pipette 100 μL of digested adult thymus cells (5×10^6 cells; see Subheading 3.3.1, step 7) into separate wells of a 96-well plate. Pellet via centrifugation at $360\times g$ for 3 min at 4°C.
2. Flick out the supernatant and resuspend thymus cells in desired antibody cocktail. To analyze TEC subsets, we stain cells with CD45, MHC class II, EpCAM, UEA1, and Ly51 antibodies (see Subheading 2.4, item 3) in separate color channels leaving a channel free for PI.
3. Incubate cells with Ab at 4°C for 15 min, wash cells with FACS-buffer and centrifuge at $360\times g$ for 3 min at 4°C to collect cells. Repeat procedure with appropriate secondary antibody if required.
4. For compensation, resuspend 5×10^6 cells with each individual fluorochrome. As stromal cells are relatively rare, we routinely compensate with a marker of increased frequency. For example, if we use Ly-51 in the PE channel, we would stain cells separately with both Ly51-PE as well as CD45-PE (which is present on all hematopoietic cells) to make sure we have correct compensation.
5. After staining, resuspend cells in 0.3–0.5 mL of FACS-buffer and run sample on flow cytometer. Add PI 30 s before running sample. Run entire sample to collect as many stromal events as possible. If applicable, the file size can be reduced to a more manageable size by storing only CD45 negatively gated events.
6. For analysis, a wide FSC/SSC gate is used to ensure inclusion of larger stromal cells, whilst excluding small debris and RBC (see Fig. 2a). A PI-negative gate is then drawn to exclude visibly dead cells (see Fig. 2b and Note 11). Whole TSC are defined as CD45 negative (see Fig. 2c); TEC are positive for both MHCII and EpCAM, whilst non-TEC are classified as negative for both MHCII and EpCAM (see Fig. 2d). Non-TEC can be further classified into $CD31^+$ endothelium and $MTS15^+/PDGFR\alpha^+$ fibroblast (data not shown; (8)). TEC can be further divided into $UEA1^+$ medullary thymic epithelium (mTEC) and $Ly51^+$ cortical thymic epithelium (cTEC)

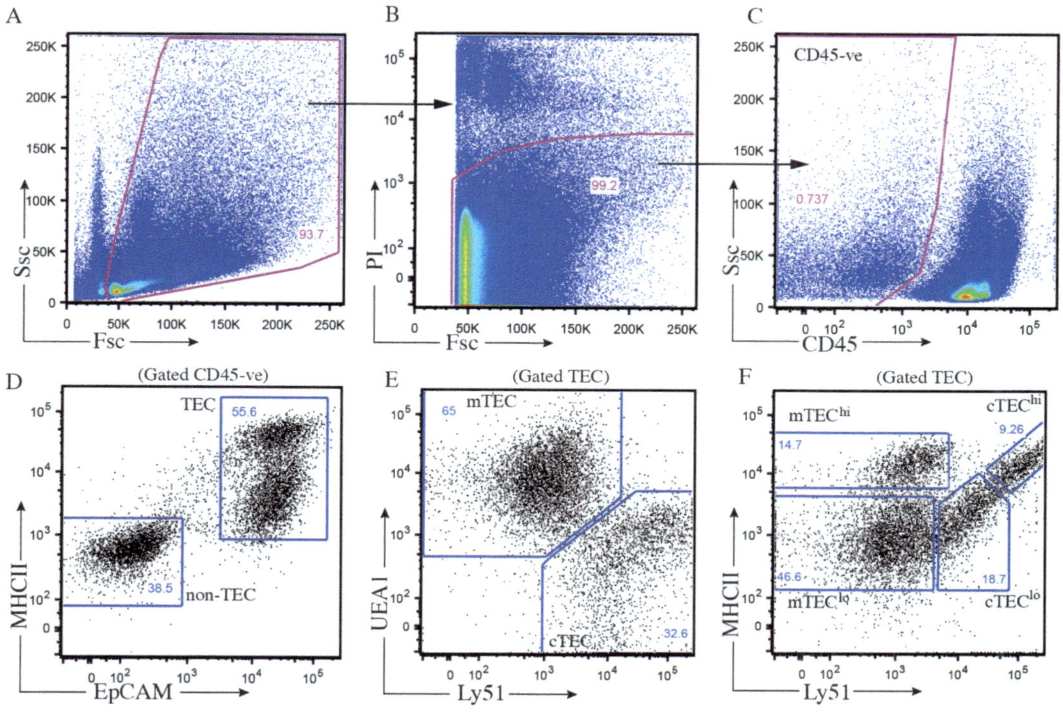

Fig. 2. Flow cytometric analysis of adult stromal cells: Individual thymus digests of adult mice (4–8 weeks) were stained with surface antibodies and analyzed for thymic stromal cell (TSC) subsets using flow cytometry. (**a**) A large forward and side scatter gate (FSC and SSC respectively) is set to include all TSC. (**b**) Viable cells are gated using propidium iodide (PI) to exclude dead cells. (**c**) TSC are gated as CD45 negative (–) and are divided into MHCII/EpCAM positive (+) thymic epithelial cells (TEC) and MHCII⁻ EpCAM⁻ Non-TEC. (**d**) (**e**) TEC can be divided into UEA1⁺ medullary epithelium (mTEC) and Ly51⁺ cortical epithelium (cTEC). (**f**) Further TEC subsets are based on high (hi) and low (lo) levels of MHCII expression: $mTEC^{hi}$ ($Ly51^-MHCII^{hi}$), $mTEC^{lo}$ ($Ly51^-MHCII^{lo}$), $cTEC^{hi}$ ($Ly51^+MHCII^{hi}$) and $cTEC^{lo}$ ($Ly51^+MHCII^{lo}$). Dot plots are representative; $n = 5$.

(see Fig. 2e). Further subsets of both medullary and cortical TEC can be based on high (hi) and low (lo) level expression of MHCII (see Fig. 2f): $mTEC^{hi}$ ($CD45^-$ $EpCAM^+$ $Ly51^-$ $MHCII^{hi}$), $mTEC^{lo}$ ($CD45^-$ $EpCAM^+$ $Ly51^-$ $MHCII^{lo}$), $cTEC^{hi}$ ($CD45^-$ $EpCAM^+$ $Ly51^+$ $MHCII^{hi}$) and $cTEC^{lo}$ ($CD45^-$ $EpCAM^+$ $Ly51^+$ $MHCII^{lo}$).

3.5. Depletion of CD45 Positive Cells from Pooled Adult Thymus Digestion

Depletion of CD45 positive cells from the thymus allows removal of the majority of thymocytes and thus enriches for the more rare CD45 negative TSC fraction. This depletion allows for more efficient sorting of TSC subsets. The following protocol is a modified version of the manufacturer's instructions for use of CD45 MicroBeads and AutoMACS® separation (see Subheading 2.5, item 3).

1. From the completed pooled thymus digest (see Subheading 3.3.2), spin down cells in a 15 mL tube and resuspend them in FACS-buffer at a concentration of 10^7 cells per

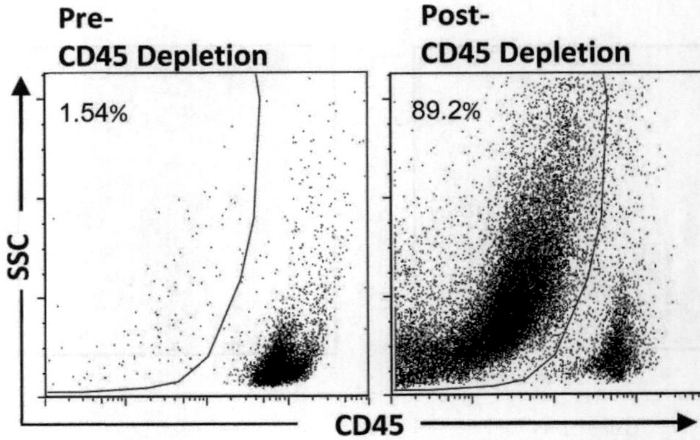

Fig. 3. Depletion of CD45 positive cells: Representative dot plots from a pooled thymus digest of ten young mice (4–8 weeks) demonstrating a low percentage of CD45 negative thymic stromal cells (TSC) pre-CD45 depletion (1.54%) and enrichment of TSC post-CD45 depletion (89.2%). Depletion was performed with CD45 MicroBeads and autoMACS® separation. Dot plots are representative; ($n=3$).

95 μL. Add CD45 MicroBeads at a concentration of 5 μL/10^7 cells. E.g., A pooled digest of ten adult mice (aged 4–8 weeks) may yield approximately 12×10^8 cells post digestion (excluding thymocytes). Therefore you would resuspend cells in 11,400 μL of buffer and add 600 μL of beads (see Note 12). Make sure you keep some cells from fraction 3 aside prior to bead depletion to use as compensation controls for the FACS sorter.

2. Incubate samples gently on a rotating mixer in the cold room at 4°C for 20 min.
3. Following the incubations, dilute any unbound beads by adding 10–20× FACS-buffer (or just fill to top of 15 mL tubes). Centrifuge cells for 10 min at $300 \times g$.
4. Aspirate the supernatant and resuspend the pellet at a concentration of 0.5×10^8 cells/mL before placing on the AutoMACS® for separation using the "Depletes" program.
5. Elute fractions into separate tubes. The negative fraction will contain your $CD45^-$ TSC. Keep the $CD45^+$ positive fraction to run through a second time and elute any remaining $CD45^-$ TSC.
6. Once all $CD45^-$ cells are eluted, pool and centrifuge the negative fractions to collect the cell pellet. Figure 3 demonstrates a representative CD45 flow cytometric profile of a pooled thymus digest pre- and post-CD45 depletion.

3.6. Fluorescence Activated Cell Sorting

3.6.1. Adult TSC Sort for RNA Analysis

1. Once the pooled thymic digests have been depleted of hematopoietic cells (see Subheading 3.5), the $CD45^-$ fraction is stained with the same Ab cocktail as per Subheading 3.4, step 2.
2. Following staining, resuspend cell sample at a concentration of 10×10^6 cells/mL of FACS-buffer for cell sorting.

Replicate	Parameter	TEC	Non-TEC	cTEC	mTEC
1	**Cell number (x10^5)**	**8.45**	**22.5**	**0.9**	**7.44**
	ng/10^{5}cells	98	40	150	370
	rRNA ratio [28s/18s]	2.11	2.4	1.8	1.87
2	**Cell number (x10^5)**	**7.8**	**20**	**1.0**	**7.5**
	ng/10^{5}cells	87	35	101	247
	rRNA ratio [28s/18s]	2.1	2.4	1.83	1.86
3	**Cell number (x10^5)**	**5.5**	**32**	**2.0**	**11.6**
	ng/10^{5}cells	96	54	79	324
	rRNA ratio [28s/18s]	1.95	2.05	1.83	1.91
Total Average	**Cell number (x10^5)**	**7.3**	**25**	**1.3**	**8.8**
	ng/10^{5}cells	94	43	110	314
	rRNA ratio range	(1.95 – 2.11)	(2.05 – 2.4)	(1.80 – 1.83)	(1.86 – 1.91)

Fig. 4. Cell number and RNA analysis from sorted thymic epithelial cell populations: Thymic epithelial cell (TEC) populations were sorted from pooled thymic digestions of ten young (4–8 week) adult mice and separated into either; whole TEC, non-TEC, cortical TEC (cTEC), and medullary TEC (mTEC) subsets (sorting gates are based on Fig. 2). Total cell number collected post sort is indicated for each subset as well as the quantity and quality of RNA extracted. The averages for the combined experimental replicates are then calculated ($n = 3$ replicates). The quantity and quality of RNA was analyzed using the Nanodrop instrument. RNA quality was determined by the ratio of the 28s/18s ribosomal RNA (rRNA), with range 1.8–2.2 depicting good quality RNA.

3. Fill 2 mL Eppendorf collection tubes with 200–500 μL of collection medium (30% FBS in RPMI-1640). Ensure all sides of the Eppendorf are coated thoroughly to ensure any cells that hit the side of the tubes will slide down into medium and not dry out. Drizzle 100 μL of 100% FBS down sides of the collection tube.
4. Run sample on cell sorting machine (see Note 13).
5. Adult TCS subsets were sorted into whole TEC, non-TEC as well as cTEC and mTEC populations based on gates shown in Fig. 2.
6. Figure 4 demonstrates the numbers of TSC subsets recovered from three separate experiments.

3.6.2. Embryonic Day 15 MTS24$^+$ TEC Sort for RTOC

1. Cells from the E15 thymus digest (see Subheading 3.3.3) are stained with MTS24 supernatant (see Subheading 2.4, item 3) and incubated at 4°C for 15 min. Wash with FACS-buffer and centrifuge at 480 × *g* for 5 min at 4°C to collect cells.
2. Resuspend sample with the appropriate secondary Ab for detection of MTS24 (see Subheading 2.4, item 3) and incubate for 15 min at 4°C in the dark. Wash with FACS-buffer and recentrifuge to collect cells.
3. Resuspend pellet with diluted normal rat serum (see Subheading 2.4, item 4) for 5 min before adding antibody cocktail containing CD45, MHCII and EPCAM (see Subheading 2.4, item 3). Incubate for 15 min at 4°C. Wash with FACS-buffer and centrifuge at 480 × *g* for 5 min at 4°C to pellet cells.

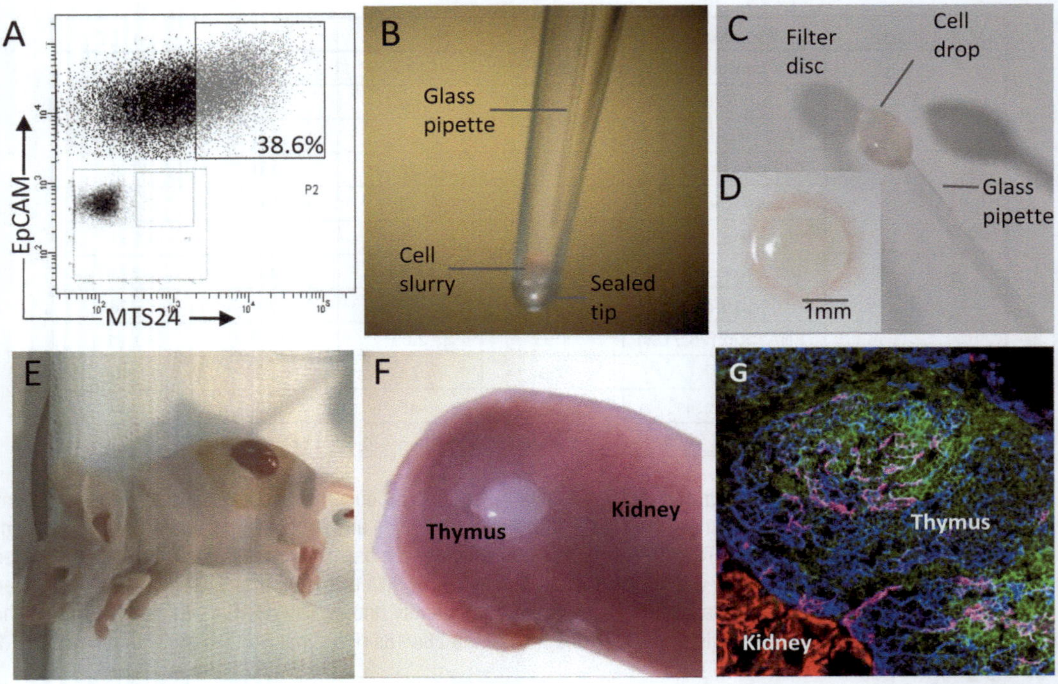

Fig. 5. Reaggregate thymic organ culture: (**a**) Dot plot showing sorted MTS24⁺ thymic epithelial cells (gated CD45⁻ MHCII⁺ EpCAM⁺) from digested E15 thymus lobes compared to isotype control (*inset*). (**b**) Sealed 200 μL tip containing pelleted cells and elongated glass pipette for cell aspiration. (**c**) Droplet of cell slurry onto filter disk for cell reaggregation. (**d**) Cells reaggregate following 24 h incubation. (**e**) Exposed kidney on a *nude* mouse for kidney grafting. (**f**) Development of an ectopic thymus graft 8 weeks post kidney capsule grafting of a 100,000 cell E15 MTS24⁺ reaggregate. (**g**) Immunohistochemical section of ectopic thymus–kidney graft. Thymus microenvironment stains with pan-cytokeratin (*blue*) and co-localization of MTS10 indicates medullary regions (*pink*). The kidney also stains with MTS10 (*red*). CD3⁺ T cell staining can be seen throughout the thymus graft (*green*).

4. For cell sorting, gate on CD45⁻ EpCAM⁺ TEC and then set the EpCAM⁺ MTS24⁺ gate compared to isotype control (see Fig. 5a).
5. Collect in Eppendorfs as per Subheading 3.6.1, step 3.
6. Use sorted MTS24⁺ embryonic TEC straight away for RTOC assay.

3.7. RNA Isolation and cDNA Synthesis from Adult TSC Subsets

1. Following cell sorting, top off the sort collection tubes with sterile nuclease-free PBS. Pellet cells by centrifugation for 5 min at 500 × *g* at 4°C. Repeat this step to ensure removal of all media. Leave ~20 μL of supernatant and resuspend pellet by flicking the Eppendorf tube gently.
2. Add lysis buffer to the sample using the appropriate RNA extraction kit (see Subheading 2.7, item 2). Vortex thoroughly to mix the cells in the lysis buffer (1–2 min) and snap freeze immediately in liquid nitrogen. Store at –80°C if unable to extract RNA immediately.

3. Extract RNA as per kit instructions (see Subheading 2.7, item 2). It is optimal to extract RNA at the same time for all replicate samples.
4. Analyze the quantity and quality of RNA from your sample using a Nanodrop-1000.
5. Figure 4 demonstrates the quantity and quality of RNA extracted from sorted TSC subsets from pools of ten thymus digests ($n=3$ experiments).
6. cDNA synthesis was performed on extracted RNA using the Superscript™ III first strand synthesis system for RT-PCR as per kit instructions (see Subheading 2.7, item 5).

3.8. Quantitative PCR

1. For qPCR analysis of TSC subsets, we use the Platinum® SYBR® Green qPCR SuperMix-UDG kit (see Subheading 2.8, item 1).
2. GAPDH (or other appropriate housekeeper gene) equilibration is performed using a 1 in 4 serial dilution of TSC cDNA template. A total reaction volume of 10 μL was used to minimize the amount of template used (see Note 14). Following GAPDH equilibration, appropriate test primers can be employed to analyze gene expression levels in TSC subsets. Test samples are run in triplicate.
3. Samples are run on a Rotor-Gene 3000 using the following protocol; 1 cycle 50°C 2 min (UDG incubation), 1 cycle 94°C 2 min (enzyme activation), 40 cycles 94°C 15 s + 60°C 30 s (denaturation, annealing, and extension), melt analysis 1 cycle 65–95°C (0.5°C increments per step, 5 s per step).

3.9. Reaggregate Organ Culture

To date, the most successful reaggregate cultures have been obtained using embryonic thymic stromal cells. Limited survival of adult thymic stromal cell reaggregates has been reported when cocultured with embryonic thymic stroma in vitro (21). Here we demonstrate the reaggregation and engraftment of FACS-sorted E15 MTS24$^+$ TEC (see Subheading 3.6.2) and their potential to form a functional thymus graft when placed under the kidney capsule.

1. Initial plate set-up: Remove gel-foam and cut into approximate 1–1.5 cm^2 pieces. Place gel foam into Petri dish containing RTOC medium and wet pieces thoroughly by submerging them in the medium using curved forceps. Ensure the majority of air-bubbles are removed from the foam.
2. Squeeze out excess liquid from gel-foam in a separate Petri dish and place gel foam into a 12-well plate containing approximately 1 mL of RTOC medium. Place filter disk on top of the gel foam. It is important to ensure that the filter disk is sitting at the liquid–air interface and is not submerged in the medium.

3. Reaggregates were performed with 50–100,000 MTS24$^+$ cells (see Note 15). Take E15 MTS24$^+$ TEC and spin down in a 1.5 mL Eppendorf tube. Remove supernatant and resuspend cells in a maximum volume of 50–100 μL RTOC medium and transfer into a sealed yellow tip (see Subheading 2.9, item 8). Centrifuge at 300 × *g* for 5 min at 4°C to pellet cells to the bottom of tip (see Fig. 5b).
4. Remove filter disk from well and rest on well-bridge to remove excess moisture (wait at least 20 s).
5. Under the microscope, use a finely drawn glass pipette, controlled with a mouth pipette (see Subheading 2.9, item 9) via a length of tubing, to aspirate the liquid from the yellow tip. Try to remove as much liquid as possible to ensure the cell slurry is not too wet. Vortex remaining pellet gently to form a slurry of cells.
6. With a new glass pipette, take up the cell slurry under microscope. This should occur mostly by capillary action and aspiration should be avoided as it may force cells too far up the pipette.
7. Visualizing the glass pipette under a microscope, slowly expel a small cell drop and touch onto filter disk (see Fig. 5c). Repeat this process and pile cells up on top of each other, pausing at least 10–15 s between each cell drop to allow sufficient time for liquid to drain away, otherwise cells will spread out too far and will not reaggregate properly.
8. Once all cells are piled on filter disk, wait another 10–15 s before carefully placing disk back onto gel-foam in culture. Cells should remain in this basic structure upon replacement into the culture well (see Note 16).
9. Incubate reaggregate for 24 h at 37°C in a humidified 5% CO_2/95% air incubator. Successfully reaggregated cells should be visible under microscope as a rounded disk-like formation (see Fig. 5d) which is able to be picked up (in one piece) by fine forceps for kidney capsule engraftment

3.10. Kidney Capsule Grafting of Reaggregate Thymus

1. Anesthetize mouse using appropriate dose of ketamine–xylazine cocktail (see Subheading 2.10, item 2) and apply eye gel.
2. Rub the incision area with Betadine® using a sterile cotton-bud. Using a scalpel, an intra-abdominal incision (~0.5 cm long) should be made running parallel but just below the spine, in between the hip and the bottom of the rib cage, to expose kidney.
3. The kidney, which is attached to a fat pad, is exposed by gently grasping the fat pad with forceps and levering the kidney out of the incision. Apply gentle pressure on either side of incision to expose kidney (see Fig. 5e).

4. Using the microscope to visualize the kidney capsule, grasp the capsule with one pair of forceps and pierce the capsule with the other pair. Move forceps gently around to make space under capsule for graft. Gently blot the capsule at incision site with sterile cotton bud to remove excess moisture.
5. Remove the filter disk containing the reaggregate and place on well-bridge. Allow it to dry for 10–15 s. Using fine forceps, carefully slide one arm under reaggregate and pick up gently. Insert the reaggregate into the space under the capsule.
6. Replace the kidney into the abdominal cavity and suture muscle tissue before stapling skin closed. Sprinkle wound area with antibiotic powder (see Subheading 2.10, item 13) and replace mice in cage with sunflower seeds. Place cage on heat-pad.
7. Mice are given Caprofen for pain (see Subheading 2.10, item 16) and monitored daily. After 8 weeks, the mice are sacrificed and the whole kidney-thymus graft removed for histology.

3.11. Analysis of Ectopic Thymus Graft

1. The ectopic thymus graft can be visualized on the kidney as a white tissue mass (see Fig. 5f).
2. For immunohistochemical analysis, remove the kidney and cut in half through the mid section (avoiding thymus graft) and place in OCT with the cut section facing down. Cutting cross-sections should allow both the kidney tissue and thymus graft to be visualized.
3. Freeze by placing in a liquid-nitrogen and isopentane slurry. Store at −80°C for further use.
4. Cut 8–10 μm sections onto glass slides using cryostat. Place cut sections in 4°C cold room with circulating air for 1 h to dry sections.
5. Stain sections with primary antibody cocktail to visualize thymic microenvironment. In this instance, we visualized the graft by using pan-cytokeratin (labels the thymic epithelial microenvironment), MTS10 (labels medullary epithelium) and CD3 (labels T cells). Incubate for 20 min at room temperature. Wash slides in 1× PBS. Incubate with secondary antibodies as required. Capture florescent images via 3-color microscope.
6. Figure 5g demonstrates an ectopic thymus graft under the kidney capsule. The organized thymus structure can be visualized by pan-cytokeratin staining (blue). The development of medullary regions was confirmed by co-staining of keratin and MTS10 (pink) and the presence of CD3+ T cells (green) can be seen in both medullary and surrounding cortical regions. MTS10 also stains the murine kidney tissue.

4. Notes

1. Two female mice are placed in a cage overnight with one male breeder. Mice are separated in the morning and gestational day designated as 0.5 in females with vaginal plug.
2. Liberase TM and TH may cleave some cell surface epitopes. We recommend testing individual antibodies before use.
3. For individual adult thymus digest, approximately 4 mL total enzyme/RPMI mix is required per thymus. For pooled digestions of $n = 10$, 15 mL total enzyme mix is required.
4. This kit contains the following items which can be purchased separately as required: Oligo(dT)$_{20}$ (50 mM), 10 mM dNTP mix, Superscript™ III Reverse Transcriptase (200 U/mL), RNaseOUT™ (RNase inhibitor)(40 U/mL), *E.coli* RNase H (2 U/mL).
5. For E15 reaggregate TEC experiments, time-mate between 10 and 20 pregnant mice. This should yield approximately $0.5–1 \times 10^6$ TEC from the pooled and sorted E15 thymic lobes. Approximately 38% of TEC are MTS24$^+$ (see Fig. 5a).
6. This is a fairly tedious procedure and requires some practice. After elongation, do not bend the glass pipette to break, as the ends will be jagged. A smooth end is required to place cell droplets optimally on the filter disk. This is achieved by simply pulling the pipette ends apart after elongation. Do not make the glass pipettes too fine or else the cells may become lodged.
7. It is technically difficult to isolate the thymus from younger embryos, therefore we recommend practising isolating lobes from embryonic day 16 embryos first.
8. If the thymocyte wash is too vigorous or prolonged, some loss of cortical epithelium may occur.
9. With aged murine thymus tissue (>6 months) it may not be possible to digest all visible thymic tissue due to increased fibrosis.
10. Pooling all fractions from the thymic digest will give an unbiased representation of all stromal cell types by flow cytometry. Fractions can be analyzed separately, however, later fractions will be enriched for particular TSC subsets (9).
11. TSC are inherently autofluorescent. To set initial live gate, run sample without PI and backgate on EpCAM$^+$ TEC to see where they fall on the PI plot. Following the addition of PI, any TEC, which then fall outside this gate, are considered to be true dead or dying cells.
12. You do not have to wait until the pooled thymus digest is finished before you start the CD45$^+$ cell depletion step. Once the initial digest fractions are completed, it is suggested you

combine fractions 1–2 to incubate with beads and put through the AutoMACS®, then continue with fractions 3–5 once they are ready. Combine all CD45$^-$ cells at the end for staining with Ab and FACS.

13. For optimal sorting and viability of thymic epithelium, we suggest using a minimal sort pressure (20 psi) and a large nozzle size (100 μm); this has been optimized for the BD Influx and may vary from machine to machine. The cells should be run at approximately 2,000–5,000 cells per second ensuring the sorting sample and collection tubes are kept cold throughout sort time.
14. A 10 μL reaction contains; 5 μL of 2× Platinum® SYBR® Green qPCR SuperMix-UDG, 1 μL 10× primer set (forward + reverse), 0.02 μL ROX reference dye, 0.98 μL nuclease-free water and 3 μL of diluted template.
15. We recommend initially using 500,000 whole CD45-E15 TSC to practise reaggregation and engraftment, as the reaggregation of lower cell numbers is more difficult.
16. If the cell pile does not hold its structure and disperses upon replacement into culture, it may indicate too much initial liquid in slurry or require longer time between drops and/or replacement of disk into the well to drain excess fluid.

Acknowledgments

The authors would like to thank Jade Homann for her expertise in kidney grafting and Joanna Lim for the embryonic photos, Professor Richard Boyd for general support, and the National Health and Medical Research Council and the Australian Stem Cell Centre for funding.

References

1. Anderson G, Jenkinson EJ (2001) Lymphostromal interactions in thymic development and function. Nat Rev Immunol 1:31–40
2. Anderson G, Owen JJ, Moore NC, Jenkinson EJ (1994) Thymic epithelial cells provide unique signals for positive selection of CD4+CD8+ thymocytes in vitro. J Exp Med 179:2027–2031
3. Oosterwegel MA, Haks MC, Jeffry U, Murray R, Kruisbeek AM (1997) Induction of TCR gene rearrangements in uncommitted stem cells by a subset of IL-7 producing, MHC class-II-expressing thymic stromal cells. Immunity 6:351–360
4. Anderson MS, Venanzi ES, Klein L, Chen Z, Berzins SP, Turley SJ, von Boehmer H, Bronson R, Dierich A, Benoist C, Mathis D (2002) Projection of an immunological self shadow within the thymus by the aire protein. Science 298:1395–1401
5. Gallegos AM, Bevan MJ (2004) Central tolerance to tissue-specific antigens mediated by direct and indirect antigen presentation. J Exp Med 200:1039–1049
6. van Ewijk W, Wang B, Hollander G, Kawamoto H, Spanopoulou E, Itoi M, Amagai T, Jiang Y-F, Germeraad WTV, Chen W-F,

Katsura Y (1999) Thymic microenvironments, 3-D versus 2-D? Semin Immunol 11:57–64
7. Gray DH, Seach N, Ueno T, Milton MK, Liston A, Lew AM, Goodnow CC, Boyd RL (2006) Developmental kinetics, turnover, and stimulatory capacity of thymic epithelial cells. Blood 108:3777–3785
8. Gray DH, Tull D, Ueno T, Seach N, Classon BJ, Chidgey A, McConville MJ, Boyd RL (2007) A unique thymic fibroblast population revealed by the monoclonal antibody MTS15. J Immunol 178:4956–4965
9. Gray DH, Fletcher AL, Hammett M, Seach N, Ueno T, Young LF, Barbuto J, Boyd RL, Chidgey AP (2008) Unbiased analysis, enrichment and purification of thymic stromal cells. J Immunol Methods 329:56–66
10. Jenkinson EJ, Owen JJ (1990) T-cell differentiation in thymus organ cultures. Semin Immunol 2:51–58
11. Hare KJ, Jenkinson EJ, Anderson G (1999) In vitro models of T cell development. Semin Immunol 11:3–12
12. Jenkinson EJ, Anderson G, Owen JJ (1992) Studies on T cell maturation on defined thymic stromal cell populations in vitro. J Exp Med 176:845–853
13. Anderson G, Jenkinson EJ (2007) Investigating central tolerance with reaggregate thymus organ cultures. Methods Mol Biol 380: 185–196
14. Depreter MG, Blair NF, Gaskell TL, Nowell CS, Davern K, Pagliocca A, Stenhouse FH, Farley AM, Fraser A, Vrana J, Robertson K, Morahan G, Tomlinson SR, Blackburn CC (2008) Identification of Plet-1 as a specific marker of early thymic epithelial progenitor cells. Proc Natl Acad Sci U S A 105:961–966
15. Godfrey DI, Izon DJ, Tucek CL, Wilson TJ, Boyd RL (1990) The phenotypic heterogeneity of mouse thymic stromal cells. Immunology 70:66–74
16. Gill J, Malin M, Hollander GA, Boyd R (2002) Generation of a complete thymic microenvironment by MTS24(+) thymic epithelial cells. Nat Immunol 3:635–642
17. Nijhof JG, Braun KM, Giangreco A, van Pelt C, Kawamoto H, Boyd RL, Willemze R, Mullenders LH, Watt FM, de Gruijl FR, van Ewijk W (2006) The cell-surface marker MTS24 identifies a novel population of follicular keratinocytes with characteristics of progenitor cells. Development 133:3027–3037
18. Bennett AR, Farley A, Blair NF, Gordon J, Sharp L, Blackburn CC (2002) Identification and characterization of thymic epithelial progenitor cells. Immunity 16:803–814
19. Rossi SW, Chidgey AP, Parnell SM, Jenkinson WE, Scott HS, Boyd RL, Jenkinson EJ, Anderson G (2007) Redefining epithelial progenitor potential in the developing thymus. Eur J Immunol 37:2411–2418
20. Ueno T, Liu C, Nitta T, Takahama Y (2005) Development of T-lymphocytes in mouse fetal thymus organ culture. Methods Mol Biol 290:117–133
21. Gray D, Abramson J, Benoist C, Mathis D (2007) Proliferative arrest and rapid turnover of thymic epithelial cells expressing Aire. J Exp Med 204:2521–2528

Chapter 16

Enrichment and Clonal Culture of Hepatic Stem/Progenitor Cells During Mouse Liver Development

Akihide Kamiya and Hiromitsu Nakauchi

Abstract

Liver regenerates after hepatectomy or chemical-induced injury. In contrast to cells in other tissues that can regenerate, mature cells (hepatocytes), but not undifferentiated stem cells, are mainly responsible for acute liver regeneration. Liver stem cells take part in liver regeneration in some forms of chronic liver injury, when the proliferative ability of differentiated hepatocytes is impaired. During liver development, both hepatocytes and cholangiocytes are differentiated from common precursor cells, called hepatoblasts. By combining fluorescence-activated cell sorting (FACS) and an in vitro clonal culture system for stem/progenitor cells, we established a method to isolate stem/progenitor cells prospectively from mouse fetal and adult livers. FACS clone-sorted single $CD45^-Ter119^-c\text{-}kit^-CD13^+CD133^+$ cells (from fetal mid-gestational livers) or $CD45^-Ter119^-c\text{-}kit^-Sca1^-CD13^+CD49f^+CD133^+$ cells (from adult livers) can form a colony containing both albumin-positive hepatocytes and cytokeratin 19-positive bile ductal cells, indicating that these cells have the characters of liver stem/progenitor cells (proliferative capability and bipotency for hepatic and for biliary epithelial differentiation). These cells can maintain these capabilities for several months in culture.

Key words: Liver stem/progenitor cells, Clonal culture system, Prospective isolation, Flow cytometry, Colony assay

1. Introduction

Stem cells have self-renewal capacity and multi-potency. They can differentiate into cells of several lineages. In regenerative tissues such as hematopoietic cells, skin, and hair, somatic stem cells contribute to maintain and to repair the tissues in which they are found (1). Regeneration after hepatectomy or chemical-induced injury is one of the most remarkable abilities of the liver. In contrast to mature cells in other regenerative tissues, mature hepatocytes are important for acute liver regeneration (2, 3). However, when the proliferative ability of mature hepatocytes is impaired, as in some

Scott H. Randell and M. Leslie Fulcher (eds.), *Epithelial Cell Culture Protocols: Second Edition*, Methods in Molecular Biology, vol. 945, DOI 10.1007/978-1-62703-125-7_16, © Springer Science+Business Media, LLC 2012

types of chronic liver injury, liver stem/progenitor cells are still required for liver regeneration. When serious liver injury induced by treatment with several chemicals such as retrorsine or 2-acetylaminofluorine is combined with partial hepatectomy, oval cells in periportal regions increase in number (4). Oval cells exhibit high proliferative activity and bi-potency for differentiation into hepatocytes and cholangiocytes, indicating that oval cells are candidate hepatic progenitor cells. The source of oval cells in normal livers remains unknown; however, a few candidate liver stem/progenitor cells might exist in normal livers, and indeed cells that can be cultured for long periods, a class of candidate liver stem/progenitor cells, are found at a low frequency when parenchymal or nonparenchymal cell fractions derived from murine and human livers are cultured (5–7).

Unlike adult liver stem/progenitor cells, embryonic liver stem/progenitor cells derived from mid-fetal livers are well described (8). In developing livers, hepatoblasts (common stem/progenitor cells) differentiate into both hepatocytes and cholangiocytes (9). A subset of sorted hepatic cells gives rise to relatively large colonies in suitable culture conditions. We designated an individual member of this class of cells as a "hepatic colony-forming unit in culture (H-CFU-C)" (10). Dlk, E-cadherin, and Liv-2 are identified as cell surface markers expressed on hepatoblasts, which can make H-CFU-C colonies (11–13). Isolated single cells in mid-fetal livers, identified by CD13 and CD133 expression, can form H-CFU-C colonies containing both albumin-positive hepatocytes and cytokeratin 19 (CK19)-positive cholangiocytes, indicating that CD13 and CD133 are also cell surface markers expressed on hepatoblasts (14). Replating assays of these colonies derived from hepatoblasts revealed self-renewal capability in progeny of such cells (15). These results suggested that embryonic liver has several hepatic stem/progenitor cells having both self-renewal capacity and multi-potency to differentiate into cells of several lineages.

We recently established an efficient method of culturing clone-sorted postnatal-liver progenitor cells and fetal liver stem/progenitor cells (16, 17). We found that Y-27632, the Rho-associated kinase (ROCK) inhibitor, induced efficiency of colony formation derived from postnatal-liver stem cells. Liver stem cells are numerous in the $CD45^{-}Ter119^{-}c\text{-}Kit^{-}Sca1^{-}CD13^{+}CD49f^{+}CD133^{+}$ subpopulation of nonparenchymal cells derived from non-injured postnatal livers. Expansion and reclone-sorting of postnatal-liver stem cells confirmed that these cells have the capacity for bi-potent differentiation and for self-renewal activity (17).

2. Materials

2.1. Purification of Adult Liver Progenitor Cells

1. C57BL/6CrSlc mice: 4-week-old to 16-week-old mice are used. All animal experiments comply with national and institutional regulations.
2. Solution for anesthetization: Diethyl ether or locally approved replacement e.g., isoflurane.
3. Liver perfusion medium: To approximately 850 mL Milli-Q water (Millipore) in a 1,000-mL beaker add NaCl 8.0 g, KCl 0.4 g, $Na_2HPO_4/12H_2O$ 0.121 g, KH_2PO_4 0.06 g, $NaHCO_3$ 0.35 g, glucose 1.0 g, HEPES 2.383 g, and EGTA/2Na 0.190 g. Adjust pH to 7.5 with NaOH. Adjust to 1,000 mL and filter with a 0.22-μm sterile filter system (Millipore).
4. Liver digestion medium: To approximately 850 mL Milli-Q water into a 1,000-mL beaker add NaCl 8.0 g, KCl 0.4 g, $Na_2HPO_4/12H_2O$ 0.121 g, KH_2PO_4 0.06 g, $MgSO_4/7H_2O$ 0.2 g, $CaCl_2/2H_2O$ 0.735 g, $NaHCO_3$ 0.35 g, glucose 1.0 g, and HEPES 2.383 g. Adjust pH to 7.5 with NaOH. Adjust to 1,000 mL. Just before use, add 0.5 g collagenase (Yakult Pharmaceutical Industry) into 1,000 mL of the liver digestion medium and warmed (37°C). Dissolve completely and filter with a 0.22-μm sterile filter unit (Millipore).
5. Surflash catheter: Surflo® 24 G 3/4.
6. Surgery Instruments: Dissecting scissors and forceps. For dissociation of fetal livers, INOX 5 forceps are used.
7. Surgery sutures: VICRYL 4-0 27″.
8. 10× Dulbecco's Phosphate Buffered Saline (PBS, Sigma-Aldrich®): Add 9.6 g of powder PBS to 100 mL MilliQ water. Autoclave, store at 4°C.
9. 1× PBS: Add 9.6 g of powder PBS to 1,000 mL MilliQ water. Autoclave and store at 4°C.
10. Density buffer: A mixture of 7.5 mL DMEM/10% FBS, 2.5 mL Percoll, and 250 μL 10× PBS.
11. Propidium iodide solution (Sigma-Aldrich®): Dissolve in PBS (1 mg/mL).
12. Fetal bovine serum (FBS): Inactivate by incubation at 55°C for 30 min.
13. Staining medium: 1× PBS with 3% FBS.
14. Dexamethasone stock solution: Add 39.2 mg dexamethasone (Sigma-Aldrich®) to 10 mL ethanol (10^{-2} M stock solution). Dilute to 100× (10^{-4} M) with ethanol. Store at −20°C until use.

15. Collagen-coated dishes: Dilute Cellmatrix® type I-C (Nitta gelatin) tenfold with HCl (pH 3.0). Add to culture dishes and keep at RT for 1 h. Remove collagen solution, dry completely, wash two times with PBS, and use.
16. Gelatin-coated dishes: Add 0.5 g gelatin (Sigma-Aldrich®) to 500 mL PBS, to yield 0.1% gelatin/PBS. Autoclave and store at 4°C. Add to culture dishes 0.1% gelatin/PBS and keep at RT for 1 h, wash with PBS, and use.
17. Hepatocyte culture medium: DMEM supplemented with 10% FBS, 1× penicillin/streptomycin/L-glutamine (Sigma-Aldrich®), 1× Minimum essential medium nonessential amino acid solution (Life Technologies™) and 10^{-7} M dexamethasone.
18. Hemolysis buffer: Dissolve Tris–HCl 0.5 g and NH_4Cl 1.4 g in 250 mL MilliQ water; sterilize with a 0.22-μm filter system.
19. E14.5 liver-derived conditioned medium: Using micro-forceps, dissect livers from E14.5 embryos, completely removing gut and other connective tissues. Mince livers, using surgical scissors, into 1 mm^3-blocks and incubate in liver digestion medium (37°C, 15 min) to dissociate cells. Wash cells with DMEM/10% FBS and centrifuge (150×*g*, 3 min); decant wash medium; repeat washing. Add 7 mL hemolysis buffer to cell pellets and incubate (4°C, 8 min) to remove contaminating red blood cells. Add 7 mL DMEM/10% FBS and centrifuge (150×*g*, 3 min). Wash cell pellets with DMEM/10% FBS. Filter cell suspension through a 70-μm cell strainer into 15-mL tubes. Count cells; inoculate onto 60 cm^2 gelatin-coated dish (2.0×10^6 cells/dish). At 1, 3, and 5 days of culture, change culture medium (10 mL/dish); at 5 and 7 days of culture, harvest culture medium, pass medium through 0.45-μm sterile filters and store at −20°C until use.
20. Standard culture medium: DMEM/F-12 (Sigma-Aldrich®) supplemented with 10% FBS, 1× Insulin-Transferrin-Selenium X (Life Technologies™), 10 mM nicotinamide (Sigma-Aldrich®), 10^{-7} M dexamethasone, 2.5 mM Hepes, 1× penicillin/streptomycin/L-glutamine, and 1× Minimum essential medium nonessential amino acid solution. Mix 1:1 with conditioned medium derived from E14.5 liver cells (see Subheading 2.1, item 19). Add 40 ng/mL human hepatocyte growth factor (HGF, Peprotech) and 20 ng/mL mouse epidermal growth factor (EGF, Peprotech) into standard culture medium just before use.
21. Y-27632 (ROCK inhibitor; Calbiochem): Dissolve in MilliQ water to yield 10 mM solution (500× stock solution). Store at −20°C.
22. 150-μm nylon filter net: PP-150n. Sterilize by autoclaving.

Table 1
Antibody lists for flow cytometry and immunocytochemistry

Antigen	Characteristic	Source	Catalogue number
CD13	FITC-conjugated	Pharmingen (San Jose, CA)	558744
CD133	APC-conjugated	eBioscience (San Diego, CA)	17-1331
CD45	PE-Cy7-conjugated	eBioscience	25-0451
CD49f	PE-conjugated	Pharmingen	555736
c-Kit	PE-Cy7-conjugated	eBioscience	25-1171
Sca-1	PE-Cy7-conjugated	eBioscience	25-5981
Ter119	PE-Cy7-conjugated	Pharmingen	557853
Dlk	FITC-conjugated	Medical & Biol. Lab. (Nagoya, Japan)	D187-4
Albumin	Purified	Bethyl Lab. (Montgomery, TX)	A90-135
CK19	Purified	Gift from Prof. Miyajima (University of Tokyo)	
Alexa546-conjugated donkey anti-goat IgG antibody	2nd antibody	Life Technologies	A11056
Alexa488-conjugated donkey anti-rabbit IgG antibody	2nd antibody	Life Technologies	A21206

FITC fluorescein isothiocyanate, *PE* phycoerythrin, *APC* allophycocyanin

23. Antibodies for flow cytometry: (see Table 1).
24. 70 μm cell strainer.
25. FACS tubes.
26. 96 well plates.

2.2. Establishment of Normal-Liver-Derived Stem-Like Cells and Induction of Hepatic Maturation of NLS Cells

1. Cell digest solution: 1× Trypsin-EDTA (Sigma-Aldrich®).
2. Solution for hepatic maturation: Mouse oncostatin M (OSM, R&D systems). Dissolve in DMEM/10% FBS to yield 20 μg/mL solution (1,000× stock solution). Store at −20°C.
3. Matrigel™ (BD Biosciences).
4. Solution for RNA purification: Trizol® (Life Technologies™).

2.3. Purification and Clonal Culture of Fetal Liver Stem/Progenitor Cells

1. Animals: Pregnant female C57BL/6CrSlc mice carrying embryonic-day (E) 13.5 fetuses.
2. Antibodies for immunocytochemistry: (see Table 1).

Table 2
Primer lists for RT-PCR

Protein encoded	Direction	Sequence
Albumin	F R	CATGACACCATGCCTGCTGAT CTCTGATCTTCAGGAAGTGTAC
CK19	F R	GTCCTACAGATTGACAATGC CACGCTCTGGATCTGTGACAG
G6Pase	F R	GCGCAGCAGGTGTATACTATGT ATCAACTCAACCTGGGATGG
TAT	F R	CAATGACAGCCTACCTTCAGTGC ACCTCAATCCCCATAGACTCAGC
GAPDH	F R	CTTCACCACCATGGAGAAGGC GGCATGGACTGTGGTCATGAG

CK19 cytokeratin 19, *G6Pase* glucose-6-phosphatase, *TAT* tyrosine aminotransferase, *GAPDH* glyceraldehyde 3-phosphate dehydrogenase, *F* forward primer, *R* reverse primer

2.4. Characterization of Adult and Fetal Liver Stem/Progenitor Cells

1. Solution for cDNA synthesis: Primescript™ II first strand cDNA synthesis kit (Takara).
2. Probe for quantitative RT-PCR: Rodent GAPDH TaqMan™ Probe (Life Technologies™).
3. Solution of cell fixation: 4% Paraformaldehyde in PBS.
4. Solution of cell blocking: Donkey serum (Sigma-Aldrich®). Inactivate by incubation at 55°C for 30 min. Dilute to yield 5% donkey serum/PBS. Filter (0.45-μm pore) and store at −20°C.
5. PCR primers: (see Table 2).
6. Agarose gels: Dilute 1.0 g agarose to 100 mL 1× Tris-acetate-EDTA Buffer (40 mM Tris-acetate, 1 mM EDTA, pH 8.3). Warm using microwave oven, and dissolve agarose completely. When the agarose solution has cooled to about 50°C, add 2 μL of ethidium bromide and pour solution directly into the casting tray.
7. Methanol.

3. Methods

3.1. Purification and Clonal Culture of Adult Liver Progenitor Cells

1. Set up a perfusion system using a peristaltic pump and a warm water bath (see Fig. 1a). Incubate liver perfusion and digestion media (see Subheading 2.1, items 3 and 4). Fill silicone perfusion tubing with liver perfusion medium before the experiment (see Note 1).

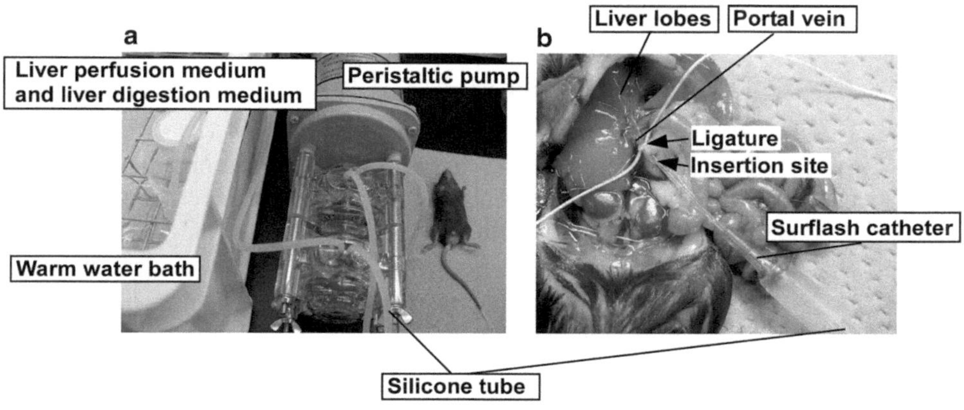

Fig. 1. Perfusion of the mouse adult liver. (**a**) Setting up the perfusion system. Liver perfusion and digestion medium is warmed in a water bath before use. A silicone tube and a catheter (use only a plastic head) are set up with a peristaltic pump. (**b**) Position of the inserted catheter in the portal vein.

2. Anesthetize adult mice (4–16 weeks old) with diethyl ether (see Subheading 2.1, item 2).
3. Using surgical scissors, open the abdominal wall into the peritoneal cavity. Identify the portal vein.
4. Incise the portal vein, using surgical scissors, and insert the tip of Surflash catheter containing liver perfusion medium (see Subheading 2.1, item 3) into the portal vein (see Fig. 1b). Ligate the catheter in place using surgical suture material.
5. Start perfusion at a flow rate of 2 mL/min.
6. Cut the inferior vena cava and wash out the blood from the liver. To prevent an increase of intra-hepatic pressure, the vena cava must be cut as soon as perfusion starts.
7. After 10 min of perfusion, turn the pump off and exchange the liver perfusion medium for the liver digestion medium (see Subheading 2.1, item 4).
8. Turn the pump on and start digestion at a flow rate of 2 mL/min.
9. After 20–30 min of perfusion, stop the flow and excise the liver.
10. Finely mince the liver in a sterilized Petri dish.
11. Add 20 mL of DMEM/10% FBS. Dissociate cells by very gentle pipetting using a 25-mL pipet. Filter the cell suspension through a 150-μm nylon filter into a 50-mL tube. Store at 4°C.
12. Add non-dissociated cells on the filter into a fresh 20 mL liver digestion medium and incubate (37°C, 15 min).
13. Dissociate cells by pipetting; filter suspension through a 150-μm nylon filter into 50-mL tubes.

14. Combine harvested cells in steps 11 and 13; centrifuge (50 × *g*, 4°C, 1 min). Collect supernatants and transfer to new tubes.
15. Centrifuge (200 × *g*, 5 min). Wash cell pellet with DMEM/10% FBS.
16. Centrifuge (50 × *g*, 1 min). Collect supernatants and transfer to new tubes. Repeat four to five times to remove contaminated mature hepatocytes.
17. Collect cells by centrifugation (200 × *g*, 5 min). Remove cell debris by density centrifugation (200 × *g*, 5 min) in density buffer (see Subheading 2.1, item 10).
18. Wash cell pellets with DMEM/10% FBS. Centrifuge (200 × *g*, 5 min). Repeat this step.
19. Add 50 μL of staining medium and fluorescent-conjugated antibodies to cell pellets. (Antibody aliquots: 1 μL fluorescein isothiocyanate (FITC)-conjugated anti-CD13 antibody, 4 μL phycoerythrin (PE)-conjugated anti-CD49f antibody, 1 μL allophycocyanin (APC)-conjugated anti-CD133 antibody, and 0.5 μL PE-Cy7-conjugated anti-CD45, -Ter119, -Sca1, and -c-Kit antibodies, respectively) (see Table 1).
20. Incubate cells (4°C, 1 h). Wash with 5 mL staining medium and centrifuge (200 × *g*, 5 min). Add cell pellets to 1 mL staining medium with 1 μL propidium iodide solution (see Subheading 2.1, item 11). Resuspend cells. Filter cell suspension through a 70-μm cell strainer into 5-mL FACS tubes.
21. Analyze cells by flow cytometry. Gates are shown in Fig. 2a. Exclude doublets by forward scatter (FSC) and side scatter (SSC). Exclude dead cells by propidium iodide staining.
22. Add 100 μL of standard culture medium (see Subheading 2.1, item 20) supplemented with 20 μM Y-27632 (see Subheading 2.1, item 21) to individual wells of collagen type I-coated 96-well plates (see Fig. 3a, Note 2).
23. Sort single $CD45^-Ter119^-c\text{-}Kit^-Sca1^-CD13^+CD49f^+CD133^+$ cells (adult liver progenitor cells) into individual wells of 96-well plates.
24. Culture individual single cells for 10–13 days to yield colonies in the 5% CO_2 incubator at 37°C. Two types of colonies will appear (see Fig. 3b, Note 3).

3.2. Establishment of NLS Cells and Induction of Hepatic Maturation of NLS Cells

1. After 10–13 days of culture (37°C, 5% CO_2), several hepatocytic colonies will have formed from individual liver progenitor cells.
2. Wash colonies with PBS and add 100 μL trypsin-EDTA (see Subheading 2.2, item 1) to individual wells of 96-well plates. Plates are incubated at 37°C for 5 min. Trypsin-EDTA is removed by aspiration. Incubate plates (37°C, 5 min).

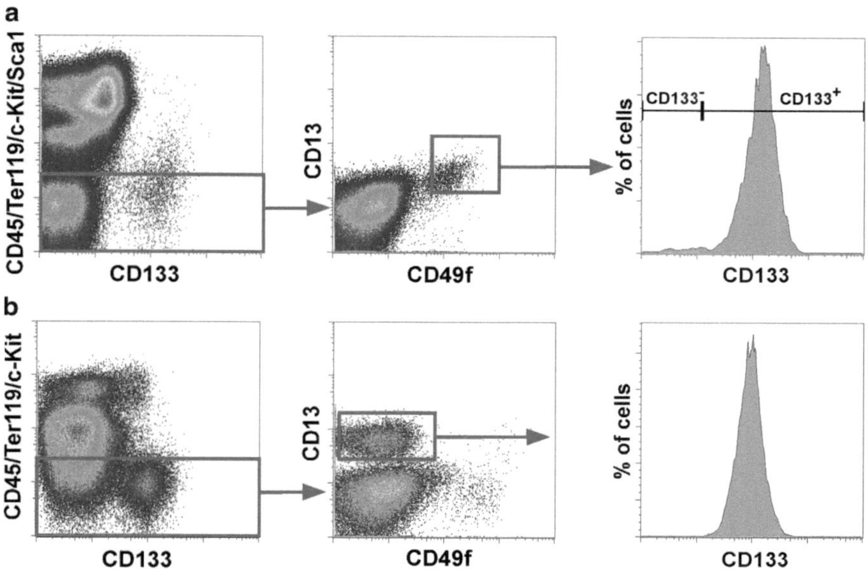

Fig. 2. Gates for prospective isolation of fetal and adult liver stem/progenitor cells. (**a**) Adult $CD45^-Ter119^-c\text{-}Kit^-Sca1^-CD13^+CD49f^+$ cells are fractionated by CD133 expression. (**b**) Fetal $CD45^-Ter119^-c\text{-}Kit^-CD13^+CD49f^{-/low}CD133^+$ cells are fractionated as fetal liver progenitor cells. (Reproduced from ref. 17 with permission from Elseiver).

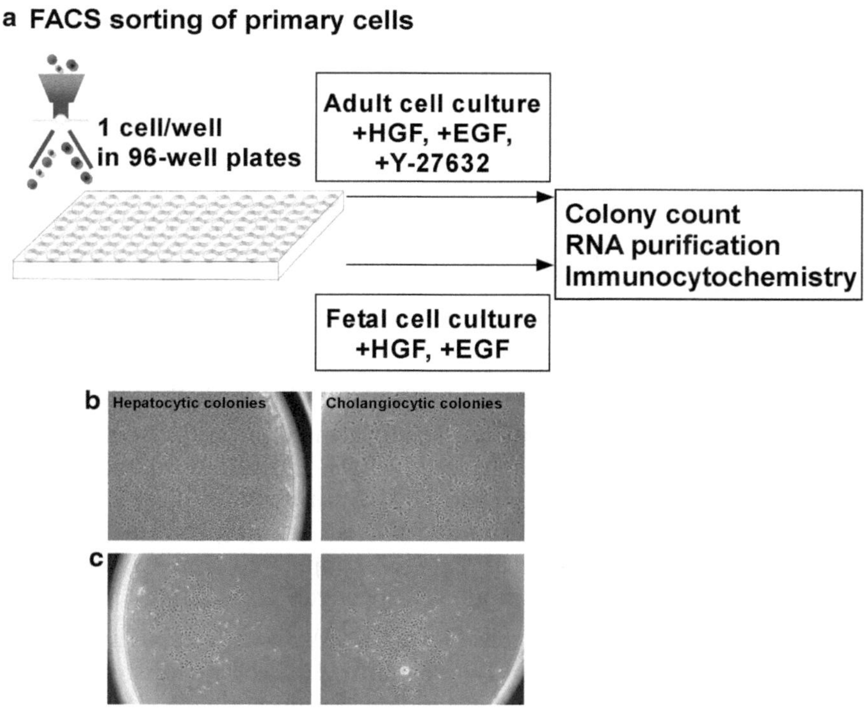

Fig. 3. Clonal culture of fetal and adult liver stem/progenitor cells. (**a**) Schemata of culture systems of fetal and adult liver progenitor cells. Cells are clone-sorted into individual wells of collagen type-I coated 96-well culture plates. Colonies were counted and expression of liver-cell markers was analyzed using RT-PCR and immunocytochemistry. (**b**) Representative views of colonies derived from clone-sorted $CD45^-Ter119^-c\text{-}Kit\text{-}Sca1^-CD13^+CD49f^+CD133^+$ cells. Both hepatocytic and cholangiocytic colonies are shown. (**c**) Representative views of colonies derived from clone-sorted fetal liver stem/progenitor cells ($CD45^-Ter119^-c\text{-}Kit^-CD13^+CD49f^{-/low}CD133^+$ cells). (Reproduced from ref. 17 with permission from Elsevier).

3. Add 100 μL standard culture medium (see Subheading 2.2, item 20) to individual wells. Dissociate cells by pipetting.
4. Inoculate cells into individual wells of 48-well culture plates in 500 μL standard culture medium.
5. Culture for several days (yield: almost $3–4 \times 10^4$ cells/well); Cells are dissociated with Trypsin/EDTA and transferred to 12-well culture plates.
6. Culture to semi-confluent state (yield: almost 1×10^5 cells/well); Cells are dissociated with Trypsin/EDTA and transferred to 6-well culture plates.
7. Repeat step 6, for continued growth. These cells are termed normal-liver-derived stem-like (NLS) cells.
8. To induce hepatocytic maturation, culture NLS cells to semi-confluent state in 6-well culture plates.
9. Wash cells twice with PBS and changed culture medium to hepatocyte culture medium supplemented with 20 ng/mL OSM and 20% Matrigel™ (18, 19).
10. After 3 days, wash with PBS to remove Matrigel™ overlaying on cells. Add 1 mL Trizol® into individual wells to lyse cells.
11. Purify total RNA and analyze gene expression by RT-PCR. (see Subheading 3.4.1).

3.3. Purification and Clonal Culture of Fetal Liver Stem/Progenitor Cells

1. Using micro-forceps, dissect livers from E13.5 embryos, completely removing gut and other connective tissues (almost ten embryos per one pregnant female mouse).
2. Mince livers, using surgical scissors, into 1 mm³-blocks and incubate in liver digestion medium (37°C, 15 min) to dissociate cells.
3. Wash cells with DMEM/10% FBS and centrifuge ($150 \times g$, 3 min); decant wash medium; repeat washing.
4. Add 50 μL of staining medium and fluorescent-conjugated antibodies to cell pellets. (Antibody aliquots: 1 μL FITC-conjugated anti-CD13 antibody, 4 μL PE-conjugated anti-CD49f antibody, 1 μL APC-conjugated anti-CD133 antibody, and 0.5 μL PE-Cy7-conjugated anti-CD45, -Ter119, and -c-Kit antibodies, respectively) (see Table 1 and Note 4).
5. Incubate cells (4°C, 1 h). Wash with staining medium and centrifuge ($150 \times g$, 5 min). Add cell pellets to staining medium with propidium iodide. Resuspend cells. Filter cell suspension through a 70-μm cell strainer into 5-mL FACS tubes.
6. Analyze cells by flow cytometry. Gates are shown in Fig. 2b. Exclude doublets by FSC and SSC. Exclude dead cells by propidium iodide staining.

7. Add 100 μL of standard culture medium to individual wells of collagen type I-coated 96-well plates (see Fig. 3a).
8. Sort single $CD45^{-}Ter119^{-}c\text{-}Kit^{-}CD13^{+}CD49f^{-/low}CD133^{+}$ cells (fetal liver stem/progenitor cells) into individual wells of plates.
9. Culture individual single cells for 6 days to yield colonies (37°C, 5% CO_2) (see Fig. 3c).

3.4. Characterization of Adult and Fetal Liver Stem/Progenitor Cells

3.4.1. Analyses of Expression of Hepatic and Cholangiocytic Marker Genes Using RT-PCR

1. Culture colonies derived from fetal and adult liver stem/progenitor cells in individual wells of 96-well culture plates; (see Subheadings 3.1 and 3.3). Add 100 μL of Trizol® to individual wells. For analyses of hepatic maturation, culture NLS cells with OSM and Matrigel™ and lyse with Trizol®; (see Subheading 3.2).
2. Extract total RNA using Trizol® according to manufacturer's protocol.
3. Synthesize first-strand cDNA (PCR amplification template) using the Primescript™ II first strand cDNA synthesis kit (see Subheading 2.4, item 1) according to manufacturer's protocol.
4. Using the TaqMan™ probe, normalize cDNA samples by number of glyceraldehyde 3-phosphate dehydrogenase (GAPDH) copies.
5. Use RT-PCR to detect expression of a hepatocyte marker (albumin), a cholangiocyte marker (CK19), and of mature hepatocyte markers (glucose-6-phosphatase (G6Pase) and tyrosine aminotransferase (TAT)). As PCR templates, apply aliquots of cDNA samples that contain equal numbers of GAPDH copies. Anneal primers at 55°C for 30 s and repeat amplification for 27 ~ 40 cycles.
6. Separate amplified products by electrophoresis on a 1.5% agarose gel; stain with ethidium bromide (see Fig. 4a–c).

3.4.2. Immunocytochemical Analyses of Hepatic and Cholangiocytic Marker Expression

1. Culture sorted fetal and adult liver stem/progenitor cells (fetal cells: 5–6 days; adult cells: 9–10 days).
2. Fix cells with 4% paraformaldehyde in PBS at RT for 10 min. Wash cells with PBS twice.
3. Permeabilize cells using cold methanol at −20°C for 10 min. Wash cells with PBS twice.
4. Block nonspecific reactions by incubating cells with 5% donkey serum/PBS (see Subheading 2.4, item 4) at RT for 1 h. Add rabbit anti-mouse CK19 antibody (diluted to 1:2,000 with 5% donkey serum/PBS) and goat anti-mouse albumin antibody (diluted to 1:500 with 5% donkey serum/PBS). As a negative control, incubate cells (4°C, overnight) with rabbit and goat IgG fractions in 5% donkey serum/PBS.

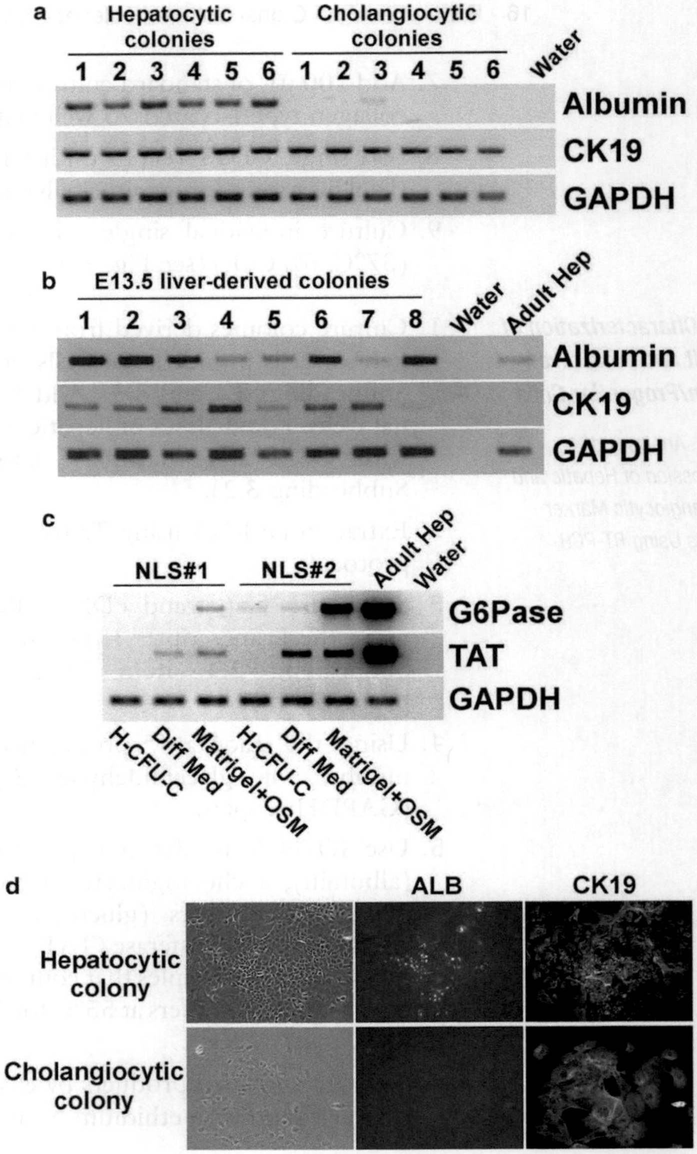

Fig. 4. Characterization of adult and fetal liver stem/progenitor cells. (**a**) Characterization of hepatocytic and cholangiocytic colonies using RT-PCR. Total RNAs derived from six colonies of hepatocytic and cholangiocytic cells were purified. A hepatocyte-marker gene (*Albumin*) and a cholangiocyte-marker gene (*CK19*) are detected. Glyceraldehyde 3-phosphate dehydrogenase (GAPDH) was used as an internal control. (**b**) Total RNAs derived from eight fetal liver stem/progenitor colonies were purified. A hepatocyte-marker gene (Albumin) and a cholangiocyte-marker gene (CK19) were detected. Adult Hep; cDNA derived from adult hepatocytes. (**c**) Induction of hepatic maturation in NLS cell culture. NLS cells were cultured for 2 days and expression of G6Pase and TAT was induced by change of culture medium. After 3 days of culture, cells were collected. Under normal culture conditions (standard culture medium), NLS cells barely expressed mature-hepatocyte genes (G6Pase and TAT). In contrast, these genes were induced under hepatocyte culture medium conditions (Diff Med), in particular, in the presence of Matrigel™ and OSM (Matrigel™ + OSM). Adult Hep; cDNA derived from adult hepatocytes. (**d**) Cells were stained with anti-albumin and -CK19 antibodies. Signals were detected using Alexa546- and Alexa488-conjugated antibodies (against albumin and CK19, respectively). (Reproduced from ref. 17 with permission from Elsevier).

5. Wash cells with PBS three times. Add Alexa546-conjugated donkey anti-goat IgG antibody (diluted to 1:500 with 5% donkey serum/PBS) and Alexa488-conjugated donkey anti-rabbit IgG antibody (diluted to 1:500 with 5% donkey serum/PBS). Incubate cells (RT, 1 h).
6. Wash cells with PBS three times (see Fig. 4d).

4. Notes

1. The perfusion step is particularly important for purification of adult liver progenitor cells. Collagenase activity is best at 37°C. The water bath must be adjusted.
2. The addition of ROCK inhibitor Y-27632 to adult liver progenitor cells in culture is strictly required for colony formation (17). In contrast, Y-27632 is not required for colony formation from E13.5 fetal liver stem/progenitor cells. The mechanisms regulating proliferation of adult liver progenitor cells may differ from those operating in fetal liver cells.
3. When $CD45^-Ter119^-c\text{-}Kit^-Sca1^-CD13^+CD49f^+CD133^+$ cells derived from adult livers are cultured, two morphologically different types of colonies are detected. Colonies that form a monolayer with most cells resembling hepatocytes are called "hepatocytic," These colonies express both albumin and CK19, suggesting that these colonies arise from a liver stem/progenitor cell. Other colonies express only CK19, suggesting that these colonies arise from a cholangiocyte precursor cell.
4. In addition to CD13 and CD133, Dlk also is useful as a surface marker of fetal liver stem/progenitor cells (11).

Acknowledgments

This work was supported in part by Grants-in-Aid for Scientific Research from the Ministry of Education, Culture, Sports, Science, and Technology in Japan. Some figures are reproduced from ref. 17 with permission from Elsevier.

References

1. Mimeault M, Batra SK (2006) Concise review: recent advances on the significance of stem cells in tissue regeneration and cancer therapies. Stem Cells 24:2319–2345
2. Fausto N, Campbell JS, Riehle KJ (2006) Liver regeneration. Hepatology 43:S45–S53
3. Cantz T, Manns MP, Ott M (2008) Stem cells in liver regeneration and therapy. Cell Tissue Res 331:271–282
4. Shafritz DA, Oertel M, Menthena A, Nierhoff D, Dabeva MD (2006) Liver stem cells and

prospects for liver reconstitution by transplanted cells. Hepatology 43:S89–S98

5. Herrera MB, Bruno S, Buttiglieri S, Tetta C, Gatti S, Deregibus MC, Bussolati B, Camussi G (2006) Isolation and characterization of a stem cell population from adult human liver. Stem Cells 24:2840–2850
6. Duret C, Gerbal-Chaloin S, Ramos J, Fabre JM, Jacquet E, Navarro F, Blanc P, Sa-Cunha A, Maurel P, Daujat-Chavanieu M (2007) Isolation, characterization, and differentiation to hepatocyte-like cells of nonparenchymal epithelial cells from adult human liver. Stem Cells 25:1779–1790
7. Tsuchiya A, Heike T, Baba S, Fujino H, Umeda K, Matsuda Y, Nomoto M, Ichida T, Aoyagi Y, Nakahata T (2007) Long-term culture of postnatal mouse hepatic stem/progenitor cells and their relative developmental hierarchy. Stem Cells 25:895–902
8. Lemaigre F, Zaret KS (2004) Liver development update: new embryo models, cell lineage control, and morphogenesis. Curr Opin Genet Dev 14:582–590
9. Dabeva MD, Petkov PM, Sandhu J, Oren R, Laconi E, Hurston E, Shafritz DA (2000) Proliferation and differentiation of fetal liver epithelial progenitor cells after transplantation into adult rat liver. Am J Pathol 156: 2017–2031
10. Suzuki A, Zheng Y, Kondo R, Kusakabe M, Takada Y, Fukao K, Nakauchi H, Taniguchi H (2000) Flow-cytometric separation and enrichment of hepatic progenitor cells in the developing mouse liver. Hepatology 32:1230–1239
11. Tanimizu N, Nishikawa M, Saito H, Tsujimura T, Miyajima A (2003) Isolation of hepatoblasts based on the expression of Dlk/Pref-1. J Cell Sci 116:1775–1786
12. Nitou M, Sugiyama Y, Ishikawa K, Shiojiri N (2002) Purification of fetal mouse hepatoblasts by magnetic beads coated with monoclonal anti-e-cadherin antibodies and their in vitro culture. Exp Cell Res 279:330–343
13. Watanabe T, Nakagawa K, Ohata S, Kitagawa D, Nishitai G, Seo J, Tanemura S, Shimizu N, Kishimoto H, Wada T et al (2002) SEK1/MKK4-mediated SAPK/JNK signaling participates in embryonic hepatoblast proliferation via a pathway different from NF-kappaB-induced anti-apoptosis. Dev Biol 250: 332–347
14. Kakinuma S, Ohta H, Kamiya A, Yamazaki Y, Oikawa T, Okada K, Nakauchi H (2009) Analyses of cell surface molecules on hepatic stem/progenitor cells in mouse fetal liver. J Hepatol 51:127–138
15. Suzuki A, Zheng YW, Kaneko S, Onodera M, Fukao K, Nakauchi H, Taniguchi H (2002) Clonal identification and characterization of self-renewing pluripotent stem cells in the developing liver. J Cell Biol 156:173–184
16. Kamiya A, Kakinuma S, Onodera M, Miyajima A, Nakauchi H (2008) Prospero-related homeobox 1 and liver receptor homolog 1 coordinately regulate long-term proliferation of murine fetal hepatoblasts. Hepatology 48:252–264
17. Kamiya A, Kakinuma S, Yamazaki Y, Nakauchi H (2009) Enrichment and clonal culture of progenitor cells during mouse postnatal liver development in mice. Gastroenterology 137:1114–1126, 1126, e1111–1114
18. Kamiya A, Kinoshita T, Ito Y, Matsui T, Morikawa Y, Senba E, Nakashima K, Taga T, Yoshida K, Kishimoto T et al (1999) Fetal liver development requires a paracrine action of oncostatin M through the gp130 signal transducer. EMBO J 18:2127–2136
19. Kamiya A, Kojima N, Kinoshita T, Sakai Y, Miyaijma A (2002) Maturation of fetal hepatocytes in vitro by extracellular matrices and oncostatin M: induction of tryptophan oxygenase. Hepatology 35:1351–1359

Chapter 17

Long-Term Culture and Coculture of Primary Rat and Human Hepatocytes

Maria Shulman and Yaakov Nahmias

Abstract

The liver is the largest internal organ in mammals, serving a wide spectrum of vital functions. Loss of liver function due to drug toxicity or viral infection is a major cause of death in the United States. The development of Bioartificial Liver (BAL) devices and the demand for pharmaceutical and cosmetic toxicity screening require the development of long-term hepatocyte culture techniques. However, primary hepatocytes rapidly lose their cuboidal morphology and liver-specific functions over a few days in culture. Accumulation of stress fibers, loss of metabolic function, and cell death are known phenomena. In recent years, several techniques were developed that can support high levels of liver-specific gene expression, metabolic and synthetic function for several weeks in culture. These include the collagen double-gel configuration, hepatocyte spheroids, coculture with endothelial cells, and micropatterned cocultures with 3T3-J2 fibroblasts. This chapter covers the current status of hepatocyte culture techniques, including: hepatocyte isolation, media formulation, oxygen supply, heterotypic cell–cell interactions, and basic functional assays.

Key words: Liver, Hepatocytes, Metabolism, Oxygen, Coculture, Culture medium, Non-parenchymal cells

1. Introduction

The liver is the largest internal organ in the human body, ascribed with over 500 functions. Among those functions are embryonic hemopoiesis, protein synthesis (albumin, fibrinogen), bile acid production, glycogen storage, and xenoboitic metabolism, as well as metabolic homeostasis by regulating carbohydrate, lipid, and amino acid levels. One of the most unique features of the liver is its capability to fully regenerate even after 80% mass loss (1). In spite of its ability to regenerate from various insults, loss of liver function

Scott H. Randell and M. Leslie Fulcher (eds.), *Epithelial Cell Culture Protocols: Second Edition*, Methods in Molecular Biology, vol. 945, DOI 10.1007/978-1-62703-125-7_17, © Springer Science+Business Media, LLC 2012

due to drug toxicity or viral infection is a major cause of death in the United States, reaching over 27,000 individuals in 2006 (2). Despite recent advances in split graft transplantation, there is an extreme shortage of human organs, resulting in high prices and low availability of human hepatocytes for clinical, research, or pharmaceutical application. This shortage is aggravated by the inability to expand primary hepatocytes in vitro (3). For these reasons, most human hepatocytes are isolated from marginal livers (fatty, fibrotic) or those rejected for transplantation. Several companies sell freshly isolated human hepatocytes or cryopreserved cells that can be plated for a variety of research, clinical, or pharmaceutical applications.

One such research direction is the development of a Bioartificial Liver (BAL) system, an extracorporeal assist device that could extend the life of those waiting for transplantation by providing critical metabolic and synthetic functions (4). Pharmaceutical drug screening is another major field. Drug development is currently estimated at $400 million/drug with the majority spent on pre-clinical screening (5). However, as animal studies are inadequate to evaluate drug toxicity because of species-specific variations (6), liver drug metabolism models use primary human hepatocytes for ADME/Tox (absorption, distribution, metabolism, excretion and toxicity) screening. Hepatocyte cultures are also used in the study of metabolism (7, 8), liver development (9), regeneration (10), viral infection (11), and inflammation (12).

Reliable culture techniques have become available over the last two decades (13, 14). The main problem with the culture of liver cells is that the cells rapidly lose their cuboidal morphology and liver specific functions during standard culture. The cells accumulate actin stress fibers, described as becoming "fibroblast-like," lack bile canaliculi and die within a few days. Recent developments allow preserving cell function and structure for several months in culture under several widely different configurations. Culture in sandwich configuration or spheroids relies on the assembly of 3D-like structures, while coculture with endothelial cells or 3T3-J2 fibroblasts depends on heterotypic interactions (15). Finally, oxygen supply and medium formulation are critical aspects in the design of the optimal hepatocyte microenvironment (16).

This chapter covers the current status of hepatocyte culture, problems and limitations, and the most common culture techniques. We describe, primary hepatocyte isolation (see Subheading 3.1), hepatocyte culture media (see Subheading 3.2), and hepatocyte culture techniques (see Subheading 3.3). We also cover heterotypic interactions (see Subheading 3.4), and the importance of oxygen supply (see Subheading 3.6). Finally, we describe standard albumin and urea quantification methods (see Subheading 3.7) to characterize liver-specific function.

2. Materials

2.1. Rat Hepatocyte Isolation

1. $CaCl_2$ solution: Add 2.7 g $CaCl_2 \cdot H_2O$ to 196 mL H_2O. Sterilize solution using a 0.22 μm filter and store at 4°C.
2. Krebs Ringer Buffer (KRB): Add to 4 L Ultrapure water 28.54 g NaCl, 3.98 g D-Glucose, 8.40 g $NaHCO_3$, 1.68 g KCl, and 19.06 g HEPES stored at 4°C, pH 7.2–7.4, filter through 0.22 μm filter.
3. KRB with EDTA: Add 1.7 mL of 0.5 M EDTA to 500 mL of KRB buffer on the day of the procedure. Store at room temperature.
4. Collagenase solution: Add 0.105 g collagenase type IV from *clostridium histolyticum* (Sigma-Aldrich®) to 150 mL of KRB (see Notes 1 and 2). Add 9 mL $CaCl_2$ (see Subheading 2.1, item 1) to activate the enzyme and sterile filter. Use within minutes to a few hours after preparation.
5. Percoll® solution (Sigma-Aldrich®).
6. Ketamine 100 mg/mL, Xylazine 20 mg/mL.
7. Instruments: Scalpel blade, blunt-tipped fine forceps, scissors, hemostat, two cotton buds and 2–3 sets of sterile 2 × 2 pads.
8. 250 and 60 μm pore mesh nylon filter.
9. Peristaltic pump.
10. 70% (v/v) Ethanol.
11. 3.0 or 5.0 sutures.
12. 18 or 20 G catheter.
13. 10 cm dish.
14. 50 mL conical tubes.

2.2. Culture Media

1. Thaw Fetal Bovine Serum (FBS): Heat-inactivate the FBS by placing the thawed aliquot in water bath set at 60°C for 30 min.
2. L-Glutamine, 200 mM (see Note 3).
3. Epidermal Growth Factor (EGF) from murine submaxillary gland 0.1 mg (Sigma-Aldrich®): Prepare stock solution by adding 1.0 mL Ultrapure water in sterile conditions, aliquot to 100 μL, and store at −20°C.
4. Penicillin–Streptomycin (10,000 units penicillin and 10 mg streptomycin/mL, Sigma-Aldrich®). Aliquot to 5 mL and store at −20°C.
5. Hydrocortisone sodium succinate for injection, USP, 100 mg (Solu–Cortef®, Pfizer): Prepare stock solution by adding 2 mL Ultrapure water in sterile conditions. Store at 4°C.

6. Insulin, Transferrin, Selenium (ITS) Liquid Media Supplement (Sigma-Aldrich®): 100× stock.
7. Hepatocyte culture medium: Under sterile conditions add 0.2 mM of L-glutamine, 100 μL of EGF stock solution, 75 μL hydrocortisone, 5 mL of liquid ITS, 5 mL of penicillin–streptomycin solution, and 50 mL of FBS, to 440 mL of high glucose Dulbecco's Modified Eagle's Medium (DMEM). Hepatocyte culture medium should be stored at 4°C and used within a few weeks (see Notes 4–6).
8. Rat tail collagen type I (BD) store at 4°C.
9. 10× PBS: Add 80.0 g NaCl, 2.0 g KCl, and 14.4 g Na_2HPO_4 to 1 L Ultrapure water.
10. 1N NaOH: Add 4.0 g NaOH to 10 mL Ultrapure water.

2.3. Albumin and Urea Quantification

1. Purified rat albumin (MP Biomedicals): Prepare 10 mL of 50 μg/mL purified rat albumin in PBS. Store aliquots at −20°C.
2. HRP conjugated Rabbit anti-rat albumin antibody (MP Biomedicals).
3. Phenylenediamine dihydrochloride (OPD) pill, 10 mg.
4. PBS-Tween: Add to 975 mL Ultrapure water, 8.0 g NaCl, 0.2 g KCl, 0.2 g KH_2PO_4, and 2.89 Na_2PO_4.
5. 30% H_2O_2.
6. Blood urea nitrogen (BUN) kit (Stanbio Labs).
7. 96-well plates.
8. Citrate-phosphate buffer: Dissolve 5.1 g Citric Acid Mono-Hydrate and 13.78 g $Na_2HPO_4 \cdot 7H_2O$ in 975 mL Ultrapure water.
9. 8N H_2SO_4: Mix 22.2 mL of concentrated sulfuric acid with 77.8 mL Ultrapure water.

3. Methods

3.1. Primary Hepatocyte Isolation

Primary hepatocytes do not proliferate in vitro and therefore need to be freshly isolated for each experiment. Rats are common source, with 100 to 400 million cells routinely isolated from each animal with >90% viability. The two-step enzymatic digestion technique was first established by Seglen in 1976 (17), and was slightly modified (18). It allows for the isolation of a relatively pure population with high viability. All animals need to be maintained in accordance with National Research Council guidelines and experimental protocols approved by the appropriate institutional Animal Care Committee.

3.1.1. Preparation

1. Open a set of sterilized instruments in a laminar flow hood.
2. Place KRB with EDTA buffer (see Subheading 2.1, item 3) and collagenase solution (see Subheading 2.1, item 4) in water bath and connect to a peristaltic pump. Perfuse at a speed of 17 mL/min.
3. Set a bubble trap and ensure that the line is free of bubbles (see Note 7).
4. Turn on water bath to 40°C.

3.1.2. Surgical Procedure

Rats should normally be 140–180 g body weight. Anesthesia should be administered using an approved protocol and closely monitored. Commonly anesthesia is administered as an intraperitoneal injection of ketamine and xylazine (50–80 mg/kg and 5–10 mg/kg respectively).

1. Once the animal is anesthetized, place and fixate the animal on a surgical table in a dorsal position. Make surgery preparation in a sterile fashion by shaving the abdominal area from the pubis to the xiphoid process and wiping the shaved area with 70% ethanol.
2. Make a median abdominal skin incision. Extend the abdominal incision through the linea alba into the abdominal cavity.
3. Extend the cut to the inferior edge of the xiphoid process. Make two incisions laterally from the midpoint of the abdomen and fixate the flaps with a hemostat.
4. Deflect the intestines to the rat's left. Identify the portal vein, pancreatic and bile ducts, abdominal aorta, and inferior vena cava.
5. Divide the falciform ligament on the ventral and anterior aspects of the liver.
6. Identify the esophagus and the esophageal–gastric junction.
7. The accessory lobes of the rat liver lie posterior to the stomach, nestled in the lesser curvature of the stomach and inferior to the esophageal–gastric junction. Pick the lesser omentum and divide it to fully expose the ventral surface of the accessory lobes. Displace the accessory lobes upwards and identify any posterior ligaments attaching the accessory lobe to the stomach. Divide these. Using a gentle rolling motion, retract these lobes through the window within the lesser omentum.
8. Place two ligatures around the portal vein.
   (a) Proximal ligature: Use a blunt-tipped fine forceps to create a window in the omental triangle formed by the portal vein, the pancreatic bile duct, and the splenic vein. Grasp the omentum adjacent to the portal vein and retract gently upwards. Pass a suture through this window, position it immediately superior to the splenic vein, and ligate loosely with a half-hitch.

(b) Distal ligature: Create a second window between the mesenteric (cranial (superior) and caudal (inferior)) veins, approximately 1.5 cm distal to the proximal ligature. Identify the posterior vessels. Pass a suture through this window and ligate loosely with a half-hitch.

9. Measure an 18 or 20 G catheter lead to the approximate location of the cannula inlet.
10. Grasp the omentum about 1 cm or less, inferior to the distal ligature. Retract gently ventrally and inferiorly, and insert the catheter into the cranial (superior) mesenteric vein, 5 mm inferior to the distal ligature retracting the needle immediately.
11. Advance the catheter until it is 5 mm superior to the proximal ligation. Draw the ligations securely (distal before proximal). Blood should immediately fill the catheter and begin to drip out.
12. Attach the catheter lead to the cannula, and immediately divide the inferior vena cava using the scalpel. The success of the perfusion should be immediately apparent. Blood will drain from the liver, replaced by perfusate, so the color of the liver will change from red to light brown.
13. Make a horizontal incision in the diaphragm ventral to the vena caval foramen. Cut the superior vena cava.
14. Tie the proximal suture, and then the distal suture.
15. Using small scissors cut the ligaments that anchor the liver to the surrounding tissue and the bile duct.
16. Remove the liver and place it in a 10 cm dish for the duration of the perfusion.

3.1.3. Perfusion-Digestion

1. KRB with EDTA solution (see Subheading 2.1, item 3) should be perfused in a single pass, until the liver becomes homogenously light brown.
2. Add the entire collagenase solution (see Subheading 2.1, item 4) in a manner that ensures that no bubbles are allowed into the line.
3. Collagenase digestion will vary with each isolation and batch. Successful enzymatic digestion will be marked by: swelling, lobes "bubbling," and the emergence of cellular debris.
4. Once digestion has been complete, disconnect the perfusion system and immediately place the liver in a petri dish containing 5 mL of ice-cold KRB.

3.1.4. Cell Purification

1. Move the tissue to a sterile biological safety cabinet. The metabolic activity of primary hepatocytes demands that all following steps be carried out on ice to prevent hypoxia.
2. Add ice-cold KRB to the dish, grasp the portal system using forceps, disrupt the capsule and shake the liver to remove the

cells. Add additional KRB buffer as required to wash the cells from the liver.

3. Using a plastic pipette, aspirate the cell suspension slowly and pass it through a 250 μm pore mesh filter to remove tissue fragments.
4. Aspirate the cell suspension several times and pour it slowly through a 60 μm pore mesh filter to attain a single cell suspension.
5. Divide the suspension into several 50 mL conical tubes and centrifuge at $40 \times g$ for 5 min. Centrifuge should be set at 4°C.
6. Re-suspend each cell pellet with 10 mL KRB by gently inverting the tube. Combine cell suspensions to 25 mL volume in 50 mL conical tubes and add additional 25 mL of salt-balanced Percoll solution. Mix by gently inverting the tubes and centrifuge at $500 \times g$ for 5 min.
7. Re-suspend each cell pellet with culture medium (see Subheading 2.2, item 7). Combine all fractions into 50 mL tube and centrifuge at $40 \times g$ for 5 min. Re-suspend cells and assess viability using hemocytometer.

3.2. Hepatocyte Culture Medium

While primary hepatocytes can be maintained in several types of culture medium, there is a clear distinction between serum-free and serum-containing medium formulations. Serum-free formulations are often based on Williams' Medium E and are ideal for short-term cultures, up to 10 days. One such formulation that yields excellent results is Hepatocyte Culture Medium, available commercially from Lonza. Serum-containing formulations were developed for long-term cultures (several weeks) and require an adaptation period (19, 20). One such formulation is detailed above (see Subheading 2.2).

3.3. Hepatocyte Culture Techniques

One common technique to culture primary hepatocytes is to seed the cells on a single layer of collagen gel. Under these conditions, hepatocytes secrete albumin and urea, and show minimal cytochrome P450 (CYP450) activity. Unfortunately, these liver specific functions decline within the first week, suggesting that significant survival factors are missing (see Fig. 1). In 1989, Dunn et al., suggested culturing hepatocytes in a collagen sandwich configuration (20). Adding the second collagen layer induces the formation of distinct apical and lateral membranes. Cellular cuboidal morphology was maintained for 42 days, with albumin and urea secretion slowly rising and stabilizing after 10 days in culture. Interestingly, cells that were cultured on a single layer of collagen for 1 week and after which a second layer of collagen was added, resulted in recovery of albumin secretion levels.

Hepatocytes can also be cultured in the form of hepatic spheroids. When cultured on soft or non-adhesive extracellular matrix,

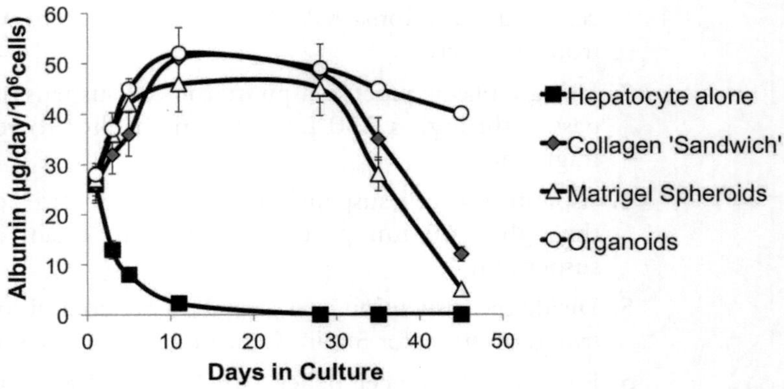

Fig. 1. Long-term albumin production in cultures of primary rat hepatocytes. Primary hepatocytes cultured alone (Hepatocytes alone) rapidly lose albumin secretion in vitro. Cells seeded on Matrigel™ (Matrigel Spheroids), or in a collagen sandwich configuration (Collagen Sandwich) stabilize over 7–10 days and maintain liver specific function for 28 days. Self-assembled hepatic organoids (Organoids) maintain albumin secretion for over 50 days in culture.

cells form spherical aggregates during the first 48 h. Cells in these aggregates maintain their morphology and liver specific functions for over a month (see Fig. 1). The disadvantages of this technique include lack of control over spheroid size, and thus, variation in the transport of metabolites. Another disadvantage is the formation of a necrotic core in big aggregates (1).

Cell morphology in collagen sandwich and spheroid culture configurations is different (see Fig. 2). While hepatocytes entrapped in collagen matrix show a cuboidal morphology (21), spheroids are round, and form closely associated aggregates, which are not found in the mature liver. One assumption is that spheroids resemble the organization of liver during regeneration (22, 23). Unfortunately, hepatocytes cultured in either configuration cannot generate a large quantity of cells, and show little to no proliferation capacity (24) (see Fig 3).

3.3.1. Hepatocyte Sandwich Technique

1. All solutions must be ice-cold.
2. Determine the final volume of collagen solution needed: for a 10 cm² plate, use 0.5 mL collagen solution
3. Determine volume of 10× PBS (see Subheading 2.2, item 9) by dividing the final volume calculated by 10.
4. Calculate the volume of collagen needed by dividing the final volume by collagen concentration in the bottle (see specification sheet)
5. The volume of 1N NaOH (see Subheading 2.2, item 10) is given by the formula:

   Volume of collagen × 0.023 mL = mL NaOH
6. Calculate the volume of Ultrapure water needed, by subtracting from the final volume calculated in step 1 the volume of 10× PBS (step 2), collagen (step 3), and 1N NaOH (step 4)

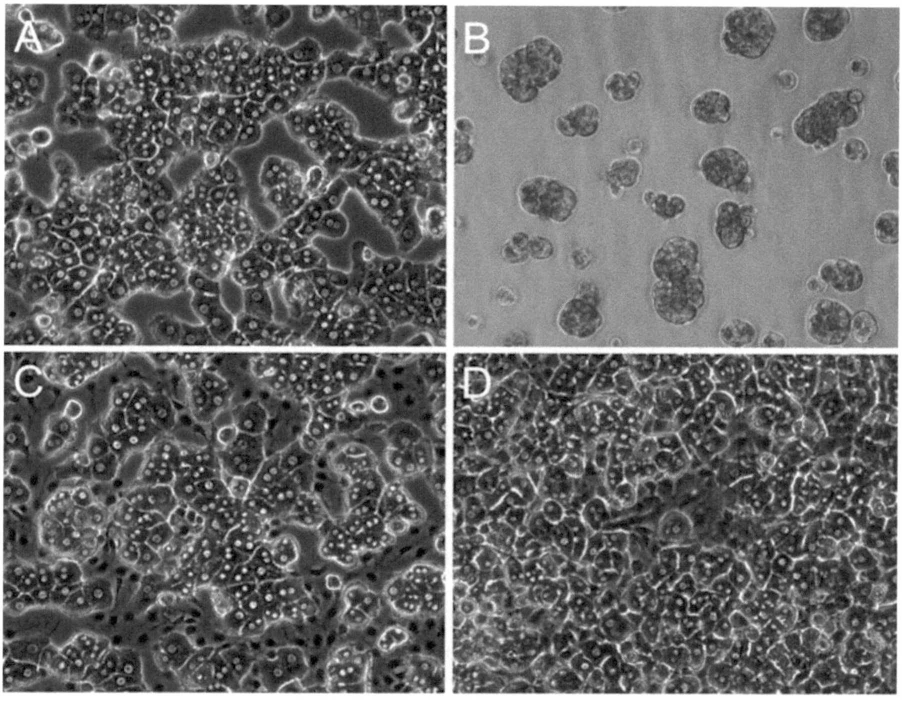

Fig. 2. Phase images of hepatocyte morphology in culture and coculture. (**a**) Hepatocytes cultured in a collagen double gel configuration form plate-like structures and stabilize synthetic and enzymatic activity a week after isolation. Cells in double gel were shown to exhibit native cell–cell contacts such as E-cadherin and bile canaliculi (45), but do not express sinusoidal receptors such as EGF-R (45) and LDL-R (11). (**b**) Hepatocyte spheroids formed on Matrigel exhibit significant synthetic and enzymatic activity. Spheroids were shown to express E-cadherin (45), form extensive bile canaliculi (46), and show sinusoidal surface markers at the interface between cells in the spheroid (46). (**c**) Hepatocytes cocultured with liver sinusoidal endothelial cells show both traditional polarity markers (33, 34), and express a high level of sinusoidal receptors (EGF-R, LDL-R) at the interface between the hepatocytes and the LSEC. At least part of the interaction between hepatocytes and LSEC has been shown to be mediated by growth factors (47, 48). (**d**) Hepatocytes cocultured with 3T3 fibroblasts grow in distinct clusters and exhibit hepatic cell–cell contacts such as connexin-32 (49) and bile canaliculi (50), but also do not express the EGF-R and LDL-R. At least part of the interaction between hepatocytes and 3T3 fibroblasts was shown to be mediated by N-cadherin and decorin (51).

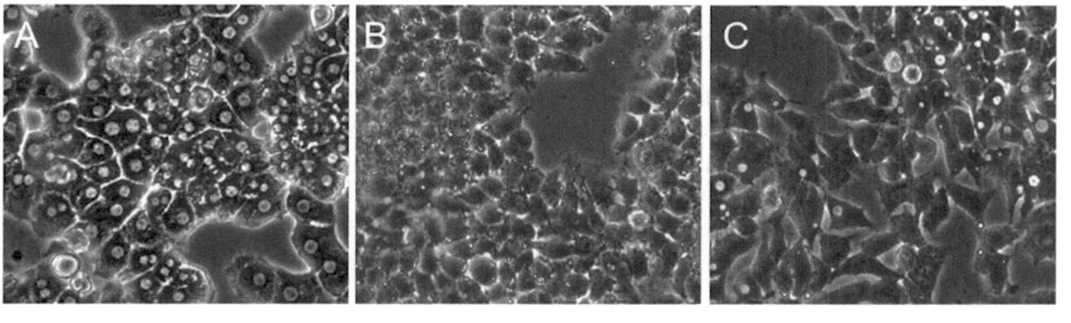

Fig. 3. Morphological differences between primary hepatocytes and hepatocellular carcinoma cell lines. (**a**) Primary rat hepatocytes, (**b**) Huh7 cells (P53 mutant), (**c**) HepG2 cells (P53 wt).

7. In sterile conditions, on ice, add the calculated volume of 1N NaOH to the calculated volume of PBS, and water.
8. Add the calculated volume of collagen and mix gently.
9. Add 0.5 mL mixture into each 10 cm^2 well, and distribute evenly.
10. Incubate at 37°C for 30–40 min.
11. Seed 100,000 cells/cm^2 in a final volume of 1 mL/10 cm^2. This roughly translates to 1×10^6 hepatocyte in each well of a 6-well plate.
12. Incubate cells overnight at 37°C and 5% CO_2.
13. Prepare fresh collagen solution (steps 1–7).
14. Aspirate culture medium and add 0.5 mL of collagen mixture into each well
15. Incubate for 30–40 min.
16. Add 1 mL of hepatocyte culture medium to each well.

3.4. Heterotypic Interactions

Cellular functions are influenced by a sum of extracellular factors, which include neighboring cells, extracellular matrix, soluble factors, and physical forces (25). Neighboring non-parenchymal liver cells play a major role in the regulation of hepatocyte function (26). To mimic these cell–cell interactions, several groups cocultured hepatocytes with non-parenchymal cells, demonstrating improved maintenance of hepatic function (27). The most successful configurations to date include the coculture of primary hepatocytes with microvascular endothelial cells or 3T3-J2 mouse fibroblasts (19, 28). Other work showed that small hepatocytes can marginally proliferate in culture and form large aggregates, in the presence of epithelial cells and fibroblasts (29, 30). In a later study, those aggregates were formed on a collagen mesh and showed liver-specific structure and function (31). When hepatocytes and non-parenchymal cells are cultured in roller bottles, they self-organize and form simple epithelial structures consisting of a superficial layer from biliary epithelial cells, middle layer of hepatocyte and connective tissue, and an inner layer of endothelial cells (32).

In 1997, Bhatia and colleagues created micropatterns of hepatocytes and 3T3-J2 mouse fibroblasts (see Table 1) using lithographic techniques. It was demonstrated that the fibroblasts maintain urea and albumin secretion through a combination of cell contact and short acting diffusible substances. Only hepatocytes that were in close contact to the 3T3-J2 fibroblasts maintained albumin expression, while those distant from the fibroblasts quickly lost function. In a more recent report, Khetani and Bhatia used a polydimethylsiloxane (PDMS) stencil to pattern islands of collagen. Selective human hepatocyte adhesion to these domains yields islands that were subsequently surrounded by 3T3-J2 mouse fibroblasts. Liver specific functions stabilized within 7–10 days and the cells attained synthetic function, mRNA expression, and CYP450 activity, which were similar to in vivo. The authors report optimal function for the configuration of 500 μm islands with 1,200 μm center-to-center spacing. Cell morphology and gene expression were preserved for 4–6 weeks (28).

Endothelial cells were also shown to have a similar supportive function. Early work by Morin and Normand showed that liver sinusoidal endothelial cells (LSEC) stabilize hepatic urea and albumin secretion for up to a month in vitro (33, 34). Nahmias and colleagues demonstrated that micropattering LSEC using laser guided direct writing into capillaries, allows hepatocytes to assemble into sinusoid-like structures which maintained liver specific function for over a month in vitro (35). Later, the same group demonstrated that LSEC induced hepatic expression of basal receptors for low density lipoproteins (LDL) and epidermal growth factor (EGF) as well as the uptake of HCV-like particles (11). LDL uptake is an important function of hepatocytes and was dramatically elevated following coculture of hepatocytes with LSEC. More recent work from our group demonstrated that under the culture medium and oxygen conditions, primary human hepatocytes cocultured with endothelial cells attain synthetic function, mRNA expression, CYP450 activity and drug clearance equivalent to in vivo (19).

Finally, Kupffer cells, the liver's resident macrophages are known to become activated in response to a foreign stimulus and secrete reactive oxygen and nitrogen species (ROS/RNS), which causes hepatic damage. Cocultures of hepatocytes with Kupffer cells have been shown to mimic in vivo damage due to ischemia or drug toxicity (36). While LSEC and Kupffer cells can be difficult to isolate and do not proliferate in vitro, there are several commercially available sources of cells. Non-parenchymal cell sources and culture specifications are listed in Table 1.

3.5. Dynamic Flow Cultures

The culture of hepatocytes under flow is thought to mimic certain aspects of the physiological environment found in vivo. For example, the flow of hormones, nutrients and oxygen across a flat plate bioreactor readily forms gradients due to cellular uptake, such gradients are thought to induce metabolic zonation in vivo. In fact, one group suggested that oxygen gradients formed across the reactor induce metabolic zonation (37), with albumin production high at the entrance (periportal) and CYP450 activity higher at the exit (perivenous). However, it is possible that the loss of synthetic function and gain of CYP450 activity were the result of hypoxia, as oxygen content in the bioreactor was much lower than in vivo. Other groups, however, used similar designs to demonstrate the negative effects of shear forces >5 dyn/cm^2 on hepatocyte function (38), and a higher rate of drug metabolism in perfused cells due to enhanced mass transfer (39).

Table 1
Non-parenchymal cell source and cell culture

Endothelial cells
Liver sinusoids are composed of a highly specialized type of microvascular endothelial cells, which play an important role in lipid metabolism, coagulation, cellular growth, differentiation, immune, and inflammatory response. Liver sinusoidal endothelial cells (LSEC) are usually purified from the nonparenchymal fraction of the liver by a two-step Percoll gradient separation (44). Alternatives for LSEC include rat heart microvaslcular endothelial cells (Vec Technologies) for coculture with rat hepatocytes, and human lung microvascular endothelial cells (Lonza) for coculture with human hepatocytes. In all three cases the endothelial cells are cultured in EGM2mv (Lonza). Add 5 ng/mL of VEGF to hepatocyte culture medium if coculturing with endothelial cells
Kupffer cells
Kupffer cells are the liver resident macrophage, thought to originate from the bone marrow. Kuppfer cells ingest and degrade old erythrocytes, bacteria, various endotoxins, and play an important role in iron metabolism. Due to their similar density and size it is difficult to separate Kupffer cells from LSEC. However, relatively pure populations can be purified using centrifugal elutriation. The ED2 antibody is specific for Kupffer cells. Kupffer cells do not proliferate and rapidly activate in culture. A mouse Kupffer cell line is available (KC13-2)
Fibroblasts
Stellate Cells (Ito cells, fat-storing cells) are vitamin A storing pericytes (fibroblasts), which decorate the liver's sinusoids. Stellate cells are the main matrix-producing cell in the liver and play an important role in regeneration, differentiation, and inflammation. Stellate cells are thought to become activated during liver fibrosis, increasing collagen-synthesis and acquiring a fibroblast-like phenotype. Low density of the cells allows for simple purification using density centrifugation. CD95 and Desmin II are specific markers for stellate cells. HSC become rapidly activated in culture and readily proliferate. An immortalized rat liver stellate cell line (HSC-T6) is available for coculture. One popular cell line shown to support long-term liver specific function is the 3T3-J2 mouse fibroblast cell line available from Prof. Howard Green (Harvard Medical School). 3T3-J2 readily proliferate when cultured in DMEM culture medium supplemented with 10% fetal bovine serum

3.6. Oxygen Supply

Oxygen is an important component in the hepatic microenvironment, which could be derived from the fact that a single hepatocyte contains over 1,500 mitochondria. The oxygen consumption rate of hepatocytes ranges from 0.3 to 0.9 nmol/s/10^6 cells. In vivo, in order to supply cells with this amount of oxygen, the liver is connected to a highly oxygenated arterial network, in addition to the portal circulation, delivering close to 1.29 nmol/s/10^6 cells. Solubility of oxygen in aqueous media is low. That is compensated in vivo by the presence of oxygen-binding hemoglobin (40). In vitro, oxygen is supplied by diffusion from the air-liquid interface, severely limiting the cell density that can be seeded in a given culture area (24).

Several groups suggested that high oxygen tensions stimulate production of free radicals possibly damaging cultured cells (41, 42). However, our group demonstrated that hepatic damage at high oxygen tensions is a result of serum adaptation at the early stages of culture. Removing serum completely from the hepatocyte culture medium, while increasing oxygen tension to 95%, caused a dramatic threefold increase in albumin synthesis and 74% increase in CYP450 activity (19). Gene expression, functional polarization, and drug metabolism were similarly enhanced in both rat and human primary hepatocytes. Remarkably, these oxygenated cocultures showed an ability to predict in vivo hepatic clearance rates of both rapid and slow clearing drugs, such as carbamazepine and antipyrine, with R^2 of 0.92.

3.7. Albumin and Urea Quantification

Albumin synthesis and urea production are classical assays for the study of liver-specific function. Albumin is quantified using enzyme-linked immunosorbent assay (ELISA) and several kits for the quantification of human albumin are available commercially (Bethyl Labs). Standard blood urea nitrogen (BUN) kit for all species is available (Stanbio Labs).

3.7.1. Rat Albumin ELISA

1. Prepare 10 mL purified rat albumin in PBS just before use (see Subheading 2.3, item 1). Add 100 μL albumin solution to each well of a 96-well plate and cover with adhesive plate sealer. Incubate at 4°C overnight. Coated-plates can be stored at 4°C for several days.
2. Using serial dilution, prepare the following rat albumin standards in hepatocyte culture media: 200, 100, 50, 25, 12.5, and 0 μg/mL.
3. Thaw 5 μL of HRP conjugated rabbit anti-rat albumin antibody (see Subheading 2.3, item 2) and dilute 1 to 10,000 in PBS-Tween.
4. Wash albumin-coated plate four times with 100 μL of PBS-Tween
5. Add 50 μL of sample or standard to each well in triplicate, and 50 μL of antibody solution to each well. Incubate overnight at 4°C or alternatively for 3 h at 37°C.
6. Dissolve one 10 mg *o*-Phenylenediamine dihydrochloride (OPD) pill in 25 mL citrate-phosphate buffer (see Subheading 2.3, item 8) at room temperature.
7. Wash plate four times with 100 μL of PBS-Tween.
8. Incubate overnight at 4°C or alternatively for 3 h at 37°C. Wash plate four times with 100 μL of PBS-Tween.
9. Add 10 μL of 30% H_2O_2 to OPD solution.
10. Add 100 μL/well of OPD solution using multi-well pipettor. Incubate for 5 min at room temperature and add 50 μL/well of 8N H_2SO_4 using a multi-well pipettor at the same rate as above.
11. Read OPD absorbance on a plate reader at 405 nm or 450 nm.

3.7.2. Urea Quantification

1. Using serial dilution, prepare the following urea/nitrogen standards in hepatocyte culture media: 200, 100, 50, 25, 12.5, and 0 μg/mL.
2. Add 10 μL of standard or sample to each well of a 96-well plate.
3. Mix 5 mL of BUN color reagent with 10 mL of BUN acid reagent (see Subheading 2.3, item 6).
4. Add 150 μL of the BUN reagent mixture into each well.
5. Cover with adhesive plate sealer and incubate at 60°C for 90 min.
6. Cool plate on ice for 5–10 min before reading.
7. Read plate absorbance at 520 nm.

4. Notes

1. Adjust collagenase concentration to the specific activity of collagenase per batch.
2. Collagenase activity is temperature dependent and fails rapidly below 37°C.
3. L-glutamine rapidly degrades at temperatures above 4°C and should be added before use
4. Human cells require ascorbic acid, absent from the above formulation. Add ascorbic acid if using human hepatocytes. Prepare stock solution (100 mM) by adding 1.76 g ascorbic acid to 100 mL Ultrapure water, aliquot, and store at −20°C. Add 5 mL stock solution to 500 mL hepatocyte culture medium.
5. Phenol red is known to be a weak estrogen (43). Change to phenol red-free basal medium is this is an experimental concern.
6. EGF is not required if hepatocytes are cocultured with 3T3-J2 fibroblasts which are known to secrete this growth factor.
7. Reusable glass bubble traps can be obtained through a commercial vendor (i.e., Radnoti Glass Technology), while single use plastic traps can be fitted from an IV line.

Acknowledgments

The authors wish to thank Ms. Candice Calhoun for technical advice. This work was supported by the National Institute of Diabetes and Digestive and Kidney Diseases (K01DK080241) and the European Research Council Starting Grant (TMIHCV 242699). Resources were provided by the Silberman Institute of Life Sciences and the Center for Bioengineering in the Service of Humanity.

References

1. Nahmias Y, Berthiaume F, Yarmush ML (2007) Integration of technologies for hepatic tissue engineering. Adv Biochem Eng Biotechnol 103:309–329
2. Heron M, Hoyert DL, Murphy SL, Xu J, Kochanek KD, Tejada-Vera B (2009) Deaths: final data for 2006. Natl Vital Stat Rep 57: 1–134
3. Block GD, Locker J, Bowen WC, Petersen BE, Katyal S, Strom SC, Riley T, Howard TA, Michalopoulos GK (1996) Population expansion, clonal growth, and specific differentiation patterns in primary cultures of hepatocytes induced by HGF/SF, EGF and TGF alpha in a chemically defined (HGM) medium. J Cell Biol 132:1133–1149
4. Allen JW, Hassanein T, Bhatia SN (2001) Advances in bioartificial liver devices. Hepatology 34:447–455
5. DiMasi JA, Hansen RW, Grabowski HG (2003) The price of innovation: new estimates of drug development costs. J Health Econ 22:151–185
6. Pritchard JF, Jurima-Romet M, Reimer ML, Mortimer E, Rolfe B, Cayen MN (2003) Making better· drugs: decision gates in non-clinical drug development. Nat Rev Drug Discov 2:542–553
7. Rodrigues MA, Gomes DA, Andrade VA, Leite MF, Nathanson MH (2008) Insulin induces calcium signals in the nucleus of rat hepatocytes. Hepatology 48:1621–1631
8. Hewitt NJ, Lechon MJ, Houston JB, Hallifax D, Brown HS, Maurel P, Kenna JG, Gustavsson L, Lohmann C, Skonberg C, Guillouzo A, Tuschl G, Li AP, LeCluyse E, Groothuis GM, Hengstler JG (2007) Primary hepatocytes: current understanding of the regulation of metabolic enzymes and transporter proteins, and pharmaceutical practice for the use of hepatocytes in metabolism, enzyme induction, transporter, clearance, and hepatotoxicity studies. Drug Metab Rev 39:159–234
9. Kamiya A, Kinoshita T, Ito Y, Matsui T, Morikawa Y, Senba E, Nakashima K, Taga T, Yoshida K, Kishimoto T, Miyajima A (1999) Fetal liver development requires a paracrine action of oncostatin M through the gp130 signal transducer. EMBO J 18:2127–2136
10. Michalopoulos GK, Khan Z (2005) Liver regeneration, growth factors, and amphiregulin. Gastroenterology 128:503–506
11. Nahmias Y, Casali M, Barbe L, Berthiaume F, Yarmush ML (2006) Liver endothelial cells promote LDL-R expression and the uptake of HCV-like particles in primary rat and human hepatocytes. Hepatology 43:257–265
12. Richards CD, Brown TJ, Shoyab M, Baumann H, Gauldie J (1992) Recombinant oncostatin M stimulates the production of acute phase proteins in HepG2 cells and rat primary hepatocytes in vitro. J Immunol 148:1731–1736
13. Dunn JC, Tompkins RG, Yarmush ML (1991) Long-term in vitro function of adult hepatocytes in a collagen sandwich configuration. Biotechnol Prog 7:237–245
14. Koide N, Shinji T, Tanabe T, Asano K, Kawaguchi M, Sakaguchi K, Koide Y, Mori M, Tsuji T (1989) Continued high albumin production by multicellular spheroids of adult rat hepatocytes formed in the presence of liver-derived proteoglycans. Biochem Biophys Res Commun 161:385–391
15. Bhatia SN, Balis UJ, Yarmush ML, Toner M (1999) Effect of cell-cell interactions in preservation of cellular phenotype: cocultivation of hepatocytes and nonparenchymal cells. FASEB J 13:1883–1900
16. Nishikawa M, Kojima N, Komori K, Yamamoto T, Fujii T, Sakai Y (2008) Enhanced maintenance and functions of rat hepatocytes induced by combination of on-site oxygenation and coculture with fibroblasts. J Biotechnol 133:253–260
17. Seglen PO (1976) Preparation of isolated rat liver cells. Methods Cell Biol 13:29–83
18. Berry MN, Grivell AR, Grivell MB, Phillips JW (1997) Isolated hepatocytes—past, present and future. Cell Biol Toxicol 13:223–233
19. Kidambi S, Yarmush RS, Novik E, Chao P, Yarmush ML, Nahmias Y (2009) Oxygen-mediated enhancement of primary hepatocyte metabolism, functional polarization, gene expression, and drug clearance. Proc Natl Acad Sci U S A 106:15714–15719
20. Dunn JC, Yarmush ML, Koebe HG, Tompkins RG (1989) Hepatocyte function and extracellular matrix geometry: long-term culture in a sandwich configuration. FASEB J 3:174–177
21. Berthiaume F, Moghe PV, Toner M, Yarmush ML (1996) Effect of extracellular matrix topology on cell structure, function, and physiological responsiveness: hepatocytes cultured in a sandwich configuration. FASEB J 10:1471–1484
22. Taub R (2004) Liver regeneration: from myth to mechanism. Nat Rev Mol Cell Biol 5: 836–847
23. Michalopoulos GK, DeFrances MC (1997) Liver regeneration. Science 276:60–66
24. Yarmush ML, Toner M, Dunn JC, Rotem A, Hubel A, Tompkins RG (1992) Hepatic tissue engineering. Development of critical technologies. Ann N Y Acad Sci 665:238–252

25. Bhatia SN, Yarmush ML, Toner M (1997) Controlling cell interactions by micropatterning in co-cultures: hepatocytes and 3 T3 fibroblasts. J Biomed Mater Res 34:189–199
26. Kmiec Z (2001) Cooperation of liver cells in health and disease. Adv Anat Embryol Cell Biol 161:III–XIII, 1–151
27. Strain AJ (1999) Ex vivo liver cell morphogenesis: one step nearer to the bioartificial liver? Hepatology 29:288–290
28. Khetani SR, Bhatia SN (2008) Microscale culture of human liver cells for drug development. Nat Biotechnol 26:120–126
29. Mitaka T, Sato F, Mizuguchi T, Yokono T, Mochizuki Y (1999) Reconstruction of hepatic organoid by rat small hepatocytes and hepatic nonparenchymal cells. Hepatology 29:111–125
30. Mitaka T (2002) Reconstruction of hepatic organoid by hepatic stem cells. J Hepatobiliary Pancreat Surg 9:697–703
31. Harada K, Mitaka T, Miyamoto S, Sugimoto S, Ikeda S, Takeda H, Mochizuki Y, Hirata K (2003) Rapid formation of hepatic organoid in collagen sponge by rat small hepatocytes and hepatic nonparenchymal cells. J Hepatol 39:716–723
32. Michalopoulos GK, Bowen WC, Mule K, Stolz DB (2001) Histological organization in hepatocyte organoid cultures. Am J Pathol 159:1877–1887
33. Goulet F, Normand C, Morin O (1988) Cellular interactions promote tissue-specific function, biomatrix deposition and junctional communication of primary cultured hepatocytes. Hepatology 8:1010–1018
34. Morin O, Normand C (1986) Long-term maintenance of hepatocyte functional activity in co-culture: requirements for sinusoidal endothelial cells and dexamethasone. J Cell Physiol 129:103–110
35. Nahmias Y, Schwartz RE, Verfaillie CM, Odde DJ (2005) Laser-guided direct writing for three-dimensional tissue engineering. Biotechnol Bioeng 92:129–136
36. Milosevic N, Schawalder H, Maier P (1999) Kupffer cell-mediated differential down-regulation of cytochrome P450 metabolism in rat hepatocytes. Eur J Pharmacol 368:75–87
37. Allen JW, Khetani SR, Bhatia SN (2005) In vitro zonation and toxicity in a hepatocyte bioreactor. Toxicol Sci 84:110–119
38. Tilles AW, Baskaran H, Roy P, Yarmush ML, Toner M (2001) Effects of oxygenation and flow on the viability and function of rat hepatocytes cocultured in a microchannel flat-plate bioreactor. Biotechnol Bioeng 73:379–389
39. Arno WT, Harihara B, Partha R, Martin LY, Mehmet T (2001) Effects of oxygenation and flow on the viability and function of rat hepatocytes cocultured in a microchannel flat-plate bioreactor. Biotechnol Bioeng 73:379–389
40. Fisher R, Peattie R (2007) Controlling tissue microenvironments: biomimetics, transport phenomena, and reacting systems. Adv Biochem Eng Biotechnol 103:1–73
41. Fariss MW (1990) Oxygen toxicity: unique cytoprotective properties of vitamin E succinate in hepatocytes. Free Radic Biol Med 9:333–343
42. Martin H, Sarsat JP, Lerche-Langrand C, Housset C, Balladur P, Toutain H, Albaladejo V (2002) Morphological and biochemical integrity of human liver slices in long-term culture: effects of oxygen tension. Cell Biol Toxicol 18:73–85
43. Berthois Y, Katzenellenbogen JA, Katzenellenbogen BS (1986) Phenol red in tissue culture media is a weak estrogen: implications concerning the study of estrogen-responsive cells in culture. Proc Natl Acad Sci U S A 83:2496–2500
44. Smedsrod B, Pertoft H (1985) Preparation of pure hepatocytes and reticuloendothelial cells in high yield from a single rat liver by means of Percoll centrifugation and selective adherence. J Leukoc Biol 38:213–230
45. Moghe PV, Berthiaume F, Ezzell RM, Toner M, Tompkins RG, Yarmush ML (1996) Culture matrix configuration and composition in the maintenance of hepatocyte polarity and function. Biomaterials 17:373–385
46. Abu-Absi SF, Friend JR, Hansen LK, Hu W-S (2002) Structural polarity and functional bile canaliculi in rat hepatocyte spheroids. Exp Cell Res 274:56–67
47. Davidson AJ, Zon LI (2003) Love, honor, and protect (your liver). Science 299:835–837
48. LeCouter J, Moritz DR, Li B, Phillips GL, Liang XH, Gerber H-P, Hillan KJ, Ferrara N (2003) Angiogenesis—independent endothelial protection of liver: role of VEGFR-1. Science 299:890–893
49. Sugimachi K, Sosef MN, Baust JM, Fowler A, Tompkins RG, Toner M (2004) Long-term function of cryopreserved rat hepatocytes in a coculture system. Cell Transplant 13:187–195
50. Bhandari RN, Riccalton LA, Lewis AL, Fry JR, Hammond AH, Tendler SJ, Shakesheff KM (2001) Liver tissue engineering: a role for coculture systems in modifying hepatocyte function and viability. Tissue Eng 7:345–357
51. Khetani SR, Szulgit G, Rio JAD, Barlow C, Bhatia SN (2004) Exploring interactions between rat hepatocytes and nonparenchymal cells using gene expression profiling. Hepatology 40:545–554

Chapter 18

Tissue Culture Correlational Study of Genetic Cholangiopathy of Autosomal Recessive Polycystic Kidney Disease

Yasuni Nakanuma, Yasunori Sato, and Kenichi Harada

Abstract

Cholangiocytes are epithelial cells that line the biliary tract and are also known as biliary epithelial cells (BECs). In vitro culture studies of BECs in correlation with tissue section examination may give us a comprehensive analysis of biliary tract diseases. Herein, we discuss genetic cholangiopathy of autosomal recessive polycystic kidney disease (ARPKD), mainly using a polycystic kidney (PCK) rat, an animal model of ARPKD. The hepatobiliary lesions in ARPKD patients (Caroli's disease and congenital hepatic fibrosis) and in PCK rats are speculated to be related to mutations to polycystic kidney and hepatic disease 1 (*PKHD1*) which have been recently demonstrated, though the exact causal relation between these mutations and hepatobiliary pathology remain to be clarified. Recently we clarified that BECs of PCK rat showed increased cell proliferation followed by irregular dilatation of intrahepatic bile ducts. We also identified the essential involvement of the MEK5-ERK5 pathway in the abnormal proliferation of BECs in the PCK rat. The degradation of laminin and type IV collagen (basal membrane components of bile ducts) was closely related to the biliary dysgenesis and cystogenesis in the PCK rats. BECs also showed mesenchymal phenotype followed by progressive portal tract fibrosis, indicating TGF-β1 may be involved in this acquisition of mesenchymal phenotype. Detailed tissue culture correlation studies of ARPKD and PCK rats are mandatory to evaluate the pathogenesis of this genetic cholangiopathy.

Key words: Biliary tract, Biliary epithelial cells, Cholangiopathy, Polycystic liver, MAPK kinase, MEK5, Epidermal growth factor, Collagen gel, Basement membrane

1. Introduction

Cholangiocytes, also called as biliary epithelial cells (BECs), are lining epithelial cells of the biliary tract which comprises a complex three-dimensional network of conducts from the intrahepatic small bile ducts to extrahepatic bile duct (1, 2). BECs are specialized secretory/excretory epithelial cells involved in the modification of hepatic bile and are increasingly recognized as biologically and pathologically important epithelial cells because of the diverse array of cellular processes in which they participate (3). As for the biliary

Scott H. Randell and M. Leslie Fulcher (eds.), *Epithelial Cell Culture Protocols: Second Edition*, Methods in Molecular Biology, vol. 945, DOI 10.1007/978-1-62703-125-7_18, © Springer Science+Business Media, LLC 2012

anatomy, the biliary tree is divided into the intrahepatic, hilar (right and left hepatic duct), and extrahepatic bile duct (EHBD) (1). The intrahepatic bile duct is a branch of the right or left hepatic bile duct, and the first to third branches of both hepatic ducts are known as the intrahepatic large bile duct. Intrahepatic small bile ducts are classified into septal and interlobular bile ducts, and bile ductules are located at the periphery of the portal tracts. The biliary tree is covered by a layer of BECs, and the large bile ducts have their own dense fibrous wall. The BECs of the large bile duct are usually columnar, while BECs of the bile ductules and small intrahepatic bile ducts are small and cuboidal cells.

There have been many studies on cholagiocyte biology and biliary tract diseases such as immune-mediated, inflammatory, and neoplastic cholangiopathies. In this chapter, we will focus on recent progresses in genetic cholangiopathy (4, 5). Autosomal recessive polycystic kidney disease (ARPKD) and autosomal dominant polycystic kidney diseases (ADPKD) are representative genetic cholangiopathy showing polycystic changes in the liver (5). Caroli's disease and congential hepatic fibrosis (CHF) are well-known hepatobiliary lesions of ARPKD (4), and the polycystic kidney (PCK) rat has been established as an animal model of ARPKD (6). Mutations to *PKHD1* have been identified in ARPKD, including Caroli's disease with CHF as well as the PCK rats (7). Particularly, the gene responsible for human ARPKD (the *PKHD1* gene) was identified on human chromosome 6 as fibrocystin. In the PCK rat, a spontaneous splicing mutation of *Pkhd1* initiates the development of cystic dilatation of the bile ducts. *PKHD1* is a large gene from which multiple transcripts may be generated by alternative splicing. Developing and mature intrahepatic bile ducts express the *PKHD1* protein, fibrocystin, whereas bile ducts of ARPKD patients do not. In the PCK rat, BECs of cystically dilated bile ducts possess short and malformed cilia that do not express fibrocystin, the *Pkhd1* protein. It seems likely that the liver and kidney lesions in ARPKD patients and PCK rats are related to these mutations to *PKHD1*. However, the exact causal relations, including the pathogenesis between such mutations to *PKHD1* and the pathologic conditions including the hepatobiliary dysgenesis in ARPKD patients and the PCK rats, are not fully understood.

Herein, the methods for the analysis of pathologic BECs of ARPKD and their results will be discussed using mainly PCK rats (6). So far the studies on the cholangiopathies of ARPKD using PCK rats have been mainly done in our group and Mayo's group (7–13). The representative methods for the cell culture method including three-dimensional cell culture in collagen gel and the preparation and isolation of BECs from the biliary tract will also described. The tissue culture correlational study will give us a comprehensive analysis of biliary dysgenesis and cystogenesis of ARPKD (9).

2. Materials

2.1. Autosomal Recessive Polycystic Kidney Disease

Caroli's disease and CHF are a hepatobiliary manifestation of human ARPKD (5). The former is characterized by a congenital, multiple, and saccular dilatation of the intrahepatic bile ducts and the latter by tortuous and abnormally shaped dilated bile ducts and ductules in fibrously enlarged portal tracts at the microscopic level (see Note 1). These biliary lesions are regarded as a failure of remodeling of the ductal plate after birth ("ductal plate malformation").

2.2. PCK Rat

PCK rat, a mutant animal established from a Crj:CD/SD strain and a rodent model of ARPKD, is found to show characteristic hepatobiliary features identical to Caroli's disease with CHF in addition to renal cystic lesions (8–13). The PCK rat has also been established as an animal model of human ARPKD (see Note 2).

2.3. Culture Solution for Rat Intrahepatic BECs

1. Ca^{2+}- and Mg^{2+}-free Hanks' Balanced Salt solution (HBSS, pH 7.2).
2. Collagenase solution: Dulbecco's modified Eagle's medium and Ham F-12 (DMEM/F-12, Gibco®) containing 0.04% collagenase (collagenase S-1, Nitta Gelatin) and 0.22% dispase (Life Technologies™), sterile filter.
3. Standard medium for cultured rat BECs: DMEM/F-12 medium containing 10% Nu-Serum (Becton Dickinson), 1% insulin-transferrin-selenium (ITS, Becton Dickinson), 5 μM/L forskolin, 1 μM/L dexamethasone (Sigma-Aldrich®), 5 μM/L triiodo-thyronine (Sigma-Aldrich®), 5 mg/mL glucose (Sigma-Aldrich®), 25 mM/L sodium bicarbonate, 1% antibiotics-antimycotic (Life Technologies™), and 20 ng/mL epidermal growth factor (EGF, Upstate Biotechnology).
4. EGF: Dissolve 1 mg/mL in DMEM/F-12, store at −20°C in single-use aliquots.
5. Forskolin: Dissolve 4.11 mg/L in HBSS, store at 4°C, filter-sterilize.
6. Dexamethasone: Dissolve 7 mg/L in HBSS, store at 4°C, filter-sterilize.
7. Triiodo-thyronine: Dissolve 10 mg/mL in HBSS, store at 4°C, filter-sterilize.
8. Sodium bicarbonate: Dissolve 2.17 g/L in culture solution, store at 4°C, filter-sterilize.
9. Type I collagen gel matrix: At 8:1:1 ratio of solution A (Cellmatrix Type 1-A, Nitta Gelatin), solution B (10× concentrated DMEM/F-12), and solution C (0.05N sodium hydroxide, 260 mM/L sodium bicarbonate, and 200 mM/L HEPES).

10. Sodium hydroxide: Dissolve 1.1 g/50 mL in distilled water.
11. HEPES: Dissolve 2.39 g/50 mL in distilled water, filter-sterilize.
12. 24 G cannula.
13. Surgical tools required to open the abdomen: scissors and tweezers, sterilized.
14. Ether or a locally acceptable replacement, i.e., isoflurane.
15. 50 mL centrifuge tube.
16. Dispase (1,000 U/mL): 2 g/L in DMEM/F-12, filter-sterilize.
17. 6-well and 96-well plates.
18. 60 mm dish.
19. Cell scraper.
20. 10% Neutral-buffered formalin.
21. Gefitinib: Dissolve 10 mM in DMSO, stored at −20°C.
22. Matrigel™ (BD Biosciences): Stored at −20°C.
23. Growth factor reduced Matrigel™ matrix (BD Biosciences): Stored at −20°C.
24. Sirius red reagent (Biocolor Ltd.): Sirius red dye in pycric acid, provided in the Sircol collagen assay kit.
25. α2-antiplasmin (Calbiochem): Dissolve 1 mg/mL in sterile water, stored at −20°C.
26. TGF-β1 (2 ng/mL; R&D Systems, Inc.): Dissolve in sterile 4 mM HCl containing 1 mg/mL BSA, stored at −20°C in single-use aliquots.
27. Sircol collagen assay kit (Biocolor Ltd.): A quantitative dye-binding method designated for the analysis of acid-soluble collagens.

3. Methods

In addition to histologic, molecular, and genetic studies of the tissue from the biliary tract in various diseases, variable approaches to pathophysiology of the biliary tree are available using cultured BECs (2, 13, 14). The use of monolayer and three-dimensional cell culture systems has become an accepted method in hepatobiliary research (2, 8, 13, 14). Depending on the diseases affecting the biliary tract, the aim and design of the experiment should guide the selection of the most appropriate isolation and cultivation technique. Herein, the characteristics of cultured BECs with respect to biliary dysgenesis and cystogenesis in ARPKD will be reviewed by referring to in vivo tissue studies, mainly using PCK rats. According to our recent studies (6–11), the increased proliferation of BECs of PCK rat led to irregular dilatation of intrahepatic bile ducts. The essential involvement of the MEK5-ERK5 pathway was pointed out in the abnormal proliferation of BECs of the PCK rat. The degradation of laminin and type IV collagen was closely related to the biliary dysgenesis in the PCK rats. BECs also showed mesenchymal phenotype followed by progressive portal tract fibrosis, and TGF-β1 may be involved in this transformation. In vivo studies of tissue sections of the biliary tract combined with in vitro culture studies of BECs give us a comprehensive analysis of biliary tract pathophysiology.

3.1. Establishment of Cultured BECs

Because the dilatation of the intrahepatic large bile ducts has been regarded as an essential feature of Caroli's disease, BECs were isolated from the intrahepatic large bile duct from 8-week-old PCK rats and from those of control rats (8, 14) and cultured for analysis.

3.1.1. Isolation and Monolayer Cell Culture of BECs

Under ether anesthetization, the abdomen was opened and a 24 G cannula was inserted into the inferior vena cava as an inflow and the portal vein was cut to make an outflow. The livers were first perfused with HBSS via the portal vein trunk, and then with collagenase solution (see Subheading 2.3, item 2) for 20 min (2, 5, 6). Several tissue fragments were cut from the parts of the biliary tree (see Note 3), and floated for 48 h in standard medium for cultured rat BECs (see Subheading 2.3, item 3) in a 5% CO_2 incubator. Type 1 collagen gel matrix (see Subheading 2.3, item 3) was prepared and poured into 60 mm dishes. The tissue fragments were then placed as an explant on type I collagen gel, and cultured at 37°C in a 5% CO_2 incubator.

The dishes were periodically observed under a phase-contrast microscope for 2 weeks. BECs proliferated on collagen gels and spread as a sheet from the edge of these explants. Some areas of these sheets, which consisted only of BECs with no mesenchymal contamination, were selectively cut with scissors and placed on other collagen gels as secondary explants (subculturing). Subcultured BECs on gels were fed with standard medium in a 5% CO_2 incubator, and passaged every 3 weeks by cutting them with scissors. After four passages, the cells were scraped off using a cell scraper and placed in a collagenase solution (see Subheading 2.3, item 2). After centrifugation $200\times g$ for 5 min at room temperature, the pellets were added to HBSS and cells were set onto collagen-1-coated plastic dishes for 3 weeks until confluent. Isolated and cultured intrahepatic BECs from the livers of both control and PCK rats proliferated and spread as an epithelial cell sheet on the collagen gels (see Fig. 1a). Histologically, the cultured BECs after four passages, which were used for the subsequent experiments, showed a monolayer of cuboidal epithelial cells on the collagen gel.

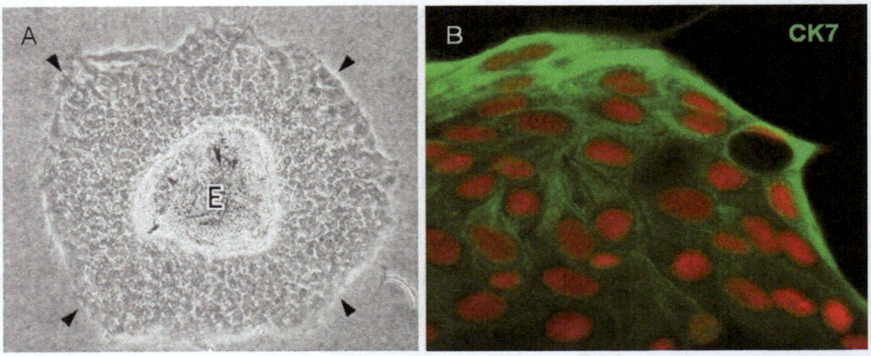

Fig. 1. Cultured intrahepatic biliary epithelial cells (BECs) obtained from polycystic kidney (PCK) rats. (**a**) BECs formed an epithelial sheet (*arrowheads*) spreading from the isolated epithelial cell explant (E). (**b**) Confocal immunocytochemistry for cultured BECs derived from a PCK rat. The cytoplasm of BECs were positive for cytokeratin 7 (*green*), and the nucleus was counterstained with propidium idodide (*red*) (modified from ref. 8).

Next, cultured cells were scraped off and stored at −80°C for the culture experiments. These cell lines were confirmed to be BECs by the expression of a biliary-type cytokeratin, CK19 (see Fig. 1b).

3.1.2. Three-Dimensional Cell Culture of BECs

The BECs (passage 4) of the PCK rats and control rats were subcultured on collagen gels for 3 weeks (2, 9, 14). These collagen gels were scraped using cell scrapers and placed into a 50 mL centrifuge tube containing 10 U/mL dispase. The tube was rocked for 40 min. BECs were dispersed and directly embedded in a fluid collagen gel matrix. The fluid collagen gel was divided into 6-well plates, which soon became gelatinous. The rat BEC cultures were then covered with 2 mL standard medium (see Subheading 2.3, item 3) and cultured at 37°C in an atmosphere of 5% CO_2.

3.2. Morphology of Cultured BECs

3.2.1. Preparation for Histologic Studies

BECs cultured on collagen gels for 72 h were fixed in 10% neutral-buffered formalin, and cut into several slices, which were embedded in paraffin. Then, several sections, 3-μm thick, were cut and deparaffinized, and stained with HE and examined morphologically.

3.2.2. Histology and Phenotypes

An oval or round nucleus was located in the middle of the cytoplasm of cultured BECs, and acid glycoprotein was clearly detected on the apical cell surface. Phenotypes of in vivo BECs such as cytokeratin (CK)-7, CK-19, and γ-GTP were expressed diffusely in the cytoplasm of cultured BECs of PCK and control rats, while vimentin was absent (2, 8, 13). These histological and phenotypic features were similar in the cultured BECs of control and PCK rats. Muff et al. (13) showed that BECs from PCK rats grown on collagen, formed a polarized monolayer with well-developed junction complexes, and distinct apical and basolateral membranes. Cultured BECs from the PCK rat retained properties of the in vivo BECs lining cystically dilated bile ducts.

3.3. Increased Cell Proliferation of BECs of PCK Rats

The normal development of intrahepatic bile ducts requires a balance between the proliferation and deletion of BECs (5). The bile ductal changes of ARPKD characterized by an irregular and multiple dilatation are thought to be caused by a combination of uneven and disproportionate overgrowth of BECs and their supporting connective tissue (6).

3.3.1. Increased Growth of BECs in Three-Dimensional Cell Culture

When BECs were seeded and cultured in a collagen gel matrix, BECs proliferated and became spherical or elliptical small cystic masses 2–3 days later under phase-contrast microscopy (9). The micro cysts gradually enlarged and transformed into multicellular cysts forming well-developed cysts. The size and number of the biliary cysts increased over a period of 2 weeks. The cultured BECs of PCK rats in collagen gels formed cystic structures which expanded progressively in their size and number in comparison with those of normal rats (9) (see Fig. 2a, b, d, e).

Muff et al. (13) also did the three-dimensional cell culture study from BECs of the PCK rat and control rats. The BECs cells seeded in collagen gels organized into cystic structures as assessed by light microscopy. In BECs of normal and PCK rats, formation of small cysts was observed within 24–48 h and the cysts continued to grow. Both cultures were maintained in this configuration for up to 3–4 weeks. However, the surface areas of the cysts formed by BECs of PCK rat were substantially larger compared to BECs of control rats. The surface areas of the cysts of PCK rats continued to grow and expand progressively up to day 21 while those of cysts formed by BECs of control rats grew up to days 3–6 until they reached the plateau around day 9 and remained the same size after day 9.

3.3.2. Molecules and Genes Related to Increased Cell Proliferation of BECs

The overall differences in gene expression between cultured BECs of control and PCK rats at the subconfluent stage were examined using a cDNA microarray analysis (8). It was found that the vast majority of the 1,081 genes examined were expressed to a similar degree in both rats. In the PCK rat, the expression of 18 genes was enhanced by at least twofold, and the expression of three genes was reduced by half or less compared with the control. Although most of the 21 differentially expressed genes identified here have not been well studied with regard to function in BECs, seven genes (the mitogen-activated protein kinases (MAPK)/ERK (extracellular signal-regulated protein kinase) kinase 5 (MEK5), thymidine kinase, junD, TGF-β3, TGF-β type I receptor (TβR-I), basic fibroblast growth factor, and plasma glutathione peroxidase) seemed to be related to cell kinetics. Among them, MEK5 is a component of the MAPK pathway, which is typically involved in epithelial cell proliferation by reacting with EGF. As described below, we identified the essential involvement of the MEK5-ERK5 pathway in the abnormal growth of BECs of the PCK rat (8).

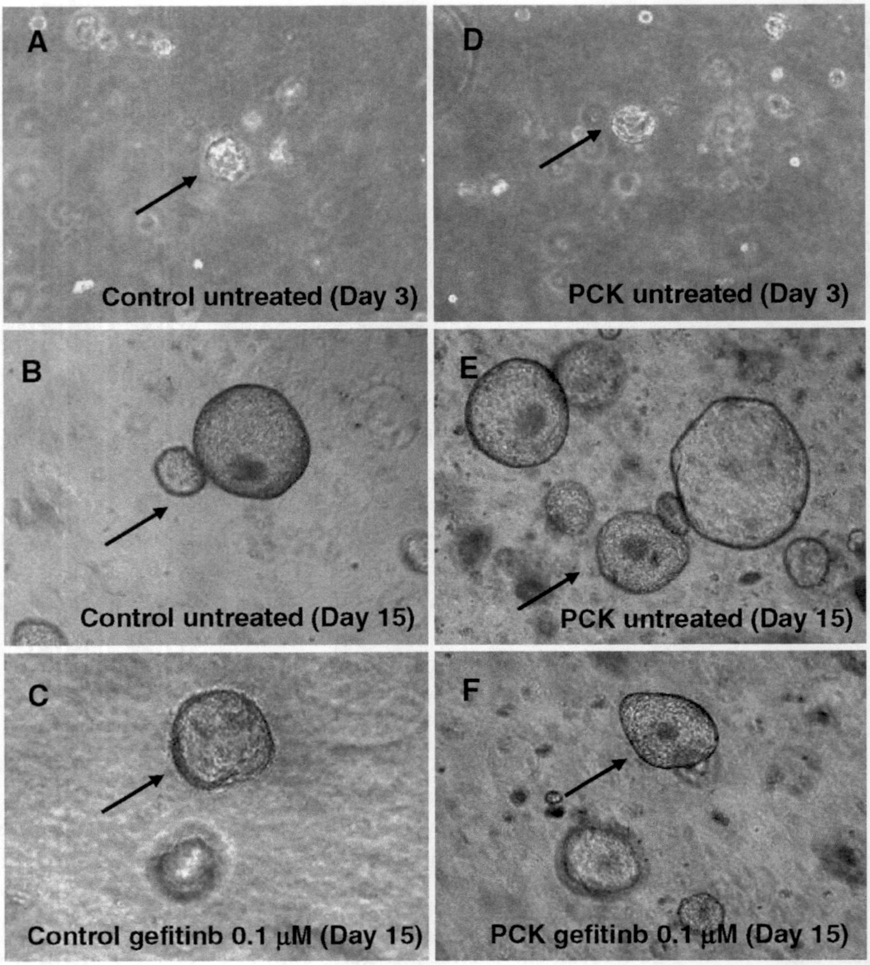

Fig. 2. Cyst formation of cultured BECs and inhibitory effects of Gefitinib on cyst formation. BECs were obtained from the control and PCK rats, and the three-dimensional cell culture system was used for the analysis. Gefitinib (0–1 m*M*) was administered at the beginning of the three-dimensional cell culture and on day 7 after the beginning of the culture, and the size and number of biliary cysts were monitored. **a** to **f**: Representative photographs of the biliary cysts in a collagen gel matrix (**a** to **c**, control rats; **d** to **f**, PCK rats). Day 15 (**c** and **f**) is equal to day 8 after Gefitinib treatment. *Arrows* indicate biliary micro cysts. When BECs were cultured in a collagen gel matrix with the medium that did not contain Gefitinib, the cells became spherical or elliptical small cystic masses 2–3 days later under phase-contrast microscopy (**a**, **d**). The micro cysts gradually enlarged and transformed to multicellular cysts forming well-developed cysts. The size and number of the biliary cysts increased over a period of 15 days (**b**, **e**). Gefitinib had inhibitory effects on the size and number of the cysts for both the control and PCK rats (**c** and **f**) (9).

By binding to EGF receptor (EGFR), EGF activates the MAPKs. The MAPK pathway consists of three protein kinases that act sequentially within a pathway: a MAPK kinase kinase, a MAPK/ERK kinase (MEK), and a MAPK (ERK). The first and best studied is the MAPK pathway which consists of Raf-1 or B-Raf, MEK1/2, and ERK1/2. Recently, the novel MEK5-ERK5 pathway has been implicated in the regulation of cellular proliferation by acting with EGF (15). That is, EGF activates ERK5, and this activation requires the direct and specific upstream activator, MEK5.

3.3.3. In Vivo Tissue Studies

EGF was immunohistochemically localized in the mononuclear cells around the relatively large bile ducts in PCK rats, and more EGF-positive cells were accumulated around dilated large bile ducts. EGFR was expressed on almost all BECs, mainly on their basolateral location, in PCK and normal rats. Double immunostaining of EGFR and Ki-67 expression disclosed that while almost all BECs of both rats expressed EGFR, a considerable number of these BECs were positive for Ki-67 in PCK rats. This data suggests that abnormal expression of EGF and EGFR in the liver has been implicated in biliary cystogenesis in PCK rats (6) (see Note 4).

While only a few positive BECs show positive nuclear signals for phosphorylated (p)-ERK1/2 in PCK rats, p-ERK5 were diffusely expressed in the nuclei of BECs in PCK rats, and the signal intensity in BECs was remarkably higher in PCK rat. The Ki-67 labeling index of BECs was continuously higher in the PCK rats, with a value of 10.5, 8.2, and 13.5% at 3 weeks, 2 months, and 10 months of age, respectively, when compared to the index of the control rats (2.1, 0.5, and 0.3% at 3 weeks, 2 months, and 10 months of age), suggesting that p-ERK5 overexpression was involved in increased proliferative activity of BECs in PCK rats.

3.3.4. Culture Studies

The proliferative activities of cultured BECs of PCK and control rats increased dose-dependently in response to EGF. This proliferative activity of cultured BECs was significantly increased in PCK rats. EGFR mRNA was detected in cultured BECs of control and PCK rats, suggesting that BECs obtained from PCK rats were hyper responsive to EGF.

3.3.5. Western Blot Analysis

The expression of MAPK kinases (MEK1/2, p-MEK1/2, and MEK5) and MAPKs (ERK1/2, p-ERK1/2, ERK5, and p-ERK5) in cultured BECs was examined. Under the stimulation with EGF, overexpression of MEK5 was observed in cultured BECs of the PCK rat, which corresponded with its gene expression shown by a cDNA microarray analysis (8). The expression of p-MEK1/2 and p-ERK1/2 was faint or invisible. In contrast, the expression of p-ERK5 was increased in PCK rats (see Fig. 3), suggesting that increase proliferation of BECs was accompanied by overexpression of the MEK5, and subsequent phosphorylation of ERK5, and that activation of the MAPK pathway may be critically involved in the development of biliary dysgenesis of PCK rat. Thus, the involvement of the MEK5-ERK5 pathway seems essential in the abnormal growth of BECs of PCK rats.

3.3.6. Inhibition Studies

Western blot analysis confirmed that short interfering RNA (siRNA) against MEK5 mRNA blocked the protein expression in the cultured BECs of PCK rats (9). The EGF-induced increased proliferative activity of cultured BECs of PCK rats was significantly inhibited by the transfection of siRNA against MEK5 mRNA. An EGFR tyrosine kinase inhibitor, gefitinib ("Iressa"), also significantly inhibited the increased proliferative activity of BECs of PCK and normal rats. Both PD98059 and U0126, inhibitors for MEK1/2, were less effective in the inhibition of BECs from PCK rats, suggesting that the activation of the MEK5-ERK5 cascade plays a pivotal role in the biliary dysgenesis of PCK rats.

3.3.7. Effects of Gefitinib on Cystogenesis in Three-Dimensional Cell Culture

When Gefitinib was administered at the beginning of culture, biliary cysts were developed in the control and PCK rats and Gefitinib had dose-dependent inhibitory effects on the size and number of the cysts for both the control and PCK rats (9). When Gefitinib was administrated 7 days after the beginning of the culture, it also inhibited the increase in the size and number of the cysts for both the control and PCK rats. The dosage of 0.1 μM Gefitinib inhibited the cyst formation, which was more prominent in PCK rats than those of control rats after treatment (see Fig. 2c, f). At the dosages of 0.5 and 1 μM, the once formed biliary cysts decreased in size and number by day 12 after treatment in both the control and PCK rats (9).

3.4. Basement Membranes and Cystogenesis and Biliary Dysgenesis

Degradation of the basement membrane and extracellular matrix (ECM) constituents, and the remodeling of the ECM are important in processes of the development of the biliary tract (5). Most of the proteolytic enzymes involved in these processes belong to the matrix metalloproteinases (MMPs) and the serine proteinases, in particular the plasminogen activator (PA)/plasmin system. Generation of plasmin from the extracellular zymogen plasminogen can be catalyzed by either of two other serine proteinases: the tissue- and urokinase-type plasminogen activators (tPA and uPA). Both tPA and uPA have been shown to contribute to the plasminogen-dependent lysis of basement membrane laminin. In addition, plasmin contributes to activation of the zymogens of MMPs such as MMP-9 and MMP-13, which in turn degrade basement membrane components, including type IV collagen (16).

3.4.1. In Vivo Tissue Studies

For the development of bile ducts, the coordinated expression of basal laminar components such as laminin and type IV collagen and proteolytic enzymes is essential. Laminin and type IV collagen are invariably expressed in the basement membrane of normal bile ducts. Interestingly, the expression of laminin and type IV collagen around the dilated intrahepatic bile ducts was reduced, or discontinuous, or lost in Caroli's disease with CHF and the PCK rats (10). Additionally, the expression of tPA was increased in the BECs of PCK rats at any age.

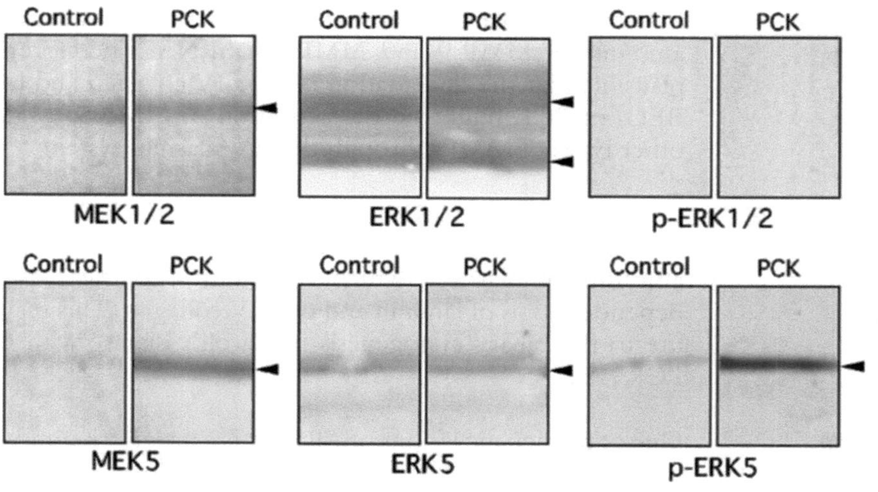

Fig. 3. Western blot analysis of the selective components of the MAPK pathway in cultured BECs of control and PCK rats. BECs were cultured in the presence of EGF (20 ng/mL), and 100 μg of total protein extracted from BECs were used. Overexpression of MEK5 and p-ERK5 was observed in cultured BECs of PCK rats in comparison with control rats. There were no differences in the expression of MEK1/2, ERK1/2, ERK5, and p-ERK1/2 was not detected in either rats (8).

3.4.2. The Collagen Degradation Assay Using Collagen Gel and Culture of BECs

To examine the ability of BECs of PCK rats to degrade ECM, BECs were cultured in Matrigel™ for 10 days, and the amounts of laminin and collagen in the gel were determined (see Note 5). BECs were embedded in a fluid gel matrix (see Subheading 2.3, item 23) in 96-well plates, and was then covered with a standard culture medium (200 μL) and incubated at 37°C in an atmosphere of 5% CO_2. Sirius red reagent (1 mL) was added to each protein extract (100 μL) and mixed for 30 min. The collagen–dye complex was precipitated by centrifugation at 15,000 × *g* for 10 min and dissolved in 0.5 M sodium hydroxide. Finally, the samples were introduced into a microplate reader and the absorbance was determined at 540 nm. The amounts of laminin and collagen in the gel used for the culture of BECs of PCK rats was significantly reduced compared with those of the control gel (gel only), suggesting that BECs of PCK rats had the ability to degrade laminin and collagen in Matrigel™.

3.4.3. Expression of Proteolytic Enzymes by BECs

Reverse transcriptase-polymerase chain reaction (RT-PCR) analysis showed that mRNA of tPA and plasminogen was overexpressed in cultured BECs of PCK rats. At the protein level, overexpression of plasminogen and tPA was also observed in cultured BECs of PCK rats. Expression levels of uPA and PA inhibitor 1 (PAI-1) mRNA were similar between BECs of normal and PCK rats.

3.4.4. Effects by Blocking Plasmin Activity

To block the activity of plasmin, α2-antiplasmin, a plasmin inhibitor, was added to the culture medium at a concentration of 1.0 μM. The addition of α2-antiplasmin, inhibited the degradation of laminin and collagen in the gel by BECs of PCK rats.

3.4.5. MMP Activities

Both cultured BECs of normal and PCK rats contained detectable amounts of MMP-9 and MMP-13 mRNA. Therefore, it seems plausible that the degradation of type IV collagen and laminin by BECs of PCK might be the activation of MMP-9, MMP-13, or other types of MMPs through the PA/plasmin system.

3.4.6. Tissue and Culture Correlation

Overexpression of plasminogen and tPA in the BECs of CHF and Caroli's disease, and PCK rats results in the generation of excessive amounts of plasmin, which contributes to the plasminogen-dependent lysis of laminin and type IV collagen. This may contribute to the biliary cystogenesis in Caroli's disease with CHF and PCK rats.

3.5. BECs with Mesenchymal Features

One of the hepatic lesions of ARPKD is a progressive and unresolving portal fibrosis. While in most of chronic liver diseases, activated myofibroblasts play major roles in hepatic fibrosis by producing ECM molecules, activation of such fibrogenetic cells were negligible in ARPKD (17). Herein, the mechanism of hepatic fibrosis of ARPKD was examined from the viewpoint of pathologic BECs.

Epithelial-mesenchymal transition (EMT), which is accompanied by a loss of epithelial cell markers such as E-cadherin and a gain of mesenchymal cell markers such as vimentin, has been implicated in a variety of biological processes, especially fibrogenesis. TGF-β is known to be the most potent inducer of EMT. A recent study has demonstrated that BECs can undergo EMT, thereby contributing to hepatic fibrosis (18).

3.5.1. In Vivo Tissue Studies

In the livers of PCK rats, two different types of intrahepatic bile ducts could be recognized; those lined by cuboidal-shaped BECs (C-type) and those lined by flat-shaped BECs (F-type). The former were a predominant phenotype in the liver, showing progressive cystic dilation. Both C- and F-type showed progressive fibrosis around them, particularly around F-type bile ducts. Immunohistochemically, C-type was positive for biliary cytokerain (CK7 and CK19) and for E-cadherin on their cell membrane. In F-type, reduction of CK7 and CK19 expression was observed, while expression of mesenchymal markers such as vimentin and fibronectin appeared to be maintained in F-type (see Fig. 4). Both C- and F-types lacked α-SMA expression.

By RT-PCR analysis, whole liver of the PCK rats expressed TGF-β1 mRNA, and the expression level increased with aging, which was in accordance with the increase in the appearance of BECs with F-type. The PCK rat also expressed a significantly high level of TGF-β1 in the whole liver. C- and F-type bile ducts and those of normal rats invariably expressed TβR-I and TGF-β type II receptor (TβR-II) to the same extent. Nuclei of many BECs of F-type showed strong immunohistochemical expression of pSmad2, suggesting the transmission of TGF-β signals from the cell surface

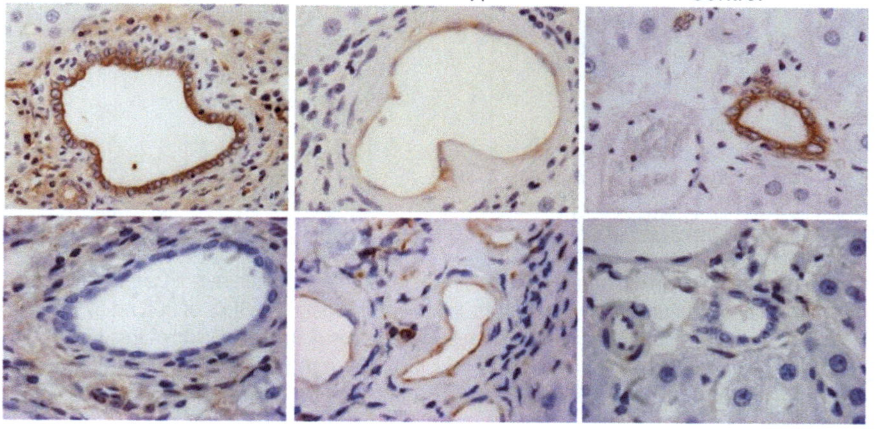

Fig. 4. Immunophenotype of intrahepatic bile ducts of the PCK rat. C-type bile duct of PCK rat and bile duct of control rat were positive for CK19 (biliary epithelial marker) but negative for vimentin (mesenchymal marker), while F-type bile duct of PCK rat was positive both for CK 19 and vimentin (11).

into the nucleus, whereas the staining was totally weak or invisible in C-type and normal bile ducts.

3.5.2. Cell Culture of BECs of PCK Rats

BECs of PCK and normal rats, subcultured between 8 and 12 times, were used for the study. At a subconfluent state on type I collagen-coated culture dishes, BECs of PCK and normal rats were incubated with the standard medium or that containing TGF-β1 (2 ng/mL) for 3 and 7 days.

3.5.3. Measurement of Collagen Content in Cell Culture Supernatant

According to the Sircol collagen assay kit (see Subheading 2.3, item 27), Sirius red reagent (50 μL) was added to each culture supernatant (50 μL) and mixed for 30 min. The collagen–dye complex was precipitated by centrifugation at 15,000 × *g* for 5 min, and the precipitate was dissolved in 0.5 M/L sodium hydroxide. Finally, the samples were read on a microplate reader, and the absorbance was determined at 540 nm.

3.5.4. TGF-β1 Treatment

Cultured BECs of PCK and normal rats were stimulated with TGF-β1 (2 ng/mL) for 3 days on type I collagen-coated 60 mm culture dishes, and the expression of mesenchymal markers was examined. Experiments using RT-PCR showed that BECs of PCK and normal rats expressed mRNA for TβR-II and also contained detectable amounts of TGF-β1 mRNA. After stimulation, an increase in the expression of mRNA for vimentin, procollagen type I, and fibronectin was observed in BECs of PCK and normal rats by RT-PCR, but α-SMA mRNA expression was unchanged (11). On immunofluorescence confocal microscopy, the majority of BECs of PCK and normal rats were negative for vimentin, and TGF-β1 increased the number of vimentin-positive BECs

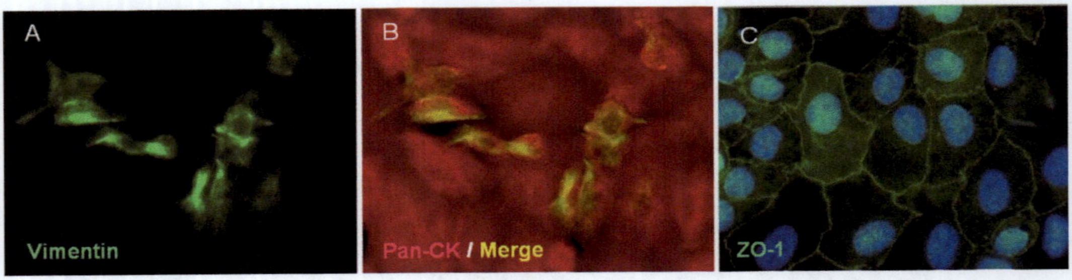

Fig. 5. Induction of mesenchymal markers in BECs of PCK rat by TGF-β1. Cultured BECs were incubated with TGF-β1 for 3 days on type I collagen-coated cell culture dishes. Induction of vimentin in BECs of PCK rat by TGF-β1. (**a**) Vimentin was visualized by Alexa-488 (*green*) under immunofluorescence confocal microscopy. (**b**) Merged images of vimentin visualized by Alexa-488 (*green*) and pan-CK visualized by Alexa-568 (*red*). BECs were *yellow*, implying expression of both pan-CK and vimentin. (**c**) ZO-1 was visualized by Alexa-488 (*green*), and nuclei were stained with 4′,6-diamidino-2-phenylindole (*blue*) (11).

(see Fig. 5a, b). In accordance with the induction of mRNA of procollagen type I and fibronectin, the supernatant of BECs of PCK rats contained significantly increased amounts of collagen and fibronectin after the TGF-β1 treatment.

As for epithelial cell phenotype, CK19 and E-cadherin mRNA expression was not affected by the TGF-β1 treatment in BECs of the PCK rats by RT-PCR analysis. Western blot analysis also showed no effects of TGF-β1 on CK19 and E-cadherin expression in BECs of the PCK rats at the protein level. Under the phase-contrast microscope, epithelial cell morphology was maintained in BECs of PCK rats after TGF-β1 treatment, and no morphological transition of the cells from an epithelial to a fibroblastic appearance was observed. In addition, tight junctions between adjacent cells were preserved after the treatment as shown by the expression of ZO-1 (see Fig. 5c). Similar results were obtained for the normal BECs.

3.5.5. Effects of TGF-β1 Treatment on Basement Membrane Components

Because cell contact with basement membrane components is an important developmental and biological microenvironment for BECs, BECs of PCK rats were cultured on type IV collagen- and laminin-coated cell culture dishes, and the effects of TGF-β1 on phenotype of BECs were examined. After a 3-day stimulation with TGF-β1, the expression of CK19 mRNA appeared to be reduced in BECs of the PCK rats cultured on type IV collagen and laminin, whereas E-cadherin mRNA expression was unchanged. Western blot analysis showed that TGF-β1 significantly reduced CK19 expression in BECs of the PCK rats cultured on type IV collagen. E-cadherin expression of the BECs of PCK rats cultured on type IV collagen and laminin was not significantly affected by TGF-β1 at the protein level, consistent with no reduction of E-cadherin expression in F-type bile ducts (see Note 6). These findings are consistent with the immunophenotype of F-type bile ducts in vivo.

3.5.6. Tissue Culture Correlation

It seems plausible that BECs of the PCK rat may acquire mesenchymal features in response to TGF-β1 and participate in progressive hepatic fibrosis. TGF-β1 may act as an inducer initiating a cascade of dedifferentiating events (see Note 7).

4. Notes

1. Caroli's disease and congenital hepatic fibrosis are rare diseases, and the liver tissue specimens are available from liver transplantation center with many transplantation services.
2. PCK rats are only maintained in Kanazawa University Graduate School of Medicine, Kanazawa, Japan, and also in Center for Basic Research in Digestive Diseases, Mayo Clinic, Rochester, Minnesota, USA.
3. To prepare the biliary fragments from the digested liver, the digested hepatic parenchyma should be gently and softly removed with a comb and soft toothbrush. Then, biliary tree thus obtained was suspended in Ca^{2+}- and Mg^{2+}-free Hanks' solution and cut into two regions: the intrahepatic large bile duct and the peripheral region. The latter contains small intrahepatic bile ducts. This approach is useful for the analysis of bile duct pathophysiology according to their size and anatomical location.
4. Double-immunostaining demonstrated that EGF-positive cells were also positive for mast cell tryptase, suggesting that these mononuclear cells are mast cells. EGF from infiltrating mast cells may be involved in the increased cell kinetics and dilation of the bile ducts of ARPKD and PCK rats.
5. Matrigel™ is composed of 61% laminin, 30% type IV collagen, and 7% entactin.
6. The F-type BECs show acquisition of mesenchymal marker but sill keep expression of reduced epithelial marker and maintain epithelial cell structure, thus a different event from EMT.
7. More than one factor are involved in cystic dilatation of intrahepatic bile ducts of the PCK rat and Caroli's disease with CHF: the increase in the proliferative activity of BECs involving the EGF/EGF receptor and the MEK5-ERK5 pathway and the degradation of laminin and type IV collagen in the basement membrane, a supportive structure of intrahepatic bile ducts.

References

1. Nakanuma Y, Hoso M, Sanzen T, Sasaki M (1997) Microstructure and development of the normal and pathologic biliary tract in humans, including blood supply. Microsc Res Tech 38:552–570
2. Katayanagi K, Kono N, Nakanuma Y (1998) Isolation, culture and characterization of biliary epithelial cells from different anatomical levels of the intrahepatic and extrahepatic biliary tree from a mouse. Liver 18:90–98
3. Chen XM, O'Hara SP, LaRusso NF (2008) The immunobiology of cholangiocytes. Immunol Cell Biol 86:497–505
4. Portman BC, Nakanuma Y (2006) Diseases of bile ducts. In: Burt AD, Portman BC, Ferrell LD (eds) MacSween's pathology of the liver, 5th edn. Churchill Livingstone, New York, pp 517–581
5. Nakanuma Y, Harada K, Sato Y, Ikeda H (2010) Recent progress in the etiopathogenesis of pediatric biliary disease, particularly Caroli's disease with congenital hepatic fibrosis and biliary atresia. Histol Histopathol 25:223–235
6. Sanzen T, Harada K, Yasoshima M, Kawamura Y, Ishibashi M, Nakanuma Y (2001) Polycystic kidney rat is a novel animal model of Caroli's disease associated with congenital hepatic fibrosis. Am J Pathol 158:1605–1612
7. Zhang MZ, Mai W, Li C, Cho SY, Hao C, Moeckel G et al (2004) PKHD1 protein encoded by the gene for autosomal recessive polycystic kidney disease associates with basal bodies and primary cilia in renal epithelial cells. Proc Natl Acad Sci U S A 101:2311–2316
8. Sato Y, Harada K, Kizawa K, Sanzen T, Furubo S, Yasoshima M et al (2005) Activation of the MEK5/ERK5 cascade is responsible for biliary dysgenesis in a rat model of Caroli's disease. Am J Pathol 166:49–60
9. Sato Y, Harada K, Furubo S, Kizawa K, Sanzen T, Yasoshima M et al (2006) Inhibition of intrahepatic bile duct dilation of the polycystic kidney rat with a novel tyrosine kinase inhibitor gefitinib. Am J Pathol 169:1238–1250
10. Yasoshima M, Sato Y, Furubo S, Kizawa K, Sanzen T, Ozaki S et al (2009) Matrix proteins of basement membrane of intrahepatic bile ducts are degraded in congenital hepatic fibrosis and Caroli's disease. J Pathol 217:442–451
11. Sato Y, Harada K, Ozaki S, Furubo S, Kizawa K, Sanzen T et al (2007) Cholangiocytes with mesenchymal features contribute to progressive hepatic fibrosis of the polycystic kidney rat. Am J Pathol 171:1859–1871
12. Vroman B, LaRusso NF (1996) Development and characterization of polarized primary cultures of rat intrahepatic bile duct epithelial cells. Lab Invest 4:303–313
13. Muff MA, Masyuk TV, Stroope AJ, Huang BQ, Splinter PL, Lee SO et al (2006) Development and characterization of a cholangiocyte cell line from the PCK rat, an animal model of autosomal recessive polycystic kidney disease. Lab Invest 86:940–950
14. Kawamura Y, Yoshida K, Nakanuma Y (1989) Primary culture of rabbit gallbladder epithelial cells in collagen gel matrix. Lab Invest 61:350–356
15. Kato Y, Tapping RI, Huang S, Watson MH, Ulevitch RJ, Lee JD (1998) Bmk1/Erk5 is required for cell proliferation induced by epidermal growth factor. Nature 395:713–716
16. Lijnen HR, Van Hoef B, Lupu F, Moons L, Carmeliet P, Collen D (1998) Function of the plasminogen/plasmin and matrix metalloproteinase systems after vascular injury in mice with targeted inactivation of fibrinolytic system genes. Arterioscler Thromb Vasc Biol 18:1035–1045
17. Ozaki S, Sato Y, Yasoshima M, Harada K, Nakanuma Y (2005) Diffuse expression of heparan sulfate proteoglycan and connective tissue growth factor in fibrous septa with many mast cells relate to unresolving hepatic fibrosis of congenital hepatic fibrosis. Liver Int 25:817–828
18. Xia JL, Dai C, Michalopoulos GK, Liu Y (2006) Hepatocyte growth factor attenuates liver fibrosis induced by bile duct ligation. Am J Pathol 168:500–512

Chapter 19

Primary Mouse Small Intestinal Epithelial Cell Cultures

Toshiro Sato and Hans Clevers

Abstract

The intestinal epithelium is the most rapidly self-renewing tissue in adult mammals. We have recently shown that Lgr5 (Leucine-rich repeat-containing G protein-coupled receptor) is expressed in intestinal stem cells by an in vivo genetic lineage tracing strategy. In the past, extensive efforts have been made to establish primary small intestinal culture systems. However, no defined, reproducible and robust culture system had been developed. To establish such a system, we screened for optimal growth factor combinations based on genetic evidence of self-renewal regulation, differentiation, and carcinogenesis of intestinal stem cells. Here, we describe methods that we have established for the isolation and culture of primary small intestinal epithelial stem cells. In this culture system, isolated crypts form "organoid structures" with a histological hierarchy recapitulating in vivo small intestinal epithelium. Single isolated Lgr5+ intestinal stem cells also form these organoid structures, in which stem cells are maintained by self-renewal and give rise to all lineages of the intestinal epithelium. This culture system is particularly useful for studying the regulation of intestinal stem cell self-renewal and differentiation.

Key words: Small intestinal crypts, Intestinal stem cells, Wnt signal, Lgr5, Flow cytometry, Confocal immunofluorescence

1. Introduction

The small intestine consists of two histological compartments: crypts and villi. Villi are finger-like protrusions that point toward the lumen, and crypts are invaginations into the submucosa. The function of the adult small intestinal epithelium is mediated by four different types of mature cells: enterocytes (the most abundant cell type, absorbing water and nutrients), goblet cells (the second major cell type and secreting mucus), enteroendocrine cells (secreting intestinal hormones), and Paneth cells (secreting

Scott H. Randell and M. Leslie Fulcher (eds.), *Epithelial Cell Culture Protocols: Second Edition*, Methods in Molecular Biology, vol. 945, DOI 10.1007/978-1-62703-125-7_19, © Springer Science+Business Media, LLC 2012

antibacterial substances). Undifferentiated cells reside in the crypts and produce the four types of mature cells. Enterocytes, goblet cells, and enteroendocrine cells migrate upward along the crypt–villus axis, while Paneth cells move downward and are confined at the bottom of crypts. Recently, we found that Lgr5 is specifically expressed in small crypt base columnar (CBC) cells, which are intermingled with Paneth cells. In an in vivo genetic lineage tracing experiment, we have shown that the Lgr5+ CBC cells possess intestinal stem cell properties: long-term self-renewal and multipotential differentiation (1).

Even though several intestinal epithelial culture systems were established and reported, no reproducible, robust, and defined culture method has existed for intestinal stem cells (2–4). In most culture systems, intestinal cells transiently divide, yet cease proliferation within 1–2 weeks. Since we failed to observe Lgr5+ cells in those culture systems, we concluded that the culture conditions were not sufficient to maintain intestinal stem cells. To establish a new culture system for intestinal stem cells, we tested combinations of growth factors, selected on genetic evidence of intestinal stem cell self-renewal, differentiation and carcinogenesis. Subsequently, we found that the combination of growth factors, epidermal growth factor (EGF), noggin, and R-spondin, is essential to maintain intestinal stem cells in vitro (5). Interestingly, each growth factor signal controls a signaling pathway that is frequently mutated in human colorectal cancer (6), i.e., EGF activates *KRAS* signaling (activating mutations occur in *KRAS*), noggin blocks BMP (bone morphogenic protein) signaling (loss-of-function mutations occur in *BMPRIA* or *SMAD4*), and R-spondin activates the Wnt pathway (pathway-activating mutations occur in *APC* or *CTNNB1*). Intestinal epithelial cells also require basement membrane-derived extracellular matrix (Matrigel™) as a scaffold for three dimensional growth.

In the culture system, crypts form "organoids" which mimic crypt–villus structures. The organoids are composed of a central cyst structure and surrounding crypt-like budding structures. Intestinal stem cells and Paneth cells reside at the bottom of the crypt-like structures, while post-mitotic enterocytes migrate toward the central cyst structure and end up being shed into the lumen. The organoids produce all four types of mature cells, and interestingly, their frequency and localization patterns are preserved (5). It was believed that intestinal stem cell self-renewal and differentiation were regulated by subepithelial mesenchymal niche cells. However, we clearly showed that a single intestinal stem cell forms crypt–villus structures in vitro without a mesenchymal cellular niche. Therefore, intestinal epithelium "epithelial-cell-autonomously" organize their morphology, differentiation, and self-renewal.

2. Materials

2.1. Small Intestinal Crypt Isolation

1. Phosphate-buffered saline without Ca^{2+} and Mg^{2+} (PBS) (see Note 1).
2. Crypt isolation buffer: 2 mM EDTA in PBS.
3. Basal culture medium: Advanced DMEM/F12 (Life Technologies™) supplemented with 2 mM GlutaMax (Life Technologies™), 10 mM HEPES, and 100 U/mL penicillin/100 μg/mL streptomycin (see Note 2).
4. 70-μm cell strainer.
5. BSA coated 50 mL tube: 50 mL tube coated with 1% (w/v) bovine serum albumin (BSA)/PBS.

2.2. Small Intestinal Crypts Culture

The stock solutions (1–9) should be stored at −20°C.

1. Matrigel™, basement membrane matrix, growth factor reduced (GFR), phenol red-free (see Note 3).
2. N2 supplement, 100× (Life Technologies™).
3. B27 supplement (Life Technologies™), 50× (see Note 4).
4. *N*-Acetylcysteine (500× stock, Sigma-Aldrich®): 81.5 mg/mL in distilled water (500 mM) (see Note 4).
5. 0.1% BSA/PBS: 0.1% (w/v) BSA in PBS, sterile filter 0.2 μm filter.
6. Murine recombinant EGF, 10,000× stock (Life Technologies™): 500 mg/mL in 0.1% BSA/PBS.
7. Murine recombinant Noggin, 1,000× stock (PeproTech): 100 μg/mL in 0.1% BSA/PBS.
8. Human recombinant R-spondin1, 1,000× stock (see Note 5): 1 mg/mL in 0.1% BSA/PBS.
9. Complete crypt culture medium: Basal culture medium with N2 supplement (1×), B27 supplement (1×), and 1 mM *N*-acetylcysteine, 50 ng/mL EGF, 100 ng/mL Noggin, 1 μg/mL R-spondin (see Note 6).

2.3. Small Intestinal Stem Cell Culture

1. *Lgr5-GFP-ires-CreER* mouse (The Jackson Laboratory).
2. 40-μm cell strainer.
3. 20-μm cell strainer.
4. FACS tube.
5. BSA coated collecting tube: 1.5 mL Eppendorf tube coated with 1% BSA/PBS.
6. ROCK inhibitor, 1,000× stock (Y-27632, Sigma-Aldrich®): 10 mM in distilled water, store at 4°C.

7. Propidium iodide (Molecular Probes®): Reconstitute the reagent with distilled sterile water for a final concentration, 1.5 μM.
8. Jagged-1 peptide, 1,000× stock (Ana Spec): 1 mM in distilled sterile water, store at −20°C.
9. Single cell dissociation medium: Basal culture medium (see Subheading 2.1, step 3) including 1× N2, 1× B27, and 10 μM Y-27632.
10. Single cell culture medium: Crypt culture medium (see Subheading 2.2, step 9) including 10 μM Y-27632.

2.4. Imaging for Intestinal Organoids

1. Glass bottom dish, size 3.5 cm.
2. Paraformaldehyde (PFA): Prepare 4% (w/v) solution in PBS, heat (65°C) to dissolve. Adjust pH to 7.2 with NaOH.
3. 0.2% Triton™-X: Prepare a 0.2% (w/v) solution in PBS.
4. Blocking solution: Power Block 10× (BioGenex), reconstitute in distilled sterile water.
5. DAPI (4′,6-diamidino-2-phenylindole; Molecular Probes®): final concentration, 1 μg/mL.
6. Mounting medium: ProLong® Gold (Life Technologies™), reconstitute with distilled sterile water.

3. Methods

Isolated crypts form round-shaped structures immediately (1–2 h) after culture initiation and start to bud 2–3 days afterward. The budding structure splits every 1.5–2 days and ends up forming organoids with 30–50 crypt-like domains (see Fig. 1). The organoid structure can be disassembled into crypt-like domains and can be passaged for over a year.

Intestinal epithelial cells are vulnerable to single cell dissociation and quickly undergo detachment-induced-apoptosis, "anoikis," and this phenomenon hampered the establishment of single intestinal epithelial stem cell culture. To reduce anoikis, we preferentially use mechanical cell dissociation rather than digestive enzymes for single intestinal stem cell culture. It has been reported that ROCK inhibitor prevents anoikis in dissociated human ES cells (7), and this finding was applied to intestinal stem cell culture (5). In addition, we added synthetic Notch ligand in Matrigel™ to avoid loss of Notch signaling, which is indispensable for intestinal stem cell maintenance. Combining these modifications, we managed to establish the single intestinal stem cell culture system. A single intestinal stem cell forms a symmetric cyst structure within

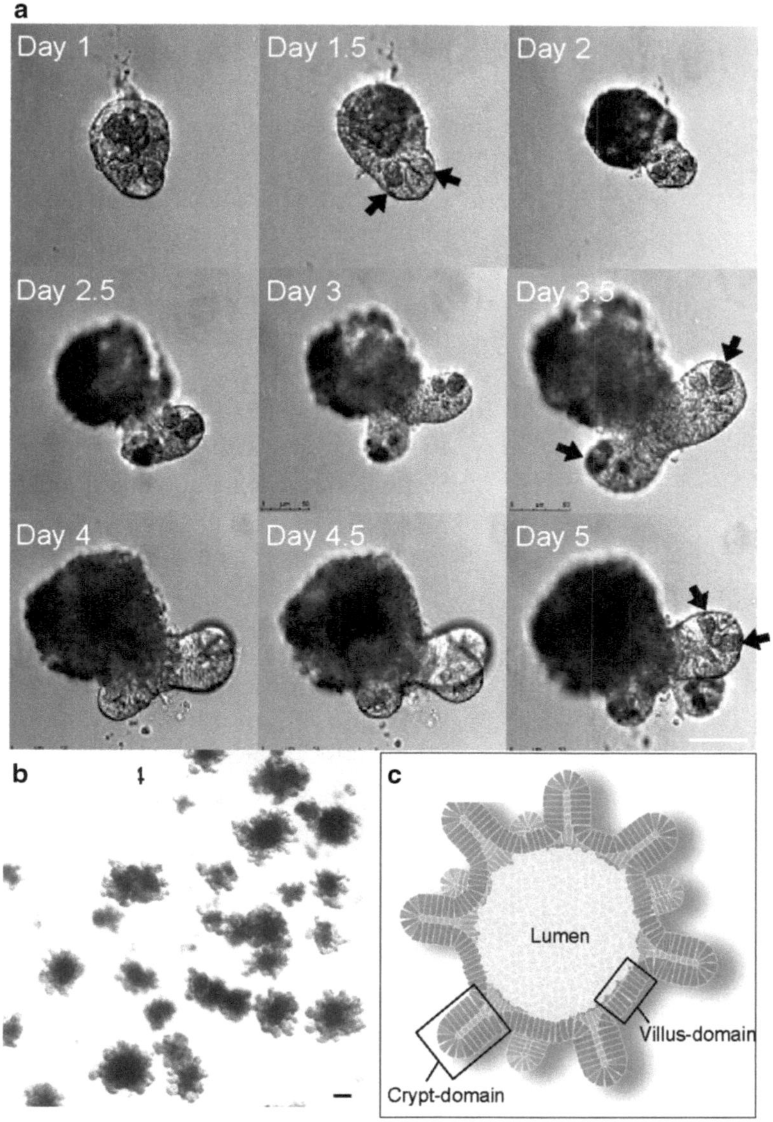

Fig. 1. (**a**) Time course of an isolated single crypt growth. Differential interference contrast image reveals granule-containing Paneth cells at crypt bottoms (*arrows*). (**b**) Single isolated crypts efficiently form large crypt organoids 14 days after culture initiation. (**c**) Schematic representation of a crypt organoid, consisting of a central lumen lined by villus-like epithelium and a number of surrounding crypt-like domains. Scale bar; 50 μm. Reproduced from Sato et al. (5) with permission from Nature press.

2 days after culture initiation, followed by formation of budding structures. Thereafter, single intestinal stem cell-derived-organoids show identical morphology and time course to whole crypt-derived organoids over a year.

3.1. Isolation of Small Intestinal Crypts

1. Thaw aliquots of Matrigel™ on ice before isolation. Pre-incubate 24-well plate in a CO_2 incubator (5% CO_2, 37°C).

2. Isolate proximal half of mouse small intestine and open longitudinally. Wash the intestine with ice-cold PBS until most of the luminal contents are cleared.
3. Scrape off the villi using a coverslip and wash with ice-cold PBS (see Note 7).
4. Cut the intestine into 2–4 mm pieces with scissors and transfer them to a 50-mL tube.
5. Add 30 mL ice-cold PBS and wash the fragments. Gently pipette the fragments up and down with a 10 mL pipette and discard supernatant after settling down. Repeat this step 5–10 times until the supernatant is almost clear.
6. Add 30 mL ice-cold crypt isolation buffer (see Subheading 2.1, step 2) and gently rock the tube at 4°C for 30 min.
7. Settle the fragments down and remove the supernatant.
8. Add 20 mL of ice-cold PBS and pipette up and down. After settling the fragments down, the supernatants are inspected to see whether they are enriched with villi (villous fraction) or crypts (crypts fraction) by inverted microscopy. Repeat this procedure until most of crypts are released (see Note 8). Villous fractions are discarded and crypts fractions are passed through a 70-μm cell strainer and collected into BSA coated 50 mL tube(s) (see Subheading 2.3, step 5).
9. Spin down the crypt fractions at 300 × *g* 5 min.
10. The pellets are resuspended with 10 mL of ice-cold basal culture medium (see Subheading 2.1, step 3) and transferred to a 15 mL Falcon tube. The tube is centrifuged with lower speed (150–200 × *g*, 2 min), to remove single cells (mostly lymphocytes). Repeat this washing step 2–3 times until most of single cells are cleared.

After the washing steps, the number of crypts is calculated. Take 10 μL from 10 mL crypts suspension and count numbers of crypts by inverted microscopy. The total number of crypts = the number of crypts counted × 1,000.

3.2. Small Intestinal Crypt Culture

1. Centrifuge small intestinal crypts, and remove supernatant as much as possible to avoid dilution of Matrigel™ in the next step. Keep the tube at 4°C.
2. Resuspend the crypts pellet with Matrigel™ (200–500 crypts/50 μL).
3. Apply 50 μL of the resuspended Matrigel™ into the pre-warmed 24-well plate. Matrigel™ should be applied on the center of the well so that it can form a hemispherical droplet. The plate should be transfer to a CO_2 incubator (5% CO_2, 37°C) as quickly as possible after the seeding (see Note 9).
4. Allow the Matrigel™ solidify for 5–10 min and add 500 μL complete culture medium/well (see Subheading 2.2, step 9).

5. The sample is cultured in the CO_2 incubator.
6. Exchange culture medium with new crypt culture medium every 4–6 days.
7. Crypt organoids can be passaged after 7–14 days after seeding. Remove the culture medium and gently break up the Matrigel™ by pipetting with p1000 pipette. The organoids are resuspended with 1–2 mL basal culture medium (see Subheading 2.1, step 3) and transferred to a 15 mL tube. Gently disrupt the organoids using a fire-polished Pasteur pipette. To remove released dead cells and single cells, wash the organoids with 10 mL basal culture medium as in Subheading 3.1, step 10.
8. Apply Matrigel™ and culture in a new 24 culture plate, as in steps 1–5.

3.3. Small Intestinal Stem Cell Culture

1. Isolate crypts from *Lgr5-GFP-ires-CreER* mice as in Subheading 3.1, steps 1–9 (see Note 10).
2. Resuspend the pellet with 2 mL of pre-warmed single cell dissociation medium (see Subheading 2.3, step 9) for 30–45 min until cell integrity becomes loose. During incubation the crypts suspension is pipetted every 5–10 min.
3. The sample is mechanically pipetted with fire-polished Pasteur pipette until most of the crypts are dissociated into single cells or 2–3 cell clusters.
4. The sample is passed through a 40-μm cell strainer, followed by a 20-μm cell strainer.
5. Resupend with basal culture medium and centrifuge at $300 \times g$, 4°C. Repeat this washing step twice.
6. The pellet is resuspended with single cell dissociation medium (see Subheading 2.3, step 9) (around 5×10^6 cells/mL) and transferred to FACS tube. Add propidium iodide (see Subheading 2.3, step 7) to stain dead cells.
7. A single cell population is gated by forward scatter, side scatter, and pulse width parameter. Live cells are gated with propidium iodide. Lgr5-GFP^{hi} cells are defined as the brightest GFP population in *Lgr5-GFP-ires-CreER* mice.
8. Sort single, live Lgr5-GFP^{hi} cells into a BSA coated collecting tube including single cell dissociation medium.
9. After sorting, centrifuge the collecting tubes ($300 \times g$, 3 min) and discard the supernatant. The pellet is resuspended within Matrigel™ (1–100 cells/10 μL) including 1 μM Jagged-1 peptide 1. The sample is seeded on pre-warmed 48- or 96-well plate (25 μL/well for 48-well plate, 10 μL/well for 96-well plate). After solidification, add single cell culture medium (250 μL/well for 48-well plate, 100 μL/well for 96-well plate).

10. Add PBS to the empty wells to maintain humidity throughout the plate.
11. 4–6 h after seeding, single live cells are counted for calculation of colony forming efficiency (see Note 11).

3.4. Imaging of the Organoids

1. Remove culture medium and wash the sample with PBS 2 times. PFA (see Subheading 2.4, step 2) is then added for 10–30 min at room temperature.
2. Discard the fixative and mechanically break the fixed Matrigel™ up using a p1000 pipette. The organoids are transferred into 15 mL tube (preferably pre-coated with BSA), washed twice with PBS, and permeabilized by incubation in 0.2% Triton™-X (see Subheading 2.4, step 3) for 30 min at 4°C.
3. The samples are washed twice with PBS, and blocked by incubation in blocking solution for 1 h at 4°C.
4. The blocking solution is removed and replaced with primary antibody in blocking solution over night at 4°C (preferably with gentle rocking).
5. The primary antibody is removed and the samples are washed three times for 5 min each with PBS.
6. Incubate with secondary antibody for 1 h at room temperature.
7. The secondary antibody is discarded and DAPI (see Subheading 2.4, step 6) is added for 10 min at room temperature to stain the DNA and identify the nuclei.
8. The samples are washed three times for 5 min each with PBS. Remove as much PBS as possible after the last washing.
9. Resuspend the samples with mounting medium (see Subheading 2.4, step 6), and apply to a glass bottom dish.
10. The organoids are analyzed by standard confocal microscopy.

4. Notes

1. If PBS is prepared from powder or concentrated stock, it is important to reconstitute it accurately. Slight differences in osmolarity may affect viability of isolated crypts.
2. We normally store the reconstituted basal culture medium in the dark at 4°C, and keep it up to 14 days after reconstitution.
3. Matrigel™ should be thawed on ice. It takes around 4–6 h to thaw a 10 mL vial of Matrigel™. Thawed Matrigel™ should be aliquoted into precooled, sterile cryogenic tubes and stored at −20°C. Matrigel™ can be refrozen and thawed up to three times without significant loss of culture efficiency in our hands.

Matrigel™ potentially exhibits batch-to-batch differences. We have tried more than ten different batches of Matrigel™, but we observed no discernable differences among the batches. However, once the crypt culture is established, it is advisable to use the same batch of Matrigel™ as much as possible.

4. B27 and *N*-acetylcysteine is acidic, and the culture medium becomes yellowish after addition of the reagents.
5. Nuvelo no longer provides recombinant R-spondin1. It can be substituted with recombinant protein from other sources (R&D) or with conditioned medium from R-spondin1-Fc producing cell lines (8).
6. We keep complete culture medium at 4°C for up to 4 days.
7. Removal of villi by scraping can be skipped. In that case, wash small intestinal mucosa more than ten times with PBS in Subheading 3.1, step 4.
8. Typically villous fractions appear in the first and second supernatants, and crypts appear in the third fraction and thereafter. The number of positive crypt fractions is variable, and we always inspect fractions by inverted microscopy.

   Isolated intact villi are removed by a 70-μm strainer. However, some villi fall apart and the villous fragments may pass through the strainer. Make sure that crypt fraction is enriched with crypts but not villous fragments. Crypts can be discriminated from villous fragments by inverted microscopy (optimally equipped with phase contrast or different interference contrast filters) with specific appearance of Paneth cell granules. The contaminating fragments of villi die and disintegrate within 24 h after culture.
9. It is sometimes difficult to form a hemispheric Matrigel™ droplet on the well. If this is the case, Matrigel™ will spread out on the well and organoids attach to the bottom of well and lose 3D structure. Crypts also sink through the Matrigel™ and attach to the plate when we leave samples for more than a few minutes without transferring to the incubator.
10. The single intestinal stem cells culture is of lower efficiency (2–7%) than that of whole crypts culture (80–90%) presumably due to isolation stress and insufficient growth condition. Therefore, it is important to be proficient with the whole-crypts culture technique before starting intestinal stem cell culture.
11. Half of sorted intestinal stem cells are pro-apoptotic due to isolation stress and quickly die after initiation of culture, while contaminated cell clusters evade anoikis and form organoids with higher efficiency than single cells. Therefore, to exclude those cells from colony formation efficiency calculation, we manually inspect and count single healthy cells 4–6 h after seeding.

References

1. van Barker N, Es JH, Kuipers J et al (2007) Identification of stem cells in small intestine and colon by marker gene Lgr5. Nature 449(7165):1003–1007
2. Evans GS, Flint N, Somers AS, Eyden B, Potten CS (1992) The development of a method for the preparation of rat intestinal epithelial cell primary cultures. J Cell Sci 101(Pt 1):219–231
3. Whitehead RH, Demmler K, Rockman SP, Watson NK (1999) Clonogenic growth of epithelial cells from normal colonic mucosa from both mice and humans. Gastroenterology 117(4):858–865
4. Fukamachi H (1992) Proliferation and differentiation of fetal rat intestinal epithelial cells in primary serum-free culture. J Cell Sci 103(Pt 2): 511–519
5. Sato T, Vries RG, Snippert HJ et al (2009) Single Lgr5 stem cells build crypt-villus structures in vitro without a mesenchymal niche. Nature 459(7244):262–265
6. Markowitz SD, Bertagnolli MM (2009) Molecular origins of cancer: molecular basis of colorectal cancer. N Engl J Med 361(25): 2449–2460
7. Watanabe K, Ueno M, Kamiya D et al (2007) A ROCK inhibitor permits survival of dissociated human embryonic stem cells. Nat Biotechnol 25(6):681–686
8. Ootani A, Li X, Sangiorgi E et al (2009) Sustained in vitro intestinal epithelial culture within a Wnt-dependent stem cell niche. Nat Med 15(6):701–706

Chapter 20

Isolation, Growth, and Characterization of Human Renal Epithelial Cells Using Traditional and 3D Methods

John J. Gildea, Helen E. McGrath, Robert E. Van Sciver, Dora Bigler Wang, and Robin A. Felder

Abstract

The kidney is a highly heterogeneous organ that is responsible for fluid and electrolyte balance. Much interest is focused on determining the function of specific renal epithelial cells in humans, which can only be accomplished through the isolation and growth of nephron segment-specific epithelial cells. However, human renal epithelial cells are notoriously difficult to maintain in culture. This chapter describes the isolation, growth, immortalization, and characterization of the human renal proximal tubule cell. In addition, we describe new paradigms in 3D cell culture which allow the cells to maintain more in vivo-like morphology and function.

Key words: Proximal tubule, Human renal epithelial cells, Immortalization, 3D cell culture

1. Introduction

The relevance of using renal proximal tubule cell (RPTC) lines of human origin to study human physiology and pathophysiology over using those derived from nonhuman sources, such as the LLC-PK (pig), LLC-RK1 (rabbit), NRK-52E (rat), and OK (opossum) cannot be overemphasized. Initially, primary RPTC lines were used to study renal cellular physiology since they retained many of their in vivo characteristics in vitro (1). There are some limitations to the use of primary RPTCs. For example, primary cell lines cannot be sustained in long-term culture, which makes inter-assay comparisons difficult (2, 3). Furthermore, primary cell lines representing a broad genetic background are often only available in limited supply. Therefore, transformed cell lines were generated

Scott H. Randell and M. Leslie Fulcher (eds.), *Epithelial Cell Culture Protocols: Second Edition*, Methods in Molecular Biology, vol. 945, DOI 10.1007/978-1-62703-125-7_20, © Springer Science+Business Media, LLC 2012

using immortalizing virus (e.g. SV40, human papilloma virus (HPV)) which extended their growth potential beyond the 8–15 passage limit of primary cells (4, 5). Transformed RPTCs retain many of the functional and morphological characteristics of RPTCs in primary culture (5, 6). However, they also retain any genetic aberrations contained in their genome. Thus, laboratories that intend to study RPTCs should use techniques to grow and use primary as well as immortalized RPTCs, from as wide a genetic background as possible. In addition, RPTCs should be expanded under conditions that enable them to more closely resemble their in vivo counterparts (what some are calling "3D biology") (7).

We have reserved part of this chapter to cover the new technology of growing RPTCs on a 3D, biomimetic, pipettable, polarized, transporting matrix that facilitates studying RPTCs in large numbers under relatively in vivo-like conditions. Microcarriers are becoming the mainstay of cell culture laboratories. Microcarriers that are manufactured to diameters of 75–150 μm may be pipetted using standard laboratory liquid handling devices, allowing cell culture to remain entirely in the liquid-phase throughout the growth and analysis phases. Human RPTCs grown on biomimetic substrates demonstrate an in vivo-like morphology that is not observed when these cells are cultured on 2D flat culture dishes (see Fig. 1). Total cellular expression of villin is increased dramatically when primary RPTCs are cultured on GEMS™ (global eukaryotic microcarriers) vs. 2D culture dishes (see Fig. 2). Growing RPTCs on transwells is an improvement over flat plasticware in that RPTCs are polarized and experiments can be performed to study apical or basolateral membranes. However, RPTCs grown on microcarriers have a tenfold greater number of microvilli on their surfaces than RPTCs grown on transwells (see Fig. 3). By more closely replicating in vivo-like conditions, cells provide more physiologically and biochemically relevant data. Figure 4 shows that RPTCs on 3D GEMs accumulate more sodium than their 2D counterparts.

Microcarriers also provide the maximum surface area-to-cell ratio, allowing for a high yield of cells in a minimal volume of space, thus reducing the footprint required to maintain cultures as well as resources. Paramagnetic particles placed in the microcarriers further enhance their utility since they enable relatively simple noncontact manipulation of the cultured cells while on the substrate and provide for integration into automated equipment. Furthermore, RPTCs that have reached confluence remain viable in culture for several months with proper feeding. Primary cells will remain quiescent, enabling them to be used for replicate experiments over a longer time than if they were constantly dividing and getting progressively closer to replicative senescence, where they no longer maintain a normal proximal tubule cellular phenotype.

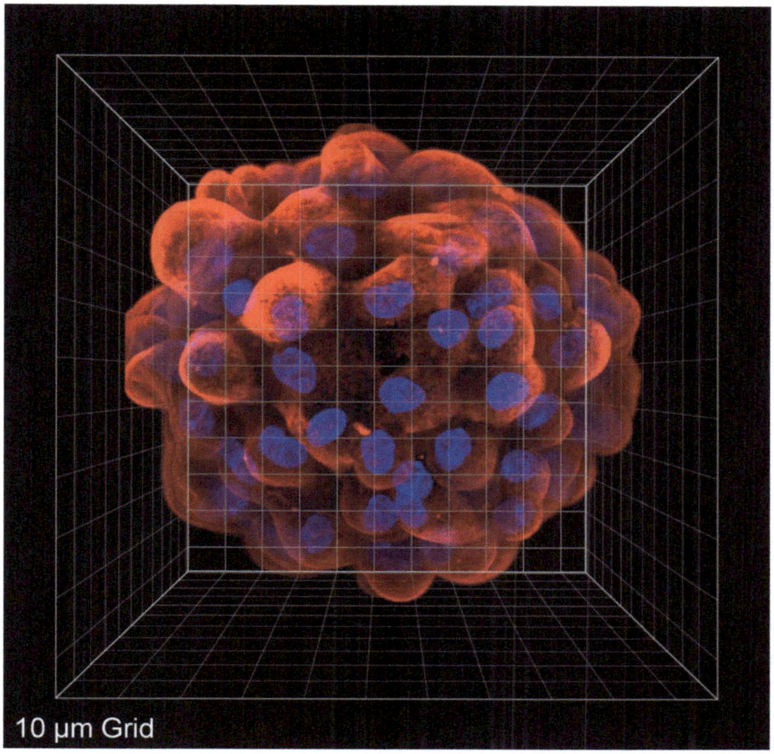

Fig. 1. Photomicrograph of primary human RPTCs cultured on a single GEM™ microcarrier and stained with WGA (10 µg/mL), and Hoechst 33342 (50 µg/mL, *blue nuclear stain*) at a magnification of 600×. Each block represents 10 µm in each dimension: *X*, *Y*, and *Z*.

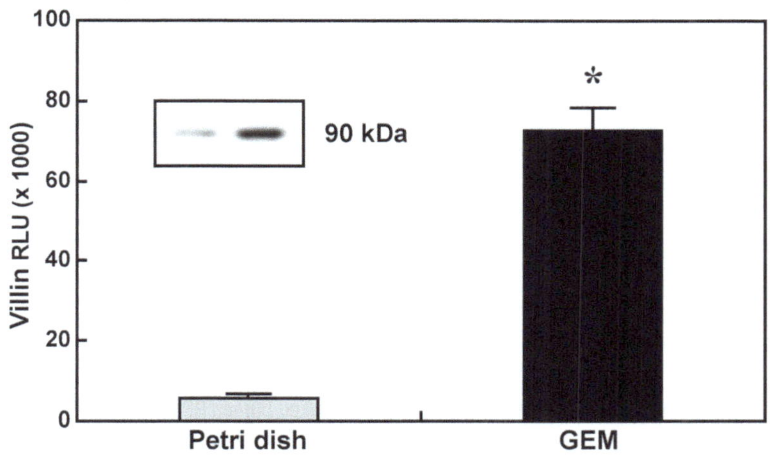

Fig. 2. Western blot analysis of villin expression after 5 day growth of primary RPTCs on GEMs™ vs. plastic cell culture dishes. Total cellular expression of villin (90 kDa, *inset*) is increased dramatically when primary RPTCs are cultured on GEMS™ and analyzed by chemiluminescent western blot (*$P < 0.001$, $N = 4$).

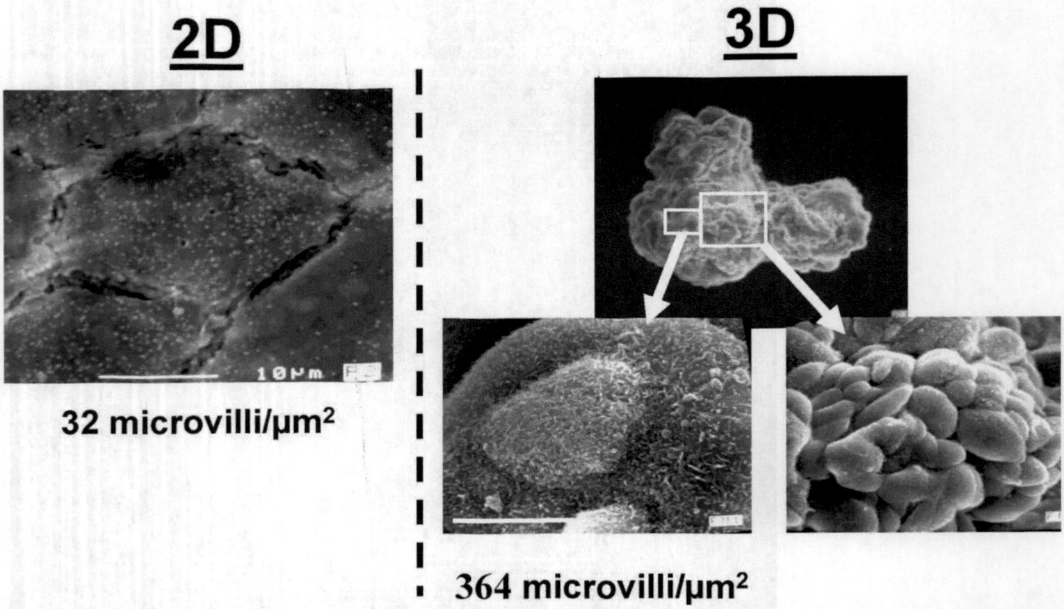

Fig. 3. Scanning electron microscopy of primary culture of RPTCs grown as a monolayer in a 2D transwell on left at 2,500× magnification, or as a 3D culture on GEMs™ on the right. The three GEMs™ pictured on the right are at 200× magnification, with *left bottom inset* at 2,500× magnification and *right bottom inset* at 500× magnification. Scale bars are 10 μm. The number of microvilli per μm^2 is tenfold greater in RPTCs grown in 3D than those grown in 2D.

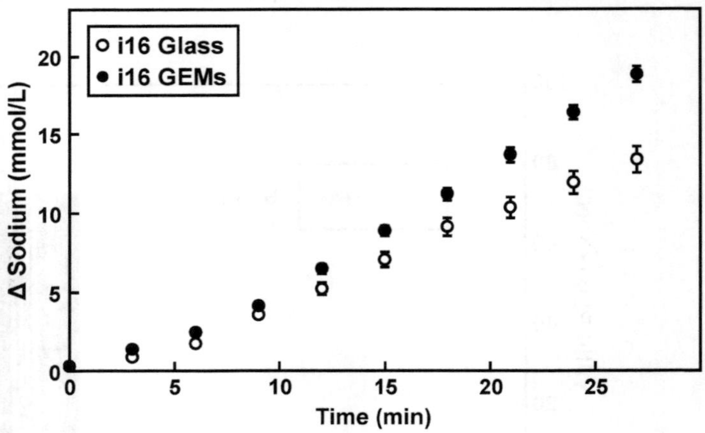

Fig. 4. Sodium influx (NHE3 activity) assay in immortalized human RPTCs (i16) grown on a collagen-coated glass plate and on GEMs™. The accumulation of sodium in response to ouabain (100 μmol/L) is greater and less variable in RPTCs cultured on 3D GEMs™ than those grown on 2D coated glass plates (GEMs™ vs. glass, $P<0.05$ for 12–24 min interval, $P<0.0001$ for >24 min).

2. Materials

2.1. Isolation of RPTCs and Primary Cell Culture

1. DMEM-F12 (Life Technologies™) is our basal cell culture medium. This formulation of DMEM-F12 is with HEPES and without phenol red. For primary cells, a serum-free, hormonally defined complete growth medium is made, using DMEM/F12 supplemented with the four ingredients (a–d) listed below. This combination was found to select specifically for epithelial over fibroblast cell proliferation from explanted cortex (8). The antibiotic mixture used in the primary media (PM) is 1× penicillin/streptomycin/amphotericin B (P/S/M, Life Technologies™). Once the primary cells have been immortalized, they are switched to DMEM-F12 media containing 2% fetal bovine serum (FBS) and 1× penicillin/streptomycin (Life Technologies™) without amphotericin B. For both primary and immortalized cell complete growth media, 2.5 μg/mL plasmocin (InvivoGen) and the following ingredients are added:
   (a) Human recombinant epidermal growth factor (EGF, Sigma-Aldrich®): Is made up with 1% low endotoxin BSA. Store in 100 μg/mL aliquots at −20°C and dilute 1:10,000 to a final concentration of 10 ng/mL in the medium.
   (b) Dexamethasone (Sigma-Aldrich®): Is water soluble at a ratio of active ingredient to solid weight (see Note 1). Store in 360 μg/mL aliquots at −20°C and dilute 1:10,000 to a final concentration of 36 ng/mL in the medium.
   (c) Triiodothyronine (Sigma-Aldrich®): Is only soluble in base. Store in 20 μg/mL aliquots at −20°C and dilute 1:10,000 to a final concentration of 2 ng/mL in the medium. Weigh 1 mg of triiodothyronine, add to 1 mL 1 N NaOH and stir to dissolve. Aliquot stock solution into 50 μL tubes and freeze at −20°C. Each 50 μL aliquot is then diluted 1:50, to make the 10,000× working stock.
   (d) Insulin–Transferrin–Selenium (ITS, Life Technologies™): Is diluted 1:100 into medium.
2. Collagenase A (Roche Diagnostics): A sterile 2 mg/mL collagenase solution is made in primary media (PM) and is used to digest cortical tissue.
3. Stacking metal sieves (Newark Wire Cloth Co): 212-μm (#70) is on top and 106-μm (#140) is on bottom (see Note 2).
4. Sterile forceps and scalpels (#10 blades work well) used for mincing tissue.
5. 100 mm tissue culture dish.
6. 75 cm^2 tissue culture flask.

2.2. Subculture and Cryopreservation of RPTCs

1. Solution of 10× trypsin/EDTA (0.5% trypsin, 0.2% EDTA, Life Technologies™) is diluted 1:10 in Dulbecco's PBS without calcium and magnesium (D-PBS) and this 1× solution is used for passaging adherent cells grown in tissue culture dishes.
2. For cryopreservation, DMSO is used at a final concentration of 10% in medium containing 50% FBS.
3. Cryovials.
4. Dulbecco's PBS without calcium and magnesium (D-PBS).

2.3. Immortalization of RPTCs

1. Lentivirus vectors HIV-7/CNPO-Tag and HIV-7/CNPO-hTERT were provided by Dr. Jiing-Kuan Yee and are described in detail by Kowolik et al. (9).
2. Adenovirus (ori-) SV40 TS A209 (10, 11) were provided by Dr. Janice Chou, NIH.
3. ViraPower™ Lentiviral Packaging Mix (Life Technologies™).
4. Lipofectamine™ 2000 (Life Technologies™).
5. HEK-293FT cells (Life Technologies™).
6. Opti-MEM® I + GlutaMax™ I media (Life Technologies™).
7. 0.45 μm Nalgene® low protein-binding syringe filter.
8. Geneticin (G418).
9. DMEM-H supplemented with 10% FBS.
10. DMEM/F12 supplemented with 0.5% FBS.

2.4. Characterization of RPTCs

All primary antibodies are diluted in PBS-T with 5% milk (see Subheading 2.4.4, step 5).

2.4.1. Proximal Tubule-Specific Markers for Immunofluorescent Staining

1. Biotinylated-*Lotus tetragonolobus* agglutinin (LTA, Vector Labs, 1:100 dilution).
2. Monoclonal antibody to γ-glutamyl transpeptidase (gGT, NeoMarkers clone 138H11, 1:100 dilution).
3. Rabbit polyclonal antibody to megalin (MEG, Santa Cruz Biotechnology®, Inc., 1:50 dilution).
4. Rabbit polyclonal antibody to aminopeptidase A (APA, Santa Cruz Biotechnology®, Inc., 1:50 dilution).
5. Monoclonal antibody to aminopeptidase N (APN, also known as CD13, clone 452, provided by Dr. Meenhard Herlyn, The Wistar Institute of Anatomy and Biology, Philadelphia, PA) is used at a 1:20 dilution of the hybridoma culture supernatant. Antibodies to CD-13 are also commercially available from Southern and Santa Cruz Biotechnologies®, Inc.

2.4.2. Markers of Other Nephron Regions Which Serve as Negative Controls

1. Tamm-Horsfall protein is synthesized by the thick ascending limb (TAL) and the rabbit polyclonal antibody (THP, Santa Cruz Biotechnology®) is used at a 1:200 dilution.

2. Sodium chloride co-transporter is expressed in the distal convoluted tubule and the rabbit polyclonal antibody (NCC, Millipore™) is used at a 1:500 dilution.
3. Aquaporin 2 is expressed in the collecting duct, and the rabbit polyclonal antibody (AQP2, Abcam®) is used at a 1:500 dilution.

2.4.3. Secondary Antibodies or Conjugates

All are from Life Technologies™ and are used at a final concentration of 2 μg/mL and are diluted in PBS-T with 2% milk (see Subheading 2.4.4, step 3).

1. Alexa 647-labeled streptavidin conjugate is used for the biotinylated LTA staining.
2. Donkey anti-rabbit Alexa Fluor® 647 IgG is used for MEG, APA, THP, AQP2, and NCC staining.
3. Donkey anti-mouse Alexa Fluor® 647 IgG is used for GGT and APN staining.

2.4.4. Other Reagents

1. Fixing of cells is done with 4% paraformaldehyde (Electron Microscopy Sciences), diluted 1:8 from 32% stock, make fresh for each use. Diluent is D-PBS (see Subheading 2.2, step 4). Triton™-X100 is added to the fixative to a final concentration of 1% in order to permeabilize the cells.
2. Fixative is stopped by washing with tris-buffered saline, TBS: 10 mM Tris, 150 mM NaCl supplemented with 0.02% Tween 20 (TBS-T).
3. Washes and the diluent for blocking and antibody incubations is PBS-T (PBS, without calcium and magnesium, plus 0.02% Tween-20).
4. Unlabelled streptavidin (Pierce, 2 μg/mL final concentration, diluted in PBS-T) and biotin (Sigma-Aldrich®, stock solution of 15 mmol/L in water, and then diluted in PBS-T to a final concentration of 4 μmol/L) are used for blocking in the LTA staining.
5. 5% Non-fat dry milk: Dissolve in PBS-T, heated to 60°C for 10 min, sonicated for 1 min and centrifuged at 9,000 × *g* for 30 min at room temperature. It is used as the blocking reagent for the rest of the staining and is the diluent for the primary antibodies.
6. Collagen Type I (BD Biosciences): Diluted 1:100 in collagen coating buffer (200 mM KCl, 50 mm Tris, pH 8.5).
7. Glass bottom 96-well plates, coated with collagen type I—Pipette 100 μL into glass bottomed 96-well plate and incubate for 1 h at 37°C. Rinse plates 3× in ddH_2O and dry. Plates are sterilized by placing on UV transilluminator for 5 min. Dried dishes can be stored at 4°C for a month.

2.5. Microcarrier 3D Cell Culture

1. Magnetized alginate microcarriers with gelatin coating, such as the Global Eukaryotic Microcarriers (GEM™, Global Cell Solutions).
2. A strong neodymium boron rare earth magnet (at least 1″×2″×0.25″).
3. Complete culture medium for primary or immortalized RPTCs (see Subheading 2.1).
4. 5% Poly-(2-hydroxyethyl methacrylate) solution in ethanol (Poly-HEMA, Sigma-Aldrich®): Prepare a 100% poly-HEMA solution by dissolving 120 mg/mL solution of poly-HEMA in 100% ethanol overnight (e.g., 3 g poly-HEMA into 25 mL 100% ethanol). Dilute this solution 1:20 to create a 5% poly-HEMA solution in 100% ethanol.
5. BioLevitator™ automated microcarrier culture system and LeviTubes™ for culturing (Global Cell Solutions, Inc.™) equipped with a hydrophobic filter cap (see Note 3).
6. Freezing medium: Culture medium supplemented with 20% FBS. Make fresh the day of freezing cells and keep on ice.
7. DMSO freezing wash: Culture medium supplemented with 5% DMSO and 20% FBS.

3. Methods

Human kidneys may be obtained from various sources. The most convenient source is kidneys surgically removed from patients with renal cell carcinoma or Wilms' tumors that have affected only one pole of the kidney cortex. Another source may be transplant tissue that cannot be used, or fresh cadaveric tissue. All tissues should be obtained and used under institutional review board approval. Unaffected tissue may then be harvested from the normal pole of the surgically removed kidney. The renal cortex then becomes the source of RPTCs that are isolated using the protocol described in Subheading 3.1. Our isolation procedure uses a combination of enzyme digestion of the tissue, followed by sieving to separate the nephron segments.

We define primary cells as the first outgrowth of cells from fresh tissue. RPTCs exhibit typical epithelial cell morphology growing in monolayers exhibiting cobblestone appearance. RPTCs are notable for their high concentration of mitochondria, and their expression of short and long microvilli. Primary RPTCs will grow for 10–12 passages (before slowing down), or indefinitely when immortalized (see Subheading 3.2). Primary cells may be grown on novel alginate microcarriers that assure their polarization, more closely resembling the morphology and behavior of RPTCs in vivo (see Subheading 3.5).

3.1. Isolation of RPTCs and Primary Cell Culture

1. Make a sterile 2 mg/mL solution of collagenase (see Subheading 2.1, step 2) in primary media (PM) warmed to 37°C.
2. Decapsulate the kidney and isolate cortex; a piece approximately 1 cm × 3 cm × 0.5 cm thick is usually obtained. Mince tissue into small pieces (1–3 mm^3) using sterile forceps and scalpel blades in a 100-mm culture dish.
3. Add tissue pieces to warm collagenase (the amount of tissue described in step 2 is typically put into 20 mL collagenase solution in a 50 mL conical tube) and place in the water bath for 5 min. The collagenase mixture is moved into a 37°C incubator on a rocker and gently rocked for 20 min. Every 10 min, the tissue pieces are gently dispersed by pipetting with a 25-mL pipet.
4. Centrifuge tube for 3 min at 90 × *g*, then carefully remove the supernatant and resuspend the pellet in 10 mL PM using a 25-mL pipet.
5. Pipette tissue pieces onto top sieve (212-μm, pre-stacked on top of the 106-μm sieve and set on top of a 100-mm culture dish) and gently mash the pieces using rubber end of a 10-mL syringe. Add 10 mL PM onto the tissue pieces two times, mashing the tissue pieces each time. Glomeruli and large nephron segments will be caught in the 106-μm sieve.
6. Recover 30 mL of sieved cells and tubules from the dish and transfer into a 50 mL conical tube. Add 5 mL PM back onto the 100-mm dish and pipette vigorously to remove remaining tubules and add to the 50 mL tube. Centrifuge tube for 3 min at 90 × *g*.
7. Resuspend cell pellet in 10 mL PM and let settle for 10 min. To remove red blood cells and debris, carefully pipette off supernatant down to within 1 mL of the pellet. Add 30 mL of PM (scale according to size of initial cortex) and distribute resuspended tubules into 3 × 75 cm^2 flasks (10 mL each). Incubate at 37°C in a 5% CO_2/95% air humidified chamber.
8. Change the culture medium the next morning in order to remove debris and unattached nephron segments. The medium is changed every 2 days and cells will grow out (reproduce away from the explanted tissue), form monolayers and can become confluent following 6–7 days in culture.

3.2. Immortalization of RPTCs

Four different methods have been used to immortalize RPTCs. The first one uses SV40 T antigen alone, and the second uses SV40 T antigen containing a mutated temperature-sensitive SV40 (SV40 TS) (10, 12). SV40 has also been used along with hTERT (9) and the fourth method that has been successful uses hTERT alone (13). We describe two methods of immortalization, one with the hTERT and Tag (SV40) using lentivirus and the second with SV40 TS using adenovirus. However, each of the lentivirus constructs could also be used alone.

3.2.1. Lentivirus Production

Lentiviruses are produced in HEK-293FT cells by transient cotransfection with the Lentivirus vectors HIV-7/CNPO-Tag and/or HIV-7/CNPO-hTERT and the ViraPower Lentiviral Packaging Mix (an optimized mix of packaging plasmids: pLP1, pLP2, pLP/VSVG) at a 1:3 ratio, according to the manufacturer's instructions, using Lipofectamine™ 2000 (see Subheading 2.3).

1. Passage an HEK-293FT cell culture at a 1:3 ratio the day before transfection, into a 100-mm tissue culture dish in DMEM-H/10% FBS (no antibiotics) medium. It is optimum that they will be 30–40% confluent the next day.
2. For each transfection, dilute 3 μg Lentivirus vector DNA and 9 μg ViraPower Lentiviral Packaging Mix in 1.5 mL Opti-MEM® medium. Mix gently.
3. In a separate tube, mix 36 μL Lipofectamine™ 2000 with 1.5 mL Opti-MEM® medium, mix gently and incubate for 15 min at room temperature.
4. Combine the DNA mix with the Lipofectamine™ 2000 mixture, mix gently and incubate for 20 min at room temperature to allow the DNA/Lipofectamine™ complexes to form.
5. Add the DNA/Lipofectamine™ complexes drop-wise to the cells, mix gently by rocking the dish and incubate at 37°C in a humidified 5% CO_2 incubator.
6. Collect the supernatant containing the virus after 48–72 h, filter it through a low protein-binding 0.45 μm syringe filter, aliquot and store at −80°C. Viral titers using this method are between 0.5 and 2×10^6 colony forming units per mL.

3.2.2. Adenovirus Production

Infectious immortalizing adenovirus is made using a chimeric origin-defective temperature sensitive (ts209) SV40 large T antigen containing El minus adenovirus (SV40 TSA209). HEK-293 cells were originally immortalized using an adenovirus that contains the El region and therefore can produce functional adenovirus using El minus virus particles. The virus is only replication-competent in a cell that has the El region of the adenovirus, which most cells do not.

1. Plate the HEK-293 cells the day before transfection into a 100-mm tissue culture dish so that they will be 80–90% confluent the next day. Do not include antibiotics in the medium.
2. Thaw one vial of adenovirus stock and add the amount of viral stock to achieve an MOI of 0.1 (approximately 10 μL of a stock at 1×10^8 viral particles per mL for each 100-mm dish used).

3. Grow the cells until a majority of the cells have lifted off the culture surface and have lysed (56–72 h). Medium does not need changing during this time.
4. Collect the medium and remaining cells and transfer to a 15-mL centrifuge tube.
5. Freeze and thaw this tube three times by placing the tube at −80°C until frozen, and then bringing to room temperature in a water bath.
6. Centrifuge the tube at 900 × *g* for 5 min at 25°C to pellet debris; collect the supernatant and place in a fresh centrifuge tube.
7. Filter-sterilize the supernatant through a low protein-binding 0.45 μm syringe filter, aliquot, and store at −80°C. Viral titers using this method are between 1 and 2×10^8 viral particles per mL.

3.2.3. hTERT and Tag (SV40) Transduction

1. To establish immortalized clones, primary RPTCs from third and fourth passage are transduced with both HIV-7/CNPO-Tag and HIV-7/CNPO-hTERT at an MOI of 1 for 18–20 h.
2. 48 h after infection with the virus, the cells are put under G418 selection (250 μg/mL) in immortalized cell medium (see Subheading 2.1).
3. Using this method, RPTCs continue to grow past the passage number where the cells normally begin to show signs of senescence (between passages 10 and 15).

3.2.4. SV40 TS Transduction

1. RPTCs are transduced with adenovirus SV40 TSA209 by incubating early passage primary RPTCs with adenovirus at an MOI of 100.
2. After incubating for 2 h, the medium is removed and fresh adenovirus is added again but this time in immortalized cell medium (see Subheading 2.1). Continue to culture in that medium at 33°C.
3. Between passages 10 and 15, the appearance of senescent cells starts and small colonies of immortalized cells begin to appear. After 5 more passages, the appearance of senescent cells is minimal.
4. In order to stop the growth of the TS immortalized RPTCs, switch the culture conditions to just DMEM/F12 with 0.5% FBS and place the culture at 39°C for 3 days to turn off the SV40 TS large T antigen. 39°C is the nonpermissive temperature for the adenovirus and it becomes unstable.

3.3. Subculture and Cryopreservation of RPTCs

3.3.1. Passaging Primary or Immortalized Cells

1. Pre-warm medium, $DPBS^{-/-}$, and 1× trypsin–EDTA (final concentration of 0.05% trypsin, and 0.02% EDTA) to 37°C.
2. Add $DPBS^{-/-}$ to tissue culture flasks for 5–10 min.
3. Remove DPBS and add 1× trypsin–EDTA. Put in 37°C incubator for up to 3 min.

4. When cells are detached, pipette with 3 volumes of medium with 2% FBS to inactivate trypsin and disperse cells.
5. Centrifuge cells for 4 min at 90 × *g*.
6. Resuspend cell pellets in fresh medium and distribute to new flasks at a 1:3 subculture ratio.

3.3.2. Cryopreservation of Primary or Immortalized Cells

1. When cells are 80–90% confluent in a 75 cm^2 flask, there are approximately five million cells. It is optimum to freeze them at a minimum concentration of two million cells per cryovial.
2. Primary or immortalized cells are trypsinized, centrifuged, and resuspended in medium with 50% FBS.
3. Keeping cells cold, add DMSO drop-wise into swirling tube, starting with 1 drop and doubling the number of drops consecutively until reaching the final concentration of 10% DMSO (see Subheading 2.2, step 2).
4. Aliquot cells into cryovials on ice. Place in Styrofoam boxes in −80°C freezer overnight, then put in liquid nitrogen for long-term storage.

3.3.3. Thawing of Primary or Immortalized Cells

1. Add 10 mL pre-warmed medium into an empty 75 cm^2 flask or 100-mm culture dish.
2. Take cryovial from liquid nitrogen, wipe exterior with 70% ethanol.
3. Thaw rapidly with agitation in water bath for 1 min.
4. Immediately pipette cells from cryovial into flask (there is no need to spin cells first); rinse out vial using media from the flask or dish.
5. The next morning, check for cell adherence and change media to remove any residual DMSO.
6. Feed cells by changing media every 2 days until they reach 80–90% confluence and then passage.

3.4. Characterization of RPTCs

1. Cells are placed on collagen I-coated glass bottom 96-well plates and are washed twice with D-PBS.
2. They are fixed in D-PBS containing 4% paraformaldehyde and 1% Triton™-X100 for 5 min and washed three times with TBS-T for 5 min per wash.
3. The fixed cells are blocked overnight in 5% non-fat dry milk.
4. Cells to be used with biotinylated lectin are first blocked with 2 μg/mL unlabelled streptavidin and washed in PBS-T. 4 μmol/L biotin is added, and cells are washed and blocked with bovine serum albumin. Lectin (biotinylated LTA, 20 μg/mL) staining is performed in 1% bovine serum albumin and detected with 2 μg/mL Alexa 647-labeled streptavidin.

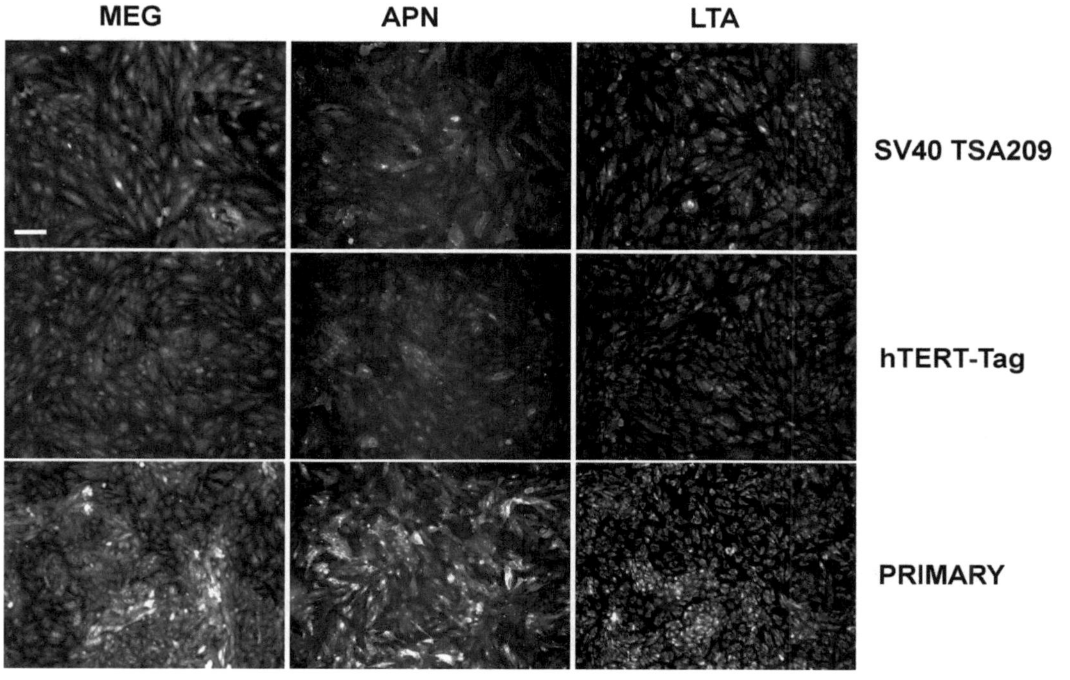

Fig. 5. Immunofluorescent staining of RPTCs in primary culture or immortalized with either hTERT-Tag or SV40 TS. The three cell lines were stained for the RPTC markers; megalin (MEG), aminopeptidase N (APN), and LTA. All are 100× magnifications, scale bar (*white-upper left image*) is 100 μm.

5. Other primary antibodies (GGT, MEG, APA, APN, THP, NCC, and AQP2) are added at the concentrations listed in Subheading 2.4.
6. Perform all primary antibody incubations overnight with gentle rocking at 4°C, followed by washing three times (5 min each) in PBS-T (see Subheading 2.4.4, step 3).
7. The secondary incubation is carried out in 2% milk with the corresponding Alexa-647-conjugated secondary antibodies (see Subheading 2.4.3) for 90 min at room temperature.
8. Cells are washed three times in PBS-T and imaged using a fluorescence microscope. See examples of stained cells in Fig. 5, which show staining for MEG, APN, and LTA. GGT and APA also stain positive, and the staining for THP, NCC, and AQP2 is completely negative.
9. Controls for nonspecific fluorescent streptavidin binding and fluorescent secondary antibody staining are performed for each cell type and threshold levels for background staining are established. Exposure times are identical for each cell type across each individual antibody or lectin.

3.5. Microcarrier 3D Cell Culture

3.5.1. Preparing the GEMs

1. Hold the vial of GEM™ substrate over a neodymium boron rare earth magnet for 2 min to attract the GEMs into the bottom of the vial. The solution should appear in two distinct layers: a clear GEMs-free layer on the top and a packed layer of GEMs on the bottom.
2. Carefully suction off the storage buffer, being careful not to aspirate GEMs.
3. Add 1 mL of the appropriate culture medium (for either immortalized or primary RPTCs) at room temperature, as prepared in Subheading 2.1.
4. Cap the vial and invert to mix the GEMs and medium.
5. Repeat steps 1–4 for a total of four washes.
6. Aspirate off the final wash and add 1 mL of culture medium. The GEMs are now ready for use.

3.5.2. Preparing RPTCs for Loading onto the GEMs

1. For manual loading of cells on GEMs, coat a few sterile 100-mm culture dishes with 5.8 mL of 5% poly-HEMA solution per dish and let them dry overnight in a laminar flow hood (*see* Subheading 2.5, step 4). For automated loading of cells on GEMs, turn on the BioLevitator™ (37°C, 5% CO_2).
2. Trypsinize cells (80–90% confluent) in a 75 cm^2 flask, as described in Subheading 3.3.
3. Pipette 8 mL 2% FBS culture medium onto the cell layer to recover the trypsinized cells into a 50-mL tube.
4. Centrifuge the tube of cells at 90 × *g* for 3 min.
5. Suction off the supernatant and resuspend the cells in 30 mL culture medium, pipetting vigorously to ensure a single cell suspension.
6. Invert the vial of GEM™ slurry to resuspend the GEMs.

3.5.3. Manual Loading a Culture of RPTCs onto GEMs

1. Transfer 10 mL of the resuspended cells to each poly-HEMA coated 100-mm dish.
2. Pipette the GEM™ mixture several times to create an even, heterogeneous mixture.
3. Transfer 100 μL GEM™ slurry to each 100-mm dish.
4. Cover the dishes and incubate for 4 h, gently stirring them once every 30 min.
5. Place the dishes on a rocking mixer in a conventional laboratory incubator (37°C, 5% CO_2) for 24 h.

3.5.4. Growing RPTCs on GEMs in an Automated Mixing Levitating Incubator

1. Transfer the 30 mL of resuspended cells to a LeviTube™.
2. Pipette the GEM™ mixture several times to create an even, heterogeneous mixture.

3. Transfer 300 μL GEM™ slurry to LeviTube™ with cells and culture medium.
4. Seal the LeviTube™ with the filter cap and place in the BioLevitator™.
5. Set the BioLevitator™ parameters for the "Inoculation" and "Culture" protocols according to the manufacturer's instructions.

3.5.5. Feeding the GEM™ Cultures

1. 24 h after inoculating cells, visually inspect them under a microscope. The GEMs have a faint outline and the cells themselves will appear as bubbles on the surface of the GEM™.
2. On day 2, feed the cultures by pulling the GEMs to the bottom corner of the dish or LeviTube™ with the magnet.
3. Suction off half of the spent media (5 mL for the 100-mm culture dishes, 15 mL for the LeviTubes™).
4. Replenish GEM™ suspension with pre-warmed culture media, bringing the total volume back up to 10 mL per dish or 30 mL per LeviTube™.
5. Feed the cells as described in steps 2–4 every 3–4 days.
6. After approximately 6 days, cells should reach confluence and be ready for experiments or cryopreserving.

3.5.6. Freezing Cells on GEMs

1. Prepare DMSO freezing wash and freezing media (see Subheadings 2.5, steps 6 and 7) and keep solutions on ice.
2. If growing GEMs in a LeviTube™, proceed to the next step. If growing GEMs in a 100-mm culture dish, transfer the GEM™ solution from the 100-mm culture dishes to a 50 mL tube.
3. Place the LeviTube™ or 50 mL tube on the magnet to pull the GEMs to the bottom.
4. Gently suction off supernatant, slowly add the DMSO freezing wash and mix gently (see Subheading 2.5, step 7).
5. Pull the GEMs to the bottom with a magnet, then suction off the DMSO freezing wash.
6. Repeat steps 3–4, taking care not to aspirate any of the GEM™ layer.
7. Gently add freezing media and mix gently.
8. With a 1 mL pipet, add DMSO drop-wise into the swirling tube, doubling the number of drops each time until the final concentration of DMSO is 10% of the freezing media.
9. Still swirling, aliquot 1 mL of the GEM™ solution into pre-labeled 2 mL cryovials.
10. Immediately store the cryovials at −80°C in a double-sided Styrofoam container.
11. After 24 h at −80°C, move the cryovials to liquid nitrogen.

3.5.7. Thawing Frozen Cells on GEMs

Depending on the type of experiments you have planned, the frozen cells on GEMs may be immediately ready for assaying once thawed.

1. Thaw a cryovial of frozen cells on GEMs in a 37°C water bath with gentle agitation until just thawed.
2. Pull the GEMs to the bottom of the vial with the magnet.
3. Gently aspirate the freezing media and resuspend the GEMs in 5 mL pre-warmed culture media in a 15 mL tube. This is just enough for one half of a 96-well plate at the correct concentration of GEMs (100 μL/well). For continuing culture, resuspend the GEMs in 5 mL pre-warmed culture media in a poly-HEMA treated100-mm culture dish or LeviTube™.
4. Place the culture dish in the incubator or place the LeviTube™ in the BioLevitator™ and run the experiment with the parameters defined in Subheading 3.5.4.
5. The cell line may be expanded at this point by adding 25 mL of pre-warmed culture medium, and 250 μL GEM™ slurry to the LeviTube™ as outlined in the application note.

4. Notes

1. For example, if the activity of a certain lot of dexamethasone is 64.5 mg/g, in order to make a stock solution of 360 μg/mL (10,000× stocks), one must multiply by 15.5 (because active ingredient is 1/15.5 of weight). So weigh out 55.8 mg into 10 mL D-PBS for actual 10,000× stock, and then add 50 μL per 500 mL bottle for a final concentration of 36 ng/mL.
2. After decontaminating and cleaning sieves (may need to sonicate in order to fully remove the tissue debris), place sieves inside a UV light box to sterilize them for 10 min/side.
3. The BioLevitator™ and LeviTubes™ are optional and only necessary for automated loading of cells onto GEM™ microcarriers.

Acknowledgments

This work was supported by NIH grants HL074940, HL023081, DK039308, HL068686, and HL092196.

Disclosure

Two of the authors of this chapter, Felder RA and Gildea JJ, invented the 3D method described here in order to grow human RPTCs for NIH-sponsored research. This technology has been subsequently commercialized and these two authors have an equity position in the company (Global Cell Solutions, Charlottesville, VA, USA).

References

1. Baer PC, Bereiter-Hahn J, Schubert R, Geiger H (2006) Differentiation status of human renal proximal and distal tubular epithelial cells in vitro: differential expression of characteristic markers. Cells Tissues Organs 184:16–22
2. Wilson PD, Dillingham MA, Breckon R, Anderson RJ (1985) Defined human renal tubular epithelia in culture: growth, characterization, and hormonal response. Am J Physiol 248:F436–F443
3. Detrisac CJ, Sens MA, Garvin AJ, Spicer SS, Sens DA (1984) Tissue culture of human kidney epithelial cells of proximal tubule origin. Kidney Int 25:383–390
4. Racusen LC, Monteil C, Sgrignoli A, Lucskay M, Marouillat S, Rhim JG, Morin JP (1997) Cell lines with extended in vitro growth potential from human renal proximal tubule: characterization, response to inducers, and comparison with established cell lines. J Lab Clin Med 129:318–329
5. Ryan MJ, Johnson G, Kirk J, Fuerstenberg SM, Zager RA, Torok-Storb B (1994) HK-2: an immortalized proximal tubule epithelial cell line from normal adult human kidney. Kidney Int 45:48–57
6. Racusen LC, Wilson PD, Hartz PA, Fivush BA, Burrow CR (1995) Renal proximal tubular epithelium from patients with nephropathic cystinosis: immortalized cell lines as in vitro model systems. Kidney Int 48:536–543
7. Justice BA, Badr NA, Felder RA (2009) 3D cell culture opens new dimensions in cell-based assays. Drug Discov Today 14:102–107
8. Taub M, Sato G (1980) Growth of functional primary cultures of kidney epithelial cells in defined medium. J Cell Physiol 105: 369–378
9. Kowolik CM, Liang S, Yu Y, Yee JK (2004) Cre-mediated reversible immortalization of human renal proximal tubular epithelial cells. Oncogene 23:5950–5957
10. Loghman-Adham M, Nauli SM, Soto CE, Kariuki B, Zhou J (2003) Immortalized epithelial cells from human autosomal dominant polycystic kidney cysts. Am J Physiol Renal Physiol 285:F397–F412
11. Lei KJ, Gluzman Y, Pan CJ, Chou JY (1992) Immortalization of virus-free human placental cells that express tissue-specific functions. Mol Endocrinol 6:703–712
12. Orosz DE, Woost PG, Kolb RJ, Finesilver MB, Jin W, Frisa PS, Choo CK, Yau CF, Chan KW, Resnick MI, Douglas JG, Edwards JC, Jacobberger JW, Hopfer U (2004) Growth, immortalization, and differentiation potential of normal adult human proximal tubule cells. In Vitro Cell Dev Biol Anim 40: 22–34
13. Wieser M, Stadler G, Jennings P, Streubel B, Pfaller W, Ambros P, Riedl C, Katinger H, Grillari J, Grillari-Voglauer R (2008) hTERT alone immortalizes epithelial cells of renal proximal tubules without changing their functional characteristics. Am J Physiol Renal Physiol 295:F1365–F1375

Chapter 21

Human Female Reproductive Tract Epithelial Cell Culture

Rachel W.S. Chan, Abby S.C. Mak, William S.B. Yeung, Kai-Fai Lee, Annie N.Y. Cheung, Hextan Y.S. Ngan, and Alice S.T. Wong

Abstract

The female reproductive system is a complex system. Epithelia of the female reproductive system including the ovaries, the oviduct, and the uterus are important sites for follicular development, ovulation, fertilization, implantation, and embryo development. They are also able to synthesize and secrete various hormones, growth factors, and cytokines, which are essential to women's health, sexuality, and reproduction. Conversely, their dysfunction has been implicated in disorders such as infertility, endometriosis, and many other gynecological diseases, as well as cancer. In this chapter, we describe detailed procedures for establishing and maintaining primary cultures of human ovarian surface epithelium, oviductal epithelium, and endometrium. We also provide protocols for cell immortalization, clonal isolation, and in coculture with stromal cells. These cultures can be useful models for investigating the molecular and cellular functions of these epithelia in both normal and pathological states.

Key words: Ovary, Oviduct, Endometrium, In vitro culture, Reproduction

1. Introduction

The epithelium of the reproductive tract covers the surface of the ovary, the inner lining of the oviduct and the endometrium. While there is no doubt that the tissue plays important roles in reproductive events, its improper function can be the underlying cause of many gynecological diseases as well as cancer.

The ovarian surface epithelium (OSE) is the part of the pelvic peritoneum that overlies the ovary. During each reproductive cycle, the OSE takes part in the cyclical ovulatory rupture and repair. This inconspicuous epithelium is also of major importance in gynecologic pathology, because it is considered to give rise to epithelial ovarian carcinomas, which are the most common and most lethal of all human ovarian neoplasms (1). In spite of the clinical importance, studies of human OSE and of early events in ovarian

Scott H. Randell and M. Leslie Fulcher (eds.), *Epithelial Cell Culture Protocols: Second Edition*, Methods in Molecular Biology, vol. 945, DOI 10.1007/978-1-62703-125-7_21, © Springer Science+Business Media, LLC 2012

carcinogenesis have been hindered by the minute amounts of OSE cells that can be obtained from the ovary and by their limited growth potential in culture (2).

The growth potential of cultured OSE can be increased by the introduction of viral genes simian virus 40 (SV40) T antigen or human papilloma virus (HPV) E6/E7 (3, 4). As an alternative, telomerase reverse transcriptase (TERT) has been known to bypass senescence and to increase lifespan. Such cells are non-tumorigenic, but they resemble preneoplastic phenotype in many ways, including reduced serum requirement, loss of contact inhibition, and abnormal growth control and apoptosis (3). Given the frequent alterations of p53 and hTERT found in ovarian carcinomas (4–7), immortalized OSE cell lines with these genetic changes make them invaluable for studying the biology of OSE and ovarian carcinogenesis.

The oviduct has long been considered to be passive in reproduction serving as the site of fertilization and a passage for transporting the gametes and preimplantation embryos. Accumulating evidence suggest that the oviduct also modulates the functions of spermatozoa (8) and promotes the development of preimplantation embryos (9). These latter functions of the oviduct are mediated significantly by its epithelium, which consists of ciliated cells and secretory cells. The proportion of these cell types in different regions of the oviduct changes in the reproductive cycle. The epithelium especially in the ampullary region is actively synthesizing and secreting various molecules into the luminal fluid of the oviduct, including complement protein-3 (10) and demilune cell and parotid protein (11) that are known to stimulate the development of preimplantation embryos. In addition, recent studies suggest that some serous ovarian carcinomas may arise from the epithelium of the oviduct (12).

The endometrium is the inner lining of the uterus. It is covered by two types of epithelium, namely the luminal epithelium and the glandular epithelium. The former lines the uterine lumen and is the site of implantation, where the blastocyst attaches to and eventually invades through it into the endometrial stroma. The glandular epithelium secretes a number of molecules such as glycodelin, mucin-1, and granulocyte-macrophage colony-stimulating factor in a hormone-dependent manner (13–15). These molecules modulate the implantation process (16, 17) and the subsequent placentation (18) and cytokine environment in the decidua (19).

The interaction of the gametes/embryos with the epithelium of the female reproductive tract is complex and can be dissected by using coculture models, i.e., culture of the gametes/embryos with the epithelial cells of the reproductive tract. Such studies will enable us to understand how the microenvironment within the reproductive tract affects fertilization, preimplantation embryo development, and implantation in normal as well as in pathological conditions such as ectopic pregnancy and infertility.

The human endometrium undergoes cyclical growth and regression in each reproductive cycle. The endometrium regenerates from the lower basalis layer, a germinal compartment that persists after menstruation to give rise to the new upper functionalis layer. Since the endometrium is a highly dynamic regenerative tissue, it has been hypothesized that stem/progenitor cells are responsible for this remarkable capacity. The hypothesis is supported by the identification of clonogenic human endometrium stem/progenitor cells with self-renewal and multipotent differentiation capacity (20). As a well-accepted endometrial stem cell-specific marker is unavailable, several functional approaches have been used to identify rare populations of adult stem/progenitor cells in human endometrium (21).

2. Materials

2.1. Preparation of Purified Human OSE, Oviductal, and Endometrial Epithelial Cells

1. 35 and 100 mm tissue culture dishes.
2. 75 cm^2 culture flask.
3. 15 mL centrifuge tubes: Sterile, for preparing medium and incubation.
4. Table top centrifuge.
5. Incubator: Humidified, 37°C, 5% CO_2 in air.
6. Shaking water bath at 37°C.
7. Surgical instruments (autoclaved, sterile): Cell lifter scrapers (Corning) for scraping the OSE cells, one pair of scissors and a pair of forceps for opening of the oviduct, and two pair of small fine forceps for peeling of oviductal epithelium, surgical scalpels for mincing of the endometrial tissue.
8. Nylon cell strainer with pore size of 40 μm.

2.2. Isolation of OSE, Oviductal, and Endometrial Cells

1. Medium 199 (M199).
2. MCDB 105 Medium.
3. Ca^{2+}- and Mg^{2+}-free phosphate-buffered saline (PBS).
4. 0.05% (w/v) trypsin/0.5 mM ethylenediamine tetraacetic acid (EDTA) solution (Gibco®).
5. 10 mg/mL hydrocortisone(Sigma-Aldrich®) stock: Dissolve 100 mg hydrocortisone powder in 10 mL absolute ethanol.
6. Bovine serum albumin (BSA; 200×; GE Healthcare Amersham Biosciences) 200 mg/mL: 1 g in 5 mL of double-distilled water (final volume); add water only up to 5 mL once solid dissolves.
7. Collagenase Type 1A (0.1%, w/v; Sigma-Aldrich®) and deoxyribonuclease type I (DNase 0.001%, w/v; Sigma-Aldrich®) in PBS.

2.3. Immortalization

1. SV40 and hTERT plasmids were kind gifts of Dr. N. Auersperg (University of British Columbia, Vancouver, B.C., Canada) and Dr. G. Tsao (University of Hong Kong, Hong Kong), respectively.
2. G418.
3. Lipofectamine™ 2000 (Life Technologies™).
4. OPTI-MEM® I Reduced Serum Media (Life Technologies™).
5. Mouse monoclonal antibody anti-SV40 T antigen mAb (PAb419) (DAKO) and goat anti-mouse IgG labeled with Texas Red dye (Jackson ImmunoResearch).
6. Polybrene (Sigma-Aldrich®).
7. TRAPeze® Telomerase Detection Kit (Chemicon).

2.4. Cryopreservation

1. Cell culture medium.
2. Dimethyl sulphoxide (DMSO, Sigma-Aldrich®).
3. Cryogenic vials.

2.5. Preparation of Single Cell Suspension of Purified Human Endometrial Epithelial and Stromal Cells

1. Collagenase Type III (Worthington Biochemical Corp.): Dissolve in PBS for a final concentration of 2.5 mg/mL; stored in single use aliquots at −20°C.
2. Deoxyribonuclease type I (DNase, Worthington Biochemical Corp.): Dissolve in PBS for a final concentration of 4 mg/mL; stored in single use aliquots at −20°C.
3. Collection Medium: DMEM/F12 (Gibco®) supplemented with 1% fetal bovine serum (FBS), 1% penicillin/streptomycin (Gibco®), and 1% L-glutamine (Gibco®).
4. Ficoll-Paque™ (GE Healthcare).
5. Cell strainer 40 μm.
6. Dynabeads® Epithelial Enrich (Ber-EP4 coated Dynabeads®, Life Technologies™).
7. Dynabeads® CD45 (Life Technologies™).
8. Magnetic particle collector (Life Technologies™).
9. Gelatin (Sigma-Aldrich®): Dissolve gelatin in PBS to make 0.2% gelatin for coating of cultureware.
10. 60-mm Petri dishes.
11. 10% Formalin.
12. Toluidine Blue O (Sigma-Aldrich®): Dissolve in distilled water at 10 mg/mL.
13. 25 mm Thermanox coverslips (Nalge Nunc International) for 6-well plates. Coverslips were coated with 0.2% gelatin for 30 min and washed with PBS before use.

2.6. Culture Media

1. OSE cell culture medium: To 500 mL M199:MCDB105 (1:1, v/v), 100 U/mL penicillin, and 100 μg/mL streptomycin supplemented with 10% heat-inactivated FBS. All components are sterile filtered.
2. Oviductal cell culture medium: In 450 mL Dulbecco's modified Eagle's medium Nutrient Mixture Ham F-12 (DMEM/F12 1:1, v/v; Sigma-Aldrich®), add 50 mL FBS or human preovulatory serum, supplemented with L-glutamine (0.1%, w/v), penicillin (50 IU/mL), and streptomycin (50 IU/mL).
3. Endometrial gland culture medium: DMEM/F12 supplemented with 10% (v/v) heat-inactivated FBS, insulin (5 μg/mL; Sigma-Aldrich®), transferrin (5 μg/mL; Sigma-Aldrich®), selenium (5 ng/mL; Sigma-Aldrich®), L-glutamine (0.1%, w/v), penicillin (50 IU/mL), and streptomycin (50 IU/mL).
4. Oviductal cell-embryo coculture medium: Chatot-Ziomek-Bavister (CZB) (22): 81.62 mM NaCl, 4.83 mM KCl, 1.18 mM KH_2PO_4, 1.18 mM $MgSO_4{\cdot}7H_2O$, 25.12 mM $NaHCO_3$, 1.70 mM $CaCl_2{\cdot}2H_2O$, 31.30 mM sodium lactate, 0.27 mM sodium pyruvate, 0.11 mM EDTA, 1 mM L-glutamine, 5 mg/mL BSA, 100 IU/mL penicillin, and 0.7 mM streptomycin.
5. Endometrial epithelial and stromal clonal culture medium: DMEM/F-12 supplemented with 10% FBS, 1% penicillin/streptomycin, and 1% L-glutamine.
   (a) Endothelial cell growth factor supplement (ECGS) is dissolved at a concentration of 20 μg/mL in 0.1% PBS/BSA and stored in single use aliquots at −20°C. ECGS is required for clonogenic endometrial epithelial cells.
   (b) Epidermal growth factor (EGF, Millipore) is dissolved at a concentration of 10 μg/mL in distilled water and stored in single use aliquots at −20°C. EGF is required for clonogenic endometrial epithelial cells.

3. Methods

3.1. Isolation, Growth, and Characteristics of OSE, Oviductal and Endometrial Epithelia

3.1.1. Culture of Human OSE

The first method was developed in Dr. Nelly Auersperg's laboratory in 1980s (23). For the explant method, however, there is the risk of contamination of the cultures by stromal cells/fibroblasts and other components of the ovary. The scrape method which takes advantage of the characteristically loose attachment of human OSE to underlying structures is a simple and reliable method to isolate a predominant population of epithelial OSE cells (2). Moreover, effects observed of these cells in vitro appear to mimic those occur in vivo (24–26).

1. Human OSE can be obtained from ovarian biopsies at abdominal or pelvic surgery or at laparoscopic procedures for nonmalignant gynecological conditions (see Note 1).
2. For biopsy specimens, hold the entire specimen surface down over a 35 mm culture dish containing 2 mL of culture medium (see Subheading 2.6, step 1), and scrape the ovarian surface gently 2–3 times with a rubber scraper (see Note 2). Alternatively, OSE can also be obtained by the surgeon in situ at the beginning of laparoscopic procedures.
3. Primary cultures are incubated at 37°C in 5% CO_2:95% air and left undisturbed for 48 h. When OSE outgrowths begin to 70% confluency, the cells can be passaged with a split ratio of 1:3 after dissociation with 0.05% trypsin/EDTA.
4. Growth patterns and morphology of OSE can be classified as: Compact epithelial cells forming a cobblestone pattern, comprising the majority of all primary cultures (see Fig. 1a); atypical, fibroblast-like cells found in passage 3 or beyond as a part of the wound healing process in culture (see Fig. 1b), and elongated, irregularly shaped, overlapping cells that sometimes

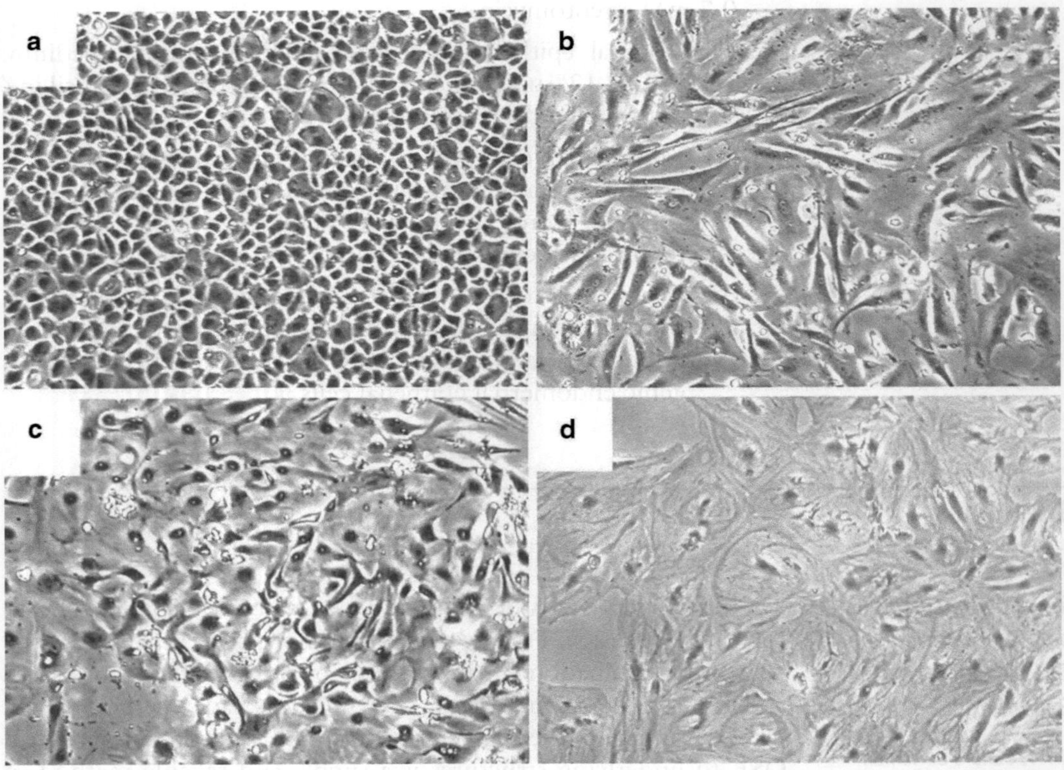

Fig. 1. OSE cell morphology and growth characteristic in culture. (**a**) Compact epithelial cells forming a cobblestone pattern. (**b**) Atypical fibroblast-like cells. (**c**) Elongated cells that sometimes appear as whirls, which signifies metaplastic OSE. (**d**) Large, flattened senescent cells.

form colonies appearing as whorls (see Fig. 1c), which are present in primary culture and are rarely cultured beyond passage 1. These colonies appear to be derived from metaplastic OSE on the basis of their morphological and functional resemblance to cultured Mullerian duct-derived epithelia (24).

5. The culture usually proliferates for 3–4 passages and then senesce, which are characterized by large flat cells that do not reach confluence over 1 month (see Fig. 1d). The addition of 10 ng/mL EGF and 1 μg/mL hydrocortisone can increase the growth rate and growth potential of OSE cells (see Note 3).
6. There is no single marker specific for OSE. A combination of markers is used to distinguish these cells from potential contaminating cells. These include keratins 7, 8, 18, 19, which separate OSE from other cell types; while additional markers such as laminin help to distinguish OSE from the underlying stroma. 17β-OH steroid dehydrogenase and mucin can be used to isolate OSE from the extraovarian mesothelium. A lack of factor VIII is useful to differentiate OSE from endothelial cells (27).

3.1.2. Culture of Human Oviductal Epithelial Cells

1. Human oviduct can be obtained from women admitted for tubal ligation or hysterectomy for benign gynecological conditions.
2. Keep the fresh human oviduct in oviductal cell culture medium or PBS as listed above.
3. Cut open the oviduct longitudinally and tease off the oviductal mucosa by fine sterile forceps (see Note 4). Mince the epithelium with a pair of scissors in minimal amount of PBS.
4. Digest the epithelium with 5 mL 0.05% typsin/EDTA solution with shaking at 37°C for 45 min. Collect the cells by centrifugation at 300 × *g* for 2 min at room temperature.
5. Rinse the dispersed oviductal cells twice with oviductal cell culture medium (see Subheading 2.6, step 2).
6. Culture 5×10^5 cells in the same medium in 75 cm^2 flasks at 37°C in an atmosphere of 5% CO_2 in air (see Note 5).
7. Monitor the growth of the cells daily by phase contrast microscopy thereafter. Change the culture medium every 48 h. The epithelial cells usually take 4–6 days to reach confluence. The purity of the epithelial cell preparation can be determined by standard immunohistochemical staining for cytokeratin (see Note 6).
8. The epithelial cells can be subcultured by standard trypsin/EDTA method for another cycle before experimentation or cryopreservation. The subcultured cells grow faster, and take shorter time to reach confluence (see Note 7).

3.1.3. Culture of Human Endometrial Glandular Epithelial Cells

1. Human endometrium biopsies can be obtained from women at different phases of the menstrual cycle by endometrial aspirate (see Note 8).
2. Mince the endometrial samples with sterile scalpels into tiny fragments (about 1 mm^2) in a tissue culture dish.
3. Transfer the endometrial fragments into a tube containing 0.1% collagenase Type IA and 0.001% DNase and allow the cells to disperse in a shaking water bath at 15 rpm at 37°C for 45–60 min.
4. Separate the stromal cells and the epithelial glands by passing the digest through a 40 μm nylon cell strainer.
5. Rinse the glands retained on the strainer with PBS and backwash into a 15 mL centrifuge tube for further washing with endometrial culture medium (see Subheading 2.6, step 3).
6. Seed the glands at a density of 200 endometrial glands/cm^2 and culture the glands in endometrial culture medium at 37°C in an atmosphere of 5% CO_2 in air. Do not disturb the glands in the first 48 h (see Note 9).
7. Monitor the growth of the cells daily by phase contrast microscopy thereafter. Change the culture medium every 48 h. The cells will reach 60–70% confluence in 3–5 days (see Note 10).

3.2. Establishment of Preneoplastic Immortalized OSE

There are several methods for immortalizing OSE cells in culture. One method is to use the SV40 T antigen, which inactivation of the tumor suppressor genes Rb and p53 avoids senescence and leads to immortalization.

3.2.1. SV40 T Antigen

1. Expression of SV40 T antigen is achieved by transfection with the SV40 T antigen-containing plasmid, PX-8 (28).
2. Use as early a passage as possible. When cells are approximately 80% confluent in a 35 mm culture dish, transfect the cells with 1 μg of SV40 T antigen-containing plasmid DNA in 6 μL Lipofectamine™ 2000 in 94 μL OPTI-MEM™ (see Note 11).
3. Leave the cells undisturbed for 1 week. Once they become confluent, split 1:5. Colonies of small cells begin to appear 4–6 days, and cells can be either trypsinized with cloning cylinders or scraped off and transferred to new dishes. There is no selectable marker included in this case, since the untransfected cells are lost through senescence. Split them a couple of more times, the untransfected ones will die and only the SV40 T antigen-transfected cells will be alive.
4. The expression of large T antigen is confirmed by immunocytochemical examination with mouse monoclonal antibody PAb419 and goat anti-mouse IgG labeled with Texas Red dye (see Subheading 2.3).

3.2.2. hTERT

An alternative method of cell immortalization is through the expression of TERT. The exogenous expression of hTERT enables the cells to bypass replicative senescence through telomere synthesis. Importantly, hTERT cells are able to maintain a stable genotype that also retains critical phenotypic traits (29).

1. Constitutive expression of hTERT is achieved by transduction using a retroviral hTERT expression vector, pLXIN-hTERT (30).
2. The day before transduction, remove the culture medium, plate cells (see Subheading 2.6, step 1) at 30–50% confluence in 6-well plate and incubate at 37°C overnight.
3. On the day of transduction, thaw the hTERT retrovirus supernatant in a 37°C water bath.
4. Infect the subconfluent cultures of OSE cells with 2 mL/well viral supernatant in the presence of 5 μg/mL polybrene (see Subheading 2.3 and Note 12) and incubate at 37°C overnight for 16–20 h.
5. The following day, remove the medium containing virus and replace with fresh cell culture medium (see Subheading 2.6, step 1).
6. After 72 h incubation, subculture the cells into 2 × 100 mm dishes and add fresh culture medium containing 200 μg/mL G418 for selection.
7. Check telomerase activity by the TRAPeze® Telomerase Detection Kit (see Subheading 2.3) according to manufacturer instructions (see Fig. 2).

It is worth noting that OSE cells overexpressing either SV40 T antigen or hTERT are not truly immortal, the cells eventually enter crisis and die. Recent research has found that coexpression of the

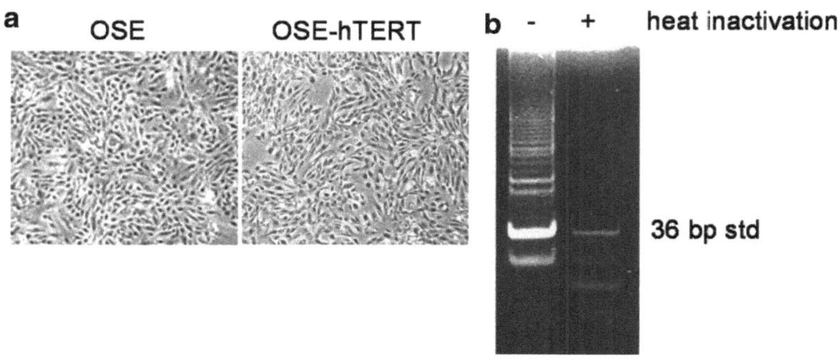

Fig. 2. hTERT immortalization of OSE cell lines. (**a**) hTERT-transduced cells. (**b**) TRAP assay showing telomerase activity (a characteristic ladder with six base increments starting at 50 bp, with a 36 bp internal standard). Heat inactivation of the telomerase extract was used to confirm the specificity of the TRAP assay.

hTERT with either Rb or p53 siRNA can permanently immortalize OSE cells, providing an indefinite source of nontumorigenic human OSE cells (29, 31).

3.3. Embryo-Oviductal Cell Coculture System

Coculture requires both the oviductal cells and the embryos simultaneously. As the availability of the donated oviduct is limited and often cannot be controlled by the researchers, it is more practical to use cryopreserved oviductal cells for coculture study.

1. Cryopreservation of the oviductal epithelial cells can be done when the cells reach confluent growth. The cells to be cryopreserved are trypsinized and cryopreserved in DMEM/F12 containing 10% DMSO. They are cooled successively at 4°C for 2 h, at –20°C for 4 h and at –70°C overnight before storing in liquid nitrogen until use.
2. One day before coculture, thaw the frozen oviductal cells in a water bath at 37°C and rinse the cells in PBS three times before seeding at a density of 2×10^4 cells/well in a 96-well culture plate.
3. Culture the cells in DMEM/F12 medium (see Subheading 2.6, step 2) for 24 h (see Note 13).
4. After washing the cells with CZB medium (see Subheading 2.6, step 4), the cells can be used for coculture in CZB medium or for the production of conditioned medium (see Note 14).
5. For coculture, transfer the embryos onto the monolayer of epithelial cells and culture for the duration according to the experimental design.
6. Change at least half of the CZB medium every other day (see Note 15).

3.4. Retrieval of Endometrial Stem Cells in Human Endometrium

3.4.1. Endometrial Tissue Collection and Preparation

To isolate endometrial stem/progenitor cells, it is important to obtain full thickness of the endometrial tissue from women undergoing hysterectomy. Tissue samples comprising endometrium attached to myometrium of thickness 5–10 mm are collected in the collection medium (see Subheading 2.5, step 3) and can be processed immediately. Viable endometrial cells can also be obtained by storing the sample at 4°C overnight, if the isolation method cannot be conducted on the day of sample collection.

3.4.2. Preparation of Single Cell Suspensions of Purified Human Endometrial Epithelial and Stromal Cells

A single cell suspension of endometrial cells can be obtained by enzymatic digestion and mechanical means as described (32).

1. The endometrium is scraped off the underlying myometrium, finely minced and dissociated in a tube containing Ca^{2+}- and Mg^{2+}-free PBS (pH 7.4) containing 300 µg/mL collagenase type III and 40 µg/mL DNase type I in a shaking water bath at 150 rpm at 37°C.

2. Every 15 min interval, the digest is pipetted vigorously and the dissociation of single endometrial cells is monitored microscopically (see Note 16).
3. A fine slice of the myometrium that formed the endometrial-myometrial junction is also digested separately to ensure that any penetrating endometrial glands and stromal tissues are dissociated.
4. After 45–60 min, the cell suspensions are filtered through the 40 μm cell strainer to separate single cells from debris and undigested myometrial tissue fragments.
5. Further dissociation of the filtrate is prevented by the addition of collection medium and washing of the cell suspension by centrifugation at $100 \times g$ for 5 min at room temperature.
6. The strainers are back washed with PBS to obtain myometrial and glandular fragments, which are further dissociated for 45–60 min as described above, to yield more single cell suspensions.
7. The myometrial fragments are checked microscopically at 15-min interval to ensure all the penetrating glands are dissociated. The cell suspensions are filtered as above and combined with the cells isolated from the endometrial fragments.
8. To remove erythrocytes and dead cells, 2.5 mL of Ficoll-Paque™ is loaded underneath the cell suspensions in 10 mL of collection medium and centrifuged at $400 \times g$ for 10 min at room temperature (see Note 17).
9. The endometrial cells are removed from the Ficoll-Paque™ medium interface, washed and resuspended in a 1 mL volume of collection medium (see Note 18).
10. Endometrial epithelial cells are obtained by positive selection using Ber-EP4-coated magnetic Dynabeads®. Ber-EP4 antibody has been shown to be specific for luminal and glandular epithelium in full thickness endometrium. The required number of Ber-EP4 beads is calculated assuming 50% of the cells in the suspension are epithelial cells and four beads/epithelial cell are required (see Note 19). The beads are incubated with the cell suspensions for 30 min at 4°C with end-over-end rotation.
11. The beaded epithelial cells are recovered and washed three times in collection medium using a magnetic particle collector and the supernatant containing the stromal cells and the leukocytes is also collected.
12. The stromal cells are then negatively selected using anti-CD45 antibody coated Dynabeads for removal of the leukocytes, assuming 50% of the remaining cells are leukocytes (see Note 20).

3.4.3. Endometrial Epithelial and Stromal Clonal Culture

1. Beaded endometrial epithelial cells and purified endometrial stromal cells are seeded in triplicate at clonal density, 500 and 300 cells/cm^2, respectively, into 60-mm Petri dishes. All dishes are coated with a thin layer of 0.2% gelatin for 30 min at room temperature and washed with PBS before used. For epithelial cells, culture medium is supplemented with 20 μg/mL of ECGS (see Subheading 2.6, step 5a) and 10 μg/mL of EGF (see Subheading 2.6, step 5b).
2. All cultures are incubated for 15 days in a humidified CO_2 incubator at 37°C in 5% CO_2:95% air and medium is changed every 7 days.
3. Colonies are monitored microscopically on a daily basis to ensure they are derived from single cells (see Note 21).
4. After 15 days in culture, colonies can be depicted (see Fig. 3a). Large colonies (>4,000 cells, see Fig. 3b) are considered to be derived from endometrial stem/progenitor cells whereby small colonies (<4,000 cells, see Fig. 3c) are derived from transit amplifying cells (see Note 22).
5. To determine the cloning efficiency, cultures can be terminated at 15 days by fixation in 10% formalin for 10 min, wash with tap water, stain with Toluidine Blue O for 4 min (see Note 23), rinse with tap water and evaporation to dryness (see Fig. 3a).

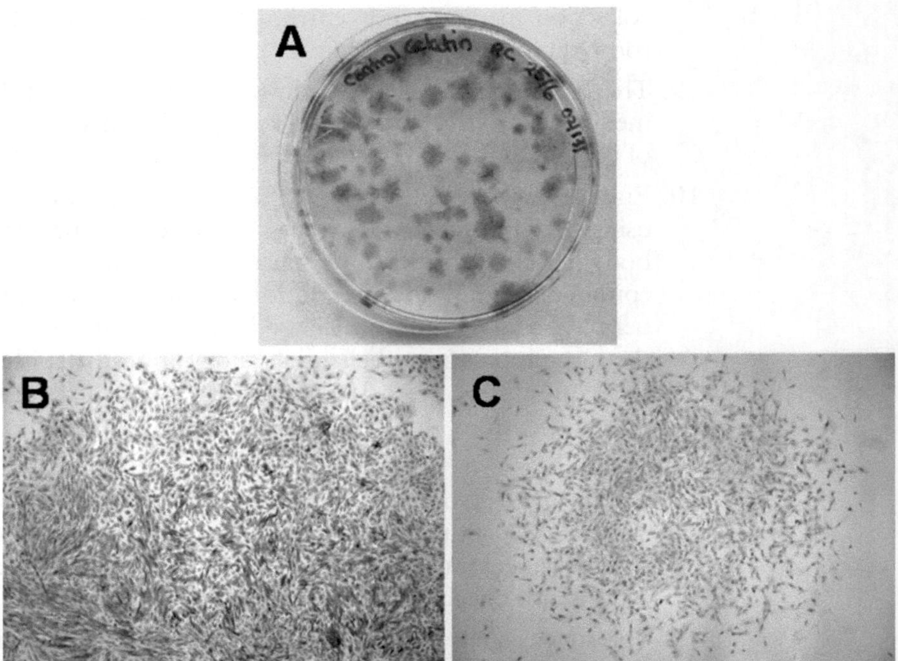

Fig. 3. Representative phase contrast photomicrograph. (**a**) Of a culture plate after 15 days in culture, showing the distribution of colonies and colony size variation. (**b**) Of a typical large epithelial colony containing small, densely packed cells with a defined border. (**c**) Of a typical small epithelial colony containing loosely packed cells.

Clusters of cells are considered colonies when they are visible macroscopically and contain more than 50 cells. The number of colonies formed is counted. The cloning efficiency (CE) can be determined from the formula CE (%) = (number of colonies/ number of cells seeded) × 100.

6. Alternatively, for phenotypic examination of the endometrial clones by immunohistochemistry, the endometrial epithelial and stromal cells are seeded onto 25-mm gelatin-coated Thermanox coverslips in six well dishes and cultured for 15 days as described above. Coverslips can then be used for immunohistochemical analysis using the standard procedure for culture cells (see Note 24).

4. Notes

1. Request that they scrape off the cells first thing after opening the abdomen or right after laparoscope insertion to avoid blood contamination. Immerse cells in culture medium (see Subheading 2.6, step 1) right away because OSE cells are on the ovarian surface and drying occurs rapidly after exposure to air.
2. OSE is very loosely attached. Scraping too hard can result in removal of the underlying stromal cells.
3. This treatment will not only enhance growth of OSE cells, but may also change their phenotype by initiating epithelial to mesenchymal transition (EMT), which likely mimics a repair response following ovarian rupture (33).
4. The protocol described produces a cell preparation with enriched oviductal epithelial cells. The amount of oviductal stromal contamination depends on the skill in teasing off the oviductal mucosa by fine forceps. Cleaner preparation can be obtained when using only the longitudinal folds of the oviductal mucosa. The contaminated stromal fibroblasts can be reduced by the standard differential attachment method, i.e., culture of the enzyme-dispersed cell suspension in a culture flask for 30–60 min allowing the attachment of the fibroblasts, followed by transfer of the supernatant containing the enriched epithelial cells into another flask for culture.
5. The primary oviductal epithelial cells attach slowly to the culture plastic. To facilitate attachment these cells, it is important not to disturb the culture flask in the first 48 h post-seeding.
6. A protocol likely to produce purer oviductal epithelial cell preparation has been described (34). The protocol uses rubber cell scraper to dislodge the epithelial cells by scrapping the

oviductal epithelium. However, the yield of this protocol is low. In contrast to the protocol described in this chapter, which can work on small (1–2 cm in length) oviductal samples such as those obtained from tubal ligation, the scrapping method requires much larger pieces of oviductal tissue as starting material.

7. The oviductal epithelial cells gradually lose their cilia and reduce their secretory activity in culture. Prolonged culture of the oviductal epithelial cells leads further to change in the morphology of the cells to fibroblast-like. Therefore, only primary epithelial cells or early passaged cells should be used for experimentation. Alternatively, the epithelial cells can be immortalized to preserve the functions of the cells (35).
8. The techniques for culture of the endometrial glands are similar to that of the oviductal epithelial cells. For instance, stromal cell contamination in glandular preparation can be further reduced by selective attachment onto to plastic culture dish for 20–30 min.
9. Dispersed endometrial epithelial cells in suspension can be collected by using antibody-coated beads as described for isolation of the endometrial progenitor cells (see Subheading 3.4.2).
10. Sophisticated endometrial culture systems have been described in the literature including a three-dimensional reconstituted endometrial model, which reconstructs the endometrium from dispersed endometrial stromal cells and isolated endometrial epithelial glands (36).
11. Since OSE has a very limited life span and senesce after a finite number of passages, it is recommended to use an early passage of these cells for transfection.
12. Polybrene is a polycation that promotes the binding of viruses and the cellular membrane by neutralizing the anionic charge at the cell surface. The optimal concentration of polybrene for virus transduction varies with cell type, but is usually in the range of 4–8 μg/mL.
13. The oviductal cells detach and die after confluent growth. The death of the cells as well as the metabolic waste of the living cells accumulated in the medium will affect the development of the cocultured embryos. Therefore, coculture should be started when the cells are 50–60% confluent and appropriate change of the culture medium is necessary during the coculture experiment.
14. As the development of early preimplantation embryos in DMEM/F12 is poor, coculture has to be done in embryo culture medium (see Subheading 2.6, step 4). Apart from CZB medium, other embryo culture medium such as potassium

simplex optimized medium supplemented with amino acids (KSOMaa) (37) can be used.

15. The protocol described can be adapted for endometrial cell-embryo coculture.
16. Place a small drop of the cell suspension onto a culture plate and check under an inverted microscope.
17. IMPORTANT: When layering the sample do not mix Ficoll-Paque™ with the cell suspension.
18. Differential migration during centrifugation results in the formation of layers containing different cell types. The bottom layer (red pellet) contains erythrocytes, above (clear color) contains mostly granulocytes and is the Ficoll-Paque™ layer. Endometrial cells are found at the interface (white buffy coat) between the collection medium and Ficoll-Paque™. The size of the interface layer depends on the number of endometrial cells isolated. If the endometrial sample contains a lot of red blood cells, this step may need to be repeated to ensure all erythrocytes are removed.
19. The number of epithelial cells isolated with the protocol described depends on the phase of the menstrual cycle of the hysterectomy sample. A higher number of epithelial cells are expected from the secretory phase. This should be considered when calculating the number of Dynabeads required for cell separation (see Subheading 3.4.2).
20. All reagents and cell suspensions are to be kept on ice when not in use during experimentation.
21. The formation of colonies after 15 days in culture needs to be initiated from a single cell as it represents the proliferative potential of the cell (a stem cell property). Hence, low seeding density of endometrial cells and regular microscopic monitoring of the cultures are important. If no colonies form, it could be due to patient variation and its pathological condition.
22. The ability of a single cell to initiate a colony of cells indicates the high proliferative potential of the cell and its ability to generate a line of genetically identical cells (38). Although this technique examines one stem cell property, it does not provide clues on how to locate putative stem/progenitor cells in tissues as it detects stem cell activity retrospectively.
23. Only a thin layer of Toludine Blue O is required to stain the colonies on the cultureware.
24. The protocol can be adapted to isolate stem/progenitor cells from ovarian endometriotic samples.

Acknowledgments

We thank the Department of Obstetrics and Gynecology, University of Hong Kong, which kindly provided OSE, oviductal, and endometrial samples. This work was supported by grants from the Hong Kong Research Grant Council to A.S.T.W. and (HKU 782209M) W.S.B.Y. and from the HKU Outstanding Young Researcher Award to A.S.T.W.

References

1. Auersperg N, Wong AS, Choi KC, Kang SK, Leung PC (2001) Ovarian surface epithelium: biology, endocrinology, and pathology. Endocr Rev 22:255–288
2. Kruk PA, Maines-Bandiera SL, Auersperg N (1990) A simplified method to culture human ovarian surface epithelium. Lab Invest 63: 132–136
3. Maines-Bandiera SL, Kruk PA, Auersperg N (1992) Simian virus 40-transformed human ovarian surface epithelial cells escape normal growth controls but retain morphogenetic responses to extracellular matrix. Am J Obstet Gynecol 167:729–735
4. Tsao SW, Mok SC, Fey EG, Fletcher JA, Wan TS, Chew EC, Muto MG, Knapp RC, Berkowitz RS (1995) Characterization of human ovarian surface epithelial cells immortalized by human papilloma viral oncogenes (HPV-E6E7 ORFs). Exp Cell Res 218: 499–507
5. Kohler MF, Kerns BJ, Humphrey PA, Marks JR, Bast RC Jr, Berchuck A (1993) Mutation and overexpression of p53 in early-stage epithelial ovarian cancer. Obstet Gynecol 81: 643–650
6. Oishi T, Kigawa J, Minagawa Y, Shimada M, Takahashi M, Terakawa N (1998) Alteration of telomerase activity associated with development and extension of epithelial ovarian cancer. Obstet Gynecol 91:568–571
7. Baykal A, Thompson JA, Xu XC, Hahn WC, Deavers MT, Malpica A, Gershenson DM, Silva EG, Liu J (2004) *In situ* human telomerase reverse transcriptase expression pattern in normal and neoplastic ovarian tissues. Oncol Rep 11:297–302
8. Yao YQ, Ho PC, Yeung WSB (1999) Effects of human oviductal cell coculture on various functional parameters of human spermatozoa. Fertil Steril 71:232–239
9. Yeung WSB, Xu JS, Lee CKF (2002) The oviduct and preimplantation embryo development. Reprod Med Rev 10:21–44
10. Lee YL, Lee KF, Xu JS, He QY, Chiu JF, Lee WM, Luk JM, Yeung WSB (2004) The embryotrophic activity of oviductal cell-derived complement C3b and iC3b, a novel function of complement protein in reproduction. J Biol Chem 279:12763–12768
11. Lee KF, Xu JS, Lee YL, Yeung WSB (2006) Demilune cell and parotid protein from murine oviductal epithelium stimulates preimplantation embryo development. Endocrinology 147:79–87
12. Levanon K, Crum C, Drapkin R (2008) New insights into the pathogenesis of serous ovarian cancer and its clinical impact. J Clin Oncol 26:5284–5293
13. Seppälä M, Taylor RN, Koistinen H, Koistinen R, Milgrom E (2002) Glycodelin: a major lipocalin protein of the reproductive axis with diverse actions in cell recognition and differentiation. Endocr Rev 23:401–430
14. Horne AW, Lalani EN, Margara RA, White JO (2006) The effects of sex steroid hormones and interleukin-1-beta on MUC1 expression in endometrial epithelial cell lines. Reproduction 131:733–742
15. Robertson SA (2007) GM-CSF regulation of embryo development and pregnancy. Cytokine Growth Factor Rev 18:287–298
16. Domínguez F, Gadea B, Esteban FJ, Horcajadas JA, Pellicer A, Simón C (2008) Comparative protein-profile analysis of implanted versus non-implanted human blastocysts. Hum Reprod 23:1993–2000
17. Dharmaraj N, Gendler SJ, Carson DD (2009) Expression of human MUC1 during early pregnancy in the human MUC1 transgenic mouse model. Biol Reprod 81:1182–1188
18. Lam KK, Chiu PC, Chung MK, Lee CL, Lee KF, Koistinen R, Koistinen H, Seppala M, Ho PC, Yeung WSB (2009) Glycodelin-A as a modulator of trophoblast invasion. Hum Reprod 24:2093–2103
19. Lee CL, Chiu PC, Lam KK, Chan RW, Chu IK, Koistinen R, Koistinen H, Seppälä M, Lee KF,

Yeung WSB (2010) Glycodelin-A modulates cytokine production of peripheral blood natural killer cells. Fertil Steril 94:769–771

20. Gargett CE, Schwab KE, Zillwood RM, Nguyen HP, Wu D (2009) Isolation and culture of epithelial progenitors and mesenchymal stem cells from human endometrium. Biol Reprod 80:1136–1145
21. Gargett CE, Chan RW, Schwab KE (2007) Endometrial stem cells. Curr Opin Obstet Gynecol 19:377–383
22. Chatot CL, Ziomek CA, Bavister BD, Lewis JL, Torres I (1989) An improved culture medium supports development of random-bred 1-cell mouse embryos in vitro. J Reprod Fertil 86:679–688
23. Auersperg N, Siemens CH, Myrdal SE (1984) Human ovarian surface epithelium in primary culture. In Vitro 20:743–755
24. Wong AS, Maines-Bandiera SL, Leung PC, Auersperg N (1998) Metaplastic changes in cultured ovarian surface epithelium. In Vitro Cell Dev Biol 34:668–670
25. Wong AS, Maines-Bandiera SL, Rosen B, Wheelock MJ, Johnson KR, Leung PC, Roskelley CD, Auersperg N (1999) Constitutive and conditional cadherin expression in cultured human ovarian surface epithelium: influence of family history of ovarian cancer. Int J Cancer 81:180–188
26. Wong AS, Pelech SL, Woo MM, Yim G, Rosen B, Ehlen T, Leung PC, Auersperg N (2001) Coexpression of hepatocyte growth factor-Met: as an early step in ovarian carcinogenesis? Oncogene 20:1318–1328
27. Auersperg N, Maines-Bandiera SL, Dyck HG, Kruk PA (1994) Characterization of cultured human ovarian surface epithelial cells: phenotypic plasticity and premalignant changes. Lab Invest 71:510–518
28. Fromm M, Berg P (1982) Deletion mapping of DNA regions required for SV40 early region promoter function in vivo. J Mol Appl Genet 1:457–481
29. Yang G, Rosen DG, Colacino JA, Mercado-Uribe I, Liu J (2007) Disruption of the retinoblastoma pathway by small interfering RNA and ectopic expression of the catalytic subunit of telomerase lead to immortalization of human ovarian surface epithelial cells. Oncogene 26: 1492–1498
30. Horikawa I, Cable PL, Afshari C, Barrett JC (1999) Cloning and characterization of the promoter region of human telomerase reverse transcriptase gene. Cancer Res 59: 826–830
31. Yang G, Rosen DG, Mercado-Uribe I, Colacino JA, Mills GB, Bast RC Jr, Zhou C, Liu J (2007) Knockdown of p53 combined with expression of the catalytic subunit of telomerase is sufficient to immortalize primary human ovarian surface epithelial cells. Carcinogenesis 28:174–182
32. Chan RW, Schwab KE, Gargett CE (2004) Clonogenicity of human endometrial epithelial and stromal cells. Biol Reprod 70:1738–1750
33. Salamanca CM, Maines-Bandiera SL, Leung PC, Hu YL, Auersperg N (2004) Effects of epidermal growth factor/hydrocortisone on the growth and differentiation of human ovarian surface epithelium. J Soc Gynecol Investig 11:241–251
34. Bongso A, Ng SC, Sathananthan H, Ng PL, Rauff M, Ratnam SS (1989) Establishment of human ampullary cell cultures. Hum Reprod 4:486–494
35. Lee YL, Lee KF, Xu JS, Wang YL, Tsao SW, Yeung WSB (2001) Establishment and characterization of an immortalized human oviductal cell line. Mol Reprod Dev 59:400–409
36. Lalitkumar PG, Lalitkumar S, Meng CX, Stavreus-Evers A, Hambiliki F, Bentin-Ley U, Gemzell-Danielsson K (2007) Mifepristone, but not levonorgestrel, inhibits human blastocyst attachment to an in vitro endometrial three-dimensional cell culture model. Hum Reprod 22:3031–3037
37. Ho Y, Wigglesworth K, Eppig JJ, Schultz RM (1995) Preimplantation development of mouse embryos in KSOM: augmentation by amino acids and analysis of gene expression. Mol Reprod Dev 41:232–238
38. Pellegrini G, Golisano O, Paterna P, Lambiase A, Bonini S, Rama P, De Luca M (1999) Location and clonal analysis of stem cells and their differentiated progeny in the human ocular surface. J Cell Biol 145:769–782

Chapter 22

Primary Culture and Propagation of Human Prostate Epithelial Cells

Birunthi Niranjan, Mitchell G. Lawrence, Melissa M. Papargiris, Michelle G. Richards, Shirin Hussain, Mark Frydenberg, John Pedersen, Renea A. Taylor, and Gail P. Risbridger

Abstract

Basic and translational (or preclinical) prostate cancer research has traditionally been conducted with a limited repertoire of immortalized cell lines, which have homogeneous phenotypes and have adapted to long-term tissue culture. Primary cell culture provides a model system that allows a broader spectrum of cell types from a greater number of patients to be studied, in the absence of artificially induced genetic mutations. Nevertheless, primary prostate epithelial cell culture can be technically challenging, even for laboratories experienced in immortalized cell culture. Therefore, we provide methods to isolate and culture primary epithelial cells directly from human prostate tissue. Initially, we describe the isolation of bulk epithelial cells from benign or tumor tissues. These cells have a predominantly basal/intermediate phenotype and co-express cytokeratin 8/18 and high molecular weight cytokeratins. Since prostatic stem cells play a major role in disease progression and are considered to be a therapeutic target, we also describe a prospective approach to specifically isolate prostatic basal cells that include both stem and transit-amplifying basal populations, which can be studied independently or subsequently differentiated to supply luminal cells. This approach allows the study of stem cells for the development of new therapeutics for prostate cancer.

Key words: Prostate, Basal, Luminal, Stem cells, Primary, Human, Epithelial cell culture

1. Introduction

The human prostate is composed of a number of cell types that can be isolated and cultured. Prostatic homeostasis depends on cross-talk between the epithelial and stromal cell compartments, as shown in Fig. 1. Although prostate cancer originates in the epithelium, the stroma plays a major role in directing epithelial differentiation and proliferation, as well as the progression to carcinogenesis (1). The antigenic markers that define the lineage

Scott H. Randell and M. Leslie Fulcher (eds.), *Epithelial Cell Culture Protocols: Second Edition*, Methods in Molecular Biology, vol. 945, DOI 10.1007/978-1-62703-125-7_22, © Springer Science+Business Media, LLC 2012

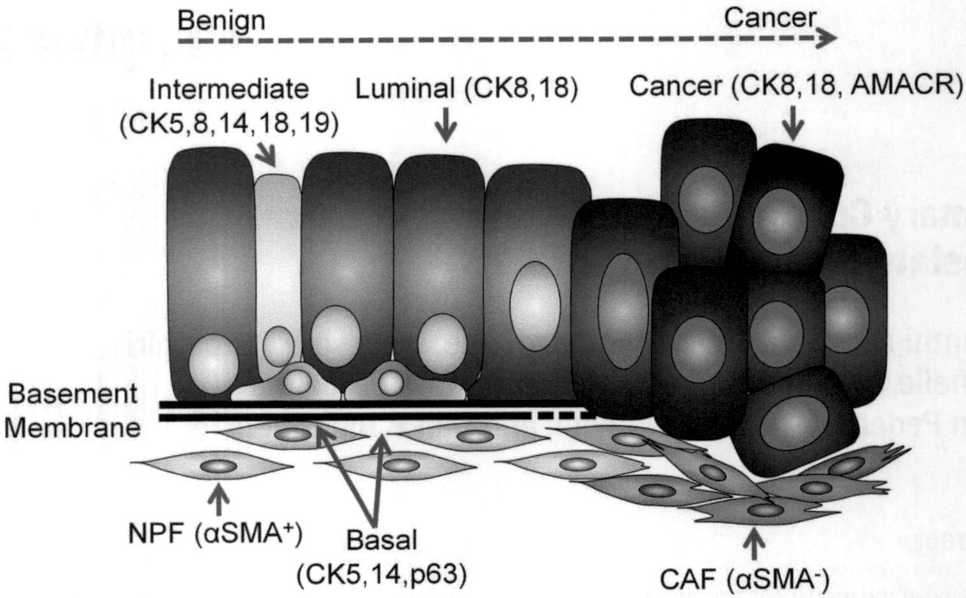

Fig. 1. Diagrammatic representation of normal and malignant prostate tissue. In the benign prostate, luminal and basal cells, including intermediate cells, are separated from neighboring normal prostatic fibroblasts (NPFs) by the basement membrane. The rare neuroendocrine cells (normally found in a suprabasal location) are not shown. In prostate cancer, the basal cells are lost and tumor cells are in direct contact with cancer-associated fibroblasts (CAFs).

and derivation of prostate epithelial cell types are less well developed than for some other tissues, but they include intracellular markers such as cytokeratins and nuclear transcription factors, as well as cell surface proteins. There are two commonly described epithelial cell types, including basal and luminal cells, although an intermediate cell has also been identified that expresses markers common to both cell types (2).

Male sex hormones regulate the development and maintenance of the human prostate. Castration leads to involution of the gland, leaving a residual, basal epithelial cell type, which can regenerate an intact prostate after restoration of androgen levels (3, 4). The luminal androgen-responsive epithelial cells in normal prostate are terminally differentiated, but highly metabolically active (5). Regeneration and homeostasis in normal glands depend on the longer lived basal cells. When normal (and indeed malignant) prostate specimens are cultured in vitro under androgen-free conditions, the cultures have an almost exclusively basal phenotype. Prostate epithelial stem cells also display a basal phenotype and are located in the basal compartment (6).

This chapter provides the basic techniques required to achieve epithelial cultures from fresh primary human tissues. We describe a method to bulk culture primary prostate epithelial cells and a modified protocol to preferentially select basal cells. The key to maximizing the growth potential of primary prostate epithelial cells is to use media that is serum-free and low in calcium, although this may also select against the growth of prostate cancer cells (7). It has also been found that addition of cholera toxin, leukemia inhibitory factor, GM-CSF, bovine pituitary extract, epidermal growth factor, and stem cell factor are beneficial to keeping the whole cell population basal and maintaining the undifferentiated phenotype.

If a more differentiated luminal cell type is required, it can be obtained by culturing the cells in differentiating medium (8), resulting in an overgrowth of cells from the monolayer (as described in Subheading 3.3.4). However, the secretory phenotype of the luminal cells in 2D cultures is limited; more complex 3D culture in a matrix such a Matrigel™ is required for a true bilayered epithelium mimicking the prostate gland. Full polarity and secretory activity in the epithelium also require cross-talk with prostatic stroma (9).

The key to maintaining prostate epithelial cells in culture is therefore to eliminate the cues from human prostate stromal cells that promote differentiation, while maintaining the properties of stroma that support growth, hence the use of inactivated and undifferentiated murine stromal cells in the protocols below. The same principles apply when culturing the epithelial components from benign and malignant prostate.

2. Materials

2.1. Establishing Primary Cultures

2.1.1. Sources and Collection of Tissue Specimens

1. Transport medium: RPMI 1640, 5% FBS, 1% antibiotic/antimycotic solution.
2. Vials for blood collection (K_2EDTA/K_3EDTA or ACD-A tubes) (BD Biosystems).

2.1.2. Preparation of Tissue for Primary Epithelial Cell Culture

1. Virkon® disinfectant or 10% bleach.
2. Digital balance for weighing tissue pieces.
3. 10 cm Petri dish.
4. Phosphate-buffered saline (PBS): 0.01 M phosphate buffer, 0.0027 M potassium chloride, and 0.137 M sodium chloride, pH 7.4. The solution should be sterilized by autoclaving it or passing it through a 0.25 μm filter.
5. Processing medium: RPMI 1640, 10% FBS, 25 mM HEPES, 1% penicillin and streptomycin, 0.5 μg/mL fungizone, 100 μg/mL gentamicin.
6. Digestion medium: Processing medium with 225 U/mL Collagenase Type I (Sigma-Aldrich®) and 125 U/mL Hyaluronidase Type II (Sigma-Aldrich®).

7. R10+ medium: RPMI 1640, 10% FBS, 2 mM L-glutamine, and 1% penicillin and streptomycin.
8. Keratinocyte serum-free medium (KSFM) with epidermal growth factor (EGF; 5 ng/mL) and bovine pituitary extract (BPE; 50 μg/mL) (Life Technologies™).
9. 10% Formalin solution, neutral buffered (Sigma-Aldrich®).
10. Sterile serological pipettes (5–25 mL).
11. 50 mL tubes with screw-top lids.
12. Tweezers, disposable scalpel (#20), and a small pair of scissors (all sterile).
13. Sterile syringe and 18 and 21 G blunt needles.
14. 0.1% Trypsin in PBS with 1 mM EDTA (filtered).
15. Serum-coated T75 tissue culture flasks: Coated with 5 mL fetal bovine serum for 20 min at 37°C.
16. Orbital shaker.

2.2. Preparation of Cells on STO Feeder Support

1. STO mouse embryonic fibroblast cells (ATCC).
2. T150 tissue culture flasks.
3. 0.1% Trypsin in PBS with 1 mM EDTA (filtered).
4. Mitomycin C solution: 1 mg/mL in sterile PBS (filtered).
5. D10 medium: Dulbecco's modified Eagle's medium, 2 mM L-glutamine, 10% fetal bovine serum, FBS (v/v).
6. Stem cell medium (SCM): KSFM with supplements (see Subheading 2.1.2, step 8), 2 mM L-glutamine, 0.05 μg/mL cholera toxin, 2 ng/mL stem cell factor (First Link, Ltd.), 1 ng/mL colony stimulating factor (First Link, Ltd.) and 2 ng/mL leukocyte inhibitory factor (Sigma-Aldrich®).
7. 15 mL tubes with screw-top lids.
8. Collagen I coated 10 cm plates (Biocoat, Becton Dickinson).
9. X-ray irradiation source.

2.3. Maintenance of Cells

1. Versene (PBS with 1 mM EDTA, pH 7.4).
2. 0.1% Trypsin in PBS with 1 mM EDTA (filtered).
3. 1% Trypsin (10×) in PBS with 1 mM EDTA (filtered).
4. Growth medium: R10+ and KSFM or SCM (see Subheading 2.1.2).
5. Freezing medium: RPMI 1640, 10% dimethyl sulfoxide (DMSO), 20% FBS.
6. Cryovials.
7. Freezing chamber (e.g., Mr. Frosty, Nalgene).
8. 15 mL conical tubes.
9. T25 or T75 flasks.

10. Inactivated STO feeder cells (see Subheading 3.2).
11. D-H10 medium: Dulbecco's modified Eagle's medium and Ham's F12 (1:1), supplemented with 10% FBS, 10 nM DHT (5α-dihydrotestosterone), and 2 mM L-glutamine.

2.4. Cell Characterization

2.4.1. Detecting Phenotypic Markers with Immunohistochemistry

1. Millicell® EZ slides (Millipore™).
2. Phosphate-buffered saline (PBS).
3. 4% Formalin: 10% Formalin is diluted in PBS.
4. Permeabilization solution: 0.3% Triton™ X-100 in PBS.
5. EnVision+ System-HRP kit (Dako). This kit contains peroxidase block, anti-mouse labeled polymer-HRP, DAB+ substrate buffer, and DAB+ chromogen. A working solution of DAB (3,3′-diaminobenzidine) is prepared by adding 1 drop of chromogen solution to 1 drop of substrate buffer.
6. CAS-Block (Life Technologies™).
7. Antibody diluent (Dako).
8. Hematoxylin.
9. 70, 95, and 100% ethanol.
10. Clearing agent: Xylene or alternative solvent such as 3B-2026 (HiChem).
11. DPX non-aqueous mounting medium (Merck).
12. Coverslips.
13. Primary antibodies including CK8/18 and HMWCK (see Table 1).

Table 1
Primary antibodies for the characterization of prostate epithelial cultures

Antigen	Application	Species/isotype	Clone	Supplier	Cell type
Cytokeratin 8/18	IHC	Mouse IgG1	5D3	Novocastra	Luminal and intermediate cells
High molecular weight cytokeratin	IHC	Mouse IgG1	34βE12	Dako Cytomation	Intermediate and basal cells
α Smooth muscle actin	IHC	Mouse IgG2a	1A4	Sigma	Fibroblasts
Trop2-APC	Flow	Mouse IgG2a	77220	R&D Systems	Epithelial cells
CD49f-PE	Flow	Rat IgG2a	GoH3	BioLegend	Epithelial cells
CD49a	Flow	Mouse IgG1	TS2/7	BioLegend	Fibroblasts

Clone information, supplier and the cell type defined by each marker is also shown

2.4.2. Detecting Cell Surface Antigens by Flow Cytometry

1. TrypLE™ Select (Life Technologies™).
2. Flow cytometry buffer: PBS, 2 mM EDTA, 0.5% bovine serum albumin (BSA).
3. Sterile 1.5 mL tubes.
4. Fluorochrome-conjugated primary antibodies and isotype controls (see Table 1).
5. SYTOX® blue viability dye (Life Technologies™).
6. Propylene or polystyrene tubes with 35 μM cell strainer caps.
7. Flow cytometer (for example BD LSR II Flow Cytometer System).

2.5. Confirming the Origin of Patient Cells

2.5.1. Isolation of Patient DNA and Genotyping Primary Cells

1. K_2EDTA/K_3EDTA or ACD-A tubes (BD Biosystems).
2. Sterile 1.5 mL tubes.
3. DNeasy-Blood and Tissue Kit (Qiagen).
4. PowerPlex 1.2 System (Promega).
5. AmpliTaq Gold DNA polymerase (Applied Biosystems).
6. Hi-Di™ formamide (Applied Biosystems).
7. 96-Well PCR plates (Applied Biosystems).

3. Methods

3.1. Establishing Primary Cultures

The limited lifespan of primary epithelial cells emphasizes the importance of efficiently isolating them from prostate tissue and culturing the cells in an appropriate defined medium that promotes proliferation. This protocol follows the guidelines reported by Chaproniere and McKeehan (10, 11).

3.1.1. Sources and Collection of Tissue Specimens

Urologists may provide tissue for propagating prostate epithelial cells once appropriate human ethics approval has been granted and patients have provided informed consent. Specimens in this study were acquired from the Australian Prostate Cancer BioResource with approval from the Cabrini (10-13-12-04; 03-14-04-08) and Monash University (2005/442MC; 2004/145) Human Research Ethics Committees. The two most common types of surgery that provide primary prostate tissue samples are radical prostatectomy and transurethral resection of the prostate (TURP). Both procedures yield surplus tissue not required for pathology. A prostatectomy removes the entire prostate gland, usually to treat patients with prostate cancer. Pathologists can initially identify regions of benign and malignant tissue based on histopathology reports from previous biopsies and palpation of the prostate, since tumor foci are usually firmer than surrounding tissue. Once the gland is dissected,

the presence and size of tumor foci should be confirmed using frozen sections stained with hematoxylin and eosin. The pathologist can then remove about 500 mg of tissue from the largest tumor foci and an uninvolved benign region. These samples are kept in transport medium for up to 24 h before tissue processing in the laboratory; however, the best results are obtained when cells are isolated immediately. The procedure is more straightforward for specimens from TURP, which removes prostate tissue that is obstructing the urethra. All tissue "chips" not required for pathology can be directly stored in transport medium. This typically yields at least 4–5 g of tissue. Since TURP is commonly performed to relieve the symptoms of benign prostatic hyperplasia (BPH), it is a good source of benign prostate cells. Occasionally, aggressive prostate cancer cells can be acquired from palliative TURP samples.
Retrospective pathology reports are important for both prostatectomy and TURP specimens, as "tumor" biopsies can lack actual tumor cells and vice versa. Blood may be taken from patients at the time of surgery to provide lymphocytes for genotyping (see Subheading 3.5).

3.1.2. Preparation of Tissue for Primary Epithelial Cell Culture

1. After weighing the tissues, they are transferred to 10 cm Petri dishes in a Class II biological safety cabinet. Transport medium (see Subheading 2.1.1, step 1) is aspirated and the tissue is washed in PBS (see Subheading 2.1.2, step 4) and then placed in 1–2 mL of processing medium (see Subheading 2.1.2, step 5).
2. The tissue is diced into 2–3 mm^3 pieces with two sterile scalpels (see Note 1). Remove 2–3 pieces and fix in 10% formalin (see Note 2).
3. Aspirate the medium before placing the diced tissue into a sterile 50 mL tube. For each 200 mg of radical prostatectomy tissue, 5 mL of digestion solution (see Subheading 2.1.2, step 6) is needed, while for TURP tissue each 200 mg needs 2.5 mL digestion solution (see Note 3). The digestion medium is added and the sample is placed at 37°C on an orbital shaker overnight for not more than 16 h (see Note 4).
4. Next morning the digestion mixture is triturated by repeated pipetting with a 5 mL serological pipette and centrifuged at 216 × *g* for 5 min. The supernatant is discarded and the tissue pieces are washed in 10 mL of processing medium. The sample is centrifuged again at 216 × *g* for 5 min to sediment the cells.
5. The cell pellet is resuspended in 10 mL R10+ (see Subheading 2.1.2, step 7) and the cells are centrifuged at 100 × *g* for 1 min (preferred) or are allowed to settle for 5 min at room temperature. Epithelial (acini) cells will settle to the bottom of the tube and the supernatant will contain stroma.

6. Remove the supernatant and repeat step 5 twice. Pool all epithelial cell pellets and pass through a blunt needle attached to a syringe to break the sample up further.
7. These epithelial acini can be seeded onto serum-coated flasks (see Subheading 2.1.2, step 15) in KSFM medium (see Subheading 2.1.2, step 8). If single cells are needed, then the acini are rinsed with PBS (5–10 mL) to remove serum before the trypsin step.
8. The acini are spun at 216 × *g* for 5 min. The PBS is removed and the acini are resuspended in 0.1% trypsin–EDTA (5 mL) and incubated at 37°C for 30 min in an orbital shaker. The sample will become stringy. If large pieces still remain, the sample can be passed through a blunt needle using a syringe to break it up further. The trypsin reaction is stopped by adding 5 mL of R10+. The sample is spun at 216 × *g* for 5 min. The pellet is washed in PBS (5 mL) and centrifuged again before resuspending in 5 mL of warm (37°C) KSFM medium. The epithelial cells are plated in serum-coated flasks in KSFM medium with supplements. The flasks are placed in a 37°C, 5% CO_2 humidified incubator.

3.2. Propagation of Cells on STO Feeder Support

As an alternative method to culturing the epithelial cells directly on tissue culture plastic, a feeder layer of STO fibroblasts, inactivated by either mitomycin C treatment or irradiation, can be used to support the growth of undifferentiated basal cells.

3.2.1. Mitomycin C Treatment of STO Cells

1. Mitomycin C solution (see Subheading 2.2, step 4) is added to D10 medium (see Subheading 2.2, step 6) to give a final concentration of 10 μg/mL.
2. A 70–90% confluent T150 flask of STO cells is washed with PBS.
3. 10 mL of D10 + mitomycin C is added and the cells are incubated in a 37°C, 5% CO_2, humidified incubator for 2–3 h.
4. Mitomycin C is removed and the cells are washed twice with PBS (10 mL/wash).
5. Cells are trypsinized from the flask.
6. The flask is washed with D10 and the resulting cell suspension is placed in a screw-top tube.
7. The flask is washed again with PBS and this is added to the cell suspension.
8. The cells are centrifuged at 216 × *g* for 5 min.
9. The supernatant is removed and the cells are resuspended in 15 mL SCM (see Subheading 2.2, step 7).
10. Mitomycin-treated feeder cells are stored at 4°C and can be used for up to 3 days.

3.2.2. Irradiation of STO Cells

1. A 70–90% confluent T150 flask of STO cells is washed with PBS.
2. Cells are trypsinized from the flask.
3. The flask is washed with D10 and the resulting cell suspension is placed in a screw-top tube.
4. The flask is washed again with PBS and this is added to the cell suspension.
5. The cells are centrifuged at $216 \times g$ for 5 min.
6. The supernatant is removed and the cells are resuspended in 15 mL SCM.
7. Cells are irradiated at 3,000 rad for 5–6 min.
8. Cells are stored at 4°C and can be used for up to a week.

3.2.3. Plating Prostate Epithelial and STO Cells

1. Prostate epithelial cells are prepared as per Subheading 3.1.2, step 8 and then resuspended in 5 mL of warmed (37°C) SCM (see Note 5). The epithelial cells are plated onto a collagen I-coated 10 cm dish.
2. Irradiated STO feeder cells (20,000 cells/cm^2) are added to form a confluent monolayer.
3. The cells are placed in a 37°C, 5% CO_2, humidified incubator. Plates are checked the following day to determine the confluency of the feeder layer.
4. If required, additional STO cells (2,000 cells/cm^2) cells are added. It is important to keep the feeder layer at least 80% confluent until colonies of epithelial cells are apparent.

3.3. Maintenance of Cells

From initial tissue digestion and plating, it may take several days to several weeks before colonies of epithelial cells are apparent. Generally, if no colonies are observed by week 4, it is unlikely that any epithelial cells will be recovered. Media is changed every 2–3 days, with PBS rinses as required to remove cell debris. If cells are cultured with a feeder layer, fresh STO cells are added as needed to keep an 80–100% confluent monolayer. When epithelial cells have reached 80% confluency, they can be passaged using the following protocol.

3.3.1. Subculture and Propagation of Cells

1. Medium is aspirated.
2. Cells are washed once with versene (see Subheading 2.3, step 1) for 5 min at 37°C to loosen the cells and remove debris.
3. 1 mL of 0.1% trypsin–EDTA (see Subheading 2.3, step 2) is added to a 10 cm plate.
4. Cells are incubated at 37°C, 5% CO_2 in a humidified incubator until they have lifted off the plate (5 min). If all cells have detached, they are resuspended in R10+ (5–10 mL) (serum is

required to inactivate the trypsin). If some cells are still attached to the plate, it is desirable to collect them because they are proliferative. There are two options: the original plate is reused (see Note 6) or 1–2 mL of 1% trypsin (10×) (see Subheading 2.3, step 3) is added to the plate and incubated at 37°C, 5% CO_2, in a humidified incubator for 5–10 min (maximum 10 min, see Note 7). Cells are resuspended in 5–10 mL additional R10+. Cells are then combined from both trypsin steps.

5. Cells are centrifuged at 216×*g* for 5 min and washed once with PBS.
6. The pellet is resuspended in 10 mL KSFM or SCM and 5 mL of cell suspension is added to each new plate (see Note 8). If required, irradiated feeder cells are added at 20,000 cells/cm^2 to form a confluent monolayer. The cells are grown in a 37°C, 5% CO_2, humidified incubator.

3.3.2. Freezing Cells

1. To freeze cells, harvest cells from plate(s) using the procedure listed in Subheading 3.3.1.
2. At step 6, instead of KSFM or SCM, cells are resuspended in 1 mL of freezing medium (see Subheading 2.3, step 5) per 1×10^6 cells and transferred to a cryovial.
3. The vials are placed at −80°C in a container that lowers the temperature 1°C/min.
4. After 24 h (and a maximum of 3 months), cells are transferred from −80°C to liquid nitrogen for long-term storage (>3 months).

3.3.3. Thawing of Cells

1. 10 mL of pre-warmed R10+ (37°C) is placed in a 15 mL conical tube.
2. Cells are thawed quickly in a 37°C water bath (<2 min).
3. The cryovial is dried and the outside sterilized with 70% ethanol. The cells are transferred into the 15 mL conical tube with R10+. The inside of the cryovial is rinsed with medium to retrieve all cells.
4. The cells are centrifuged at 216×*g* for 5 min. The cells are washed once in PBS and centrifuged again as above.
5. Cells are resuspended in 15 mL of warm KSFM for bulk epithelial culture in T75 flasks or 5 mL of SCM for stem cell culture and seeded as follows (see Note 9).
6. The cells are added to a collagen I-coated 10 cm plate with enough irradiated feeder cells to form a confluent monolayer and placed at 37°C, 5% CO_2, in a humidified incubator.
7. KSFM or SCM is changed the following day and additional feeder cells are added to maintain them at approximately 80% confluency.

3.3.4. Differentiated Cell Culture

Primary prostate basal epithelial cells can be induced to differentiate into secretory luminal cells for short-term experiments. Basal cells are grown to (or plated at) 100% confluence on STO feeder cells and the culture medium is changed from SCM to D-H10, which contains 10 nM DHT (see Subheading 2.3, step 11). Morphological changes are observed after 48 h. After 5–10 days, a bilayer may be formed, consisting of an upper layer of more differentiated luminal-like cells and a lower layer of more basal-like cells (see Note 10).

3.4. Cell Characterization

3.4.1. Morphology

Prostate epithelial cells and other cell types (such as irradiated mouse feeder cells and potential stromal contaminants) can be distinguished using a phase contrast light microscope. Basal epithelial cells resemble small, compact cobblestones in a monolayer (see Fig. 2a, arrows point to epithelial cells) and luminal cells appear larger with a flattened appearance (see Fig. 2b). Secretory vacuoles are often apparent.

3.4.2. Detecting Phenotypic Markers with Immunohistochemistry

Prostate epithelial cells in standard culture generally exhibit an intermediate phenotype, as they express both high molecular weight basal cytokeratins (CK5/14/19) and luminal cytokeratins (8/18). Subpopulations of cells express CK19 and CK14 but not together (6, 12, 13). Markers of secretory luminal cells such as androgen receptor and prostate-specific antigen (PSA) are not expressed unless the cells are cultured under differentiating conditions (see Subheading 3.3.4). In this case, there is reduced expression of basal markers and increase expression of the luminal markers. Stromal-specific markers such as smooth muscle α-actin can be used to identify contaminating non-epithelial cells.

The prostate epithelial cell cultures described in this chapter contain basal cell markers, whether they are cultured from BPH or prostate cancer tissue (see Fig. 3). The presence of basal cells in prostate cancer specimens is unexpected, since prostate cancer is defined as being negative for basal cell markers. There are two possible explanations for the presence of basal cells in primary cultures: (1) that they play a major role in the initiation of prostate cancer (14, 15) or (2) that they are derived from benign glands that lay adjacent to cancer foci in tissue specimens. Regardless of their origin, basal cells are commonly detected in cultures of primary epithelial cells from tumor specimens. The following immunohistochemistry protocol can be used to identify prostate epithelial cells and determine whether they have a basal or luminal phenotype.

1. Approximately 4×10^4 cells are seeded into each well of 8-well Millicell® EZ slides and grown overnight in a 37°C, 5% CO_2 humidified incubator. Extra wells should be seeded for positive and negative controls for primary and secondary antibodies.
2. The cells are washed in PBS once for 5 min (0.2 mL/well).

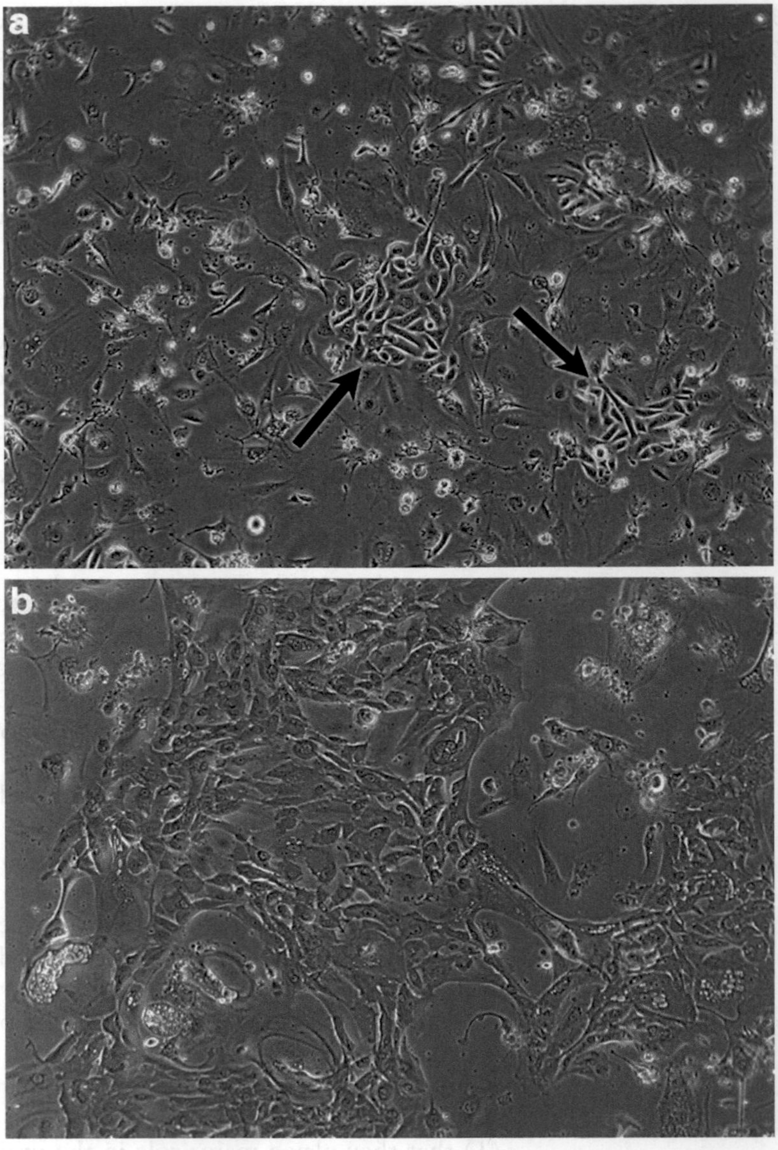

Fig. 2. Morphology of prostate epithelial cells in culture. (**a**) When grown in serum-free and low calcium medium, the predominantly basal epithelial cells resemble small, compact cobblestones in a monolayer. *Arrows* indicate epithelial cells. (**b**) High levels of serum and calcium in addition to hormones, induce luminal differentiation and cells acquire a larger, flattened appearance.

3. The cells are fixed in 4% formalin (10% formalin diluted in PBS) for 10 min at room temperature (RT) (0.2 mL/well).
4. Cells are washed with PBS, three times to remove the fixative.
5. The cells are permeabilized the with Triton™ X-100 solution (see Subheading 2.4.1, step 4) for 5 min.

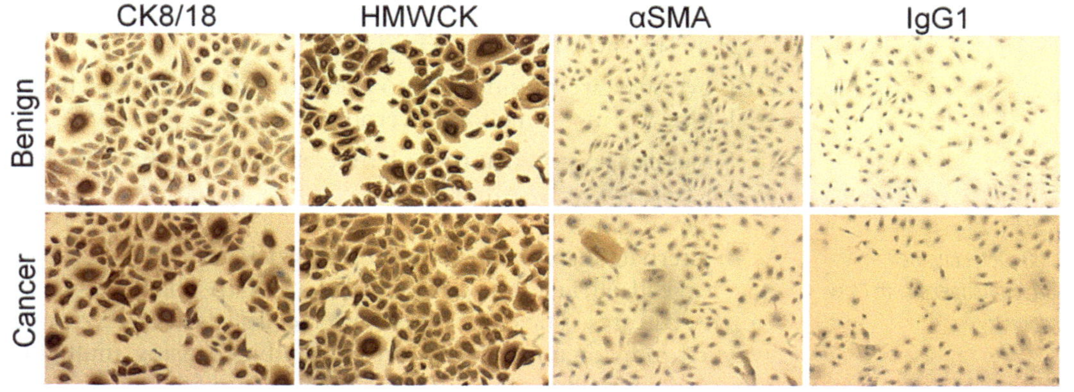

Fig. 3. Expression of cytokeratins in primary prostate epithelial cells. Immunohistochemical staining of luminal (CK8/18) and basal (high molecular weight, HMWCK) cytokeratins in cultures derived from matched benign and malignant radical prostatectomy specimens. Fibroblast contamination was identified using α smooth muscle actin (αSMA). The specificity of staining was verified using an isotype negative control (mouse IgG1).

6. The cells are washed with PBS three times for 5 min.
7. The cells are incubated in peroxidase block (see Subheading 2.4.1, step 5) for 15 min at RT.
8. The cells are washed three times with PBS.
9. CAS block (see Subheading 2.4.1, step 6) is added and incubated for 15 min at RT.
10. Primary antibodies are diluted in Ab diluent and then incubated for 1 h at RT or overnight at 4°C.
11. Cells are washed with PBS, three times.
12. Anti-mouse labeled polymer-HRP (see Subheading 2.4.1, step 5) is added for 15 min at RT.
13. Cells are washed three times with PBS for 5 min.
14. Color is developed by adding DAB+ substrate:chromogen solution (see Subheading 2.4.1, step 5) for 3–5 min or until the color change is observed under a microscope. Reactions are stopped just before background appears by washing in distilled water. At this point, samples can be stored in water overnight.
15. Cells are counterstained by adding hematoxylin for 10 s and washing off the excess with warm tap water.
16. The slides are washed again with distilled water.
17. Samples are dehydrated by immersion in 70% ethanol for 5 min with gentle rocking.
18. The slides are transferred to 95% ethanol for 5 min with gentle rocking.
19. The slides are then washed with 100% ethanol three times for 5 min each, with gentle rocking.

20. The slides are rinsed in clearing solvent (see Subheading 2.4.1, step 10) three times for 5 min each.
21. Coverslips are mounted onto the slides using DPX, which is left to dry in a fumehood overnight.
22. Slides can be stored at room temperature and analyzed by light microscopy.

3.4.3. Detecting Cell Surface Antigens with Flow Cytometry

As an alternative approach to immunohistochemistry, flow cytometry for cell surface antigens can be used to characterize the primary cultures. The expression of each cell surface marker can be analyzed individually, or they can be combined using multicolor flow cytometry. Prostate epithelial cells from fresh human tissue have previously been shown to express Trop2 (tumor-associated calcium signal transducer 2) and varying levels of CD49f (integrin α6) (16). Consistent with the predominantly basal phenotype of the primary epithelial cultures, most cells coexpress high levels of Trop2 and CD49f (see Fig. 4). Contaminating prostatic fibroblasts can be identified with CD49a (integrin α1) (17, 18). Other lineage

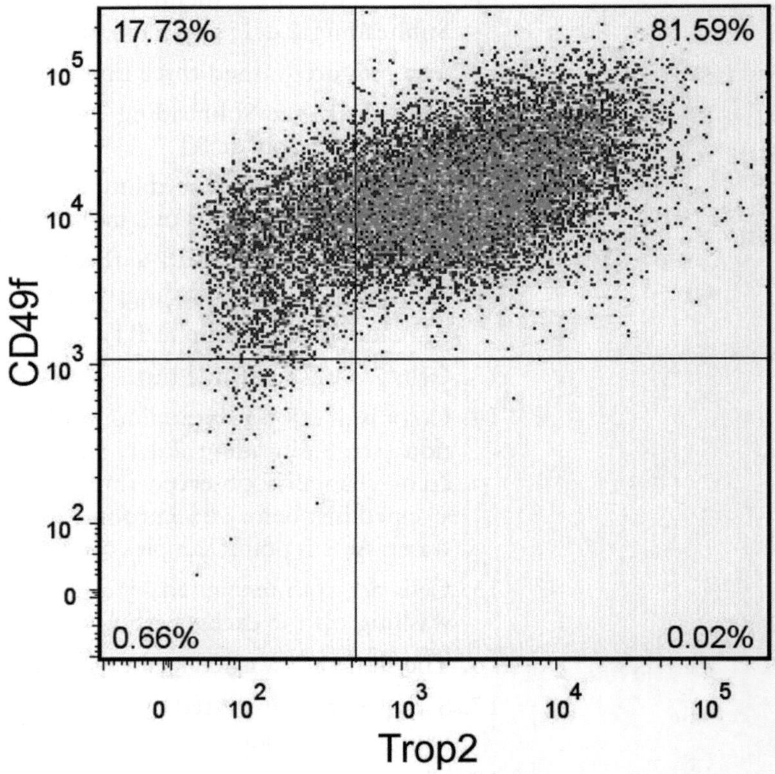

Fig. 4. Expression of cell surface antigens on primary prostate epithelial cells. A representative scatter plot of Trop2 and CD49f immunoreactivity in a primary culture derived from a benign radical prostatectomy specimen. The percentage of live cells in each quadrant is shown.

markers can also be used, such as CD31 (platelet endothelial cell adhesion molecule 1) for endothelial cells and CD45 (protein tyrosine phosphatase receptor types C) for hematopoietic cells, although we have not observed these cell types in the epithelial cultures.

1. Cells are washed with PBS and dislodged from the culture vessel using a trypsin-free solution such as versene or TrypLE™ Select (see Subheading 2.4.2, step 1).
2. The cells are counted using a hemocytometer and resuspended at 5×10^6 viable cells per milliliter in flow cytometry buffer (see Subheading 2.4.2, step 2). Cells should be kept at 4°C for all subsequent steps to reduce nonspecific labeling.
3. The cell suspension is aliquoted into sterile 1.5 mL tubes for an unstained control and each primary antibody and isotype control. If primary antibodies are combined for multicolor flow cytometry, single color compensation controls must also be prepared. Typically, $2.5–5 \times 10^6$ cells (50–100 μL of cell suspension) are sufficient for each sample, allowing for the loss of some cells during washing steps.
4. Fluorochrome-conjugated primary antibodies and isotype-matched controls (see Table 1) are added to the aliquots of cells at a final concentration of 10 μg/mL. Antibody titration curves can be used to optimize the concentration for new antibodies. The cells are then incubated at 4°C for 30 min.
5. To remove unbound antibodies, the cells are washed in flow cytometry buffer and centrifuged at $800 \times g$ for 5 min at 4°C. Repeat this step twice.
6. The cells are resuspended in buffer containing 1 μM SYTOX® Blue (see Subheading 2.4.2, step 4), or another appropriate viability dye such as propidium iodide, and then transferred to plastic flow cytometry tubes through 35 μM cell strainer caps.
7. The stained cells are analyzed on a flow cytometer, gating out dead cells and comparing samples stained with primary antibodies to corresponding isotype controls. Representative results for Trop2 and CD49f staining are shown in Fig. 4.

3.5. Isolation of Patient DNA and Genotyping Primary Cells

Blood samples are usually collected from patients prior to anesthesia and delivered with the tissue samples as described in Subheading 3.1.1. Blood samples should be processed on the day of arrival (see Note 11). This protocol assumes that the blood sample has been collected in K_2EDTA/K_3EDTA or ACD-A tubes.

1. The sample is centrifuged at $2,500 \times g$ for 10 min at room temperature. It is essential that the centrifuge brake is turned off (see Note 12). At the end of the centrifugation, the sample will have separated into three layers:

Plasma and fat (yellow-orange)—Top.

Lymphocytes/buffy coat (narrow white-gray layer)—Upper middle.

Red blood cells and granulocytes (dark red)—Bottom.

2. The upper plasma layer is carefully removed. It can be stored at −80°C in 500 μL aliquots if required for a biobank or similar archive.
3. The buffy coat is collected without disturbing the red blood cells, dispensed in 200 μL aliquots and stored at −80°C.
4. DNA from primary cell cultures and corresponding buffy coat samples can be genotyped to verify that they come from the same patient. DNA is extracted using the DNeasy Blood and Tissue Kit (see Subheading 2.5.1, step 3). It is resuspended in Hi-Di™ formamide and amplified with AmpliTaq Gold and the Powerplex 1.2 system (see Subheading 2.5.1, steps 4–7) which has a panel of eight microsatellite markers.

4. Notes

1. All instruments and disposables should be decontaminated with 1% Virkon® or 10% bleach overnight. Human tissue and blood should be handled with care and treated as potentially infectious (e.g., HIV). All tissue processing and blood handling steps should be carried out in a Class II biological safety cabinet.
2. The small pieces of tissue that are retained from each specimen can be examined using hematoxylin and eosin staining to confirm that the pathology is correct.
3. In our experience, TURP specimens can be digested more easily than radical prostatectomy specimens, so less digestion solution is required.
4. For high Gleason Grade tumors, the digestion in collagenase can be decreased to 8 h. Although the best recovery of epithelial cells occurs when processed immediately, the sample can be stored at 4°C following digestion (for up to 6 h); however, viability is compromised.
5. Primary prostate cells are to be cultured in serum free and low calcium conditions in order to maximize their longevity. In addition, antibiotics (such as penicillin and streptomycin, which are commonly used to culture cell lines) are not used. The growth of primary prostate cells in serum will significantly reduce the lifespan of the cells, in addition to reducing the

stem cell fraction (typically about 0.1% when grown in SCM) to <<0.1%.

6. It is important to inactivate any trypsin that may be remaining on the plate by thoroughly rinsing with high serum medium (R10+). It is equally important to remove all serum from the plate to prevent differentiation of the prostate cells. The plate should be rinsed well with PBS.
7. It has been observed that there are some cells that will remain adherent to the collagen plate, even after 1 h in 1% trypsin–EDTA (10×). These cells should be abandoned after 10 min in 10× trypsin–EDTA.
8. Typically, cells are split in a 1:2 ratio (one plate divided into two plates). Although plating cells at a lower density (i.e., 1:3, 1:4, and even 1:10) has been successful, this is very sample dependent and the risk is that cells may not grow at too low a density (even with feeder cells making the monolayer confluent).
9. Cells should be seeded at approximately $1–1.5 \times 10^4$ cells/cm^2 after being thawed from liquid nitrogen.
10. Although the addition of hormones, serum, and high levels of calcium are enough to cause differentiation, the addition of primary human stroma by coculturing provides the most effective and long-lasting way to form primary luminal prostate cells. The human stroma provides additional growth factors that allow the luminal cells a longer lifespan.
11. Fresh blood samples may be stored overnight at 4°C, but do not store for longer than this.
12. Using the brake during the slowing down of the centrifuge can result in disturbing the layers between the blood fractions making it difficult to separate the different components.

Acknowledgments

Norman Maitland, Anne Collins, Amanda Nicholl, Richard Birnie, and April Frazier are recognized with much appreciation for their contribution to the methods described herein. We also gratefully acknowledge the Australian Prostate Cancer Bio-Resource for the tissue specimens our laboratory uses for this technique.

The authors receive funding from the Australian National Health and Medical Research Council (Enabling Grant: APCB 614296, Fellowships: G.R. 1002648 and M.L. 1035721, Project Grant: 545931), Prostate Cancer Foundation of Australia (Project Grant: PG0709, PG0810, research infrastructure grant to APCB & Movember Young Investigator Grants to R.T., M.L.), and the Peter and Lyndy White Foundation.

References

1. Taylor RA, Risbridger GP (2008) Prostatic tumor stroma: a key player in cancer progression. Curr Cancer Drug Targets 8:490–497
2. Wang Y, Hayward S, Cao M, Thayer K, Cunha G (2001) Cell differentiation lineage in the prostate. Differentiation 68:270–279
3. English HF, Drago JR, Santen RJ (1985) Cellular response to androgen depletion and repletion in the rat ventral prostate: autoradiography and morphometric analysis. Prostate 7:41–51
4. Isaacs JT (1987) Control of cell proliferation and cell death in the normal and neoplastic prostate: a stem cell model. In: Rodgers C, Coffey D, Cunha GR, Grayhack J, Hinman FJ, Horton R(eds) Benign prostatic hyperplasia. Government Printing Office, Washington, DC, pp 85–94
5. De Marzo AM, Meeker AK, Epstein JI, Coffey DS (1998) Prostate stem cell compartments: expression of the cell cycle inhibitor p27Kip1 in normal, hyperplastic, and neoplastic cells. Am J Pathol 153:911–919
6. Richardson GD, Robson CN, Lang SH, Neal DE, Maitland NJ, Collins AT (2004) CD133, a novel marker for human prostatic epithelial stem cells. J Cell Sci 117:3539–3545
7. Litvinov IV, Vander Griend DJ, Xu Y, Antony L, Dalrymple SL, Isaacs JT (2006) Low-calcium serum-free defined medium selects for growth of normal prostatic epithelial stem cells. Cancer Res 66:8598–8607
8. Swift SL, Burns JE, Maitland NJ (2010) Altered expression of neurotensin receptors is associated with the differentiation state of prostate cancer. Cancer Res 70:347–356
9. Lang SH, Smith J, Hyde C, Macintosh C, Stower M, Maitland NJ (2006) Differentiation of prostate epithelial cell cultures by matrigel/stromal cell glandular reconstruction. In Vitro Cell Dev Biol Anim 42:273–280
10. McKeehan WL, Adams PS, Rosser MP (1984) Direct mitogenic effects of insulin, epidermal growth factor, glucocorticoid, cholera toxin, unknown pituitary factors and possibly prolactin, but not androgen, on normal rat prostate epithelial cells in serum-free, primary cell culture. Cancer Res 44:1998–2010
11. Chaproniere DM, McKeehan WL (1986) Serial culture of single adult human prostatic epithelial cells in serum-free medium containing low calcium and a new growth factor from bovine brain. Cancer Res 46:819–824
12. Robinson EJ, Neal DE, Collins AT (1998) Basal cells are progenitors of luminal cells in primary cultures of differentiating human prostatic epithelium. Prostate 37:149–160
13. Collins AT, Habib FK, Maitland NJ, Neal DE (2001) Identification and isolation of human prostate epithelial stem cells based on alpha(2) beta(1)-integrin expression. J Cell Sci 114:3865–3872
14. Goldstein AS, Huang J, Guo C, Garraway IP, Witte ON (2010) Identification of a cell of origin for human prostate cancer. Science 329:568–571
15. Lawson DA, Zong Y, Memarzadeh S, Xin L, Huang J, Witte ON (2010) Basal epithelial stem cells are efficient targets for prostate cancer initiation. Proc Natl Acad Sci U S A 107:2610–2615
16. Goldstein AS, Lawson DA, Cheng D, Sun W, Garraway IP, Witte ON (2008) Trop2 identifies a subpopulation of murine and human prostate basal cells with stem cell characteristics. Proc Natl Acad Sci U S A 105: 20882–20887
17. Goo YA, Goodlett DR, Pascal LE, Worthington KD, Vessella RL, True LD, Liu AY (2005) Stromal mesenchyme cell genes of the human prostate and bladder. BMC Urol 5:17
18. Berry PA, Birnie R, Droop AP, Maitland NJ, Collins AT (2011) The calcium sensor STIM1 is regulated by androgens in prostate stromal cells. Prostate 71:1646–1655

Chapter 23

Urothelial Cell Culture: Stratified Urothelial Sheet and Three-Dimensional Growth of Urothelial Structure

Yuanyuan Zhang and Anthony Atala

Abstract

Urothelial cells line the urinary tract, including the renal pelvis, ureters, bladder, superior urethra, and the central ducts of the prostate. They are highly specialized epithelial cell types possessing unique features, imparting important functional roles in the urinary system. They act as a permeability barrier and protect underlying muscle tissues from the caustic effects of urine while also expanding with bladder filling to adjust urine pressures. The multilayered urothelium is typically structured with differentiated, mature surface cells and less mature basal cells. The basal cell layer contains tissue-specific stem cells able to self-renew for the lifetime of the mammal and also produces a pool of maturing cells for tissue homeostasis. Maintaining regenerative basal cells in a culture facilitates urothelial cell growth in vitro. Additionally, epithelial–mesenchymal communication, epithelial–matrix interactions, and cytokines/growth factors are required to maintain the normal structure and function of mature urothelial cells in vitro and to induce stem cell differentiation into urothelial cells. These cultures are useful to study the biology and physiology of the urinary tract, particularly for the development of cell-based tissue engineering strategies in urology. This chapter describes methods for the isolation of urothelial cells and their maintenance in monolayer culture, and methods for the production of multilayer urothelial cell sheets and three-dimensional cocultures of urothelial and mesenchymal cells.

Key words: Urothelial cells, Bladder, Ureter, Tissue engineering, Scaffold, Urinary reconstruction, Epidermal growth factor, Serum-free medium

1. Introduction

Urothelial cell (UC) cultures are an in vitro model system useful to advance understanding of cellular mechanisms in urothelial development, epithelial–stromal interactions, bladder cancer, interstitial cystitis, and urinary tract infection. Additionally, autologous UCs can serve as a critical cell source for urinary tract reconstruction and regeneration with tissue engineering technology. The urothelium forms a barrier between urine and the underlying connective

Scott H. Randell and M. Leslie Fulcher (eds.), *Epithelial Cell Culture Protocols: Second Edition*, Methods in Molecular Biology, vol. 945, DOI 10.1007/978-1-62703-125-7_23, © Springer Science+Business Media, LLC 2012

tissue. It consists of multiple cell layers: a basal layer with progenitor cells that attaches to the connective tissue substratum, an intermediate cell layer that is one to two layers thick, and a superficial cell layer composed of highly differentiated umbrella cells that line the luminal surface of the bladder. A few decades ago, UCs could be harvested and maintained in culture for limited periods, but extended growth and expansion was not possible. Their growth was further hampered by inadequate medium that contained serum, which facilitated fibroblast overgrowth. Improvements in culture technology, including development of serum-free medium and scraping of the mucosa, which facilitates basal cell harvest, has enabled clonal growth of bladder and ureter UCs. Additionally, the nutrient optimization, reduction of calcium concentration, and serum omission favors proliferation to form a monolayer of UCs rather than differentiation, extending culture life, and permitting serial propagation (1–4). Recently, stratified urothelial sheets and three-dimensional (3D) growth of urothelial structures have been generated for urological tissue engineering purposes (5–7). Stratified urothelial cell sheets can be produced for potential use in demucosalized gastrointestinal flaps (8–14). Three-dimensional urothelial structures can be formed using a combination of culture techniques, i.e., coculture of UCs and smooth muscle cells (SMCs) seeded on porous scaffolds in a bioreactor under dynamic culture conditions (15–18), and these structures can be potentially used in the engineering of urinary tract organs (8, 17, 19–26).

2. Materials

2.1. Culture Medium and Supplements

1. Keratinocyte serum-free medium (KSFM, Life Technologies™): Prepare KSFM with a calcium concentration of 0.09 mmol/L, supplemented with recombinant epidermal growth factor (EGF, Life Technologies™) (5 ng/mL), bovine pituitary extract (BPE, I Life Technologies™) (50 mg/mL), cholera toxin (CT, Sigma-Aldrich®) at a final concentration of 30 ng/mL, penicillin (100 U/mL)/streptomycin (1 mg/mL) solution. All supplements are added into KSFM and filtered via a 0.22 μM filter.
2. Dulbecco's Phosphate-Buffered Saline (PBS) without Calcium and Magnesium.
3. Soy Bean Trypsin Inhibitor (STI, 10 mg/mL): Dissolve STI in PBS and filter through a 0.22 μM filter. Aliquot 1 mL of stock solution and store at 4°C.
4. Dispase II for detaching cell sheets from culture dishes: 1% Dispase II solution is made by adding 10 mg (0.6–2.4 U/mL) into 100 mL in PBS. Filter-sterilize with a 0.22 μM filter.

5. 1% Collagenase IV digestion solution: 1 mg/mL Collagenase IV is made by adding 10 mg collagenase IV into 100 mL in PBS. Filter-sterilize with a 0.22 μM filter.
6. 0.2% Ethylenediaminetetraacetic acid (EDTA, Sigma-Aldrich®).
7. 0.05% Trypsin and 0.25% trypsin with 1 mM EDTA.
8. Dulbecco's Modified Eagle's Medium (DMEM).
9. Fetal bovine serum (FBS).
10. 5% Peracetic acid (PAA): 5 mL PAA in 95 mL deionized water.
11. 0.1% PAA: 100 μL PAA in 100 mL of deionized water.
12. 0.5 M Calcium ($CaCl_2$) stock solution: 5.55 g $CaCl_2$ in 100 mL deionized water.
13. Hank's Balanced Salt Solution (HBSS).
14. A pair of microsurgery forceps and scalpel #10.
15. NIH 3T3 fibroblasts.
16. 1% Triton™ X-100: 1 mL Triton™ X-100 (Sigma-Aldrich®) in 99 mL deionized water.
17. 50 mL conical tubes.
18. 30 mm, 100 mm Tissue culture treated dishes and 6-well plates.
19. Surgical scissors: 100 mm length.
20. Cell lifter.
21. 8×8 cm^2 sterile styrofoam piece.
22. Sharp metal pins.
23. Collagen-coated culture dish (Biocompare).
24. Sodium pentobarbital.
25. Temperature-responsive culture dishes (Cell Seed).

2.2. Coculture Model and Conditioned Medium

1. 0.20 μm sterile filter.
2. T25 Primaria cell culture flasks.
3. Dynamic culture system.

2.3. Reverse Transcriptase-Polymerase Chain Reaction

1. RNA isolation from cell cultures is performed using an RNA isolation kit (5 PRIME, Inc.).
2. Reverse transcription reactions are performed using a High-Capacity cDNA Transcription Kit (Applied Biosystems™).
3. Primers for uroplaklin Ia (Locus, NM) and III (Locus, NM), CK7 (Locus, NM), CK13 (Locus, BC), CK20 (Locus, NM) were designed (1PRIMER 3 SOFTWARE Version 4.0) and purchased from Eurofins MWG Operon.
4. Taq DNA Polymerase (New England Biolabs®, Inc.).

5. PCR buffer (New England Biolabs®, Inc.).
6. dNTPs (Life Technologies™).
7. ddH_2O.
8. Agarose.
9. PCR tubes.
10. Thermocycler (Eppendorf).
11. System for electrophoresis of PCR products.

2.4. Western Blot

1. RIPA buffer: 50 mM Tris–HCl pH 7.4, 150 mM NaCl, 2 mM EDTA, 1% NP-40, 0.1% SDS.
2. SDS-PAGE gels, 10–12.5% polyacrylamide.
3. 10× TGS running buffer (Bio-Rad).
4. Transfer buffer: 48 mM Trizma Base, 39 mM Glycine, 20% methanol pH 9.2.
5. Transfer apparatus (Bio-Rad).
6. PVDF (Millipore).
7. Blocking solution: 5% Non-fat dry milk powder (Carnation®) in Tris Buffer Saline Tween 20 (TBST).
8. Washing solution (TBST) (Dako).
9. Antibodies to uroplaklin Ia (Santa Cruz) and III (R&D), CK7 (Novocastra), CK13 (Novocastra), CK19 (Novocastra), CK20 (Novocastra).
10. Secondary antibodies: Horseradish peroxidase (HRP)-conjugated secondary antibodies (Cell Signaling).
11. Chemiluminescent detection kit: Western Lightning Chemiluminescence Reagent (Perkin Elmer).

2.5. Immuno fluoresence

1. Chambered slides for cell culture (8 wells, Nunc).
2. Serum-free block solution (Dako).
3. Antibody diluent solution (Dako).
4. 3.7% Paraformaldehyde: Add 100 μL 10N NaOH (7 drops 10N NaOH) and 37 g paraformaldehyde into 980 mL RNAse-free PBS in a 1,000 mL bottle PBS and stir the solution at 62–64°C until the fixative becomes clear.
5. Antibodies to uroplaklin Ia and III, CK7, CK13, CK19, CK20, AE1/AE3 (Dako).
6. Fluorescently labeled secondary antibodies (Jackson ImmunoResearch and Vector Laboratories).
7. Antifade mounting media (Vector Laboratories).
8. 4′-6-diamidino-2-phenylindole (DAPI).
9. Propidium iodide (PI).

2.6. Biomaterials for Urothelial Cell Growth

1. Natural collagen matrices derived from porcine bladder mucosa and lamina propria (i.e., bladder submucosa, BSM) and porcine small intestine submucosa (SIS).
2. Home-made bladder submucosa matrix (BSM) (or SIS) inserts: These 1 cm disks were manufactured in a manner such that the BSM (or SIS) was suspended over a circular polypropylene frame (border 5 mm on top and 2 mm on bottom) with the mucosal surface upwards to create a double well culture dish with the BSM (or SIS) acting as the separating membrane. The mucosal surface of BSM (or SIS) forms the base of the upward facing well, while its serosal surface forms the base of the bottom well. The upper well holds 500 mL of medium and the bottom well holds 200 mL. Following seeding of cells, BSM (or SIS) disks were placed in a 12 or 6 well cell culture dish filled with medium to allow free contact of the medium with both sides of the BSM (or SIS) (8, 18).

3. Methods

Culture of UC from humans and animals requires different approaches for cell isolation, expansion, and optimal culture medium for UC growth in vitro. The processes of both monolayer and multilayer culture of UC and methods for UC growth on matrices in vitro and in vivo for tissue engineering will be introduced.

3.1. Isolation and Culture Methods for Urothelial Cells from Different Species

Urothelial cells from different species require different methods for cell isolation and cultivation with distinctive culture medium. For isolation of urothelium, UCs can be retrieved by enzymes, such as dispase, trypsin, or collagenase IV. Such methods are most fitting for smaller animals (e.g., rats or mice) in which mechanical manipulation is complex. For human and large animals (i.e., pig and dog), UCs can be obtained by a combination of the microdissection and enzymatic retrieval methods or by explant culture. Generally, human UCs are much easier to culture than cells from rodents, dogs, and pigs. Human UCs tend to grow well in KSFM for over ten passages, while large animal UCs often require the addition of 1–5% serum and stop growing by passage 3–5. It has been shown that rat UCs can proliferate for up to 18 passages when conditioned medium derived from an immortalized cell line is added (9).

3.1.1. Human Urothelial Cell Culture

All of the following culture protocols should be performed in a class II laminar flow biosafety cabinet (hood) using aseptic techniques.

1. The bladder tissue specimen (size <2 × 2 cm^2) is placed in a sterile 50 mL centrifuge tube containing 20 mL chilled PBS or KSFM containing penicillin and streptomycin on ice.
2. The tissue should be processed as soon as possible once the bladder tissues are obtained (see Note 1).
3. The bladder tissue is transferred to a sterile 100 mm Petri dish and washed with PBS or KSFM containing penicillin and streptomycin to remove red blood cells. The bladder mucosa is dissected away from the underlying submucosa with sterile surgical scissors (see Note 2).

Primary Explant Method

1. Mince urothelial mucosa into 1-mm pieces. Place six to nine pieces of tissue, mucosal side down and without medium, on a small culture dish (30 mm dish or 6-well plate), which increases the chances of establishing a successful primary urothelial culture.
2. Place the tissue into an incubator for 60 min incubation in a humidified atmosphere containing 95% air and 5% carbon dioxide (CO_2) at 37°C to allow the tissue fragments to attach to the dish.
3. Add 3 mL KSFM medium (see Subheading 2.1, item 1) to the culture dish slowly. The outgrowth of cells from each piece of bladder tissue can usually be observed after 48 h of incubation in a humidified atmosphere containing 95% air and 5% carbon dioxide (CO_2) at 37°C (see Note 3).
4. Examine the cultured cells daily by phase-contrast microscopy to assess growth and morphology and to make certain the medium is clear of any contamination.

Enzyme-Based Method

1. Mince urothelium into 2–3 mm fragments and digest in 1% collagenase type IV (with no trypsin activity) at 37°C for 1–2 h.
2. To remove collagenase, centrifuge the digested cells at 500 × *g* at room temperature for 5 min, wash them in HBSS and centrifuge again.
3. Resuspend the cells in KSFM and culture the primary UCs in 5 mL in T25 Primaria tissue culture flasks at 37°C.
4. Change the medium every 48 h. When cells are 95% confluent, they should be passaged. To subculture the cells, use cold BPS (4°C) to wash the culture after aspirating the culture medium and then add 0.2% EDTA and 0.05% Trypsin for 2–3 min. Add soy bean trypsin inhibitor solution to stop the trypsin activity and spin down the cells at 500 × *g* for 5 min (see Note 4).
5. For subculture, plate the cells at a ratio of 1:5 with a cell concentration of 200 × 10^4 cells in 1 mL medium per flask; after 24 h, when the cells have attached to the flask, add 4 mL fresh KFSM.

Scraping Method

1. Transfer the bladder specimen to a 100 mm Petri dish in KSFM. Usually for a specimen sized 1×1–2×2 cm^2, 3.5 mL of medium is adequate. Detach UCs by gentle scraping with a sterile surgical blade.
2. Transfer the cell suspension to six wells or a 24-well cell culture vessel (0.5 mL in each well) by means of a 5-mL pipette or a 5-mL syringe with No. 23 needle.
3. Add another 0.5 mL fresh KFSM to each well and incubate for 48 h, then change medium (see Note 5).

3.1.2. Isolation and Culture of Urothelial Cells Derived From Large Animals

The methods of isolation and culture of UC from large animals (such as pigs and dogs) are similar to the methods used in human UCs. The procedure is a combination of the microdissection and enzymatic techniques as described below. UC isolated from large animals have been used as an autologous cell source in the production of tissue engineered bladder or urethra in the same animals (20, 27–30).

Isolation and Primary Culture of Urothelial Cells from Large Animals

1. Wash the bladder specimen as described for the microdissection technique (see Subheading 3.1.1).
2. Microdissection technique: strip the urothelial mucosa and connective tissues from muscle tissue by a pair of microsurgery forceps and scalpel.
3. Enzymatic techniques: Transfer the mucosa to a 100 mm culture dish containing an 8×8 cm^2 sterile Styrofoam piece with ten sharp metal pins along each edge. With the mucosa side up, the tissue should be stretched across the pins and then incubated overnight at 4°C in KSFM containing 2.5 mg/mL dispase.
4. Remove the medium-dispase solution and scrape the uroepithelial cells from the underlying connective tissue with two flexible cell scrapers.
5. Transfer the scraped cells to a culture dish, resuspend them in 20 mL of 0.25% trypsin 1 mM EDTA, and incubate at 37°C for 30 min.
6. Bring the cell suspension up to 50 mL with KSFM containing 5% FBS in a sterile conical tube and spin down at $500 \times g$ for 5 min to pellet the cells and remove the trypsin.
7. Carefully remove the supernatant and wash the cells in KSFM with supplements (see Subheading 2.1, item 1).
8. Resuspend the cells in a final concentration of 6–8×10^5 cells/mL and plate them in KSFM and supplements with 2% FBS (see Note 6).

3.1.3. Rodent Urothelial Cells In Vitro

Rodents are a commonly used animal model for biomedical research. However, the use of cultured UCs from rodents, such as rat urothelial cells (RUCs), has been limited due to the difficulty of their isolation and maintenance in a long-term culture. The main approach presently used for the culture of RUC is to plate the cells in serum-containing medium (5% FBS) on a collagen-coated culture dish or on a feeder layer of lethally irradiated NIH 3T3 fibroblasts that can support growth of cultured cells in vitro. However, these techniques also present problems, as the RUC culture is often contaminated with other cell types such as fibroblasts, SMCs, or capillary endothelium. These stromal cells overgrow and finally replace UCs in the serum-containing medium after only a few passages.

Roszell's procedure for rat urothelial cell culture: Conditions developed for large-scale in vitro growth and serial cultivation of normal diploid rat bladder epithelial cells. Primary cultures were initiated by attachment of bladder mucosal explants to type I collagen gels. A rapid outgrowth of epithelial cells from the explants occurred when cultured in a hormone-supplemented medium with epidermal growth factor. These primary outgrowths were passaged by nonenzymatic dispersion with 0.1% ethylenediaminetetraacetic acid (EDTA) and replating onto new gels. The capacity for routine serial passaging and maintenance of rat bladder epithelial cells required the presence of epidermal growth factor, a requirement not observed with human urothelial cells. The characteristics of the cultured rat bladder epithelial cells were similar to human urothelial cells in: ultrastructural and phase-contrast morphologic properties, showing junctional complexes, desmosomes, stratification, and an apical glycocalyx; the absence of stromal cell contamination; and the ability to be serially passaged. Spontaneous cell-line formation was observed with the rat bladder epithelial cells, but has not been found with the human urothelial cells. With the method that we have developed, the number of rat bladder epithelial cells generated from a single bladder of a 4–6-week-old rat was increased 100-fold from about 7×10^5 cells to 7×10^7 viable cells within 3 weeks of culture. The capability of culturing normal, primary rat bladder epithelial cells on this scale has not been reported previously and will facilitate comparative studies of the biological and molecular characteristics of the mammalian urothelium. Furthermore, this culture system will be useful for carcinogenesis studies, including metabolic activation of carcinogens and cellular transformation in vitro (31).

We modified Roszell's procedure (31) and developed a simple technique to isolate rat urothelium by enzymatic release of RUCs from an "everted bladder" and to maintain the primary culture in keratinocyte serum-free medium (KSFM) (8). In this study, the conditions for the growth of RUC in long-term culture were investigated systematically. We show that a mixture of conditioned medium (CM) obtained from NIH 3T3 fibroblasts and KSFM

(CM-KSFM) (32) yielded large quantities of normal RUCs without stromal cell contamination. RUCs could be subcultured up to 18 times in CM-KSFM during an observation period of up to 5 months. Additionally, rat urothelium stratification can be induced with or without a feeder layer in vitro, which provides potential application as an autograft for urothelial replacement in bladder augmentation in a rat model. This technique also serves as a tool for research on various bladder diseases including tumorigenesis and urinary tract infection. See Subheading "Rat Urothelial Cell Isolation and Culture", which describes the method for RUCs isolation.

Rat Urothelial Cell Isolation and Culture

1. Anesthetize the rats with an intraperitoneal injection of sodium pentobarbital (40–50 mg/kg). After the whole bladder is excised, apply a modified Roszell's procedure (31) to evert the bladder to expose the urothelial surface.
2. Reinsert the bladder neck into the lumen, and surgically close to form an "everted bladder ball" so that only the urothelial surface is exposed.
3. In order to choose the optimal method for the isolation of RUC, immerse the everted bladder either in 4 mL of 1% collagenase IV at 37°C on a shaker for 60 min or in 10 mL of 0.1% disodium EDTA at 48°C for 4 h.
4. Gently scrape the bladder mucosa from the muscle tissue following digestion using a forceps with coarse tips.
5. Collect, wash, and plate UCs on T25 Primaria cell culture flasks at a cell density of 5×10^6 cells/mL in 5 mL mixed media containing KSFM as mentioned above and conditioned medium derived from a 3T3 cell culture.
6. The conditioned medium (CM) is derived from NIH 3T3 cell culture. To produce it, culture 3T3 cells in DMEM supplemented with 10% FBS and 1% P/S and then collect the culture medium when the cells reach at 60–80% confluence (12 h).
7. Filter-sterilize the conditioned medium with a 0.20 μm filter to prevent fibroblast contamination.
8. Culture the primary RUCs using a mixed media with fresh KFSM:CM-3T3 Culture (1:1 ratio).
9. Examine the cultured urothelial cells daily by phase-contrast microscopy to assess cell growth and morphology and to check for any contamination.

3.2. Stratified Urothelial Sheet Culture

Viable cultured urothelial cell sheets can be applied for urothelium regeneration on demucosalized gastrointestinal segments through grafting techniques. There are two main techniques to produce these urothelial cell sheets in culture: (1) cultured cell sheets can be gently detached from a culture dish with 2.5% dispase and (2) temperature-responsive culture dishes can be used. When using the dispase method, in vitro cultured human stratified urothelium does show complete differentiation of its superficial cells. It retains the same ultrastructure of barrier characteristics (such as microvillus and tight junctions by electron microscope observation) against principal urine components (10). The cell sheets appear to be stable in genotype and no chromosomal aberrations have been found. Cells in the sheets express the urothelial specific cell marker uroplakin and cytokeratins 7, 8, 17, and 18.

3.2.1. Production of Stratified Urothelium In Vitro

1. Plate UC into culture flasks at an average cell density of 1×10^5 cells/cm^2 in KSFM and incubate with 5% CO_2 in 95% air at 37°C.
2. Change KSFM on alternate days.
3. Culture UCs up to 100% confluence, and then enrich the serum-free medium with calcium to a final concentration of 1.5 mM.
4. When stratification is induced, usually within 10 days, remove the culture medium, and detach the in vitro urothelial cell sheet construct from the culture flask with 2.5% dispase II at 37°C for 30 min. An entire urothelial cell sheet that is about half the size of the culture dish area can usually be easily detached (see Note 7).

3.3. In Vitro Coculture Methods

Coculture of urothelial and SMCs on bio-scaffolds is usually applied in the production tissue engineered urinary tract organs (8, 19, 20), stem cell differentiation protocols (17, 33, 34), or in vitro bladder models for urinary tract infection. In order to create a layered construct that includes both a urothelial layer and a smooth muscle layer, various coculture techniques can be used, including the layered, sandwich, and mixed coculture techniques. In vitro observations with these three different coculture techniques suggest that the layered and sandwich methods of seeding have distinct advantages over the mixed coculture technique in terms of cell stratification, cell-matrix penetration, and cell differentiation. In order to induce a stratified urothelium-bladder SMC structure, cells are usually seeded on biomaterials such as BSM and SIS (see Subheading 2.6) *with* coculture techniques (8, 19, 27, 28, 35, 36).

3.3.1. Layered Coculture Method

1. Set some SIS discs in the wells of a 24-well plate. Seed bladder SMCs (1×10^5 cells/cm) onto the mucosal side of the SIS using DMEM and 10% FBS. After incubation for 1 h, UCs can be seeded on top of the SMCs at the same cell concentration (see Note 8).
2. On day 3, distinct cell sorting will be noted, in which the SMCs grow on the surface of the SIS with early cell-matrix penetrance, and the UCs will grow on top of the SMCs as a separate population of cells. This cell sorting can be confirmed by immunohistochemical analysis. UCs on the top portion of the

culture stain positive for AE1/AE3 while SMCs in the bottom portion are positive for α-smooth muscle actin, desmin, myosin and smoothlin.

3. On day 7 the culture shows further stratification and increased matrix penetrance.
4. For the remainder of the culture period cell growth and matrix penetrance is progressive while cell sorting is maintained.
5. By day 28, a well-developed, pseudo-stratified layer of UCs is present that is three to four layers thick. The SMCs layer is five to seven layers thick at this time.
6. The vast majority of SMCs will no longer be on the surface of the SIS; rather, they penetrate the matrix of SIS membrane and proliferate under its surface within the membrane. In several areas, SMCs can usually be seen traversing into deep portions of SIS membrane. (8, 27, 28)

This matrix penetrance by SMCs is distinctly different from the pattern of membrane ingress that is observed when SMCs are grown alone on SIS. The degree and pattern of cell-matrix penetrance by SMCs with the sandwich coculture technique were similar to those observed with the layered coculture technique, although some of SMCs were still located on the serosal surface of SIS.

3.3.2. Sandwich Coculture Method

1. Set SIS disks into the wells on 6-well plates. Seed SMCs on the serosal surface of the SIS and add 0.5 mL DMEM with 10% FBS on the top portion of the disk.
2. Follow by seeding the mucosal surface of the SIS with UCs in mixed culture media (KSFM:DMEM with 10% FBS, 1:1) 24 h later (see Note 6).
3. Organized, layered growth of UCs on one side of the SIS and growth of SMCs on other side of the SIS will be readily evident on day 3.
4. Cell growth will be progressive during a 28-day period of observation (as with the layered coculture technique).
5. By day 28, there will be a well-defined pseudostratified layer of UCs on the mucosal side of the SIS disk that is three to four cells thick, and there should be little to no evidence of matrix penetration by the UCs (8, 19, 37).

On the serosal surface of SIS, the SMCs layer will be about five to seven cells thick, and there is usually significant penetrance of the SIS membrane by the SMCs.

3.4. Three-Dimensional Culture of Urothelial Structures

Generation of 3D cultures of urothelial structures requires a combination of three techniques: coculture of UCs and SMCs (8, 19, 37), culture on a biomaterial with porous microstructure (17, 18, 34, 35), and the use of a bioreactor to provide dynamic culture conditions (16, 17, 34).

3.4.1. Biomaterials

A prior study, published in 2009 (18), demonstrated that recellularization of biological collagen acellular matrices with porous microstructure using cell seeding technology provides a promising option for promoting tissue regeneration. A strategy for performing tissue reconstruction using tissue engineering techniques is to repopulate a scaffold with cells isolated from the patient's own tissues to provide an autologous repair of the tissue defect. For example, in urethral reconstruction using tissue engineering techniques, a tissue-engineered tube composed of differentiated UCs on the urethral lumen side and SMCs on the submucosal side can be used. To produce such a scaffold, the acellular matrix has to be porous, contain minimal heterogeneous cellular compounds, and retain most of the extracellular matrix components to permit cell seeding.

Bladder Submucosa Matrix and Small Intestinal Submucosa

An ideal biological collagen matrix for urethral tissue engineering would have high porosity for cell seeding, be degradable, histocompatible, and have the least xenogenous cellular compounds retained within the matrix for minimum inflammatory potential. Most importantly, the matrix needs to have a three-dimensional (3-D) structure with high porosity, but at the same time, it must maintain a nearly normal tensile strength.

Two naturally derived matrix materials, i.e., BSM and SIS, meet these criteria and have been used in a number of urological applications both in vitro and in vivo. BSM is composed of a slender basement membrane with high density of collagen and a lamina propria with areolar connective tissue. BSM has been valuable in reconstruction of urethra in animal models and in humans. On the other hand, SIS is a xenogenic, acellular, collagen rich membrane with inherent growth factors that have previously been shown to promote in vivo bladder regeneration. Our pervious study evaluated in vitro use of SIS to support the individual and combined growth of bladder UCs and SMCs for potential use in tissue engineering (8). Although no in vitro cell culture substrate can fully mimic the in vivo state, SIS has significant advantages over conventional plastic and other coated surfaces because it provides a unique environment that promotes cell–cell and cell–matrix interactions. The presence of UCs significantly impacts the pattern of SMCs growth on SIS since active penetrance of the membrane only occurs when UCs are grown in conjunction with SMCs.

Biomaterial Porosity

A 3-D scaffold with higher porosity and relative larger pore size (50–200 mM) promotes cell proliferation, migration, and infiltration into the matrix, and appears to allow abundant cell loading onto the scaffold, thereby promoting in vivo tissue regeneration and wound healing in a nu/nu athymic mouse model (17, 18, 34). Such a scaffold would also allow the host cells to participate in the tissue remodeling processes by infiltration or migration into the matrix from the wound edges. Hence, a 3-D porous matrix would be potentially beneficial in the reconstruction of bladder or urethra tissues. Below is the method for obtaining porcine BSM.

1. Clean fresh porcine bladders upon receipt and manually remove muscle layers.
2. Retain and wash BSM in distilled water.
3. For decellularization, transfer BSM to a 500 mL bottle filled with distilled water and place at 4°C on a rotary shaker at 200 rpm for 2 days.
4. The distilled water is then discarded and BSM is oxidized by soaking in 5% PAA for 4 h.
5. Next, treat BSM with a solution containing 1% Triton™ X-100 for 2 days, and then wash with distilled water for another 2 days.
6. Finally, disinfect BSM using 0.1% PAA in 20% alcohol for 2 h, rinse three times with sterile distilled water for 10 min each, and store in sterile distilled water at 4°C until further use.

The methods for producing SIS scaffolds are nearly identical to those shown above for BSM.

3.5. Dynamic Culture

Dynamic culture methods using a bioreactor that can provide medium perfusion or rotation have frequently been used for seeded scaffolds. Dynamic culturing mimics the physiological environment and promotes cell adhesion, proliferation, infiltration, and differentiation. The advantages associated with dynamic culture conditions include (1) even distribution of nutrition and oxygen leads to uniform growth of cells on the matrix; (2) increased synthesis of endogenous ECM; and (3) physiologically relevant mechanical forces on the cultured cells (i.e., shear stress, pressure, and hydrodynamic compression).

Bioreactors have become an important tool to improve bladder tissue engineering under physiologic conditions. Recently, Farhat et al. developed a urinary bladder bioreactor with a hydrodynamic chamber to produce stretch and strain on cell-seeded scaffolds (15, 16). This bioreactor helps engineered tissues to better adapt to the changing environment when implanted in vivo. It enhances epithelial-stromal and cell-ECM interactions, which are necessary for building bladder tissues but which cannot be achieved using static cultures in plates. The use of a bladder bioreactor system may accelerate tissue organization and maturation in vivo and may shorten the time required to achieve a fully functioning organ.

3.6. In Vivo Urothelium Formation

Cell-based approaches to engineer human bladder tissue have been reported (23), and bioengineering has allowed creation of functional neo-bladder tissues in several animal models (19, 20, 28, 30, 33, 36). Such tissue-engineered bladder is generated from autologous cells derived from a biopsy of tissue. After bladder tissues are obtained from the biopsy, UCs and SMCs are isolated and expanded in culture and then seeded onto a biodegradable scaffold.

A clinical experience involving engineered bladder tissue for cystoplasty reconstruction was conducted starting in 1998. A small pilot study of seven patients was reported, using a collagen scaffold seeded with cells either with or without omentum coverage, or a combined PGA-collagen scaffold seeded with cells and omental coverage. The patients reconstructed with the engineered bladder tissue created with the PGA-collagen cell-seeded scaffolds with omental coverage showed increased compliance, decreased end-filling pressures, increased capacities and longer dry periods over time (23). It is clear from this experience that the engineered bladders continued their improvement with time, mirroring their continued development. Although the experience is promising in terms of showing that engineered tissues can be implanted safely, it is just a start in terms of accomplishing the goal of engineering fully functional bladders. This was a limited clinical experience, and the technology is not yet ready for wide dissemination, as further experimental and clinical studies are required. FDA Phase 2 studies have now been completed.

To identify whether urothelial cells maintain their specific phenotypes including gene, protein expression after isolation, culture and cocultured with SMCs, Reverse Transcriptase-Polymerase Chain Reaction (RT-PCR), Western Blot, and immunofluorescence analysis are performed. Urothelial cell transcripts including uroplakin (Up)-Ia, Up-III, cytokeratin (CK)-7, and CK-13 are regularly used. For RNA extraction experiments, UC are cultured in KSFM in 10 cm dishes for 14 days. Five micrograms of RNA, extracted using RNA isolation kit (see Subheading 2.3, item 1), is used for cDNA synthesis using the High-Capacity cDNA Transcription Kit (see Subheading 2.3, item 2) according to the manufacturer's instructions. Briefly, RNA is incubated with random hexamers, nucleotides, and reverse transcriptase enzyme and reaction buffer in 20 μL volume for synthesis of cDNA. One tenth of the reaction volume is taken for PCR in PCR tubes using specific primer pairs, PCR buffer, dNTPs, ddH_2O, and Taq DNA polymerase. All the amplified products were subjected to 1.5% agarose gel electrophoresis and visualized by UV light.

For Western Blot analysis, cells are harvested from culture dishes and lysed using RIPA buffer (see Subheading 2.4, item 1). Proteins extracted from whole cells are run on sodium dodecyl sulfate-polyacrylamide gel electrophoresis in TGS running buffer to separate the proteins and then transferred to PVDF membrane in transfer buffer (see Subheading 2.4, item 4) by transfer apparatus overnight at 4°C. Mouse anti-human CK7, CK13, Ck19, CK20, AE1/AE3, and uroplakin Ia/III are used as the first antibody to probe the membrane after washing by TBST and then block in the blocking solution (see Subheading 2.4, item 7) and peroxidase labeled goat anti-mouse IgG is used as the secondary antibody for detection. Protein bands are detected with an enhanced chemiluminescence assay using chemiluminescent detection kit.

For immunofluorescence, urothelial cells are assessed for the expression of seven types of cell markers. Urothelium-specific markers included uroplakin Ia and III, a tissue specific and differentiation-dependent transmembrane protein of the urothelial luminal surface. Epithelial cell markers include CKs 7, 13, 17, 19, and 20. All markers were assessed for all three urothelial layers. All surface marker assays were performed at least three times to ensure consistent results. Following urothelial cells were seeded on the chamber slides; the cells were washed with PBS twice for 2 min each and fixed with freshly prepared 3.7% paraformaldehyde (see Subheading 2.5, item 4) for 15–20 min at room temperature. After removing excess fixative with PBS washes, the cells were permeabilized with 0.1% Triton™ X-100 in PBS for 5 min and blocked in serum-free block (see Subheading 2.5, item 2) solution for 30 min in a moist chamber. Lineage-specific primary antibody was diluted in the antibody diluent solution (see Subheading 2.5, item 3) and incubated overnight at 4°C. Appropriate secondary antibodies conjugated to fluorochromes were incubated at room temperature for 1 h. The slides were mounted using antifade mounting media containing DAPI or PI and images were recorded using an upright microscope.

4. Notes

1. To enhance the success rate of primary cultures of UCs, avoid the use of electrical knives (electrocautery) when bladder tissues are harvested from patients (9). In addition, any foreign objects such as bladder stones and catheters should be removed 1 week before the tissue biopsy is procured, as these foreign bodies can interfere with the outcome of urothelial cell culture.
2. Microdissection to strip off the urothelium is one of the methods reported to be successful in large mammals such as human, pigs, and dogs. This technique, however, is difficult in small animals such as rats or mice.
3. Do not touch the culture dish or flask during the first 48 h to allow the explants attach to the plastic.
4. Use Soy Bean Trypsin Inhibitor instead of FBS to stop trypsin activity when UCs are subcultured. This can prevent any fibroblast contamination in the culture.
5. Culture of UCs is not always successful, and urothelial cells from various areas of the urinary tract respond differently to

culture protocols. For example, urothelial cells from the ureter are more easily cultured than urothelial cells obtained from bladder (9).

6. Generally speaking, UCs derived from pig and dog can be cultured up to only passage 3–4 and then they will senesce. However, human UCs can be cultured for over ten passages. The culture medium needs to be optimized for culture of UCs from larger animals.
7. Detached urothelial cell sheets can be used for histology or tissue repair. Do not keep these in the incubator for more than 3 h; otherwise, the cell sheet will shrink.
8. Air bubbles must be aspirated with a 1 mL tuberculin syringe with a fine needle.

Acknowledgment

The authors would like to thank Dr. Jennifer Olson for editorial assistance with this manuscript.

References

1. Cilento BG, Freeman MR, Schneck FX, Retik AB, Atala A (1994) Phenotypic and cytogenetic characterization of human bladder urothelia expanded in vitro. J Urol 152:665–670
2. Freeman MR, Yoo JJ, Raab G, Soker S, Adam RM, Schneck FX, Renshaw AA, Klagsbrun M, Atala A (1997) Heparin-binding EGF-like growth factor is an autocrine growth factor for human urothelial cells and is synthesized by epithelial and smooth muscle cells in the human bladder. J Clin Invest 99:1028–1036
3. Southgate J, Harnden P, Selby PJ, Thomas DF, Trejdosiewicz LK (1999) Urothelial tissue regulation. Unraveling the role of the stroma. Adv Exp Med Biol 462:19–30
4. Southgate J, Hutton KA, Thomas DF, Trejdosiewicz LK (1994) Normal human urothelial cells in vitro: proliferation and induction of stratification. Lab Invest 71:583–594
5. Atala A, Amin M, Harty JI (1992) Diethylstilbestrol in treatment of postorchiectomy vasomotor symptoms and its relationship with serum follicle-stimulating hormone, luteinizing hormone, and testosterone. Urology 39:108–110
6. Atala A, Bauer SB, Hendren WH, Retik AB (1993) The effect of gastric augmentation on bladder function. J Urol 149:1099–1102
7. Atala A (1998) Tissue engineering in urologic surgery. Urol Clin North Am 25:39–50
8. Brandt WD, Matsui W, Rosenberg JE, He X, Ling S, Schaeffer EM, Berman DM (2009) Urothelial carcinoma: stem cells on the edge. Cancer Metastasis Rev 28:291–304
9. Li HL, Chen XG, Zhang FC, Ma J, Xu CS (2008) Expression profiles of the extracellular matrix-associated genes during rat liver regeneration. Yi Chuan 30:333–340
10. Sugasi S, Lesbros Y, Bisson I, Zhang YY, Kucera P, Frey P (2000) In vitro engineering of human stratified urothelium: analysis of its morphology and function. J Urol 164:951–957
11. Ludwikowski B, Zhang YY, Frey P (1999) The long-term culture of porcine urothelial cells and induction of urothelial stratification. BJU Int 84:507–514
12. Staack A, Hayward SW, Baskin LS, Cunha GR (2005) Molecular, cellular and developmental biology of urothelium as a basis of bladder regeneration. Differentiation 73:121–133
13. Scriven SD, Trejdosiewicz LK, Thomas DF, Southgate J (2001) Urothelial cell transplantation using biodegradable synthetic scaffolds. J Mater Sci Mater Med 12:991–996
14. Feil G, Maurer S, Nagele U, Krug J, Bock C, Sievert KD, Stenzl A (2008) Immunoreactivity of p63 in monolayered and in vitro stratified human urothelial cell cultures compared with native urothelial tissue. Eur Urol 53: 1066–1072

15. Gabouev AI, Schultheiss D, Mertsching H, Koppe M, Schlote N, Wefer J, Jonas U, Stief CG (2003) In vitro construction of urinary bladder wall using porcine primary cells reseeded on acellularized bladder matrix and small intestinal submucosa. Int J Artif Organs 26:935–942
16. Farhat WA, Yeger H (2008) Does mechanical stimulation have any role in urinary bladder tissue engineering? World J Urol 26:301–305
17. Eeles RA, Kote-Jarai Z, Al Olama AA, Giles GG, Guy M, Severi G, Muir K, Hopper JL, Henderson BE, Haiman CA et al (2009) Identification of seven new prostate cancer susceptibility loci through a genome-wide association study. Nat Genet 41:1116–1121
18. Bendall SC, Hughes C, Campbell JL, Stewart MH, Pittock P, Liu S, Bonneil E, Thibault P, Bhatia M, Lajoie GA (2009) An enhanced mass spectrometry approach reveals human embryonic stem cell growth factors in culture. Mol Cell Proteomics 8:421–432
19. Song WS, Yoo JJ, Koo KH, Yoon KS, Kim YM, Kim HJ (2004) Subchondral fatigue fracture of the femoral head in military recruits. J Bone Joint Surg Am 86-A:1917–1924
20. Oberpenning F, Meng J, Yoo JJ, Atala A (1999) De novo reconstitution of a functional mammalian urinary bladder by tissue engineering. Nat Biotechnol 17:149–155
21. Bahrenberg G, Brauers A, Joost HG, Jakse G (2000) Reduced expression of PSCA, a member of the LY-6 family of cell surface antigens, in bladder, esophagus, and stomach tumors. Biochem Biophys Res Commun 275:783–788
22. Pariente JL, Kim BS, Atala A (2001) In vitro biocompatibility assessment of naturally derived and synthetic biomaterials using normal human urothelial cells. J Biomed Mater Res 55:33–39
23. Amiel GE, Komura M, Shapira O, Yoo JJ, Yazdani S, Berry J, Kaushal S, Bischoff J, Atala A, Soker S (2006) Engineering of blood vessels from acellular collagen matrices coated with human endothelial cells. Tissue Eng 12: 2355–2365
24. Hipp J, Andersson KE, Kwon TG, Kwak EK, Yoo J, Atala A (2008) Microarray analysis of exstrophic human bladder smooth muscle. BJU Int 101:100–105
25. Amiel GE, Yoo JJ, Atala A (2000) Renal tissue engineering using a collagen-based kidney matrix. Tissue Eng 6(Suppl):685
26. Kwon TG, Yoo JJ, Atala A (2008) Local and systemic effects of a tissue engineered neobladder in a canine cystoplasty model. J Urol 179:2035–2041
27. Chen LL, Trent JC, Wu EF, Fuller GN, Ramdas L, Zhang W, Raymond AK, Prieto VG, Oyedeji CO, Hunt KK et al (2004) A missense mutation in KIT kinase domain 1 correlates with imatinib resistance in gastrointestinal stromal tumors. Cancer Res 64:5913–5919
28. Andl T, Murchison EP, Liu F, Zhang Y, Yunta-Gonzalez M, Tobias JW, Andl CD, Seykora JT, Hannon GJ, Millar SE (2006) The miRNA-processing enzyme dicer is essential for the morphogenesis and maintenance of hair follicles. Curr Biol 16:1041–1049
29. Fraser M, Thomas DF, Pitt E, Harnden P, Trejdosiewicz LK, Southgate J (2004) A surgical model of composite cystoplasty with cultured urothelial cells: a controlled study of gross outcome and urothelial phenotype. BJU Int 93:609–616
30. Cartwright L, Farhat WA, Sherman C, Chen J, Babyn P, Yeger H, Cheng HL (2006) Dynamic contrast-enhanced MRI to quantify VEGF-enhanced tissue-engineered bladder graft neovascularization: pilot study. J Biomed Mater Res A 77:390–395
31. Johnson MD, Bryan GT, Reznikoff CA (1985) Serial cultivation of normal rat bladder epithelial cells in vitro. J Urol 133:1076–1081
32. Zhang YY, Ludwikowski B, Hurst R, Frey P (2001) Expansion and long-term culture of differentiated normal rat urothelial cells in vitro. In Vitro Cell Dev Biol Anim 37:419–429
33. Frimberger D, Morales N, Shamblott M, Gearhart JD, Gearhart JP, Lakshmanan Y (2005) Human embryoid body-derived stem cells in bladder regeneration using rodent model. Urology 65:827–832
34. Amante FH, Haque A, Stanley AC, Rivera Fde L, Randall LM, Wilson YA, Yeo G, Pieper C, Crabb BS, de Koning-Ward TF et al (2010) Immune-mediated mechanisms of parasite tissue sequestration during experimental cerebral malaria. J Immunol 185:3632–3642
35. Cen L, Liu W, Cui L, Zhang W, Cao Y (2008) Collagen tissue engineering: development of novel biomaterials and applications. Pediatr Res 63:492–496
36. Jack GS, Almeida FG, Zhang R, Alfonso ZC, Zuk PA, Rodriguez LV (2005) Processed lipoaspirate cells for tissue engineering of the lower urinary tract: implications for the treatment of stress urinary incontinence and bladder reconstruction. J Urol 174:2041–2045
37. Abdel-Hamid NM (2009) Premalignant variations in extracellular matrix composition in chemically induced hepatocellular carcinoma in rats. J Membr Biol 230:155–162

Chapter 24

Analysis of Tissue Interactions in Ectodermal Organ Culture

Pauliina M. Munne, Katja Närhi, and Frederic Michon

Abstract

The morphogenesis of ectodermal organs is regulated by epithelial mesenchymal interactions mediated by conserved signaling molecules. Analyzing the roles of these molecules will increase our understanding of mechanisms regulating organogenesis, and organ culture methods provide powerful tools in this context. Here we present two organ culture methods for skin and tooth development: the hanging drop method for the short-term culture of small explants and the Trowell-type method for the long-term cultures of variable size explants. The latter allows manipulations such as combining separated epithelial and mesenchymal tissues and the use of signal-releasing beads. The effects of signaling molecules on morphogenesis can be observed during culture by using tissues from GFP-reporter mice. After culture, the effects of signals on gene expression can be analyzed by in situ hybridization or quantitative RT-PCR.

Key words: Ectodermal organ, Tooth, Organ culture, qPCR, In situ hybridization

1. Introduction

The development of ectodermal organs such as hair, feather, tooth or mammary gland is characterized by sequential morphogenetic events such as epithelial thickening and mesenchymal condensation, budding, and complex epithelial morphogenesis, which are driven at the cellular level by proliferation, migration, and differentiation of various cell lineages. The dialogue between the ectodermal and mesenchymal tissues is of particular importance during the formation of ectodermal organs. Sequential interactions between cells and tissues lead to the formation of an ectodermal placode, and the subsequent organ morphogenesis is mediated by signaling molecules belonging to conserved families, such as fibroblast growth factor (FGF), Wnt, bone morphogenetic protein (BMP), and hedgehog (Hh). In addition, the tumor necrosis factor (TNF) family member ectodysplasin (Eda) and its receptor Edar regulate the development of all ectodermal organs. The genetic networks involving these morphogenetic signals

Scott H. Randell and M. Leslie Fulcher (eds.), *Epithelial Cell Culture Protocols: Second Edition*, Methods in Molecular Biology, vol. 945, DOI 10.1007/978-1-62703-125-7_24,

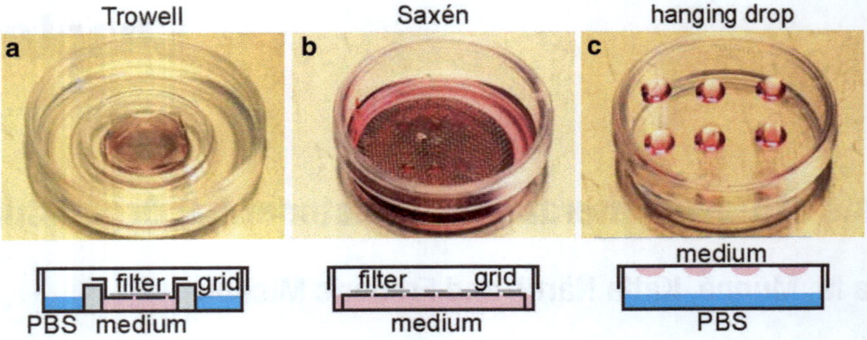

Fig. 1. Pictures of organ culture setups and their schematic representations. The original Trowell-type organ culture, the Trowell-type organ culture modified by Saxén where the metal grid supports several pieces of filters placed on the holes punched in the grid, and the hanging-drop technique.

and their specific inhibitors are particularly well documented during mouse tooth formation (1).

Various tissue culture methods have been developed to follow ectodermal organ formation: the tissue can be transplanted in vivo on the chorioallantoic membrane of a chick embryo, under the skin or the kidney capsule, or cultured in vitro. Whereas in vivo methods offer a physiological environment suitable for long-term follow-up, the in vitro culture methods allow continuous monitoring of development, enable control of the culture medium composition and manipulation of the explants.

In the Trowell-type organ culture system, the explants develop at the medium/gas interface on a filter supported by a metal grid (see Figs. 1a, b and 2) (2). This method has been used over many decades to study morphogenesis of various organs (3–8). To elucidate the molecular dialogue involved in the epithelial mesenchymal interactions during tooth formation, we have applied the Trowell technique modified by Saxén (9–18).

For short-term culture, we have recently used the so-called hanging drop culture technique (see Fig. 1c) which is commonly used to form a micromass out of a cell suspension (19). This method requires a smaller amount of medium compared to the Trowell-type culture. We used short-term hanging drop culture and quantitative real-time PCR (qPCR) to study the effects of signaling molecules on gene expression (20).

These culture methods allow different types of studies, including manipulating the amount of mesenchymal tissue during organ development (see Fig. 3), and the use of specific mutant mouse lines (see Fig. 4). Recombinant signaling molecules or their antagonists can be added to the medium, or they can be introduced using impregnated beads in order to investigate the roles of specific signal pathways in organogenesis (see Figs. 5 and 6). Epithelium–mesenchyme dissociation and reassociation between different transgenic mouse lines or different embryonic stages can be used

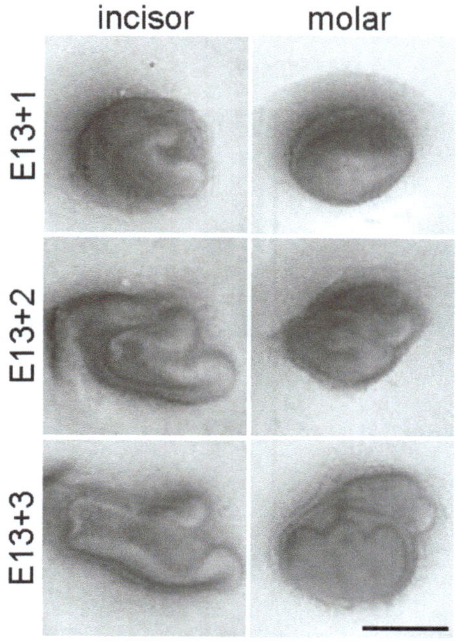

Fig. 2. Use of Trowell-type culture to follow the morphogenesis of tooth buds, molar, and incisor. The explants were dissected from E13 mouse mandibles. The explants were photographed every day using stereomicroscopy. The morphogenesis of teeth is seen as the formation of cervical loops in the incisor and the cusps in the molar. Bar: 500 μm.

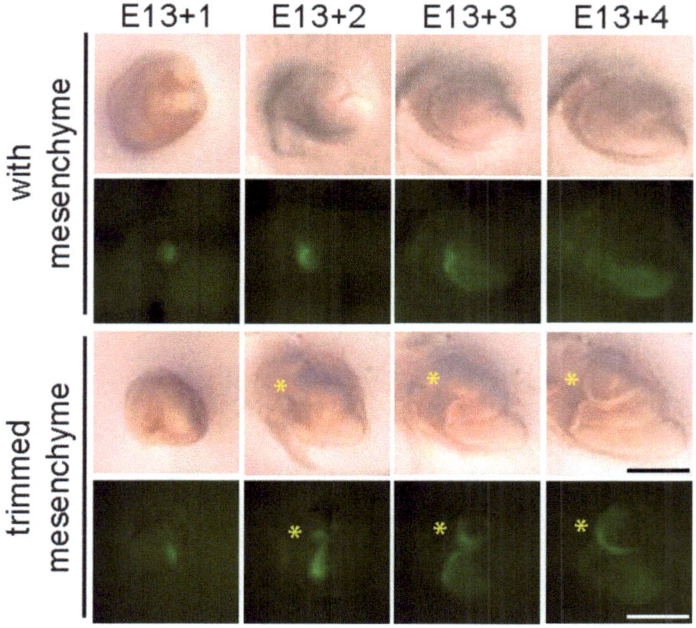

Fig. 3. The use of a transgenic mouse expressing GFP in the Shh locus allows following the formation of enamel knots (signaling centers) in the developing teeth. The trimming of the mesenchyme, surrounding the tooth bud, prior to culture, results in the formation of an extra incisor (*asterisks*) visible in fluorescent microscopy already after 2 days of culture (+2). The induction of the extra tooth results from the removal of the mesenchymally expressed BMP inhibitor Ectodin (Sostdc1) (27). Bars: 500 μm.

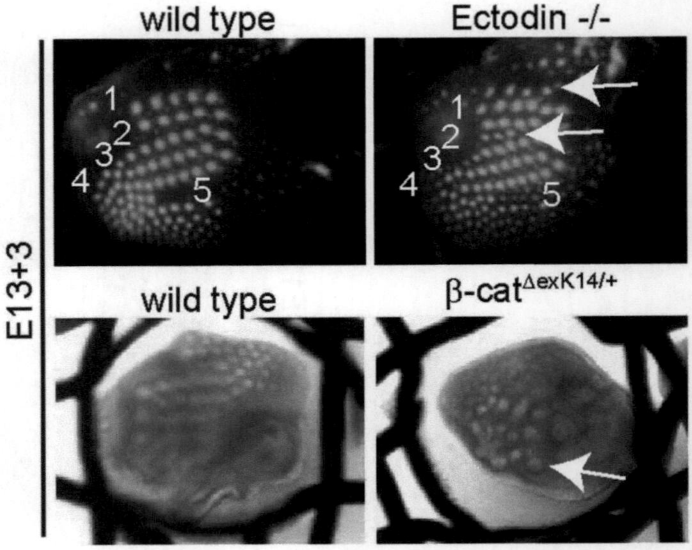

Fig. 4. The use of a transgenic mouse expressing GFP can be used to study the vibrissae development in Ectodin −/− mouse; here the GFP is in the locus of the Keratin 17. After 3 days of culture the extra vibrissae rows already formed. In light microscopy, the defect in the vibrissae patterning can be observed as well in the transgenic mouse expressing an active form of the β-catenin (β-cat$^{\Delta exK14/+}$) (29).

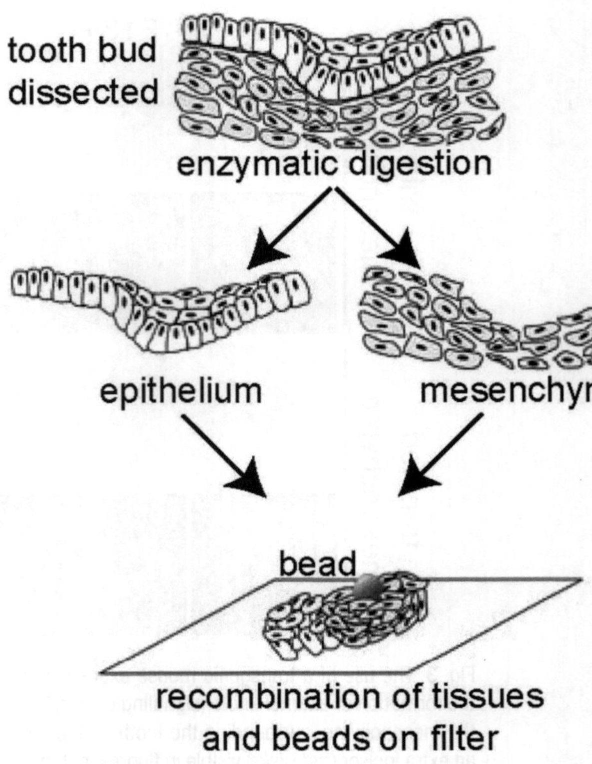

Fig. 5. Schematic representation of the separation and recombination of dental epithelium and mesenchyme and introduction of a bead releasing the factor under study.

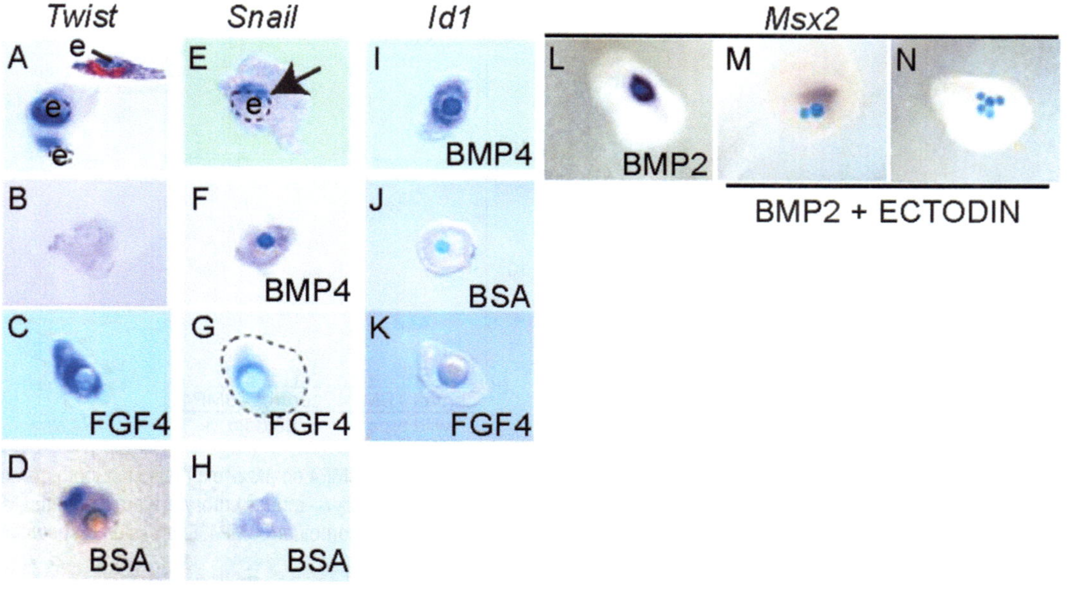

Fig. 6. Examples of recombination experiments combined to whole-mount in situ hybridization (ISH) analysis. The reassociation of E13.5 dental mesenchyme with dental epithelium leads to the expression of *Twist* (**a**) and *Snail* (**e**), whereas no Twist expression is detected without the epithelium (**b**) or with BSA (**d**). The protein-releasing beads induce the expression of specific genes: *Twist* (**c**) and *Snail* expression are by FGF4 (**g**) but not by BMP4 (**f**) or BSA (control) (**h**). *Id1* expression is induced by BMP4 (**i**) but not by BSA (control) (**j**) or FGF4 (**k**). The BMP4 induces *Msx2* expression (**l**) and this induction can be inhibited progressively by increased amounts of beads releasing the BMP inhibitor Ectodin (**m** and **n**) (**a–k** courtesy of David Rice (30) and **l–m** courtesy of Johanna Laurikkala (25)).

to elucidate the ectodermal mesenchymal timeline dialogue. Gene expression can be investigated in cultured explants by in situ hybridization (use of a riboprobe which binds the corresponding mRNA; (21)) using whole-mounts or tissue sections. Alternatively, the analysis of reporter activity, such as LacZ and GFP, in cultured tissues from transgenic mouse lines, allows gene expression analysis, cell lineage determination, and pathway activation visualization. In addition, qRT-PCR analysis allows the quantification of gene expression and its changes in response to altered signaling (see Fig. 7).

In this chapter, we describe two organ culture methods: the Trowell technique and the hanging-drop method, and tissue processing for in situ hybridization and qPCR.

2. Materials

2.1. Salt Solutions, Signals, and Culture Media

1. Phosphate-buffered saline (PBS), pH 7.4. 10× stock solution: 1.37 M NaCl, 27 mM KCl, 79 mM Na_2HPO_4, 15 mM KH_2PO_4, adjust pH if necessary with HCl. The 1× PBS working

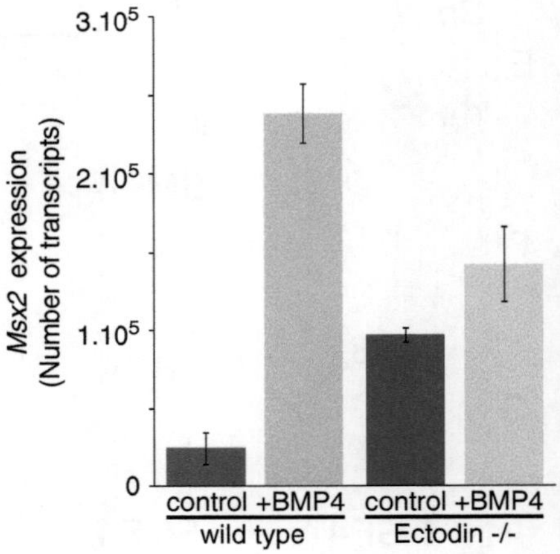

Fig. 7. Gene expression analysis of the effect of BMP4 on *Msx2* expression in tooth germs by qPCR. Tooth germs from wild-type and Ectodin –/– mouse embryos were incubated for 3 h in hanging-drop cultures with and without recombinant BMP4. Expression is shown as a number of transcripts.

solutions should be prepared by diluting 1:9 the stock in distilled water, autoclaved, and stored at room temperature.

2. Dulbecco's phosphate-buffered saline, modified (D-PBS). Store at room temperature.
3. Culture medium: Dulbecco's Modified Eagle's Medium (DMEM, Sigma-Aldrich®) supplemented with 1% (v/v) GlutaMAX-1 (Life Technologies™); aliquots stored at –20°C, 10% (v/v) heat inactivated fetal bovine serum (FBS; HyClone/Thermo Scientific, research grade EU approved); aliquots stored at –20°C, and 0.2% (v/v) penicillin 10,000 IU/mL, streptomycin 10,000 μg/mL (PS, Life Technologies™); aliquots stored at –20°C. PS is stable in medium for 1 month. Culture medium can be stored at 4°C.
4. F-12 (Ham's nutrient mixture, Life Technologies™). Store at 4°C. Use as 1:1 mixture in culture medium (see Note 1).
5. Ascorbic acid: Prepare 10 mg/mL stock solution in distilled water. Aliquots are stored at –20°C. Use 100–150 μg/mL in culture medium (see Note 1).
6. Enzyme solution for tissue separation: stock solution of Pancreatin (Sigma-Aldrich®), 1.25 g/mL; Tyrode's solution: NaCl (8.0 g), KCl (0.2 g), NaH_2PO_4 (0.05 g), glucose (1.0 g), $NaHCO_3$ (1.0 g). Adjust pH to 7.2, make up to 1 L with distilled water, and sterile filter. Store at +4°C. Dissolve 0.225 g Trypsin (Sigma-Aldrich®) in 6 mL Tyrode's solution on ice, using a magnetic stirrer. Add 1 mL Pancreatin and 20 μL PS, adjust pH to 7.4 with NaOH. Add Tyrode's solution to 10 mL and sterile filter. Aliquot 1 mL in Eppendorf tubes and store at –20°C. The enzyme solution can be stored at –20°C for 1 week.

2.2. Dissection and Culture

All glassware and metal instruments are sterilized before use. For sterilization of forceps and scissors we use Steri 250 glass bead sterilizer.

1. Dissection of tissues: 10 cm diameter plastic bacteriological Petri dishes and 10 cm diameter glass Petri dishes, small scissors, forceps, watchmaker forceps, and disposable 20- and 26-gauge needles attached to 1 mL plastic syringes (see Note 2).
2. Culture dishes: 35/10 mm plastic Petri dishes (bacteriological or cell culture dishes).
3. Metal grids: Prepare from stainless-steel mesh (corrosion-resistant, size of mesh 0.7 mm) by cutting approx. 30 mm diameter disks and bending the edges to give 3 mm height (the height of the metal grids can be altered affecting the needed amount of culture medium). Use nails to make holes in the grid to allow the analysis and photography of the explants (see Figs. 1, 2, 3, and 4). You can find commercially available organ culture dishes featuring a central well in which a metal grid (even without bent edges) can be placed (Falcon, Becton Dickinson) (see Fig. 1a).
4. Filters: 25 mm diameter Nuclepore® Polycarbonate Track-Etch Membranes (Whatman). The pore size routinely used is 0.1 μm (see Note 3). The filters are stored in 70% ethanol at room temperature.
5. Protein-releasing beads: 17–150 μm diameter Affi-Gel Blue agarose beads (Bio-Rad Laboratories) or heparin-coated acrylic beads (Sigma-Aldrich®) are divided into aliquots and stored at 4°C.
6. Glass Pasteur pipettes are used for transferring beads and tissue explants. Ideally, the diameter should be the minimal to allow free passage of the beads.
7. Stereomicroscope (e.g. Olympus SZX10) attached to a camera (e.g. Olympus DP21).
8. Pre-fixation after culture to avoid the sample to detach from the filter: Methanol, store at 4°C.
9. Fixation: 4% paraformaldehyde (PFA) (Sigma-Aldrich®) in PBS, freeze aliquots and store at −20°C. It is recommended to use fresh PFA (after thawing, store at 4°C and use within 2 days).
10. 0.1% bovine serum albumin (BSA) in PBS.

2.3. Complementary DNA Synthesis

1. RNA isolation: RNAlater (Sigma-Aldrich®; store at room temperature), Trizol® (Life Technologies™, store at 4°C), RNeasy Mini Kit (Qiagen), β-mercaptoethanol (Sigma Aldrich®, for Molecular Biology).

2. Complementary DNA (cDNA) synthesis: Random Primers (Promega; 500 μg/mL), RNase inhibitor RNasin® (Promega; 40 U/μL), dNTP mix (Finnzymes; 10 mM), Superscript™ II Reverse Transcriptase (Life Technologies™; 200 U/μL; this kit includes the 5× first strand buffer and 0.1 M DTT, as well). Store all reagents at −20°C.
3. Quantify the total RNA with UV spectroscopy, absorbance at 260 nm. Nanodrop spectrophotometer requires only 2 μL of sample and gives reliable results.

2.4. Real-Time Quantitative PCR

1. DyNAmo™ TMFlash SYBR® Green qPCR Kit (Finnzymes). Store at −20°C.
2. For miRNA real-time quantitative PCR (RT-qPCR): miScript PCR System (Qiagen).
3. Primers for sample and control cDNA (designed with Primer 3 software and ordered from Sigma-Aldrich®), store at −20°C.
4. LightCycler 480 Multiwell Plate 96 or 384 with sealing foils (Roche). Store at room temperature.
5. For running qPCR: LightCycler 480 machine (Roche).
6. For analysis of qPCR data: Lightcycler 480 software (Roche).
7. 2% agarose gels.

3. Methods

3.1. Preparation of Trowell-Type Culture Dishes

1. Take one sheet of Nucleopore filter (see Subheading 2.2, item 4.) from ethanol and rinse three times in PBS in a plastic 10 cm diameter Petri dish.
2. Cut the filter in pieces (3–5 mm² depending on the size of tissue), using small scissors and watchmaker forceps and store in PBS for several days at 4°C.
3. Place metal grids (see Subheading 2.2, item 3) in 35 mm diameter plastic Petri dishes. Add 1–3 mL culture medium (see Subheading 2.1, item 3, see Note 1) by pipetting through the grid. Avoid air bubbles. The surface of the medium should be flush with the plane of the grid. Excess medium results in floating of the filters and tissues.

3.2. Treatment of the Beads

1. Transfer agarose beads or heparin-coated acrylic beads by pipetting them into a glass Petri dish containing PBS. Count 100–200 beads under the microscope, and transfer to a non-stick Eppendorf tube with rubber bulb-controlled glass capillary pipette. Spin down the beads and remove PBS.

2. Add recombinant proteins in a small volume (10–50 μL) of 0.1% BSA (see Subheading 2.2, item 10). In general, high concentrations of proteins are used. For instance, we use 20–25 ng/μL of FGF4 and 100 ng/μL of BMP4. An equal amount of 0.1% BSA in PBS is pipetted to control beads. Incubate for 30 min at 37°C and store at 4°C. The beads can be used at least for 14 days (depending on the stability of the protein).

3.3. Dissection of Tissues

1. Place the mouse uterus (E11–E18) in a 10 cm diameter plastic Petri dish containing D-PBS, and cut open the uterine wall using small scissors and forceps. Under stereomicroscope remove the embryos from fetal membranes, and transfer them to a fresh plastic Petri dish with D-PBS. Cut off the heads using disposable needles (or with scissors when dissecting older embryos) and transfer the heads to a 10 cm diameter glass Petri dish containing D-PBS.
2. Dissect the tissue piece of interest using needles or iris knives (see Note 2): mandible, vibrissa pads, or tooth buds (see Notes 4 and 5). If tissues are cultured using hanging drop technique continue as described below (see Subheading 3.5). For hanging drop technique, it is essential to remove all the extra tissue surrounding the tissue of interest to avoid skewed data in real-time qPCR.
3. Pipet warm (37°C) culture medium (see Subheading 2.1, item 3) to a 35-mm plastic Petri dish with a grid. Transfer the dissected tissue pieces on the metal gauze by lifting with a filter piece and watchmaker forceps. Alternatively, the explants can be transferred by the capillary force between the tips of watchmaker forceps or with the Pasteur pipette or with a spatula and placed on filters lying on the metal grid.
4. Add signaling molecules (see Note 6) either directly to culture medium or introduce them with beads soaked in high concentration of molecules. Under the stereomicroscope, transfer the beads one at a time to the tissues. Depending on the experiment and tissue, 1–5 beads can be placed on one explants (see Fig. 5). Examples of BMP4-bead experiments are shown in Fig. 6 (see Note 7).

3.4. Separation of Epithelial and Mesenchymal Tissues

1. With a drawn Pasteur pipet and rubber bulb, transfer the dissected organs to a small culture dish. Remove most of the liquid. Thaw an aliquot of Pancreatin–Trypsin (see Subheading 2.1, item 6) and spin immediately for 30 s at 8,000 ×*g*. Add cold supernatant on the samples and incubate 2–10 min at room temperature or 37°C. Remove most of the liquid, add the culture medium to stop the enzymatic reaction, mix, and transfer

the organs to a glass Petri dish containing culture medium. Leave the tissues for 5–30 min at room temperature.

2. Gently separate the epithelia from the mesenchymes using needles (see Note 8). Transfer the tissues on the Nucleopore filters in culture dishes that have been prepared in advance (see Subheading 3.1, step 1). Avoid air bubbles in the pipet and avoid drawing the tissue beyond the capillary part of the pipet. Ideally, the tissues should be placed directly in their final position, but if needed, they can be gently adjusted with needles (see Fig. 2).

3.5. Hanging Drop Cultures

1. Pipet 20–50 μL drops of warm culture medium on the lid of a 35 mm plastic Petri dish. Different signaling molecules or other molecules can be added to culture medium. The culture medium used as control of the experiment should be supplemented with the solvent used to dissolve signaling molecules.
2. Transfer the dissected tissue samples carefully to the drops using the capillary force between watchmaker forceps.
3. Turn the lid quickly and place on top of the Petri dish containing 1–2 mL of sterile liquid (PBS or distilled water) in the bottom to prevent evaporation from the hanging drop.
4. Culture as described below (see Subheading 3.6 and Note 9). Hanging drop culture can be maintained only 1–3 days, whereas culture on filters can last up to 6 weeks (see Note 10).

3.6. Culture and Fixation

1. Culture the tissues in a standard incubator at 37°C, in an atmosphere of 5% CO_2 in air and 90–95% humidity. The culture medium should usually be changed every second day (see Note 10).
2. On a daily basis, photograph the explants with a camera attached to the stereomicroscope (see Note 11).
3. Concerning the hanging drop culture, replace the medium with 100 μL of PBS. Aspirate the organ with a Pasteur pipet or use a spatula to transfer it in Eppendorf tube then proceed as follows (see Subheading 3.6, step 5).
4. Concerning the Trowell culture, aspirate the medium, and pipet ice-cold methanol (pre-fixation) gently on the tissues to avoid detachment of tissues from the filters. Leave for 5 min and transfer filters by watchmaker forceps to Eppendorf tubes then proceed as follow.
5. Fix the explants for in situ hybridization, in 4% PFA for 10–24 h at 4°C. Continue with gene expression analysis.

3.7. In Situ Hybridization

1. To perform in situ hybridization (ISH) on tissue sections, process tissue explants to paraffin using standard protocols.

Cut serial sections of paraffin-embedded tissue explants and process for ISH to analyze the expression of genes of interest by using ^{35}S-UTP- or digoxigenin-labeled riboprobes. ISH is performed according to a protocol described in ref. (21) with modifications (12, 22).

2. For whole-mount ISH (with digoxigenin-labeled riboprobes) the tissue explants are rinsed in PBS and dehydrated in methanol series 25, 50, 75, and 100% (dilutions in PBS; each step 5–10 min at room temperature) after fixation. Wash with 100% methanol twice and store samples in 100% methanol at −20°C until use (can be stored for several months). Proceed for ISH according to instructions described ((13, 23, 24); see Note 12). Examples of whole-mount ISH staining are shown in Fig. 6.

3.8. Total RNA Isolation and cDNA Synthesis

1. After culture, the tissue samples are collected under microscope into Eppendorf tube containing RNAlater (see Subheading 2.3, item 1). The samples can be stored at −20°C up to 1 month without notable loss of the RNA content (see Note 13).
2. Discard the RNAlater from the tube and homogenize the samples in 100 μL of Trizol® per 5–10 mg of tissue, using 1 mL syringe and needle. Incubate 5 min at room temperature.
3. Add 20 μL of chloroform per 100 μL of Trizol®, shake the tube 15 s and incubate 3 min at room temperature prior to centrifugation at 12,000 × *g* for 10 min, 4°C.
4. In a new tube, add the upper aqueous phase and 50 μL of cold isopropanol per 100 μL of Trizol®, and after 10 min incubation, centrifuge at 12,000 × *g* for 1 h, 4°C.
5. Wash the pellet with 100 μL of 75% ethanol per 100 μL of Trizol®, vortex and centrifuge at 7,500 × *g* for 5 min, 4°C. Then, briefly dry the pellet and dissolve it in 50 μL of RNase-free water before cleaning it with the RNeasy mini kit, according to the manufacturer's instructions. Store RNA samples at −80°C after RNA quantification (see below) and avoid repeated thawing (use aliquots).
6. Quantify the total RNA with UV spectroscopy, absorbance at 260 nm (see Subheading 2.3, item 3).
7. Synthesize complementary DNA (cDNA) from total RNA according to instructions specified by manufacturer (see Subheading 2.3, item 2 and Subheading 2.4) Transcribe 50–1,000 ng of total RNA. Measure with the Nanodrop the cDNA (at this step we have a single strand DNA) obtained after the reverse transcription and dilute the cDNA to have the proper amount of cDNA in 5 μL for the quantitative PCR and freeze in aliquots at −20°C.

3.9. Quantitative Polymerase Chain Reaction (qPCR)

1. Prepare standard series (see Note 14) and 10× (e.g., 1–5 μM) primer mix of forward and reverse primers for each gene to be studied.
2. For miRNA qPCR, follow the manufacturer's instructions from the miScript qPCR System kit.
3. For mRNA qPCR, prepare master mix: 2 μL 10× primer mix, 3 μL H_2O, 10 μL 2× LightCycler 480 SYBR Green I Master. This is for one reaction, so, multiply the volumes of each reagent by number of samples (sample cDNA and standards). Also, note that you have to prepare a master mix for each gene to be studied. Mix well the prepared master mixes and keep them on ice and protected from light as they contain the SYBR Green fluorophore.
4. Pipet 5 μL of cDNA or standard into the wells.
5. Pipet 15 μL of master mix in to each well of the plate. The final volume is 20 μL in each well.
6. Seal the plate with a transparent adhesive foil and use the default PCR conditions for Lightcycler 480.
7. Normalize data against Actin or Ranbp1 for mRNA qPCR, or U6 for miRNA qPCR and analyze with Lightcycler 480 software. Gene expression is quantified by comparing the Sample data against a dilution series of PCR products (see Note 14) of the gene of interest (see Fig. 7).
8. All PCR products can be separated on a 2% agarose gel using electrophoresis to check for the correct size of the PCR product and to eliminate the possibility of primer dimers.

4. Notes

1. Depending on the tissue, the composition of the optimal culture medium varies. The culture medium in this protocol is suitable for most tissues at early stages of development, but for more advanced stages, different tissues may have special requirements. For cultures of whole tooth buds, we use culture medium composed of DMEM and F12, 1:1. For more advanced stages of tooth or calvarial bone development, ascorbic acid is added to allow deposition of collagen (8, 11).
2. Dissection can be performed with reusable tools such as scalpels and iris knives, or disposable instruments like needles. The use of needles could be preferred because there is no need to sharpen or sterilize them. The size of the needles can be chosen. In addition, the syringes do not need to be changed every time as they do not have to be absolutely sterile. To preserve

the tissue structure, dissecting should be done by determined cuts avoiding stretching and tearing of tissue. Glass Petri dishes are superior to plastic dishes because needles easily scrape and loosen pieces from the plastic surface.

3. As supporting material, lens paper may be used for large tissue pieces. Nucleopore filters with different pore sizes (0.05–8 μm) may also be used. The maximum thickness of filters is approx. 10 μm which allows good diffusion of the medium to the tissue. Small pores (0.05–0.2 μm) allow better examination of the explants in the stereomicroscope using transmitted light, but the tissues tend to detach from these filters more readily during treatments after culture. Therefore, larger pore size (up to 0.6 μm) may be preferable, depending on the experiment.
4. The preparation and dissection of tissues should be done as quickly as possible to promote survival of the tissues. Embryos waiting to be dissected should be kept on ice in a Petri dish with D-PBS and only one uterus at a time should be prepared. The dissected tissues should be transferred to the culture dishes and the incubator within 2–3 h. It should be kept in mind that the normal development could be slightly perturbed at the beginning of the culture due to the excision stress. It has been shown for example that at early stage of feather induction, the chick skin was losing the primordia and their development restarted from the beginning (19).
5. To study the reciprocal interactions between epithelium and mesenchyme, the interacting tissues need to be enzymatically separated from each other. Various manipulations can be performed after which their advancing development is followed. Separation of epithelium and mesenchyme is described under Subheading 3.4.
6. Signaling molecules include proteins, growth factors, and their antagonists (25), as well as other molecules for example retinoic acid (14).
7. Experimental and control beads can be placed on opposite halves of the jaw. Placing beads releasing different signals, or a signal and its putative antagonist, near each other on the same explant, enables monitoring antagonistic and synergistic functions of different regulatory molecules.
8. Mechanical dissociation, using a pipet with a 200 μL tip, can follow epithelial–mesenchymal separation in order to use cells in a primary cell culture. Extended treatment will lead to cell damage and RNA destruction, which could perturb gene expression analysis via quantitative RT-PCR.
9. In order to culture the organ in a hanging drop, place with a spatula the organ on the internal side of the lid, immediately pipet the warm medium on the top of it in order to form the

drop. With less than 50 µL, the drop will stay steady when the lid is turned back fast enough on the Petri dish.

10. Although organs are usually maximally cultured for 1–2 weeks, it is possible to culture tooth buds for 3–6 weeks to examine the formation of roots (26).

11. Reporter mouse lines carrying green fluorescent protein (GFP) labeled reporter constructs are useful if one wishes to follow the development of a specific structure or gene expression during advancing development in vitro. GFP expression is detected using fluorescence stereomicroscopy. For example, a mouse line expressing GFP in the Shh locus (see Fig. 3) is useful for tooth development studies because it allows visualizing the dynamics of the formation of enamel knot signaling centers, which express Shh locally (27). A mouse expressing the GFP in the K17 locus (see Fig. 4) allows following the formation of vibrissae.

12. After whole-mount ISH analysis, it is possible to analyze the gross morphology of the explants (e.g., to differentiate between epithelial vs. mesenchymal staining) by cutting vibratome sections. For this purpose explants are fixed for 24 h in 4% PFA, rinsed in PBS, embedded in gelatin/albumin, and thick sections (30–200 µm) are cut by vibratome (24).

13. In hanging drops, several small tissue pieces can be placed in one drop (depending on the size of the tissue). For instance, three to four E13 tooth germs or E11 mandibles fit well into one drop if the culture time is short. Otherwise, after 24 h of culture the explants have the tendency to fuse, leading to non-physiological development.

14. We use standard series of 10^6, 10^5, 10^4, and 10^3 amplicons in 5 µL. It has to be prepared for all the genes to be studied. To calculate the number of molecules of a PCR amplicon you need to know the length (bp) and the concentration (g/µL) of the amplicon. First, calculate the molecular weight (length × 650 g/mol) of one amplicon. Second, calculate the amount of amplicons in 1 L (concentration g/L × 10^6 µL) and then divide the product with the molecular weight to resolve the concentration of amplicon as mol/L. By using Avogadro's number (6.022×10^{23}), calculate the number of molecules in a liter and further in a microliter. For micro-RNAs qRT-PCR, due to the size of the amplicon (around 22 nucleotides), it is impossible to obtain a standard series. The most used method is the $2^{-\Delta\Delta Ct}$ method (28). This will give the relative expression level of an miRNA given compared to a small nuclear RNA, such as U6, for which you should have standard series.

References

1. Tummers M, Thesleff I (2009) The importance of signal pathway modulation in all aspects of tooth development. J Exp Zool (Mol Dev Evol) 312B(4):309–319
2. Trowell OA (1959) The culture of mature organs in a synthetic medium. Exp Cell Res 16:118–147
3. Grobstein C (1953) Inductive epithelio-mesenchymal interaction in cultured organ rudiments of the mouse. Science 118:52–55
4. Saxén I (1973) Effects of hydrocortisone on the development in vitro of the secondary palate in two inbred strains of mice. Arch Oral Biol 18:1469–1479
5. Saxén L, Lehtonen E, Karkinen-Jääskeläinen M, Nordling S, Wartiovaara J (1976) Morphogenetic tissue interactions: mediation by transmissible signal substances or through cell contacts? Nature 259:662–663
6. Nogawa H, Takahashi Y (1991) Substitution for mesenchyme by basement-membrane-like substratum and epidermal growth factor in inducing branching morphogenesis of mouse salivary epithelium. Development 112:855–861
7. Nogawa H, Ito T (1995) Branching morphogenesis of embryonic mouse lung epithelium in mesenchyme-free culture. Development 121: 1015–1022
8. Kim H-J, Rice DPC, Kettunen PJ, Thesleff I (1998) FGF-, BMP- and Shh-mediated signaling pathways in the regulation of cranial suture morphogenesis and calvarial bone development. Development 125:1241–1251
9. Saxén L (1966) The effect of tetracycline on osteogenesis in vitro. J Exp Zool 162:269–294
10. Thesleff I, Lehtonen E, Wartiovaara J, Saxén L (1977) Interference of tooth differentiation with interposed filters. Dev Biol 58:197–203
11. Partanen AM, Ekblom P, Thesleff I (1985) Epidermal growth factor inhibits tooth morphogenesis and differentiation. Dev Biol 111: 84–94
12. Vainio S, Karavanova I, Jowett A, Thesleff I (1993) Identification of BMP-4 as a signal mediating secondary induction between epithelial and mesenchymal tissues during early tooth development. Cell 75:45–58
13. Jernvall J, Aberg T, Kettunen P, Keranen S, Thesleff I (1998) The life history of an embryonic signaling center: BMP-4 induces p21 and is associated with apoptosis in the mouse tooth enamel knot. Development 125:161–169
14. Mitsiadis T, Muramatsu T, Muramatsu H, Thesleff I (1995) Midkine (MK), a heparing-binding growth/differentiation factor, is regulated by retinoic acid and epithelial-mesenchymal interactions in the developing mouse tooth, and affects cell proliferation and morphogenesis. J Cell Biol 129:267–281
15. Laurikkala J, Mikkola M, Mustonen T, Åberg T, Koppinen P, Pispa J, Nieminen P, Galceran J, Grosshedl R, Thesleff I (2001) TNF signaling via the ligand-receptor pair ectodysplasin and edar controls the function of epithelial signaling centers and is regulated by Wnt and activin during tooth organogenesis. Dev Biol 229:443–455
16. Wang X-P, Suomalainen M, Felszeghy S, Zelarayan LC, Alonso MT, Plikus MV, Maas RL, Chuong CM, Schimmang T, Thesleff I (2007) An integrated gene regulatory network controls stem cell proliferation in teeth. PLoS Biol 5:1324–1333
17. Järvinen E, Salaar-Ciudad I, Birchmeier W, Taketo MT, Jernvall J, Thesleff I (2006) Continuous tooth regeneration in mouse in induced by activated epithelial Wnt/β-catenin signaling. Proc Natl Acad Sci USA 103: 18627–18632
18. Rice D, Åberg T, Chan Y-S, Kettunen P, Pakarinen L, Maxson RE, Thesleff I (2000) Integration of FGF and TWIST in calvarial bone and suture development. Development 127:1845–1855
19. Michon F, Charveron M, Dhouailly D (2007) Dermal condensation formation in the chick embryo: requirement for integrin engagement and subsequent stabilization by a possible notch/integrin interaction. Dev Dyn 236: 755–768
20. Fliniaux I, Mikkola ML, Lefebvre S, Thesleff I (2008) Identification of dkk4 as a target of Eda-A1/Edar pathway reveals an unexpected role of ectodysplasin as inhibitor of Wnt signalling in ectodermal placodes. Dev Biol 320: 60–71
21. Wilkinson DG, Green J (1990) In situ hybridization and the three-dimensional reconstruction of serial sections. In: Copp AJ, Cockroft DL (eds) Postimplantation mammalian embryos: a practical approach. IRL, Oxford, pp 155–171
22. Vaahtokari A, Vainio S, Thesleff I (1991) Associations between transforming growth factor β1 RNA expression and epithelial-mesenchymal interactions during tooth morphogenesis. Development 113:985–994
23. Kettunen P, Thesleff I (1998) Expression and function of FGFs-4, -8, and -9 suggest functional

redundancy and repetitive use as epithelial signals during tooth morphogenesis. Dev Dyn 211:256–268

24. Mustonen T, Tümmers M, Mikami T, Itoh N, Zhang N, Gridley T, Thesleff I (2002) Lunatic fringe, FGF, and BMP regulate the Notch pathway during epithelial morphogenesis of teeth. Dev Biol 248:281–293
25. Laurikkala J, Kassai Y, Pakkasjärvi L, Thesleff I, Itoh N (2003) Identification of a secreted BMP antagonist, ectodin, interacting BMP, FGF, and SHH signals from the tooth enamel knot. Dev Biol 264:91–105
26. Tümmers M, Yamashiro T, Thesleff I (2007) Modulation of epithelial cell fate of the root in vitro. J Dent Res 86:1063–1067
27. Munne P, Tummers M, Järvinen E, Thesleff I, Jernvall J (2009) Tinkering with the inductive mesenchyme: Sostdc1 uncovers the role for dental mesenchyme in limiting tooth induction. Development 136:393–402
28. Chen Y, Gelfond JA, McManus LM, Shireman PK (2009) Reproducibility of quantitative RT-PCR array in miRNA expression profiling and comparison with microarray analysis. BMC Genomics 10:407
29. Närhi K, Järvinen E, Birchmeier W, Taketo MM, Mikkola ML, Thesleff I (2008) Sustained epithelial beta-catenin activity induces precocious hair development but disrupts hair follicle down-growth and hair shaft formation. Development 135:1019–1028
30. Rice R, Thesleff I, Rice DP (2005) Regulation of Twist, Snail, and Id1 is conserved between the developing murine palate and tooth. Dev Dyn 234:28–35

Chapter 25

Generation of Human Cell Lines Using Lentiviral-Mediated Genetic Engineering

Patrick Salmon

Abstract

Even now, most human cell lines used in research are derived from tumor cells. They are still widely used because they grow well in vitro and so far have helped answering several basic biological questions. However, as modern biology moves into more sophisticated areas, scientists now need human cell lines closer to normal primary cells and further from transformed cancerous cells. The recent identification of cellular genes involved in cell cycling and senescence, together with the development of molecular tools capable of cleanly integrating transgenes into the genome of target cells, have moved the frontier of genetic engineering. In this chapter, we present a detailed hands-on protocol, based on lentivirus-derived vectors and a combination of two native cellular genes that has proven very efficient in generating immortal cell lines from several human primary cells, while preserving most of their original properties.

Key words: Lentivirus, Vector, Lentivector, Gene transfer, Genetic engineering, Cell engineering, Cell line

1. Introduction

1.1. A Brief History of Human Cell Lines

The first deposited cell line (ATCC number CCL-1) was described in 1943 from a chemically induced mouse mammary tumor (1). The second deposited cell line (ATCC number CCL-2) is the famous HeLa cell line that was isolated in 1952 from a cervix carcinoma (2). Since then, several human cell lines have been deposited, most of them after in vitro expansion of human tumor cells. These naturally occurring tumor cells have arisen after unknown or partially known transformation events that have conferred them with the desirable features for a cancerous cell, i.e., autonomous proliferation, lack of senescence, lack of contact inhibition, anchorage-independent growth, angiogenesis, decrease or lack of MHC antigen expression, etc. The sum of these successive selection processes in vivo results in the "transformation" of a normal cell into a

Scott H. Randell and M. Leslie Fulcher (eds.), *Epithelial Cell Culture Protocols: Second Edition*, Methods in Molecular Biology, vol. 945, DOI 10.1007/978-1-62703-125-7_25,

cancerous cell. From the biologist standpoint, the main advantage is the robust growth in vitro. Indeed, these cell lines have helped define the composition of culture media, and thus they have blazed the trail of modern cell culture. The inconvenience is that, in most cases, we do not know what molecular event (or events) has occurred to render these cells tumorigenic. In almost all cases, they display an abnormal and unstable karyotype, have lost several features of their tissue of origin, and have lost the ability to differentiate. Of note, some widely used human cell lines, such as the HT-1080 cells (ATCC number CCL-121) display a near normal karyotype. This may be due to the fact that their transformation event was the activation of the Ras oncogene (3).

One of the first "genetic engineering" of human cells was the transformation of human embryonic kidney cells by transfection of DNA from Adenovirus type 5 (4). The resulting cell line is the famous 293 cell line (ATCC number CRL-1573). Although 293 cells are polyploid and tumorigenic, they are relatively stable and are used for many biological studies as well as further genetic engineering by the introduction of other transgenes. Although the procedure leading to the generation of the 293 cell line can be considered as "crude" by present standards, this type of experiment paved the way to modern genetic engineering.

The first description of "viral genetic engineering" was the transformation of rodent cells by transduction of SV40 large T antigen using a retroviral vector (5). It was rapidly followed by retroviral transduction of human cells of various origins with combinations of oncogenes, including SV40 large T, Ras and/or HPV16 E6 and E7 genes. The use of these genes generated countless cell lines from various tissues and animals, but there always seemed to be an issue of genomic or phenotypic stability. HPV16 E6 and E7 genes immortalized and induced genomic instability in human bronchial epithelial cells (6), human mesenchymal cells (7), and pancreatic duct epithelial cells (8). Since it has been shown that HeLa cells are expressing HPV16 E6 and E7 genes, this is the likely reason why they are instable and have a chromosome number ranging from 70 to 164. This genomic instability is best illustrated by SV40 large T antigen, that, when used to induce or enhance cell proliferation, invariably generated genomic instability and dedifferentiation (9–11). The immortalizing capacity of SV40 large T antigen is generally attributed to its inhibitory association with p53, which relieves the cell cycle G1 checkpoint (12). However, p53 also controls apoptosis following DNA damage (13). This could explain why introduction of SV40 large T antigen is invariably followed by genomic instability and chromosomal aberrations.

This issue of immortalization versus genomic stability and maintenance of a differentiated phenotype is crucial and is best summarized by the paradigm of human pancreatic beta-cells. Unlimited supplies of human pancreatic beta-cells would represent a major breakthrough towards the cure of diabetes. However, even though some rodent beta-cells are available, originating either from X-ray-induced tumors (INS-1) (14) or from transgenic expression of SV40 large T antigen such as MIN6 cells (15) or βTC-tet cells (16), no human beta-cell line is available yet. In early experiments, while trying to generate a human beta-cell line, Wang et al. described the isolation of an epithelial cell line from human fetal pancreas, using retroviral transduction of SV40 large T and activated Ras (17). Using the same approach, neither fetal nor adult human pancreatic cells could generate long-term insulin-producing cells. Insulin expression was gradually lost and never recovered, a likely consequence of genomic instability or dedifferentiation induced by the strong proliferative signals provided by these two oncogenes (18).

In the 1990s, the issue of cell senescence was elucidated by the discovery that the telomerase enzyme could prolong life-span of cells by delaying telomere shortening, hence "resetting" the aging clock of the cells (19, 20). Moreover, it was shown soon after that ectopic expression of telomerase alone was not associated with transformation or tumorigenicity (21, 22). Prevention of telomere shortening was, however, not sufficient for in vitro expansion of most human cells. One needed to deliver another signal to force cells to enter cell cycle or to keep them cycling.

A paradigm of "mild" cell cycle promoter was provided by the study of Jacobs et al., who showed that Bmi-1, a transcriptional repressor belonging to the Polycomb group, and downregulating expression of the ink4a locus genes p16 and p19Arf, could immortalize fibroblasts (23). Following this report, several immortal human cell lines have been created using the combination of Bmi-1 and telomerase, amongst those, vascular endothelial cells (24), myoblasts (10), cementoblast progenitors (25), mesenchymal cells (26, 27), skin, mammary, and lung epithelial cells (28, 29), skin fibroblasts (30), and glial cells (31), all with preserved and stable phenotype, and no or minimal genomic alterations, even after extended culture periods. Interestingly, Bmi-1 has also been shown to correlate with stem cell maintenance of normal and cancer cells, and its expression is linked to poor prognosis in human tumors (32–34). From a cell engineer standpoint, this latter feature has three implications. First, on one hand, it may prove useful if successful immortalization of a given cell type requires a certain level of "stemminess." Second, it may be detrimental if "stemminess" prevents full terminal differentiation of the desired cells. And last but not least, it implies that viral vectors carrying Bmi-1 must be handled with special precautions.

Taken together, it is now obvious that the generation of all these cell lines using molecular tools to introduce this specific combination of Bmi-1 and telomerase has paved the way to "molecular immortalization" of many other human cell types.

1.2. HIV-1-Based Lentiviral Vectors as Molecular Tools for Genetic Engineering

Retroviral vectors have three characteristics of a highly attractive gene delivery system. First, they integrate their genetic cargo into the chromosome of the target cell, a likely prerequisite for long-term expression. Second, they have a relatively large capacity, close to 10 kb, allowing the delivery of most cDNAs. Finally, they do not transfer sequences that encode for proteins derived from the packaging virus, thus minimizing the risk that vector-transduced cells will be attacked by virus-specific cytotoxic T lymphocytes.

An infectious retroviral particle comprises an RNA genome that carries *cis*-acting sequences necessary for packaging, reverse transcription, nuclear translocation, and integration, as well as structural proteins encoded by the gag and env genes, and the enzymatic products of the pol gene. The assembly of these components leads to the budding of the virion at the plasma membrane of the producer cell. In lentiviruses, the efficient expression of Gag and Pol requires a virally encoded posttranscriptional activator called Rev.

Conventional retroviral vectors, however, are of limited usefulness for many applications because they are derived from oncoretroviruses such as the mouse leukemia virus (MLV), and, as a consequence, cannot transduce nondividing cells. In contrast to oncoretroviruses, lentiviruses, such as the human immunodeficiency virus (HIV), are a subfamily of retroviruses that can infect both growth-arrested and dividing cells, a feature that can be essential when trying to immortalize slow or nondividing primary cells. The proof-of-principle of this concept was first provided with vectors derived from HIV-1, using the adult rat brain as an in vivo paradigm (35). Since then, gene delivery systems based on animal lentiviruses such as the simian and feline immunodeficiency viruses (SIV and FIV) and the equine infectious anemia virus (EIAV) have been described. This chapter presents exclusively the HIV-1-based vector system because it is presently the most advanced, efficient and versatile, and because, in its latest version, it offers a level of biosafety that matches, if not exceeds, that of the MLV-derived vectors currently used in the clinic.

The envelope protein (Env) mediates the entry of the viral particle into its target. HIV-1 Env specifically recognizes CD4, a molecule present on the surface of helper T cells, macrophages, and some glial cells. Fortunately, as with all retroviruses, the HIV-1 Env can be substituted by the corresponding protein of another virus. This process, which alters the tropism of the virion, is called pseudotyping. More selective tropisms were achieved by taking advantage of the natural tropisms of glycoproteins from other membrane-enveloped viruses. Although many different pseudotyped vectors have been generated using glycoproteins from other membrane-enveloped viruses, the G protein of vesicular stomatitis virus glycoprotein (VSV-G) is the most widely used envelope to pseudotype lentivectors for three reasons. First, VSV-G provides the highest vector titers. Second, it provides the highest vector stability, allowing for the concentration of vector particles by ultracentrifugation. And third, with its phospholipid receptor being ubiquitously expressed in mammalian cells, lentivectors pseudotyped with VSV-G can be used to transduce virtually all cell types. This technique is thus widely and routinely used in basic research as well as in clinical research and this chapter focuses on production of the VSV-G-pseudotyped vectors.

HIV is a human pathogen. However, its pathogenic potential stems from the presence of nine genes that all encode for important virulence factors. Fortunately, six of these genes (namely Env, Vif, Vpr, Vpu, Nef, and Tat, see Fig. 1) can be deleted from the HIV-derived vector system without altering its gene-transfer ability. The resulting multiply-attenuated design of HIV vectors ensures that the parental virus cannot be reconstituted. In addition, other improvements have been brought to achieve high levels of efficiency and biosafety.

When producing vector stocks, it is mandatory to avoid the emergence of replication-competent recombinants (RCRs). In the retroviral genome, a single RNA molecule that also contains critical *cis*-acting elements carries all the coding sequences. Biosafety of a vector production system is therefore best achieved by distributing the sequences encoding its various components over as many independent units as possible to maximize the number of recombination events that would be required to recreate a replication-competent virus. In the lentiviral vector systems described here, vector particles are generated from three or four separate plasmids, each providing either the genomic RNA containing the transgene and all the *cis*-acting sequences, or the internal structural and enzymatic proteins necessary for adequate transcription, packaging, reverse transcription and integration, or the envelope glycoprotein necessary for vector entry into the target cell. This ensures that only replication-defective viruses are produced, because the plasmids would have to undergo multiple and complex recombination events to regenerate a replication-competent entity. A diagram of the evolution of HIV-1-based systems is depicted in Fig. 1.

The first generation of lentiviral vectors was manufactured using a packaging system that comprised all HIV genes but the envelope (35). In a so-called second generation system, five of the nine HIV-1 genes were eliminated, leaving the gag and pol reading frames, which encode for the structural and enzymatic components of the virion, respectively, and the tat and rev genes, fulfilling transcriptional and posttranscriptional functions (36). Sensitive tests have so far failed to detect RCRs when this system is used. This good safety record, combined with its high efficiency and ease of use, explains why the second generation lentiviral vector packaging system is utilized for most experimental purposes. In a third generation system, geared up towards clinical applications, only gag, pol, and rev genes

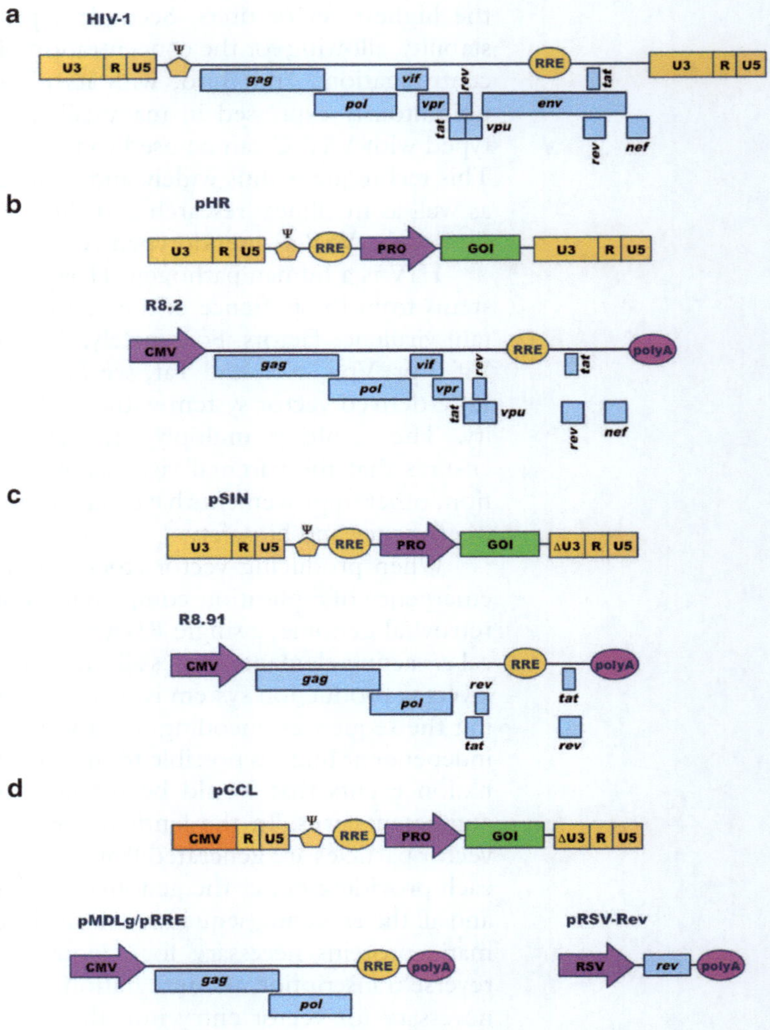

Fig. 1. Evolution in the design of HIV-1-based lentivirus vectors (LVs). HIV-1-based LVs are derived from wild-type HIV-1 (**a**) by dissociation of the *trans*-acting components (*blue boxes*) coding for structural and accessory proteins (gag, pol, env, tat, rev, vif, vpr, vpu, nef) and the *cis*-acting sequences required for packaging and reverse transcription of the genomic RNA (LTR U3-R-U5, psi, RRE) (*yellow boxes*). (**b**) First generation system. The pHR vector genome has intact 5′ LTR and 3′ LTR. The R8.2 packaging plasmid expresses all HIV-1 proteins except Env. (**c**) Second generation system. The pSIN vector genome has a self-inactivating (SIN) deletion in the U3 sequence of the 3′ LTR. The R8.91 packaging plasmid expresses only the structural and regulatory proteins of HIV-1. (**d**) Third generation system. The pCCL vector genome has a chimeric 5′ LTR that is independent of the Tat protein. The packaging system is composed of two plasmids, pMDLg/pRRE coding of the structural proteins of HIV-1 and pRSV-Rev providing the Rev protein. Note that all vector systems need the presence of complementary plasmid providing the env gene. *LTR* long-terminal repeat, *CMV* human cytomegalovirus immediate-early promoter, *RRE* rev-responsive element, *RSV* Rous sarcoma virus promoter, *polyA* polyadenylation site, *U3-R-U5* HIV-1 LTR, *psi* HIV-1 packaging signal, *PRO* virus promoter of the internal expression cassette, *GOI* transgene of interest, *deltaU3* self-inactivating deletion of the U3 part of the HIV-1 LTR.

are still present, using a chimeric 5′ LTR (long-terminal repeat) to ensure transcription in the absence of Tat.

The genetic information contained in the vector genome is the only one transferred to the target cells. Early genomic vectors were composed of the following components. The 5′ LTR, the major splice donor, the packaging signal (encompassing the 5′ part of the gag gene), the Rev-responsive element (RRE), the envelope splice acceptor, the internal expression cassette containing the transgene, and the 3′ LTR. In the latest generations, several improvements have been introduced. The Woodchuck Hepatitis Virus Posttranscriptional Regulatory Element (WPRE) has been added to increase the overall levels of transcripts both in producer and target cells, hence increasing titers and transgene expression (37). The central polypurine tract of HIV has also been added back in the central portion of the genome of the transgene RNA (38, 39). This increases titers at least in some targets. The U3 region of the 3′ LTR is essential for the replication of a wild-type retrovirus, since it contains the viral promoter in its RNA genome. It is dispensable for a replication-defective vector and has been deleted to remove all transcriptionally active sequences, creating the so-called self-inactivating (SIN) LTR (40). SIN vectors are thus unable to reconstitute their promoter and are safer than their counterparts with full-length LTRs. Finally, chimeric 5′ LTRs have been constructed, in order to render the lentiviral promoter Tat-independent. This has been achieved by replacing the U3 region of the 5′ LTR with either the CMV enhancer (CCL LTR) or the corresponding Rous sarcoma virus (RSV) U3 sequence (RRL LTR) (41). Vectors containing such promoters can be produced at high titers in the absence of the Tat HIV transactivator. However, the Rev-dependence of these third generation lentiviral vectors (lentivectors, LVs) has been maintained, in order to maximize the number of recombination events that would be necessary to generate an RCR. This latest generation represents the system of choice for future therapeutic projects. In the laboratory and in vitro genetic engineering, however, this third generation is not mandatory, and the second generation system offers a high level of safety for BSL-2 conditions. For most research applications, it is thus easier to use only three plasmids, i.e. an envelope plasmid, a second generation plasmid providing Gag, Pol, Tat, and Rev proteins, and any vector genome plasmid (second generation with native 5′ LTR or third generation with chimeric 5′ LTR) since the presence of Tat is required for optimal activity of the native LTR and does not affect the activity of the chimeric LTRs. Thus, for in vitro and vivo research, we advise to use an all-purpose packaging plasmid, such as the psPAX2 which encodes for the HIV-1 Gag, Gag/Pol, Tat, and Rev proteins.

The vector plasmid represented in Fig. 2 carries a LoxP recombination site. This LoxP sequence is duplicated during reverse

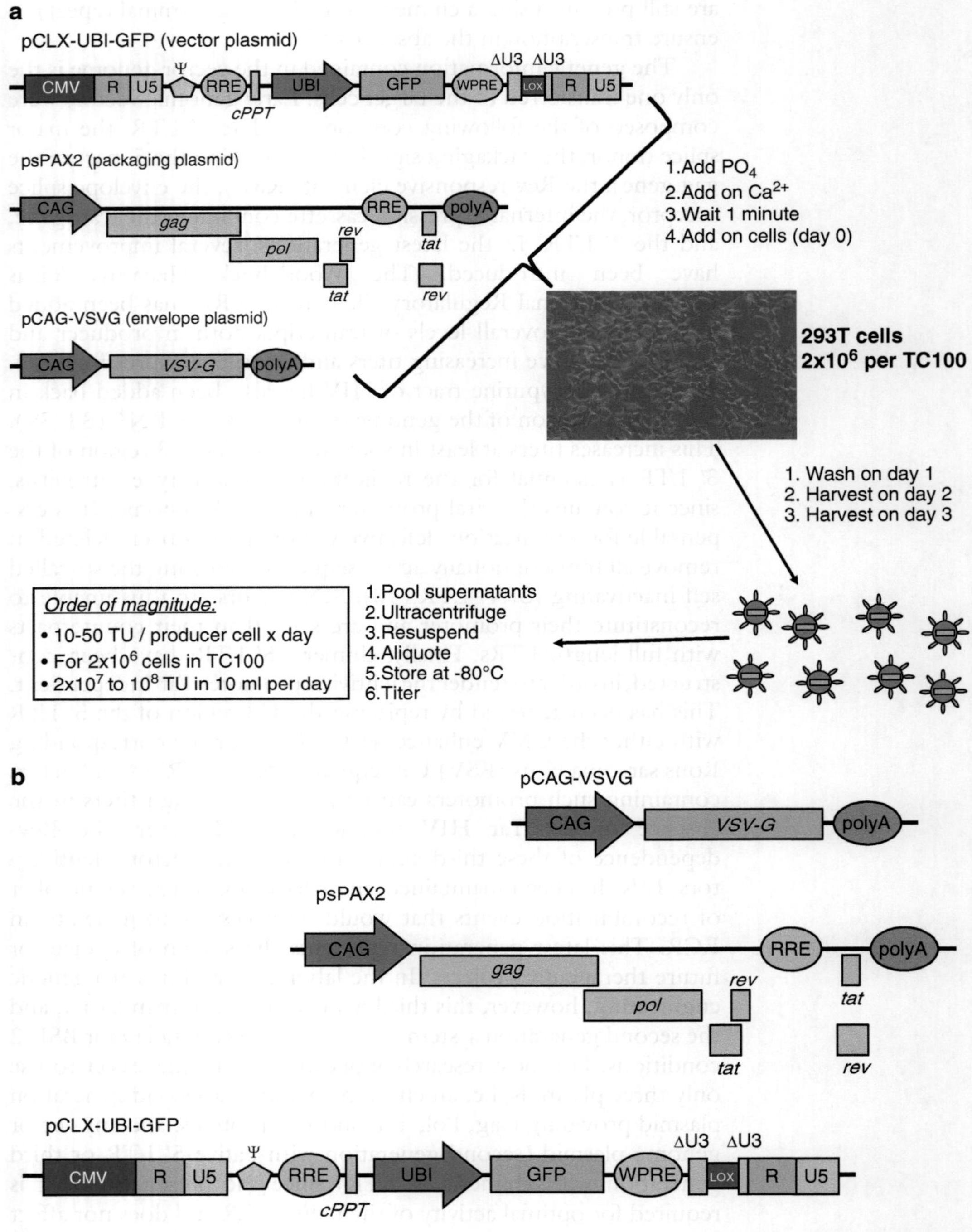

Fig. 2. Schematic diagram of lentivector production. This diagram summarizes the steps described in Subheading 3 of this chapter, to produce LVs. (**a**) The three plasmids encoding for the vector genome, the structural and accessory genes, and the envelope are mixed and introduced by transfection into 293T cells. The vectors particles are then harvested, concentrated, and titered as described in Subheading 3. The expected quantities of vectors are described in Subheading 3.8. (**b**) The pCAG-VSVG plasmid (courtesy of A. Nienhuis, (45)), providing the envelope of the LV particles is composed of the CAG compound promoter, the coding sequence of the Vesicular Stomatitis Virus Envelope protein (VSV-G), and the polyadenylation signal from the rabbit beta-globin gene. The second-generation psPAX2 packaging plasmid (P. Salmon, unpublished),

transcription and allows the integrated proviral cassette to be excised upon expression of the Cre bacterial recombinase (24). Note that, although LVs can theoretically accommodate up to 9 kb of transgenic sequence, some inserts can induce a rapid and important titer drop. This is the case, for example, for the CAG promoter in our hands. Also, the UBI promoter can be replaced by other ubiquitously active promoters, such as EF1 or EFs, or tissue-specific promoters.

Detailed information (such as maps, sequences, etc.) are available at our institutional website (http://medweb2.unige.ch/salmon/lentilab).

1.3. Current Applications and Developments

In a little more than a decade, it has become obvious that "historical" cell lines derived from tumors must be confined to "basic cell culture tasks" and that primary cells are essential to address biologically relevant questions. This is why more and more companies now propose human primary cells of various tissue origins. However, the tissue sources are limited, the life span is finite and the price can be an issue especially when up-scaling is required. Also, it is still difficult for some cell types to standardize all parameters from one preparation to another.

As a consequence, immortalization is becoming a viable alternative to provide researchers with unlimited, reproducible, and standardized human cell sources. As described above, the use of "strong oncogenes" such as SV40 large T, E6, or E7 is limited since they induce transformation and genomic instability. The final in vitro outcome is most often similar to in vivo tumorigenesis, with cells accumulating unknown molecular events, resulting in unstable and dedifferentiated cells. On the contrary, the use of "mild immortaligenes" such as Bmi-1 and telomerase, seems to preserve original phenotypes and genomic stability. Although other genes has been described and are being tested for a similar mild immortalization, such as CBX7 (42), CDK4 (43), CDK6, CyclinD1, or members of the E2F family, the Bmi-1/telomerase combination currently offers the best compromise. If the target primary cells grow well in culture, telomerase alone can suffice by providing infinite lifespan. Considering the efficiency, safety and versatility of lentivectors on one hand, and the growing list of immortalized human cell lines created using the Bmi-1/telomerase gene combination on the other hand, we propose this "molecular

Fig. 2. (continued) providing the structural and enzymatic proteins of the LV particle is composed of the CAG compound promoter, the gag, pol, tat, and rev genes, the Rev-responsive element of HIV-1 (RRE), and the polyadenylation signal from the rabbit beta-globin gene. The third-generation pCLX-UBI-GFP vector plasmid, providing the genome of the LV particles is depicted here as an example of a classical LV design. The 5′ LTR is composed of the CMV promoter and the R and U5 regions of HIV-1. This renders it tat-independent. *psi* HIV-1 packaging signal, *RRE,* rev-responsive element, *cPPT* central polypurine tract, *UBI* ubiquitin promoter, *GFP* green fluorescent protein, *WPRE* Woodchuck Hepatitis Virus Posttranscriptional Regulatory Element, $\Delta U3$ self-inactivating deletion of the U3 part of the HIV-1 LTR, *lox* Cre recombinase LoxP target sequence.

toolbox" as a start point for the immortalization of human primary cells, as well as a means for further genetic tailoring of such cells, by introduction of additional genes (30).

In this chapter, we thus describe the two lentivectors composing our "star cocktail," i.e. pLOX-CW-Bmi1 and pLOX-hTERT-iresTK, together with standard GFP vectors. Once integrated in the genome of the target cells, these lentivectors can be excised using the Cre bacterial recombinase. This can be required if removal of Bmi-1 is needed for terminal differentiation. Finally, pLOX-hTERT-iresTK encodes for the thymidine kinase of HSV-1 and thus provides a safety feature consisting in the conditional ablation of unexcised growing cells using Acyclovir or its analogs (24).

1.4. Safety Issues

The system presented here contains numerous safeguards as compared to the first-generation HIV vectors, in which genes encoding all HIV-1 proteins, except for Env, were present. A second generation was characterized by the exclusion of four accessory genes (vif, vpr, vpu, and nef). These deletions improved considerably the safety of the vector because they excluded major determinants of HIV-1 virulence. In the third-generation system, described in this unit, Gag, Pol, and Rev are the only HIV-1 proteins still present. Vectors with self-inactivating (SIN) LTR and produced with the third generation packaging system have been tested for RCR. Thus far, no RCR have been detected amongst a total of 1.4×10^{10} transducing units (44).

Since VSV-G-pseudotyped LVs have a broad tropism, both in vitro and in vivo, biosafety precautions need to take into account the nature of the transgene. LVs containing non-hazardous genes (such as GFP) should be handled in a biosafety level (BSL) laboratory as prescribed by the appropriate Institutional Biosafety Committee. We handle LVs containing non-hazardous genes (such as GFP) in a BSL-2 laboratory. We handle LVs containing genes that are themselves potentially biohazardous (such as Bmi1, hTERT, or other genes involved in cell proliferation described in this chapter) in a BSL-3 laboratory. Details on BSL-2 and BSL-3 laboratory, standard equipment and P2 safety procedures can be found at our institutional website (http://medweb2.unige.ch/salmon/lentilab).

In general, transduced cells can be fixed (using formaldehyde or paraformaldehyde as described in Subheading 3) before being taken out of the BSL-2 or BSL-3 laboratory. If a live-sorting of transduced cells is needed outside of the BSL-2 laboratory, a careful handling and decontamination of the equipment used should be performed afterwards. However, once they have cleared the RCR assay (see Subheading 3.5.4), LV-transduced cells, including LV-immortalized cells can be considered as any other human cell line.

2. Materials

All solutions and equipment coming into contact with living cells must be sterile, and aseptic technique should be used accordingly. All maps and sequences of plasmids described here are available at our institutional website (http://medweb2.unige.ch/salmon/lentilab). Common plasmids for the generation of HIV-1-based lentivectors can be obtained from www.Addgene.org. Use ultrapure or double-distilled water in all recipes.

2.1. HIV-1-Based Lentiviral Vectors

1. pCLX-CW-GFP (CMV promoter, GFP gene, excisable).
2. pCLX-UBI-GFP (Ubiquitin promoter, GFP gene, excisable).
3. pLOX-CW-Bmil (CMV promoter, mBmi-1 gene, excisable).
4. pLOX-TERT-iresTK (CMV promoter, hTERT gene, HSV-1 thymidine kinase gene, excisable).
5. pLOX-Cre (CMV promoter, Cre recombinase gene, excisable).

2.2. Production of HIV-1 Based Lentiviral Vectors by Transient Transfection of 293T Cells

1. Producer cells HEK293T/17 cells (ATCC, see Note 1).
2. Dulbecco's modified Eagle's medium (DMEM) with 4.5 g/L glucose, glutamine, pyruvate, antibiotics, and 10% FBS (abbreviated D10).
3. Serum-free Advanced DMEM (Life Technologies™) supplemented with 2 mM glutamine.
4. TE buffer: Tris–HCl 10 mM–EDTA 1 mM, pH 8.0 used to redissolve all plasmids.
5. Envelope plasmid pCAG-VSVG: dissolved at 1 μg/μL in TE buffer.
6. Packaging plasmid psPAX2 (encoding HIV-1 Gag, Pol, Tat, and Rev proteins): dissolved at 1 μg/μL in TE buffer.
7. Vector plasmid (such as pCLX-CW-GFP): dissolved at 1 μg/μL in TE buffer.
8. $CaCl_2$ Solution: 0.5 M Dissolve $CaCl_2$ (Sigma-Aldrich®, see Note 2) into 500 mL of H_2O. Filter sterilize through a 0.22-μm nitrocellulose filter. Store at −70°C in 50 mL aliquots. Once thawed, the $CaCl_2$ solution can be kept at +4°C for several weeks without observing significant change in the transfection efficiency.
9. 2× HeBS (HEPES-buffered saline): Dissolve NaCl (0.28 M final), HEPES (0.05 M final), and Na_2HPO_4, (1.5 mM final) into 800 mL of H_2O. Adjust pH to 7.0 with 10 M NaOH (see Notes 3 and 4). Add H_2O to 1,000 mL and make the final pH adjustment. Filter sterilize through a 0.22-μm nitrocellulose filter. Store at −70°C in 50 mL aliquots. Once thawed, the

2× HeBS solution can be kept at +4°C for several weeks without observing significant change in the transfection efficiency.

10. Ethanol 75% in a spray bottle.
11. PBS, pH 7.4.
12. PBS–$Ca^{2+}Mg^{2+}$, pH 7.4.
13. Sucrose 20%: Dissolve 20 g of sucrose (see Note 5) in 100 mL of PBS–$Ca^{2+}Mg^{2+}$. Filter sterilize through a 0.22-μm nitrocellulose filter. Store at +4°C.
14. Trypsin 0.25%/EDTA (Life Technologies™).
15. Bleach solution 13–14% (w/v).
16. 10-cm Tissue culture dishes.
17. 37°C Humidified incubators, 5% CO_2.
18. 1.5-mL Microcentrifuge tubes, sterile, disposable.
19. 15- and 50-mL Conical centrifuge tubes, sterile.
20. 50 mL Syringes and 0.45-μm pore size PVDF filters.
21. 30-mL Beckman Konical tubes for ultracentrifuge.
22. Ultracentrifuge (such as Beckman Optima™ L-90 K) with SW28 rotor.

2.3. Titration of Lentivectors

2.3.1. Titration by FACS

1. Target cells HT-1080 cells (ATCC).
2. D10 medium (see Subheading 2.2, item 2).
3. Trypsin/EDTA (see Subheading 2.2, item 14).
4. MW6 tissue culture plates.
5. PBS.
6. Formaldehyde 1% (w/v) in PBS: Mix 1 mL of formaldehyde (37%w/v) in 36 mL of PBS. Store at +4°C.
7. Fluorescence-activated cell sorter (FACS; Becton Dickinson with 488 nm excitation laser and green filter) and appropriate tubes.

2.3.2. Titration by qPCR

1. Target cells HT-1080 cells (ATCC).
2. D10 medium (see Subheading 2.2, item 2).
3. Trypsin/EDTA (see Subheading 2.2, item 14).
4. MW6 tissue culture plates.
5. PBS.
6. Real-time PCR machine (ABI PRISM® 7900HT Real Time PCR System, Applied Biosystems or equivalent, with a dedicated analysis program, SDS2.2.2, Applied Biosystems or equivalent).
7. Genomic DNA extraction kit (DNeasy Blood & Tissue Kit, Qiagen).
8. qPCR Master Mix (Eurogentec).

9. 96-well Optical Reaction plate (Applied Biosystems).
10. Optical caps (Applied Biosystems).
11. Filter tips (1,000, 100, 10 μL).
12. Primers and probe for quantification of human genomic sequences (10× HB2 set, see Subheading 2.6.1 and Notes 6–8).
13. Primers and probe for quantification of HIV sequences (10× GAG set) (see Subheading 2.6.2 and Notes 6–8).

2.4. Immortalization of Human Primary Cells

IMPORTANT: Immortalization of human primary cells using lentivectors pseusotyped with the VSV-G envelope and containing Bmi-1 or telomerase, as well as production, concentration, and titration of immortalizing lentivectors must be performed as per local biosafety requirements. We use BSL-3 (see Note 9).

We downgrade LV-immortalized cells (i.e., BSL3 to BSL-2 conditions) only after they have cleared the RCR assay (see Subheading 3.5.4 and Note 10).

1. Polystyrene 24 well-multidish.
2. Culture medium appropriate for the primary cells to immortalize.
3. Aerosol-barrier filter tips (1,000, 100, 10 μL).

2.5. Genetic Analysis of Lentimmortalized Cells

2.5.1. PCR Detection of Immortalizing Lentivectors

1. Standard molecular biology equipment (see Notes 11 and 12).
2. Genomic DNA extraction kit (DNeasy Blood & Tissue Kit, Qiagen).
3. REDTaq® ReadyMix™ PCR Reaction Mix (Sigma-Aldrich®).
4. 0.2 mL PCR tubes.
5. Filter tips (1,000, 100, 10 μL).
6. Primers for amplification of pLOX-CW-Bmi1 and pLOX-TERT-iresTK sequences (see Subheading 2.6 and Notes 6–8).

2.5.2. Quantification of Lentivector Copy Numbers by qPCR

1. Real-time PCR machine (ABI PRISM® 7900HT Real Time PCR System, Applied Biosystems or equivalent, with a dedicated analysis program, SDS2.2.2, Applied Biosystems or equivalent).
2. Genomic DNA extraction kit (DNeasy Blood & Tissue Kit, Qiagen).
3. qPCR Master Mix (Eurogentec).
4. 96-well Optical Reaction plate (Applied Biosystems).
5. Optical caps (Applied Biosystems).
6. Filter tips (1,000, 100, 10 μL).
7. Primers and probe for quantification of HIV sequences (10× GAG set, see Subheading 2.6 and Notes 6–8).
8. Primers and probe for quantification of human genomic sequences (10× HB2 set, see Subheading 2.6 and Notes 6–8).

2.5.3. Replication-Competent Recombinant Assay

1. All materials described in Subheading 2.5.2.
2. Genomic DNA from full HIV-1 genome-containing cells, such as 8E5 cells (ATCC) or full-length HIV-1-containing plasmid (see Notes 13 and 14).
3. Primers and probe for quantification of HIV packaging sequences (10× PRO set, see Subheading 2.6 and Notes 6–8).

2.6. Oligos

Oligos can be ordered on-line from several companies such as Eurogentec or Sigma-Aldrich®. FAM fluorescent dye can be replaced by other equivalent molecule, and TAMRA can be replaced by other quenchers.

2.6.1. Human Beta-Actin Taqman® Probe and Primers

These oligos are used to normalize for the amount of genomic DNA and are specific for the human beta-actin gene.

1. HB2-P (probe, sense) 5′-(FAM)-CCTGGCCTCGCTGTCCA CCTTCCA-(TAMRA)-3′.
2. HB2-F (forward primer) 5′-TCCGTGTGGATCGGCGGC TCCA-3′.
3. HB2-R (reverse primer) 5′-CTGCTTGCTGATCCACAT CTG-3′.

2.6.2. GAG Taqman® Probe and Primers

These oligos are used for amplification of HIV-1-derived vector sequences and are specific for the 5′ end of the gag gene (GAG). This sequence is present in all HIV-1 vectors for it is part of the extended packaging signal.

1. GAG-P (probe, antisense) 5′-(FAM)-ACAGCCTTCTGATGT TTCTAACAGGCCAGG-(TAMRA)-3′.
2. GAG-F (forward primer) 5′-GGAGCTAGAACGATTCGCA GTTA-3′.
3. GAG-R (reverse primer) 5′-GGTTGTAGCTGTCCCAGTATT TGTC-3′.

2.6.3. PRO Taqman® Probe and Primers

These oligos are used for amplification of sequences present in RCRs and are specific for the region of the pol gene coding for the HIV-1 protease (PRO).

1. PRO-P (probe, sense) 5′-(FAM)-ACAATGGCAGCAATTTC ACCAGT-(TAMRA)-3′.
2. PRO-F (forward primer) 5′-AGCAGGAAGATGGCCAGTAA-3′.
3. PRO-R (reverse primer) 5′-AACAGGCGGCCTTAACTGTA-3′.

2.6.4. CMV-Bmi Primers

These primers are for PCR detection of the pLOX-CW-Bmi1 integrated cassette in the genomic DNA of transduced cells. To avoid amplification from the endogenous Bmi-1 gene, these primers are designed to amplify a sequence comprised between the CMV

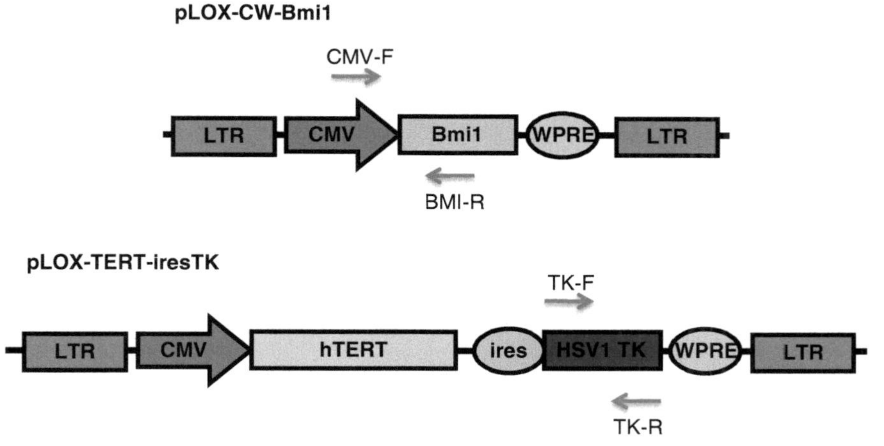

Fig. 3. Schematic diagram of PCR detection of immortalizing lentivectors. Schematic diagrams of pLOX-CW-Bmi1 and pLOX-TERT-iresTK are shown. *Arrows* indicate orientations and approximate locations of binding sites of primers described in Subheading 2.6.

promoter and the Bmi-1 gene. The corresponding amplified region is depicted in Fig. 3.

1. CMV-F (forward primer) 5′-GCAGAGCTCGTTTAG TGAACCGTC-3′.
2. BMI-R (reverse primer) 5′-CCAGACCACTCCTGAACA TAAGGTC-3′.
3. The size of the amplicon is 350 bp.

2.6.5. HSV1 TK Primers

These primers are for PCR detection of pLOX-TERT-iresTK integrated cassette in the genomic DNA of transduced cells. We could not generate PCR amplicons from CMV and hTERT sequences. Hence, detection of pLOX-TERT-iresTK is performed by specific amplification of the HSV1 thymidine kianse. The corresponding amplified region is depicted in Fig. 3.

1. TK-F (forward primer) 5′-GATGACTTACTGGCGGGTGT-3′.
2. TK-R (reverse primer) 5′-GGCCCGAAACAGGGTAAATA-3′.
3. The size of the amplicon is 620 bp.

3. Methods

3.1. Production of Lentiviral Vectors

IMPORTANT: Production, concentration, and titration of immortalizing lentivectors, must be performed as per local biosafety requirements. We use BSL-3 (see Notes 9 and 15–19).

1. Maintain 293T cells in D10 medium (see Subheading 2.2, item 2) in 10 cm tissue culture dish in a 37°C humidified incubator

with a 5% CO_2 atmosphere, and split them at ratio 1:10 using Trypsin/EDTA, three times per week (e.g., every Monday, Wednesday, and Friday).

2. The day before the transfection, seed 1–10 dishes at 2.5 million cells per dish. Cells must be approximately 1/2 to 2/3 confluent on the day of transfection. Incubate overnight in a 37°C humidified incubator with a 5% CO_2 atmosphere. On the following day, cotransfect the cells according to the following recipes.
3. For one plate of 10 cm, mix in a sterile 1.5 mL microcentrifuge tube.

Envelope plasmid	pCAG-VSVG	4 μg
Packaging plasmid	psPAX2	8 μg
Vector plasmid	pCLX-CW-GFP	8 μg

4. The vector plasmid (pCLX-CW-GFP given as example above) can be second or third generation since the psPAX2 plasmid provides Tat protein.
5. Adjust to 250 μL with sterile buffered water and mix well by pipetting.
6. Add 500 μL of 2× HeBS and mix well by pipetting.
7. Put 250 μL of 0.5 M $CaCl_2$ in a 15-mL sterile conical tube.
8. To each 15 mL tube containing the $CaCl_2$ solution, slowly transfer, dropwise, the 750 μL of DNA/HeBS mixture, while vigorously vortexing. Vigorous vortexing will ensure the formation of a fine precipitate that can be taken up efficiently by cells.
9. Leave the precipitates (1 mL final volume per tube) at room temperature for 5–30 min.
10. Add the 1 mL of precipitate dropwise to the cells in 10 mL of medium in one culture dish prepared as above. Mix by gentle swirling until the medium has recovered a uniformly red color.
11. Place the dish overnight in a 37°C humidified incubator with a 5% CO_2 atmosphere.
12. Early the next morning, aspirate the medium, wash with 10 mL of pre-warmed PBS and gently add 15 mL of fresh Advanced DMEM (see Subheading 2.2, item 3) pre-warmed to 37°C. Incubate for 24 h. If 293T cells adhere poorly, washing with PBS can be omitted.
13. Transfer the supernatant from each plate to one 50 mL centrifuge tube. Close the tubes, and spray them with 70% ethanol before taking them out of the hood. Store the supernatant at +4°C. Add another 15 mL of fresh Advanced DMEM, pre-warmed to 37°C. Incubate for another 24 h with the cell monolayer.

14. Pool the supernatants of day 1 and 2 and centrifuge for 5 min at 500 × *g*, at 4°C, to pellet detached cells and debris.
15. Filter the 30 mL of pooled supernatant (total harvest from 2 days: 30 mL/dish) with a 50 mL syringe connected to a 0.45 μm PVDF disk filter.

The LV stocks can be stored at +4°C for 1–4 days without significant titer loss, before they are used for transduction of target cells or further processing such as concentration. For longer storage, LV stocks must be kept at −80°C.

The transfection can be started late in the afternoon and the medium changed early the next morning. If you notice cell toxicity, you can transfect early in the morning and change the medium late in the afternoon the same day. The transfection procedure can be scaled up to ten culture dishes of 10 cm, or other cell culture systems with equivalent or larger surface.

3.2. Concentration of Lentiviral Vectors

1. For concentration, use 30 mL Beckman conical tubes, in a SW28 rotor in an ultracentrifuge. Put 4 mL of 20% sucrose on the bottom of the tube. Very slowly pour the supernatant on the surface of the sucrose cushion until the tube is full (allow a 2 mm dry zone to the top of the tube). Spin at 50,000 × *g* for 120 min at +16°C.
2. Remove conical tube from SW28 bucket (see Note 20).
3. Aspirate the medium with a sterile pipette down to the sucrose interface.
4. Aspirate the sucrose until you have 1–2 mL of colorless sucrose solution and then invert the tube while aspirating the remaining sucrose. Never touch the bottom of the tube where the vector pellet is.
5. Place the conical tube in a 50 mL Falcon tube and quickly add 30–100 μL of PBS–$Ca^{2+}Mg^{2+}$ on the pellet (not always visible). Do not leave the pellet dry for more than 5 min or it may result in significant titer decrease. Close the Falcon tube. You can resuspend the vector pellet of one tube in a minimal volume of 30 μL. In this case, you will achieve a ~1,000-fold concentration.
6. Vortex at half-speed for 2 s.
7. Leave the vector pellet to resuspend for 1–2 h at room temperature or 2–4 h at +4°C.
8. Vortex at half-speed for 2 s,
9. Pipet up and down 20 times and freeze at −80°C in aliquots for long time storage (see Notes 21 and 22).

3.3. Titration of Lentiviral Vectors

Titers of viruses in general and lentivectors in particular critically depend on the methods and cells used for titration. The quantification of vector particles capable of achieving every step from cell binding to expression of the transgene depends on both vector and cell characteristics. First, the cell used as target must be readily permissive to all steps from viral entry to integration of the vector genetic cargo. Second, the expression of the foreign gene must be easily monitored and rapidly reach levels sufficient for reliable quantification. Early vectors had the lacZ bacterial gene as reporter, under the control of the CMV promoter. Current vectors now have the green fluorescent protein (GFP) gene as a reporter, under the control of promoters that are active in most primary cells.

Measured titers can also vary with the conditions used for titration, i.e., volume of sample during vector-cell incubation, time of vector-cell incubation, number of cells used, etc. For several years now, numerous laboratories have been using HeLa cells as target cells for LVs. Although these cells are easy to grow and 100% susceptible to transduction by VSV-G-pseudotyped LVs, they are very unstable in terms of morphology and karyotype. For this reason, we are now using HT-1080 cells, which are stable, of human origin and give titers identical to HeLa cells.

Physical titration based on the quantification of HIV-1 capsid p24 antigen is not used anymore in our lab. Instead, our current standard procedure relies on determination of infectious titers by transduction of HT-1080 target cells. Also, we always produce a test batch of a standard GFP lentivector alongside all LV production. This test batch is used to monitor the overall efficiency of the procedure and detects any anomaly in producer cells or reagents that will result in titer drop.

Here we described a procedure that is used on a weekly basis in our lab for several years and that has been standardized in order to compare titers from one batch to another one or from one lab to another one. Changes in this procedure can be made, but one must keep in mind that, for example, reducing the cell culture surface or increasing the number of target cells will result in an increase of the final calculated titer, from the exact same vector batch.

3.3.1. General Procedure

1. On day 0, seed HT-1080 cells at 50,000 cells per well in MW6 plate in D10. Make sure that HT-1080 cells are well separated and uniformly distributed in the well.
2. On day 1, put into three independent wells 500, 50, or 5 μL of the vector suspension (either pure from unconcentrated supernatants or diluted in complete medium if it comes from a concentrated stock, i.e. 1/100 if the vector is concentrated 100-fold).
3. Polybrene can be omitted for transduction with VSV-G pseudotyped vectors since this compound does not influence permissivity of cells to VSV-G pseudotyped vectors.
4. On day 2, remove the supernatant and replace by 2 mL of fresh D10.
5. On day 5, wash the cells with 2 mL of PBS; detach them with 250 μL of Trypsin/EDTA for 1 min at 37°C.

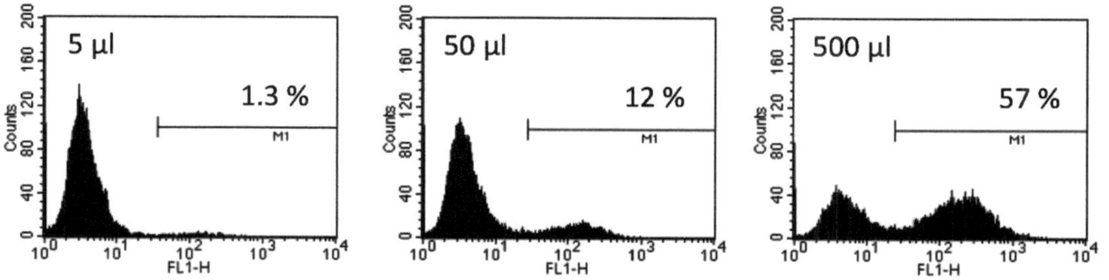

Fig. 4. A representative FACS analysis of HT-1080 cells used for titration of GFP-coding LV. HT-1080 cells (10^5) were incubated with increasing volumes of a supernatant containing an LV expressing GFP under the control of the human Ubiquitin promoter (pCLX-UBI-GFP) as described in Subheading 2.1. After 5 days, cells were detached, fixed, and analyzed by FACS for GFP fluorescence (*x* axis, 4-decade log scale, FL1) versus number of cells (*y* axis, linear scale). The percentage of GFP-expressing cells was measured by placing a marker discriminating between GFP-negative (mean of fluorescence intensity 3–4) and GFP-positive cells (mean of fluorescence intensity 200).

6. Add 250 µL of D10 and mix well to resuspend the cells. This step inactivates the trypsin and EDTA.
7. Spin cells in a microcentrifuge for 2 min at 200 × *g*. Note that if you need to run an FACS analysis and a qPCR analysis on the same sample, you must split your cells in two separate microcentrifuge tubes.

3.3.2. Titration of Lentivectors by FACS

This method can only be used to titer stocks of vectors that carry a transgene that is easily monitored by FACS (such as GFP, or any living colors, or any membrane protein that can be detected by flow cytometry), and whose expression is governed by a promoter that is active in HT-1080 cells (tissue-specific promoter-containing vector must be functionally assayed in specific cells, and titered by qPCR in HT-1080 cells (see Subheading 3.3.3).

We describe here the titration of an Ubiquitin promoter-GFP vector (pCLX-UBI-GFP, see Subheading 2.1).

1. Add 500 µL of 1% formaldehyde in PBS (see Subheading 2.3.1, item 6) to the cell pellet. This step will fix the cells and inactivate the vector particles. Samples can thus be taken out of the BSL-2 laboratory.
2. Resuspend the cells thoroughly in the well and transfer them to an FACS tube.
3. Analyze the cells in a flow cytometer. If you are not familiar with flow cytometry, you must seek help from your institutional FACS specialist.
4. Once chosen the appropriate dilution (see Fig. 4 and Notes 23–25), apply the following formula: Titer (HT-1080-transducing units/mL) = 100,000 (target HT-1080 cells) × (% of GFP-positive cells/100)/volume of supernatant (in mL).

3.3.3. Titration of Lentivectors by Quantitative PCR (qPCR)

When lentivectors contain DNAs coding for genes other than GFP or LacZ and promoter which are active only in specific primary cells and tissue, FACS titration cannot be used. Therefore, most new LVs will need an alternative method to measure the number of copies of LV stably integrated in HT-1080 target cells, after transduction as described in Subheading 3.3.2 for GFP vectors. This assay, however, only measures the number of LV copies integrated in the target cell genome. The overall functionality of the vector must be tested at least once in cells in which the promoter is active and/or with appropriate techniques to detect the expression of the transgene product. The qPCR assay proceeds as follows, using a real-time PCR machine. HT-1080 cells are transduced as for FACS analysis. Then, one half can be used if FACS analysis is performed in parallel, or target cells can be lysed directly in the plate and the DNA is extracted using a genomic DNA extraction kit and following general procedures for qPCR analysis (see Notes 26–29). Then, a fraction of the total DNA is analyzed for copy number of HIV sequences using the following real-time PCR protocol.

1. Extract target cell DNA from each individual well of an MW6 plate (see Subheading 3.3.1) using the genomic DNA extraction kit, following manufacturer recommendations. For the DNA elution step, use 100 μL of AE buffer instead of 200 μL.
2. Perform qPCR or store DNA at –20°C until use.
3. Prepare a mix containing everything but the sample DNA for the number of wells needed for the qPCR analysis, including all samples and standards in duplicates or triplicates, according to the following recipe (for one well):

qPCR Master Mix	7.5 μL
10× oligo mix (GAG or HB2, see Subheading 2.6 and Notes 6–8)	1.5 μL
DNA sample	1 μL
H_2O	5 μL

4. Distribute 14 μL of this mix into the wells of a 96-well Optical Reaction plate.
5. Add sample DNAs.
6. Close with optical caps.
7. Centrifuge the plate at 200 × *g* for 1 min to bring all liquid on the bottom of the wells.
8. Place the 96-well Optical Reaction plate in the real-time PCR machine and run the appropriate program depending on the fluorochromes and quenchers used in your Taqman probes (see Note 30).

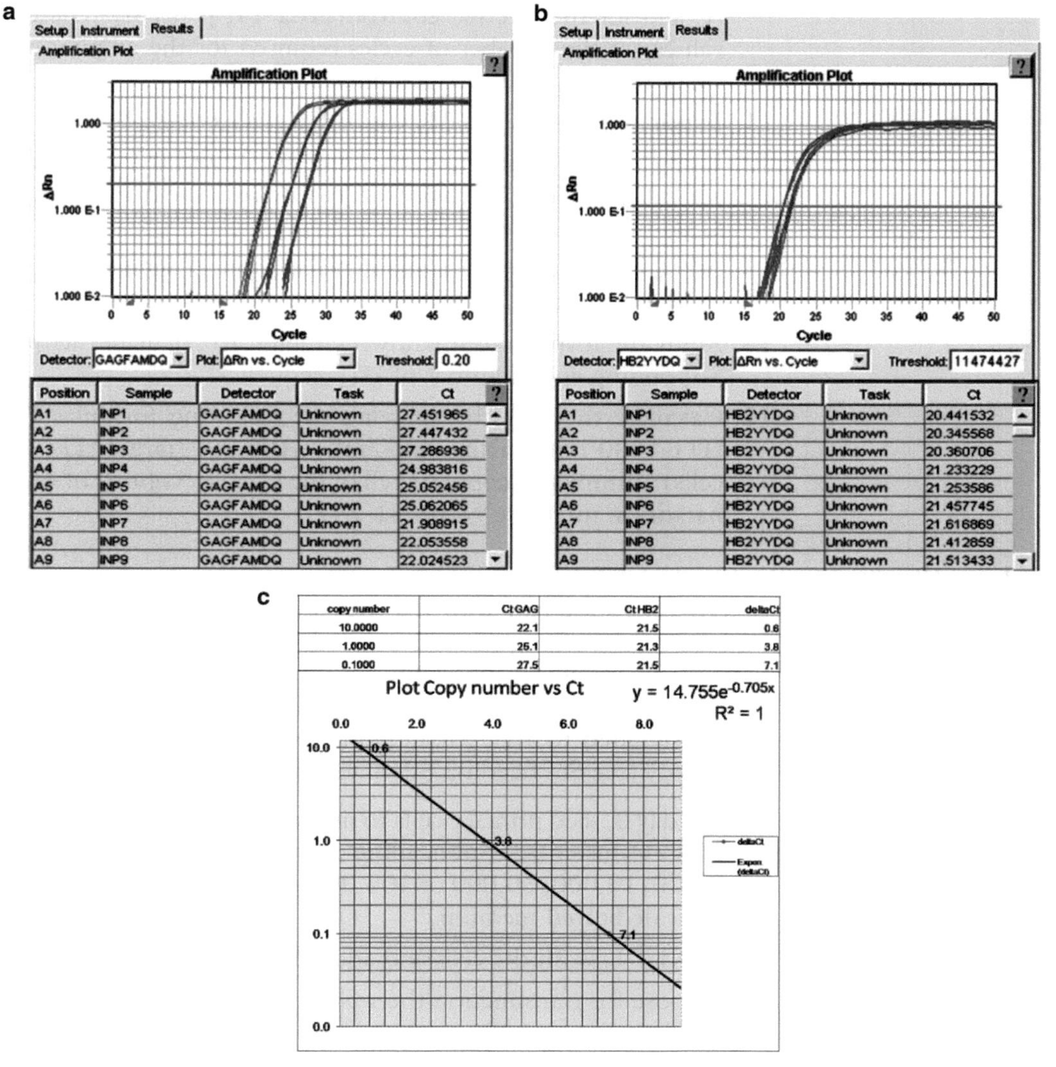

Fig. 5. A representative qPCR analysis used for titration of HIV-1-based LVs. DNA from HT-1080 cells transduced with serial tenfold dilutions of pCLX-UBI-GFP vectors was subjected simultaneously to qPCR titration analysis and FACS analysis as described in Subheading 3. A sample of each dilution was submitted to qPCR amplification and monitoring using an ABI PRISM® 7900HT Real Time PCR System (Applied Biosystems), and sets of primers and probes specific for HIV gag sequences (GAG-FAM, (**a**) or beta-actin sequences (HB2-YY, (**b**) Amplification plots were displayed and cycle threshold values (Ct) were set as described in Subheading 3.3.3. Values of GAG Ct and HB2 Ct were exported in an Excel worksheet to calculate ΔCt values (*x* axis, linear scale) and plot them against copy number values (*y* axis, log scale) (**c**) A sample giving 10% of GFP-positive cells was set as cells containing 0.1 copy of HIV sequences per cell. The regression curve can then be used to calculate GAG copy numbers (*Y* value) of unknown samples by applying the formula to ΔCt values (*X* values) of the sample.

9. Analyze results and calculate titer using the SDS 2.2.2 program. An example of amplification profiles of HIV sequences in human DNA is given in Fig. 5.
10. Set the threshold values (Ct) where the amplification curve is the steepest, both for the gene of interest (GAG-FAM, panel

A) and for the internal control (HB2-YY, panel B). These Ct values are the number of cycles required for the amplification curve to cut the absorbance threshold values.

11. Export the results as a Microsoft Excel sheet (see Note 31).
12. Using standards of cells containing 10, 1, and 0.1 copy of LV per cell (see Note 28), ask excel to calculate the ΔCt values (Ct GAG minus Ct HB2).
13. Ask Excel to display an exponential formula giving the copy number as a function of ΔCt.
14. Apply the formula to unknown samples, to calculate their corresponding copy number of HIV sequences.
15. Calculate the titers by applying the following formula: Titer (HT-1080-transducing units/mL) = 100,000 (target HT-1080 cells) × number of copy per cell of the sample/volume of supernatant (in mL) (see Notes 32–34).

3.4. Immortalization of Human Primary Cells

IMPORTANT: Immortalization of human primary cells using lentivectors pseudotyped with the VSV-G envelope and containing Bmi-1 or telomerase, as well as production, concentration, and titration of immortalizing lentivectors, must be performed in a BSL-3 laboratory (see Notes 9 and 15–19). Immortalized cells can be downgraded to BSL-2 conditions once they have cleared the RCR assay (see Subheading 3.5.4 and Note 10).

1. On day 0, seed primary cells to immortalizing in triplicate in an MW24 cluster (see Subheading 2.4) at cells densities of 1,000, 3,000, 10,000, 30,000 or 100,000 cells per well per 500 μL of their dedicated medium.
2. On day 1, wash the cells and add 500 μL of fresh medium.
3. Choose for transduction the well triplicate where the cell density is between 1/8 and 1/4 confluence. Depending on the cell size, this should amount to approximately 10,000 cells. You may add one extra well of each seeding density to detach and count cells in order to have a more precise estimation of lentivector quantities to add in further steps.
4. Add pLOX-CW-Bmi1 lentivector at an MOI of 1 (i.e. 10^4 TU) in one well and at an MOI of 10 (i.e. or 10^5 TU) in another well. The total volume of added lentivector suspension should not exceed 10% of the initial volume (500 μL of culture medium + 50 μL of vector). Keep one well as untransduced control.
5. On day 2, wash cells and add 1 mL of fresh medium.
6. Then, change medium and split cells as usual until you have enough cells to freeze, perform a CMV-BMI PCR detection and a copy number qPCR assay (see Subheadings 3.5.2 and 3.5.3) or move directly to telomerase transduction step. This should be between 1 and 2 weeks minimum, especially if a

qPCR assay of copy number is performed. Indeed, cells need to go through 3–5 cell divisions to dilute and lose unintegrated copies of lentivector DNA. Note that at this step, you may already observe a change in growth rates between transduced and untransduced cells.

7. Prepare cells in MW24 clusters and transduce them with pLOX-TERT-iresTK lentivector as described above for Bmi-1 lentivector.
8. Wash and expand cells as described above until you have enough cells to freeze and perform a TK PCR detection and a copy number qPCR assay (see Subheadings 3.5.2 and 3.5.3). Note that at this step, untransduced cells may already have ceased to grow and/or died of senescence.

3.5. Genetic Analysis of Lentimmortalized Cells

These genetic analysis assays are designed to assess for the presence of pLOX-CW-Bmi1 and pLOX-TERT-iresTK lentivectors and calculate their copy numbers. They also ensure for the absence of replication-competent recombinants in the final immortalized cells, and thus provide a biosafety label for these cells as the production, even at low levels, of infectious immortalizing vectors may represent a serious biohazard outside of BSL–3 environment. These assays should be performed at least 3 weeks after the last transduction, especially if a qPCR assay of copy number is performed. Indeed, cells need to go through 3–5 cell divisions to dilute and lose unintegrated copies of lentivector DNA.

3.5.1. Sample Preparation

1. Extract DNA from approximately 10^6 cells using the genomic DNA extraction kit, following manufacturer recommendations. For the DNA elution step, use 100 μL of AE buffer instead of 200 μL.
2. Perform PCR or store DNA at −20°C until use.

3.5.2. PCR Detection of Immortalizing Lentivectors

1. In a 200 μL PCR tube, prepare the following mix:

RedTaq reaction buffer 2×	10 μL
10× oligo mix (CMV-BMI or TK, see Subheading 2.6 and Notes 6–8)	2 μL
DNA sample	1 μL
H_2O	7 μL

2. Use DNA from Bmi+Tert immortalized cells as positive control. In case you do not have such DNA, you can prepare a stock solution at 10 pg/mL of the corresponding lentivector plasmid DNAs, diluted in control DNA from untransduced cells at 5 μg/mL (approximately equivalent to 10^6 cells/mL).
3. Use DNA from primary cells as negative control.

4. Run the PCR for 42 cycles with the following parameters:

94°C	20 s
55°C	1 min
72°C	1 min

5. Run 10 μL of PCR products in a 2% agarose gel.
6. Positive band for CMV-BMI PCR has a size of 350 bp.
7. Positive band for TK PCR reaction has a size of 620 bp.

3.5.3. Quantification of Lentivector Copy Numbers by qPCR

This assay is basically run as the qPCR titration assay described above. The specificity of this assay is that it is run with genomic DNA from immortalized cells instead of HT-1080 cells and that the copy number is directly calculated from the standard curve obtained from HT-1080 cells containing various copy numbers of LV-GFP. The normalization of LV copy numbers per genomic DNA (hence per cell) is more relevant when using standard DNA from HT-1080 cells rather than from HeLa cells. Indeed, HT-1080 cells have a close to normal diploid DNA content, whereas HeLa cells have a chromosome count ranging from 70 to 164.

3.5.4. Replication-Competent Recombinant Assay

The absence of Replication-Competent Recombinants (RCRs) is essential to downgrade the biohazard level of cells that have been transduced by retroviral vectors, including LVs. We propose here a test based on the detection (or absence of detection) in the chromosomal DNA of transduced cells of HIV sequences that are absent in the vector plasmid (vector genome) but are present in the packaging plasmid and are essential for HIV (or RCR) replication. The target sequence chosen in our assay is located in the sequence coding for the viral protease that is present in the packaging plasmid, essential for virus replication and absent in the vector genome. Although the assay described here is performed on a small number of cells, at least 3 weeks after initial transduction, it can be scaled up to meet requirements for the detection of RCR in preclinical vector batches. Other RCR tests have been described in the literature. One earlier paper describes a true RCR assay which failed to detect any RCR in vector batches produced from third generation packaging systems (44). Several other tests have been described, but they detect biological entities that need *trans*-complementation to replicate. Although these assays can measure the level of recombination during the production of lentivectors, they are not suitable to detect genuine RCR that may represent a biological hazard due to potential dissemination within primary human cells.

1. At least 3 weeks prior to assay, transduce HT-1080 cells with LV of interest and with standards (see below). This extended growth period allows for dilution of packaging DNA carried over from

vector production steps. In contrast to qPCR titration described above, two types of standards are needed in this protocol. One standard corresponds to cells containing vector sequences only (LV standard, target for GAG oligo set), and one corresponds to cells containing all HIV sequences (HIV standard, target for GAG and PRO oligo sets). The first standard is provided by cells transduced with LV as described above. The second standard is provided by cells having one copy of full-length HIV genome, such as 8E5 cells, or by full-length HIV-1-containing plasmid (see Notes 13 and 14).

2. After ≥3 weeks of cell growth, extract DNA from the transduced cells using a DNAeasy kit according to the manufacturer's instructions. Store DNA at –20°C until use. The number of cells and final volume should be such that 1 μL of the final DNA solution corresponds to 10^4 cells.
3. For each sample or standard, prepare three independent mixes containing everything but the sample DNA for the number of wells needed for the qPCR reaction, including all samples and standards in duplicates, according to the following recipe (for one well):

qPCR Master Mix	7.5 μL
10× oligo mix (GAG, PRO or HB2, see Subheading 2.6 and Notes 6–8)	1.5 μL
DNA sample	1 μL
H_2O	5 μL

4. Distribute 14 μL of this mix into the wells of a 96-well Optical Reaction plate.
5. Add sample DNAs.
6. Close with optical caps.
7. Centrifuge the plate at 200 × *g* for 1 min to bring all liquid on the bottom of the wells.
8. Place the MW96 in the real-time PCR machine and run the appropriate program depending on the fluorochromes and quenchers used in your Taqman probes.
9. Analyze as described in the qPCR titration section. In this case, however, two types of standards are used. One standard corresponds to cells containing vector sequences only (LV standard, target for GAG oligo set), and one corresponds to cells containing all HIV sequences (HIV standard, target for GAG and PRO oligo sets). The first standard is provided by cells transduced with LV as described above. The second standard is provided by cells having one copy of full-length HIV genome, such as 8E5 cells. In the case of 8E5, the DNA will contain one

copy of HIV per genome. Serial tenfold dilutions of 8E5 DNA into human DNA (up to 10^{-3} copy per genome) can be performed to provide an HIV DNA standard curve. A negative control both for LV sequences and HIV sequences will be provided by HT-1080 cells.

10. Results are expressed as Ct values for each oligo set, i.e., GAG-HB2ΔCt and PRO-HB2ΔCt. The sample DNA will be considered negative for PRO sequences and hence negative for RCR if its PRO-HB2ΔCt value is similar to the PRO-HB2ΔCt value of HT-1080 cells, with a GAG-HB2ΔCt value above the range corresponding to one copy of LV sequence per genome.

3.6. Plasmid Preparation

Plasmids containing retroviral LTRs are prone to undergo deletion in some *E. coli* strains. The Top10 strain is strongly recommended for propagating the plasmids used in this section. We recommend JetStar Kits to prepare DNAs for transfection. The last step of the DNA prep should be an additional precipitation with ethanol and resuspension in TE. Do not treat DNA with phenol/chloroform as it may result in chemical alterations. Also to avoid salt coprecipitation, do not precipitate DNA below +20°C.

3.7. Troubleshooting Lentivector Production

Transfection efficiency is the most critical parameter affecting vector titer. HEK293T/17 cells are highly transfectable using a variety of protocols. When establishing vector production procedures, it is highly recommended that the transfection protocol be optimized using a plasmid encoding GFP. Transfection efficiency should not be assessed solely on the basis of the percentage of GFP-positive cells and also on the mean fluorescence intensity, which reflects the number of plasmid copies taken up by the cells. This makes FACS analysis of the transfected cells mandatory. FACS can be done as soon as 15 h after the transfection, allowing many variables to be tested rapidly. The factors most likely to impact on the transfection efficiency are the pH of the 2× HeBS solution, the quality of the batch of fetal bovine serum used, the cell density, the total amount of DNA per plate, and the quality of DNA. A coarse precipitate will give poor transfection, whereas a fine precipitate (barely visible after application on cells) will give good transfection. As a rule of thumb, the precipitate will be coarser as the pH of 2× HeBS increases, the DNA quantity decreases, the temperature or the incubation time for precipitate formation increases.

In the case of lack of transduction of a specific cell type with a specific LV, a synoptic diagram is provided at the following URL to help addressing most of the problems that could account for it.

http://medweb2.unige.ch/salmon/lentilab/troubleshootingdiagram.pdf

3.8. Anticipated Results

When applied optimally, the procedure described here yields crude unconcentrated vector titers between 1×10^6 and 1×10^7 TU/mL. Note that pLOX-TERT-iresTK gives unconcentrated titers approximately ten times lower. After centrifugation, a yield of at least 50% is expected. A similar 50% yield is also expected after one freeze/thaw cycle. The cells produce equally during the 48 h post-transfection. You maximize the total yield by harvesting twice.

Note that there is no current procedure for purification per se, of infectious particles. The only methods available (ion exchange, centrifugation, etc.) will only concentrate the vector particles and/or wash soluble material. One must keep in mind that all other particulates generated by the producer cells, such as defective vector particles and exosomes, have a similar density to infectious vector particles and are also coated with VSV-G proteins, and will thus co-sediment or co-purify with infectious vector particles. This implies that there is no current way to enrich in infectious particles a vector stock displaying a poor infectivity index. Defective particles will be enriched alongside causing an increase in cell toxicity.

Most human cells transduced with lentivectors at MOIs ranging from 1 to 10 will end up with 1–5 copies of viral DNA in their genome. One should test transduction sensitivity of each cell type to immortalize. A sensitive cell may need only an MOI of 2. Some resistant cells may need an MOI of 10 or more to achieve one or two copies of integrated transgene. A few days to a few weeks after transduction with Bmi-1 and telomerase, immortalized cells will show sustained proliferation while untransduced control cells will undergo cell cycle arrest and senescence. PCR detection of immortaligenes lentivectors as well as qPCR quantification of their copy number will confirm molecular immortalization.

4. Notes

1. HEK293T/17 cells are highly transfectable using the $CaPO_4$ technique. Choosing other cells or other variants of 293T cells may result in lower titers. We recommend using ATCC Cat. # CRL-11268. Also, frequent passages and keeping the 293T as individual cells will ensure high transfection efficiency, hence high titers. A stock of HEK293T/17 should be frozen at early passages. When 293T producer cells in culture start showing repeated low titers of control LVs, the cells must be discarded and replaced by a fresh batch from the frozen stock.
2. For the preparation of $CaPO_4$ transfection solutions, we recommend high-quality chemicals such as SigmaUltra Cat. # C5080. Less pure chemicals may contain small amounts of contaminants that can affect transfection and viability of 293T producer cells.

3. For the preparation of 2× HeBS, we recommend high-quality chemicals such as NaCl, SigmaUltra Cat. # S7653, HEPES, SigmaUltra Cat. # H7523, and Na_2HPO_4, SigmaUltra Cat. # S7907.
4. For the 2× HeBS solution, obtaining a proper pH is very important. Below 6.95, the precipitate will not form, above 7.05, the precipitate will be coarse and transfection efficiency low.
5. For the preparation of sucrose 20% solution, we recommend high-quality chemicals such as sucrose, SigmaUltra, Cat. # S7903.
6. Stocks of probes and primers usually come lyophilized and are stored at 100 μM in water.
7. Standard concentrations in 10× oligo sets for qPCR are 1 μM of probe and 3 μM of each primer in water.
8. Standard concentrations of 10× oligo sets for classical end-point PCR are 2 μM of each primer in water.
9. BSL-3 procedures are described in our institutional website (see http://medweb2.unige.ch/salmon/lentilab). Please consult this site or your institutional biosafety officer for details about local BSL-3 procedures. In particular, personnel handling "immortaligene-containing" lentivectors must work in Class-2 laminar flows wear a double pair of gloves, a safety gown and protective glasses.
10. Cells being analyzed for the absence of RCR must be confined in a culture flask with vented cap until result of RCR analysis. If the result is negative, the biohazard level of the cells can be downgraded; after spraying the flask with 75% ethanol, it can be transferred outside of the culture laboratory.
11. These procedures require standard molecular biology equipment such as an agarose gel apparatus and a generator, a gel imager and a classical end-point PCR machine.
12. For analytical and preparative DNA visualization, we now use SYBR® Safe DNA gel stain (Cat. # S33102, Life Technologies™) together with a Safe Imager™ 2.0 Blue-Light Transilluminator (Cat. # G6600EU, Life Technologies™). This staining system related to SYBR Green dye is not carcinogenic. It is thus less biohazardous and gives higher cloning efficiencies than ethidium bromide.
13. ATCC recommends that 8E5 cells be handled in a P3 laboratory. Indeed, although they contain a full copy of noninfectious HIV, they can form syncytia with uninfected CD4+ cells.
14. A plasmid containing full-length HIV-1 DNA can be obtained from the AIDS Repository (https://www.aidsreagent.org/reagentdetail.cfm?t=molecular_clones&id=55).

15. P2 and P3 practices require that open tubes always be handled in the laminar flow hood. Tubes can be taken out of the laminar flow only when they are closed, and sprayed with 75% ethanol.
16. All solid waste and plasticware must be discarded in a trash bin in the laminar flow hood and all liquids must be aspirated into a liquid waste bottle containing fresh concentrated bleach. Refill the liquid waste bottle with fresh bleach when the color of the liquid is no longer yellow.
17. When full, bags are closed inside the laminar flow hood, then autoclaved.
18. When full, and at least 15 min after neutralization with fresh bleach, the liquid waste bottle can be emptied into a regular sink.
19. In case of a major spill of vector-containing liquid, absorb liquid with paper towels and neutralize with fresh concentrated bleach prior to disposal.
20. In case there is a leak in the SW28 buckets, remove the tubes in the hood, fill the buckets with 75% ethanol, and invert them several times. Leave under the hood for ≥20 min. Discard the 75% ethanol and remove the conical adapters under the hood. Spray the adapters with 75% ethanol and leave them under the hood for >20 min.
21. When resuspending the pellets, try to avoid bubbles since it will result in decrease of final volume and decrease of yield.
22. Try to avoid repeated freeze-thaw cycles of stored vectors. This may results in drop of titer, although the VSV-G pseudotyped particles are more resistant to this procedure than particles pseudotyped with retrovirus-derived envelopes.
23. A reliable measure of the fraction of GFP+ cells relies on the level of GFP expression. In the example shown in Fig. 4, GFP-positive and GFP-negative cells can be readily discriminated when GFP is expressed from a human Ubiquitin promoter and allowed to accumulate in cells for 4–5 days. A marker can then be set to measure the fraction of transduced versus total cells.
24. Cells fixed with formaldehyde can be stored in the dark at +4°C for several hours. A final 0.5% formaldehyde concentration is enough to fix cells and inactivate vectors. Increasing formaldehyde concentration (up to 4% final) will increase the autofluorescence of cells and decrease GFP fluorescence.
25. In a typical titration experiment, only dilutions yielding to 1–20% of GFP-positive should be considered for titer calculations. Below 1%, the FACS may not be accurate enough to reliably determine the number of GFP-positive cells. Above 20%, the chance for each GFP-positive target cell to be transduced more than once significantly increases, resulting in underestimation of the number of transducing particles.

26. Always use pipet tips containing aerosol-barrier filters when preparing solutions, mixes, samples, and plates for qPCR, to prevent cross-contamination.
27. It is advisory to run a dual titration (FACS plus qPCR) using one GFP vector alongside the other vectors, for each experiment of qPCR titration. This will help comparing the FACS titration with the qPCR titration.
28. Standards of HT-1080 cells containing 10, 1, and 0.1 copy of LV per cell can be prepared from HT-1080 cells transduced with a GFP vector, using serial tenfold dilutions. The 0.1 copy per cell standard will be provided by the sample displaying 10% of GFP-positive cells.
29. DNA typically comes from 2×10^6 HT-1080 cells (one confluent well of an MW6 plate) extracted and resuspended in 100 μL of Buffer AE (DNAeasy Tissue Kit).
30. The precise settings of a qPCR protocol depend on the real-time PCR machine used. This aspect is beyond the scope of this protocol. If you are not familiar with qPCR techniques, you should seek advice from your local qPCR expert or from the technical assistance of your real-time PCR machine.
31. A prototypic excel worksheet for calculation of qPCR titers can be downloaded from the following link: http://medweb2.unige.ch/salmon/lentilab/QPCRtitration.html.
32. Using standard DNA extraction procedures in a laboratory context where HIV sequences are often handled, you can expect a level of background contamination with HIV sequences corresponding to cells containing 1 copy per 1,000 or 100 genomes. In this case, consider higher copy numbers for calibration.
33. Vector stocks failing to give higher than 0.01 copy per genome in a qPCR assay, using the highest titration dose, must have experienced one or several problems during their design, packaging, and/or production and cannot be used. You must then refer to the troubleshooting in Subheading 3.7 to solve this issue.
34. Using careful DNA extraction procedures and standardization as described above, you can expect reproducibility within a twofold range. Ask your local qPCR expert if you need a more stringent quantification qPCR procedure.

Acknowledgment

This work is supported by the Chicago Diabetes Project (http://www.chicagodiabetesproject.org).

References

1. Earle WR (1943) Production of malignancy in vitro. IV. The mouse fibroblast cultures and changes seen in the living cells. J Natl Cancer Inst 4:165–212
2. Gey GO, Coffman WD, Tubicek MT (1952) Tissue culture studies of the proliferative capacity of cervical carcinoma and normal epithelium. Cancer Res 12:264–265
3. Chen TR, Hay RJ, Macy ML (1983) Intercellular karyotypic similarity in near-diploid cell lines of human tumor origins. Cancer Genet Cytogenet 10:351–362
4. Graham FL, Smiley J, Russell WC, Nairn R (1977) Characteristics of a human cell line transformed by DNA from human adenovirus type 5. J Gen Virol 36:59–74
5. Kriegler M, Perez CF, Hardy C, Botchan M (1984) Transformation mediated by the SV40 T antigens: separation of the overlapping SV40 early genes with a retroviral vector. Cell 38:483–491
6. Coursen JD, Bennett WP, Gollahon L, Shay JW, Harris CC (1997) Genomic instability and telomerase activity in human bronchial epithelial cells during immortalization by human papillomavirus-16 E6 and E7 genes. Exp Cell Res 235:245–253
7. Takeuchi M et al (2007) Chromosomal instability in human mesenchymal stem cells immortalized with human papilloma virus E6, E7, and hTERT genes. In Vitro Cell Dev Biol Anim 43:129–138
8. Ouyang H et al (2000) Immortal human pancreatic duct epithelial cell lines with near normal genotype and phenotype. Am J Pathol 157:1623–1631
9. Kobayashi N et al (2000) Prevention of acute liver failure in rats with reversibly immortalized human hepatocytes. Science 287: 1258–1262
10. Cudre-Mauroux C, Occhiodoro T, Konig S, Salmon P, Bernheim L, Trono D (2003) Lentivector-mediated transfer of Bmi-1 and telomerase in muscle satellite cells yields a duchenne myoblast cell line with long-term genotypic and phenotypic stability. Hum Gene Ther 14:1525–1533
11. Narushima M et al (2005) A human beta-cell line for transplantation therapy to control type 1 diabetes. Nat Biotechnol 23:1274–1282
12. Zhu JY, Abate M, Rice PW, Cole CN (1991) The ability of simian virus 40 large T antigen to immortalize primary mouse embryo fibroblasts cosegregates with its ability to bind to p53. J Virol 65:6872–6880
13. Hainaut P (1995) The tumor suppressor protein p53: a receptor to genotoxic stress that controls cell growth and survival. Curr Opin Oncol 7:76–82
14. Asfari M, Janjic D, Meda P, Li G, Halban PA, Wollheim CB (1992) Establishment of 2-mercaptoethanol-dependent differentiated insulin-secreting cell lines. Endocrinology 130: 167–178
15. Miyazaki J et al (1990) Establishment of a pancreatic beta cell line that retains glucose-inducible insulin secretion: special reference to expression of glucose transporter isoforms. Endocrinology 127:126–132
16. Efrat S, Fusco-DeMane D, Lemberg H, al Emran O, Wang X (1995) Conditional transformation of a pancreatic beta-cell line derived from transgenic mice expressing a tetracycline-regulated oncogene. Proc Natl Acad Sci USA 92:3576–3580
17. Wang S et al (1997) Isolation and characterization of a cell line from the epithelial cells of the human fetal pancreas. Cell Transplant 6:59–67
18. Itkin-Ansari P et al (2000) PDX-1 and cell-cell contact act in synergy to promote delta-cell development in a human pancreatic endocrine precursor cell line. Mol Endocrinol 14: 814–822
19. Bodnar AG et al (1998) Extension of life-span by introduction of telomerase into normal human cells. Science 279:349–352
20. Vaziri H, Benchimol S (1998) Reconstitution of telomerase activity in normal human cells leads to elongation of telomeres and extended replicative life span. Curr Biol 8:279–282
21. Jiang XR et al (1999) Telomerase expression in human somatic cells does not induce changes associated with a transformed phenotype. Nat Genet 21:111–114
22. Morales CP et al (1999) Absence of cancer-associated changes in human fibroblasts immortalized with telomerase. Nat Genet 21:115–118
23. Jacobs JJ, Kieboom K, Marino S, DePinho RA, van Lohuizen M (1999) The oncogene and Polycomb-group gene bmi-1 regulates cell proliferation and senescence through the ink4a locus. Nature 397:164–168
24. Salmon P, Oberholzer J, Occhiodoro T, Morel P, Lou J, Trono D (2000) Reversible immortalization of human primary cells by lentivector- mediated transfer of specific genes. Mol Ther 2:404–414
25. Saito M et al (2005) Immortalization of cementoblast progenitor cells with Bmi-1 and TERT. J Bone Miner Res 20:50–57
26. Zhang X et al (2006) Successful immortalization of mesenchymal progenitor cells derived from human placenta and the differentiation

abilities of immortalized cells. Biochem Biophys Res Commun 351:853–859
27. Eberhardt M et al (2006) Multipotential nestin and Isl-1 positive mesenchymal stem cells isolated from human pancreatic islets. Biochem Biophys Res Commun 345:1167–1176
28. Haga K et al (2007) Efficient immortalization of primary human cells by p16INK4a-specific short hairpin RNA or Bmi-1, combined with introduction of hTERT. Cancer Sci 98:147–154
29. Fulcher ML et al (2009) Novel human bronchial epithelial cell lines for cystic fibrosis research. Am J Physiol Lung Cell Mol Physiol 296:L82–L91
30. Unger C et al (2009) Immortalized human skin fibroblast feeder cells support growth and maintenance of both human embryonic and induced pluripotent stem cells. Hum Reprod 24:2567–2581
31. Lim F et al (2010) Reversibly immortalized human olfactory ensheathing glia from an elderly donor maintain neuroregenerative capacity. Glia 58:546–558
32. Park IK et al (2003) Bmi-1 is required for maintenance of adult self-renewing haematopoietic stem cells. Nature 423:302–305
33. Molofsky AV, He S, Bydon M, Morrison SJ, Pardal R (2005) Bmi-1 promotes neural stem cell self-renewal and neural development but not mouse growth and survival by repressing the p16Ink4a and p19Arf senescence pathways. Genes Dev 19:1432–1437
34. Glinsky GV, Berezovska O, Glinskii AB (2005) Microarray analysis identifies a death-from-cancer signature predicting therapy failure in patients with multiple types of cancer. J Clin Invest 115:1503–1521
35. Naldini L et al (1996) In vivo gene delivery and stable transduction of nondividing cells by a lentiviral vector. Science 272:263–267
36. Zufferey R, Nagy D, Mandel RJ, Naldini L, Trono D (1997) Multiply attenuated lentiviral vector achieves efficient gene delivery in vivo. Nat Biotechnol 15:871–875
37. Zufferey R, Donello JE, Trono D, Hope TJ (1999) Woodchuck hepatitis virus posttranscriptional regulatory element enhances expression of transgenes delivered by retroviral vectors. J Virol 73:2886–2892
38. Follenzi A, Ailles LE, Bakovic S, Geuna M, Naldini L (2000) Gene transfer by lentiviral vectors is limited by nuclear translocation and rescued by HIV-1 pol sequences. Nat Genet 25:217–222
39. Zennou V, Petit C, Guetard D, Nerhbass U, Montagnier L, Charneau P (2000) HIV-1 genome nuclear import is mediated by a central DNA flap. Cell 101:173–185
40. Zufferey R et al (1998) Self-inactivating lentivirus vector for safe and efficient In vivo gene delivery. J Virol 72:9873–9880
41. Dull T et al (1998) A third-generation lentivirus vector with a conditional packaging system. J Virol 72:8463–8471
42. Gil J, Bernard D, Martinez D, Beach D (2004) Polycomb CBX7 has a unifying role in cellular lifespan. Nat Cell Biol 6:67–72
43. Ramirez RD et al (2004) Immortalization of human bronchial epithelial cells in the absence of viral oncoproteins. Cancer Res 64:9027–9034
44. Escarpe P et al (2003) Development of a sensitive assay for detection of replication-competent recombinant lentivirus in large-scale HIV-based vector preparations. Mol Ther 8: 332–341
45. Hanawa H et al (2002) Comparison of various envelope proteins for their ability to pseudotype lentiviral vectors and transduce primitive hematopoietic cells from human blood. Mol Ther 5:242–251

Chapter 26

Viral Transformation of Epithelial Cells

Jennifer A. Regan and Laimonis A. Laimins

Abstract

Approximately 18% of human cancers have a viral etiology and the majority of these involve transformation of epithelial cells. Viral proteins transform epithelia by inducing alterations in the normal cell growth and differentiation pathways through the targeting of host proteins. Among the DNA viruses responsible for causing carcinomas are the human papillomaviruses as well as several members of the herpes and polyomavirus families. A number of techniques have been developed to study the mechanisms by which viruses immortalize epithelial cells and alter differentiation properties. These methods include the generation of immortalized lines by transfection or infection as well as the use of organotypic raft cultures, suspension in methylcellulose, and treatment with high calcium levels to examine how differentiation is altered.

Key words: Transformation, Epithelia, Virus, Differentiation, Human papillomavirus

1. Introduction

1.1. Viral Transformation of Epithelia

Several DNA viruses target epithelial cells for transformation and the most prominent of these are the papillomaviruses as well as several members of the herpes and polyomavirus families. The skin forms the primary barrier to infection as most viruses infect humans by entry through oral, anal, or genital portals. Only a limited number of viruses infect the stratified keratinocytes of the skin and several of these are able to induce transformation. In normal stratified epithelia, only cells in the basal layer are actively dividing. As these cells divide and leave the basal layer, they exit the cell cycle stratify and differentiate. Epithelial transformation by viruses involves deregulation of cell growth, alteration of the normal differentiation program, and induction of genetic instability.

It has been estimated that approximately 18% of human cancers are associated with an infectious etiology (1). Both DNA and RNA viruses have been linked to various human cancers. Among the RNA tumor viruses, Hepatitis B and C viruses are linked to the

Scott H. Randell and M. Leslie Fulcher (eds.), *Epithelial Cell Culture Protocols: Second Edition*, Methods in Molecular Biology, vol. 945, DOI 10.1007/978-1-62703-125-7_26, © Springer Science+Business Media, LLC 2012

formation of hepatocellular carcinoma, while human T-lymphotropic virus is associated with adult T-cell leukemia. The human DNA tumor viruses include the Epstein-Barr virus (EBV), Kaposi's sarcoma-associated herpesvirus (KSHV), the polyomaviruses, and human papillomavirus (HPV). In this review we focus on the DNA viruses that target keratinocytes and other stratified epithelia.

1.2. Herpes Viruses

EBV is a ubiquitous double-stranded DNA virus of the gamma herpes virus family, which was first implicated as the human tumor virus in 1965 (2). EBV infection is associated with the development of Burkitt's lymphoma, Hodgkin lymphoma, nasopharyngeal carcinoma (NPC), and B-lymphoproliferative disease. The association between EBV infection and NPC was first indicated by the consistent presence of viral EBV DNA in undifferentiated NPCs (3). These nasopharyngeal tumors contain clonal EBV genomes suggesting that the tumors resulted from expansion of a single EBV-infected progenitor cell (4). Development of NPC is a multifactor process that is also dependent upon the individual's human leukocyte antigen haplotype and exposure at an early age to chemical carcinogens present in specific kinds of salted fish (5, 6). EBV infections are transmitted through saliva with initial infections occurring in the oropharyngeal epithelial cells that can then be transmitted to transiting B-lymphocytes. Following entry, EBV establishes a latent infection in the B-lymphocytes that is characterized by expression of a subset of viral genes. Transformation has been shown to be dependent on the expression of six EBV-encoded gene products which include EBNA-1, -2, -3, -5, -6, and LMP-1 (7).

KSHV or human herpes virus 8 (HHV-8) is another member of the herpesvirus family that is linked to development of human cancers. KSHV has been linked to the development of Kaposi's sarcoma (KS) in AIDS patients and immunosuppressed patients as well as primary effusion lymphoma and a plasmablastic form of Castleman disease. No significant association, however, has been found with epithelial cancers. Several KSHV genes have been implicated in transformation in cell culture but the mechanism by which they contribute to malignant progression in humans is still unclear (8).

1.3. Polyomaviruses: Merkel Cell Carcinoma

Polyomaviruses are ubiquitous, non-enveloped, small double-stranded DNA viruses. The human polyomaviruses include BK virus, JC virus, Karolinska Institute (KI) virus, Washington University (WU) virus, and the recently identified Merkel cell polyomavirus (MCV). Although infections with BK and JC are common in human populations, neither is associated with the development of malignancies. In contrast, introduction of human polyomavirus genomes into mice or hamsters results in a high frequency of cancers. Feng et al. recently described a novel human polyomavirus associated with Merkel cell carcinoma (MCC), a relatively rare form of skin cancer (9). MCC is an aggressive carcinoma of cutaneous neuroendocrine cells that primarily occurs in the elderly and immunosuppressed individuals. The incidence of MCC is approximately 11-fold higher in individuals with AIDS and fivefold higher for patients with a history of organ transplantation (10, 11). MCV genomes were found at high frequency in human MCC tumors integrated into cellular DNA, while in productive infections, the viral DNA is maintained as circular episomes. These integrated viral DNAs contained mutations that result in prematurely truncated large T antigen protein abolishing the ability to replicate (12). This indicates that integration of MCV genomes into host DNAs may be linked to the development of most cases of MCC, but additional studies are needed to confirm this mechanism of action.

1.4. Human Papillomaviruses

Human papillomaviruses are small, double-stranded DNA viruses that are the causative agent of cervical and other anogenital cancers. Over 99% of cervical cancers are positive for a subset of HPV types referred to as high risk (13). Approximately 100 different types of HPVs have been identified and of these over 40 types specifically infect epithelial cells in the genital tract (14). These genital papillomaviruses are sexually transmitted and are divided into low- and high-risk types. Approximately one-third of the genital papillomavirsues are classified as high-risk types which includes types HPV-16, 18, 31, 33, and 45. In contrast, low-risk types, such as HPV-6 and 11, induce benign hyperproliferative lesions (15) and are associated with 90% of cases of genital warts (16). High-risk HPV-16 is present in approximately 60% of patients diagnosed with cervical squamous cell cancer. HPV-18 is the second most prevalent viral type associated with cancers and is detected in around 12% of patients, while HPV-31,33, or 45 are present in approximately 12% of cases (17).

The HPV productive life cycle is closely linked with the differentiation program of the host cell, the keratinocyte. HPV infection occurs by means of microabrasions that expose the undifferentiated basal cells to viral entry. Following entry, HPV genomes are established in the nucleus as episomes at approximately 50–100 copies per cell which are coordinately replicated along with the cellular DNA (17). The copy number in infected basal cells is maintained at low levels and progeny HPV virions are synthesized only in highly differentiated suprabasal cells. In this manner, the cells destroyed by virion production are suprabasal cells, while the virus is maintained in a latent state in undifferentiated basal cells. Most HPV infections in the genital tract are cleared by the immune system typically within 12 or 18 months. A small percentage of women fail to clear the viral infections and become persistently infected. The failure to clear infections by high-risk HPVs in the genital track renders these women at high risk for the development of cervical cancer. The time course between infection and development of cancer is typically several decades but can in some cases progress more rapidly.

In HPV-infected basal cells, low levels of early viral transcripts encoding E6, E7, E1, E2, E1^E4, and E5 are expressed. Due to the polycistronic nature of HPV transcripts, only low levels of factors encoded at the 3′ end of these messages, such as E1^E4 and E5, are translated into proteins in basal cells. Upon cell division, viral DNA is partitioned equally to each of the daughter cells, one of which migrates away from the basement membrane and differentiates in the suprabasal layers. Expression of viral oncoproteins enables these differentiating cells to remain active in the cell cycle. This is necessary as productive viral replication is mediated in large part by cellular enzymes resulting in the amplification of HPV genomes to several thousand copies per cell in the differentiated layers (18). The major late viral promoter is activated upon differentiation resulting in high-level expression of E1^E4 and E5 as well as the capsid proteins L1 and L2. Synthesis of HPV capsid proteins and assembly of progeny virus is induced in the uppermost layers. Interestingly, HPVs only modestly alter the expression of cellular differentiation proteins such as involucrin, filaggrin, and keratin 10. This indicates that the proliferative capacity of suprabasal and the expression of markers of differentiation are not linked.

1.5. E6 and E7 Oncoproteins

The development and maintenance of genital cancers is dependent on the expression of the viral oncoproteins E6 and E7. The E6 proteins of high-risk HPV are approximately 150 amino acids in size and localize to both the nuclei and cytoplasm of HPV-infected keratinocytes (19, 20). The E6 protein contains four C–x–x–C motifs that form two zinc-binding domains. One of the major targets of the high-risk E6 proteins is the p53 tumor suppressor protein. The E6 protein binds the cellular ubiquitin ligase E6-AP and then recruits p53 to form a trimeric complex (21–24). E6-AP then recruits other members of the ubiquitin pathway resulting in the ubiquitination of lysines on p53 and consequent proteolysis of p53 by the proteosome (22). This results in reduced levels of p53, which enables HPV-infected cells to bypass the normal growth arrest signals at the G1/S and G2/M cell cycle checkpoints. This process is also important for maintaining HPV positive cells active in the cell cycle upon differentiation.

The high-risk E6 proteins have also been shown to play an important role in the activation of telomerase, an enzyme which replicates telomeric DNA at the end of chromosomes (25). Human telomerase reverse transcriptase (hTert) is a catalytic subunit of telomerase and expression of this subunit has been shown to play a primary role in the regulation of telomerase activity (26). Studies have shown that expression of the E6 protein upregulates expression of endogenous hTert through the transcriptional activation of the hTert promoter (27, 28). The activation of telomerase is a characteristic of most human cancers.

The high-risk E6 protein has also been shown to interact with several PDZ domain-containing proteins such as hDlg (29), hScrib (30), MUPP1 (31), and MAGI-1, -2, -3 (32, 33). PDZ domains consist of approximately 90 amino acids and are most often found in cellular proteins located at areas of cell–cell contact. PDZ proteins form multiprotein signaling complexes on the inner surface of plasma membranes to control cell growth, cell polarity as well as adhesion in response to cell contact. Interaction of E6 with these PDZ domain-containing proteins leads to their ubiquitin-mediated degradation. Furthermore, the PDZ-binding domain has been shown to be necessary for some of the transforming functions of the E6 protein (34, 35).

The E7 protein is equally important in viral transformation and acts cooperatively with E6. The high-risk E7 proteins are approximately 100 amino acids in size and localize predominantly to the nucleus. Dimerization of E7 proteins has been shown to occur through a zinc finger motif located in the C terminus. The expression of E6 by itself cannot immortalize human cells; however co-expression of E7 with E6 results in efficient immortalization (35, 36). A primary target of the E7 proteins is the retinoblastoma family of proteins, which consists of Rb, p107, and p130 (37, 38). These proteins act to regulate cell cycle progression through the G1/S checkpoint by regulating the activity of the E2F family of transcription activators. E2F 1–3 are positive activators of expression of genes involved in DNA synthesis that are expressed prior to S phase entry. In G1, Rb is hypophosphorylated and binds to E2F transcription factors repressing their activity. As cells move through late G1, CDK kinases phophorylate Rb resulting in its release from E2F factors and consequent activation of expression of S-phase-specific genes. The E7 protein binds the Rb family of proteins and also induces their degradation (37, 39–41) The association of E7 with Rb family members is important not only for constitutive activation of E2F regulated transcription but also for allowing cells to remain active in the cell cycle upon differentiation. The high-risk E7 proteins have also been shown to manipulate cell cycle progression by binding to cyclin-dependent kinase inhibitors, p21 as well as p27, and deregulate their inhibitory effects (42–44).

1.6. E1^E4 Proteins

Upon differentiation, the expression of two factors, E1^E4 and E5, is induced through the activation of the late viral promoter. The E1^E4 protein is expressed as a fusion protein that includes the first five amino acids of E1 fused with the E4 ORF. E1^E4 is the most abundantly expressed HPV protein but its function is still not well understood. It is synthesized primarily in differentiated suprabasal cells and is predominantly cytoplasmic in localization though a small fraction can be detected in the nucleus. In undifferentiated cells, only low levels of E1^E4 and E5 proteins are synthesized. The E1^E4 proteins associate with cytokeratins and their high level transient expression induces the collapse of the cytokeratin network, suggesting that E1^E4 may play a role facilitating viral egress (45). Both low-risk HPV1 and high-risk HPV16 E1^E4 have been shown to bind to keratin 18 and associate with the keratin intermediate filament network through a leucine cluster (LLXLL) (46). In addition to causing cytokeratin collapse, the overexpression of E1^E4 induces a G2 arrest and in this way may facilitate productive viral replication in differentiated cells (47). In cells that maintain episomal copies of either HPV16 or HPV31 viral genomes in which the E1^E4 gene was disrupted, the level of viral DNA amplification and late gene expression in differentiated cells was significantly reduced (48–50). These results suggest that the E1^E4 protein is essential for the late productive phase of the viral life cycle in the high-risk viral types and may be linked to its role in inducing G2 arrest.

1.7. HPV E5

The E5 protein is a small, hydrophobic protein located predominantly in the endoplasmic reticulum and is co-expressed along with E1^E4 in differentiated suprabasal cells (51–53). The high-risk E5 proteins are approximately 85 amino acids in size and contain three hydrophobic domains. E5 is weakly oncogenic in heterologous expression assays and is able to stimulate proliferation of E6/E7 expressing cells but very little is known about how it functions (54, 55). In contrast to the HPV E5 protein, the role of the bovine papillomavirus type 1 (BPV-1) E5 protein has been extensively characterized and is the primary oncoprotein of the fibrosarcoma-inducing virus (56–58). The BPV-1 E5 protein induces transformation of mouse cells by binding and constitutively activating the platelet-derived growth factor (PDGF) β receptor (59). However, the amino acids necessary for the transforming ability of BPV E5 are not conserved in the HPV E5 proteins.

The epidermal growth factor receptor (EGFR) has been suggested to be a key target of HPV E5 proteins (60–62). HPV16 E5 has been reported to bind to the EGFR in COS cells, however these complexes have not been seen in other cell types (63, 64). A second target of E5 proteins is the vacuolar H^+-ATPase which is a proton pump responsible for acidifying cellular organelles such as the Golgi, lysosomes, and endosomes (51, 65, 66). The amino acids in BPV-1 E5 that are necessary for transformation, however, are not necessary for binding to the 16 kDa subunit (67). Furthermore, the vacuolar H^+-ATPase is found predominantly in endosomes, whereas the HPV E5 protein is found in the endoplasmic reticulum (51, 67). Recently, B-cell associated protein 31 (Bap31) has been identified as a binding partner of HPV16 as well as HPV31 E5. The binding of E5 to Bap31 correlates with the ability to retain the proliferative capacity of keratinocytes following differentiation (68).

Studies examining the role of E5 in the productive life cycle of HPV31 and HPV16 have demonstrated that E5 can modulate late viral functions following differentiation (69, 70). Keratinocytes that maintain wild-type HPV31 genomes remain active in the cell cycle following differentiation and retain proliferation ability as measured by colony forming ability. In contrast, cells harboring the HPV31 E5 mutant genomes showed a significant reduction in colony-forming ability following methylcellulose-induced differentiation (69). This suggests that E5 maintains cells in a proliferative competent state upon differentiation. In addition, HPV31 and HPV16 genomes containing mutations that disrupted E5 expression exhibited reduced ability to amplify viral genomes and activate late gene expression. Interestingly, the level and phosphorylation state of the EGFR were not affected in differentiating cells that maintained complete HPV31 genomes indicating there may be other targets of E5. Further work will be needed to fully clarify the role of the E5 protein in HPV pathogenesis.

Various techniques have been developed to study the mechanisms involved in viral transformation of epithelium. Generation of stable keratinocytes harboring the HPV genome transfection of cloned HPV genomes and the development of various techniques to induce differentiation has provided new tools for the study of epithelial transformation.

2. Materials

2.1. Stable Transfection

1. Keratinocyte Growth Factor Medium (KGM).
2. Fugene® 6 Transfection Reagent (Roche).
3. 0.05% Trypsin/1 mM EDTA (Life Technologies™).
4. Mitomycin C (Roche) stock solution: 0.4 mg/mL in PBS.
5. G418: 50 mg/mL stock solution.
6. T4 DNA ligase (New England Biolabs).
7. Versene: 1 mL 0.5 M sterile EDTA (Sigma-Aldrich®), pH 8 in 1 L of phosphate-buffered saline.
8. 60 mm culture dish.
9. 5× ligation buffer: 50 mM Tris–Cl pH 7.5, 10 mM $MgCl_2$ 10 mM dithiothreitol, 1 mM ATP, 25 mg/mL bovine serum albumi (BSA).
10. 5 M NaCl: solid NaCl dissolved in water.
11. HPV genome plasmid DNA (e.g. pBR322-HPV31).
12. pSV2Neo DNA.

13. Normal human foreskin keratinocytes (NHKs): isolated from human foreskins (Life Technologies™).
14. J2 mouse fibroblasts: spontaneously immortalized fibroblast cell line.

2.2. E Medium

1. E medium Supplements: 100× Cocktail: Combine 200 mL 180 mM adenine (Sigma-Aldrich®), 200 mL 5 mg/mL insulin (Sigma-Aldrich®), 200 mL 5 mg/mL transferrin (Sigma-Aldrich®), 200 mL 20 nM 3,3′,5-Triiodo-L-thyronine (Sigma-Aldrich®), 1,200 mL sterile PBS, mix well and filter-sterilize. Store in 400 mL aliquots at −20 °C.
2. 1,000× Cholera enterotoxin: 1 mg vial Cholera enterotoxin (ICN Biomedicals, Inc.) is reconstituted with 1 mL ddH_2O to make a 1×10^{-5} M stock solution. Further dilute with 99 mL ddH_2O. Store at 4 °C in the dark.
3. 1,000× Hydrocortisone: 100 g hydrocortisone (Life Technologies™) is dissolved in 20 mL 100% EtOH to make a 5 mg/mL stock. Take 19.2 mL of the stock and add 220.8 mL HEPES buffer (see Subheading 2.3, item 2). Make 6×40 mL aliquots (0.4 μg/mL) and store at −20 °C.
4. E-medium preparation: For a 40-L preparation, combine the following in a 40-L carboy: 30 L of distilled, deionized water (ddH_2O), 3 (10 L) packets of DMEM powder (Sigma-Aldrich®) dissolved in 3 L ddH_2O, 10 (1 L) packets of Hams F-12 powder dissolved in 1 L ddH_2O, 122.7 g tissue culture-grade $NaHCO_3$ dissolved in 1 L ddH_2O, 400 mL 100× cocktail E medium supplements, 400 mL penicillin–streptomycin (Life Technologies™), 40 mL 1,000× hydrocortisone, 40 mL 1,000× cholera toxin. Shake briefly to mix. Add 12.5 mL concentrated HCl (dilute in ddH_2O first), shake, and remove a small quantity to check that pH is approximately 7.1. For 5% FBS (Hyclone), add 2 L of FBS. Make up to 40 L with deionized ddH_2O water and shake. Filter-sterilize using a low-protein-binding 0.2-μM filter and aliquot aseptically. Store at 4 °C in the dark.
5. Mouse epidermal growth factor (EGF): 100 μg EGF (Life Technologies™) and 10 mg BSA (Life Technologies™), each dissolved in 10 mL ddH_2O is combined and brought up to 100 mL with ddH_2O. Filter-sterilize, aliquot and store at −20 °C. E-medium is supplemented with 5 ng/mL (1:200 dilution of 1 μg/mL) EGF immediately before use.

2.3. Differentiation of Epithelia

1. Rat Tail Collagen type I: 2 mg/mL in 0.02 N acetic acid.
2. HEPES solution 1 M stock solution is made by dissolving 71.5 g HEPES into 250 mL ddH_2O. Use 10 N NaOH to adjust the pH to 7. Bring the volume to 300 mL and filter-sterilize.

3. DMEM without sodium bicarbonate (Life Technologies™): Filter sterilize (to prevent the formation of precipitate), aliquot and store at −20 °C.
4. Stainless steel metal grids (6 cm diameter) (Williams and Mettle Co.).
5. 1.5% Methylcellulose solution: 1.5% methylcellulose solution is prepared by combining autoclaved dry methylcellulose (400 cps, Sigma-Aldrich®) with E medium and heating the solution to 60 °C for 20 min. An equal volume of E medium is then added and the solution is stirred at 4 °C overnight.
6. Lysis buffer: 400 mM NaCl, 10 mM Tris–HCl, and 10 mM EDTA at a final pH of 7.4.
7. Paraformaldehyde: 4% (w/v) solution in PBS should be prepared for each experiment. The solution should be carefully heated to dissolve and then cooled to room temperature.
8. 10× reconstitution buffer: Dissolve 2.2 g $NaHCO_3$ and 4.8 g HEPES in 100 mL 0.05 M NaOH. Filter sterilize and aliquot and store at −20 °C.
9. TE: 10 mM Tris–HCl, pH 7.5, 1 mM EDTA.
10. Phenol chloroform solution: phenol/chloroform/isoamylethanol (25:24:1).
11. RNAse A (Sigma-Aldrich®): 10 mg/mL in water stock solution.
12. Proteinase K (Sigma-Aldrich®): 10 mg/mL in water stock solution.
13. 10 cm diameter Petri dish.
14. 15 mL and 50 mL conical tubes.
15. PBS without Ca/Mg.
16. 0.2% SDS: sodium dodecyl sulfate (SDS) 0.2% (w/v) in water.
17. 18-gauge needle.
18. 3 M sodium acetate: dissolve sodium acetate powder in water and adjust pH with HCl to pH 5.2.
19. Trizol®(Life Technologies™).
20. Isopropyl alcohol.
21. Dichromate cleaning solution (Malinckrodt).
22. 6-Well culture dish.
23. BrdU: 1 mg BrdU/mL water stock solution; final working concentration 50 μg/mL.
24. 1.5 mM $CaCl_2$: dissolve solid $CaCl_2$ in water.
25. Cell scrapers.

3. Methods

3.1. Stable Transfection

1. Viral genomes are first released from bacterial plasmids following digestion with appropriate restriction enzymes in a total reaction volume of 52.5 μL at 37 °C overnight. The specific restriction enzyme to be used depends on the viral type being transfected and how it is cloned into the bacterial vector.
2. Complete digestion of plasmids is confirmed by analysis of 2.5 μL of reaction by agarose gel electrophoresis followed by heat inactivation of the restriction at 65 °C for 10 min.
3. The genomes are unimolecularly religated by adding reaction volume to 180 μL of 5× ligation buffer, 2 μL T4 DNA ligase, and 668 μL of H_2O at 16 °C overnight.
4. The DNA was then precipitated by adding 600 μL of isopropyl ethanol and 180 μL of 5 M NaCl overnight at −20 °C.
5. The following day, the DNA was isolated by 30 min centrifugation at 13,000 × *g* at 4 °C, washed with 70% ethanol, and resuspended in 15 μL of TE (*see* Subheading 2.3, item 9).
6. Ligation is confirmed by analysis of 1 μL of resuspension by agarose gel electrophoresis.
7. To establish stable HPV positive keratinocyte cell lines, an early passage normal human foreskin keratinocytes (NHKs) isolated from neonatal foreskins are plated onto a 60 mm tissue culture dish and grown in KGM to approximately 50% confluency.
8. For each plate, 6 μL of Fugene® 6 was added to 94 μL of KGM. The HPV genome (1 μg) and pSV2Neo DNA (1 μg) is then added to the KGM/Fugene® solution, mixed by gently tapping the tube, and incubated at room temperature for 45 min (see Fig. 1). One hundred microliters of the mixture is gently mixed with 4 mL of fresh KGM medium in the 60 mm plate.
9. The next day, the keratinocytes are plated onto mitomycin C-treated J2 mouse fibroblasts in E medium (see Subheading 2.2). Cells are selected with G418 at a final concentration of 200 μg/mL for the first 4 days followed by 100 μg/mL for the final 4 days (see Note 1). The drug containing medium is changed every 2 days and mitomycin treated J2 fibroblasts are added on days 3, 5, 7, and 9. The selected colonies are then pooled and expanded (see Fig. 1).

3.2. Differentiation of Keratinocytes in Semisolid Medium

1. Sub-confluent keratinocytes are removed with trypsin and suspended in a 10 cm diameter petri dish with 25 mL of methylcellulose solution (see Subheading 2.3, item 5) for either 24 or 48 h at 37 °C in a CO_2-humidified incubator (see Note 2).

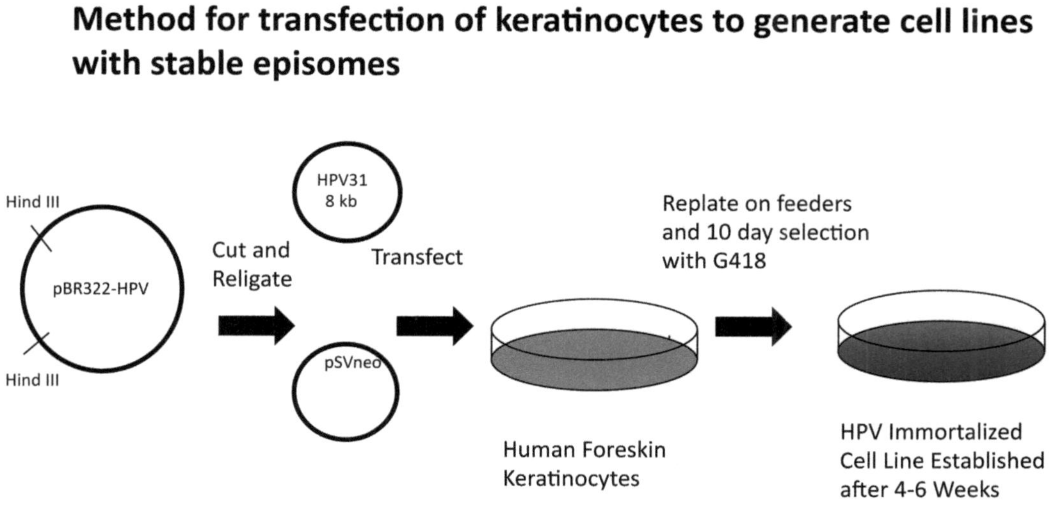

Fig. 1. Stable transfection of the HPV31 genome into normal human keratinocytes. The genomes are first released from plasmids following digestion with appropriate restriction enzymes depending upon the viral type being studied and unimolecularly religated using T4 DNA ligase. Three micrograms of DNA are then cotransfected with 1 μg of pSV2neo into 50% confluent normal human keratinocytes (NHKs) using Fugene®. The next day, the keratinocytes are plated onto mitomycin C-treated fibroblasts in E medium. Cells are selected with G418 and the selected colonies are then pooled and expanded.

2. The cells are scraped with cell scrapers into two 50 mL conical tubes, centrifuged at 5,000 × *g*, and washed four times with ice cold phosphate-buffered saline (PBS) at 4 °C. DNA and RNA are extracted from the cell pellets for further analysis.

3.3. Isolation of Viral DNA from Normal Human Keratinocytes

1. Harvest keratinocytes when they are approximately 80% confluent by first removing J2 fibroblasts with versene (see Subheading 2.1, item 7). After versene treatment, trypsinize keratinocytes (see Subheading 2.1, item 3) to detach them from tissue culture dishes and inactivate trypsinization with E-medium. Spin cells at 200 × *g* for 5 min and wash once with 10 mL of PBS.
2. Resuspend cell pellets from one 100 mm tissue culture dishes in 3 mL lysis buffer (see Subheading 2.3, item 6).
3. Add RNase A to a final concentration of 50 μg/mL, vortex and incubate at room temperature.
4. After incubation for 30 min, add Proteinase K to a final concentration of 50 μg/mL in a solution of 0.2% SDS and vortex. Incubate lysis mixture at 37 °C overnight.
5. After overnight incubation, shear cellular DNA by passing the mixture through an 18-gauge needle approximately ten times.
6. Perform a phenol chloroform extraction by adding 6 mL of phenol chloroform solution (see Subheading 2.3, item 10) and spin at 3,077 × *g* for 5 min. Repeat extraction two additional times.

7. Extract supernatant with 6 mL chloroform/isoamylethanol (24:1) two times.
8. Ethanol precipitate overnight by adding two times the volume of ethanol, one-tenth volume 3 M sodium acetate pH 5.2 and place at −20 °C.
9. Wash twice with 70% ethanol and resuspend in 100 μL TE (see Subheading 2.3, item 9). Incubate DNA in TE for 30 min at 37 °C or 15 min at 60 °C to resuspend DNA before storing.

3.4. Isolation of RNA from Normal or Virally Transformed Keratinocytes

1. Harvest RNA from monolayer cultures that are approximately 80% confluent by first removing J2 fibroblast feeders by incubation with versene for 5 min and removal of solution which contains most of the fibroblasts.
2. After versene treatment, add 7.2 mL of Trizol® to each 100 mm plate and incubate at room temperature for 1 min. Pipette the Trizol®/lysate mixture up and down using a pipetman to ensure complete lysis and then place mixture in a 15 mL falcon tube (see Note 3).
3. Following incubation at room temperature for 3–5 min, add 1.4 mL chloroform per 15 mL falcon tube and shake tubes vigorously for 15 s. Incubate tubes for additional 2–3 min at room temperature and spin at 1,730 × *g* for 15 min at 4 °C. Transfer the aqueous phase to a fresh tube (see Note 4).
4. Precipitate the RNA by adding 3.6 mL of isopropyl alcohol. Mix samples by inverting tubes gently and incubate samples at room temperature.
5. After incubation, spin down samples at 1,730 × *g* for 10 min. Precipitated RNA will form at the bottom of the tube as a gel-like pellet.
6. Remove supernatant and wash pellet once with 7.2 mL of 75% ethanol and mix by vortexing. Briefly air dry pellet and resuspend in 50 μL TE. Incubate RNA in TE at 55–60 °C for 10 minutes to resuspend RNA before storing.

3.5. Differentiation of Keratinocytes in Organotypic Raft Cultures

1. Stainless steel metal grids are first treated with dichromate cleaning solution for 1 h and then placed in a beaker and rinsed with constantly flowing tap water overnight. The following day, the grids are rinsed for an additional 3–5 h with double-distilled water. Three sides of the grid equidistant apart are bent approximately 0.5 cm and then autoclaved.
2. For each raft, make one collagen gel consisting of 2.4 mL rat-tail collagen type 1, 0.3 mL 10× reconstitution buffer, 0.3 mL DMEM without sodium carbonate, and 1–2 × 10^6 J2 mouse fibroblasts (see Note 5). The collagen is added last to prevent premature solidification. Filter sterilized 1 N NaOH may be added until the mixture has a reddish orange color.

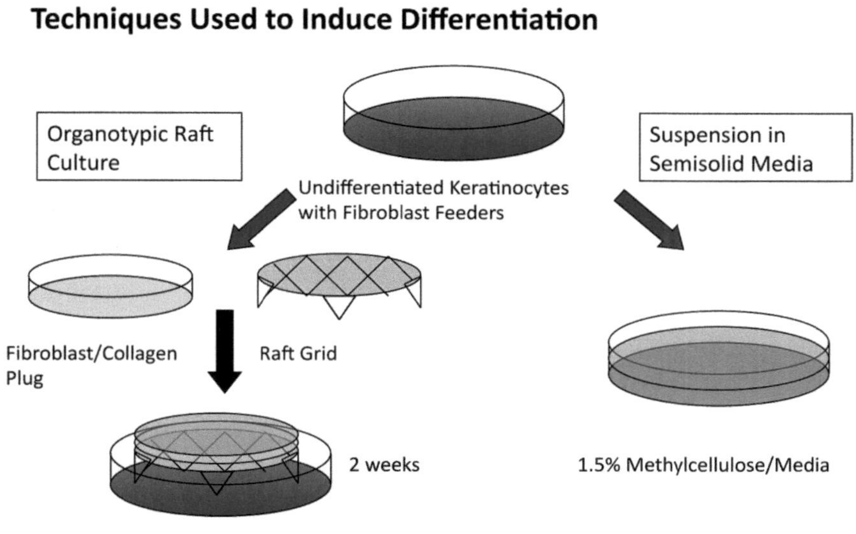

Fig. 2. Methods to induce keratinocyte differentiation to study viral effects. Keratinocyte differentiation is induced by suspension in 1.5% methylcellulose and organotypic raft culture. For methylcellulose-induced differentiation, the keratinocytes are harvested at 80% confluency and suspended in 25 mL of 1.5% methylcellulose solution for either 24 or 48 h at 37 °C. The cells are harvested into two 50 mL conical tubes, centrifuged, and washed four times with ice cold PBS at 4 °C. DNA and RNA can be extracted from the cell pellets. In order to induce differentiation in the organotypic raft culture system, cells are plated onto a solidified collagen matrix containing J2 3T3 fibroblasts. Upon confluency, the cells are transferred to a metal grid to induce differentiation. After 14 days, the raft cultures are harvested, fixed in 4% paraformaldehyde, paraffin embedded, and sectioned.

3. Three milliliters of the collagen mixture is added to a 6-well culture dish and incubated in a 37 °C incubator for 30 min to solidify. Three milliliters of E-medium with EGF (see Subheading 2.2, item 5) is then placed on top of the solidified collagen and stored at 37 °C.
4. Within 2–4 days, NHKs are versene treated, trypsinized, and $1–2 \times 10^6$ cells are plated on top of the collagen gel (see Fig. 2). Upon confluency, which typically takes 2–4 days, the cells are transferred to a 6 cm metal grid in a 10 cm dish using a sterile forceps to induce differentiation.
5. After 14 days, the raft cultures are placed in 4% paraformaldehyde (see Subheading 2.3, item 7) using tweezers, incubated for at least 4 h, paraffin embedded, and sectioned. The slides can be stained with hematoxylin and eosin for morphological analysis or examined by immunohistochemistry.
6. Proliferation was also assessed by identifying cells actively synthesizing DNA using BrdU. For this analysis, 20 μM BrdU can be added to E-medium 12 h before the raft culture was harvested.

3.6. Calcium-Induced Differentiation

1. To induce differentiation using high calcium levels, cells were grown to approximately 90% confluency in KGM with growth supplements for at least 24 h.
2. The medium was then changed to KGM without supplements containing 1.5 mM $CaCl_2$. The cells are harvested at various time points up to 96 h post-differentiation.

4. Notes

1. The effect of G418 might not be evident until several days after initial selection. If cells under selection grow too confluent, split cells at a 1:3 ratio.
2. For suspension in 1.5% methylcellulose, cells should be approximately 80% confluent. It is essential that cells are harvested during exponential growth as inadequate differentiation could occur if cells are too confluent.
3. Cell lysates in Trizol® can be stored at −80 °C until ready to continue the remaining steps of the RNA isolation protocol.
4. After centrifugation during Trizol® extraction, there should be three different phases: a lower red phenol chloroform phase, an inter-phase, and a colorless upper aqueous phase.
5. Mouse J2 fibroblasts are the preferred fibroblast feeders. While human fibroblasts can be substituted they are not as effective as mouse fibroblasts.

References

1. Carrillo-Infante C, Abbadessa G, Bagella L, Giordano A (2007) Viral infections as a cause of cancer. Int J Oncol 30:1521–1528
2. Epstein MA, Henle G, Achong BG, Barr YM (1965) Morphological and biological studies on a virus in cultured lymphoblasts from Burkitt's lymphoma. J Exp Med 121:761–770
3. Desgranges C, Wolf H, De-The G, Shanmugaratnam K, Cammoun N, Ellouz R et al (1975) Nasopharyngeal carcinoma X. Presence of epstein-barr genomes in separated epithelial cells of tumours in patients from Singapore, Tunisia and Kenya. Int J Cancer 16:7–15
4. Raab-Traub N, Flynn K (1986) The structure of the termini of the Epstein-Barr virus as a marker of clonal cellular proliferation. Cell 47:883–889
5. Busson P, Keryer C, Ooka T, Corbex M (2004) EBV-associated nasopharyngeal carcinomas: from epidemiology to virus-targeting strategies. Trends Microbiol 12:356–360
6. Chan AT, Teo PM, Johnson PJ (2002) Nasopharyngeal carcinoma. Ann Oncol 13:1007–1015
7. Tomkinson B, Roberson E, Kieff E (1993) Epstein-Barr virus nuclear proteins EBNA-3A and EBNA-3 C are essential for B-lymphocyte growth transformation. J Virol 67:2014–2025
8. Rezaee SA, Cunningham C, Davison AJ, Blackbourn DJ (2006) Kaposi's sarcoma-associated herpesvirus immune modulation: an overviews. J Gen Virol 87:1781–1804
9. Feng H, Shuda M, Chang Y, Moore PS (2008) Clonal integration of a polyomavirus in human merkel cell carcinoma. Science 319:1096–1100
10. Penn I, First MR (1999) Merkel's cell carcinoma in organ recipients: report of 41 cases. Transplantation 68:1717–1721

11. Lanoy E, Dores GM, Madeleine MM, Toro JR, Fraumeni JF, Engels EA (2009) Epidemiology of nonkeratinocytic skin cancers among persons with AIDS in the United States. AIDS 23:385–393
12. Shuda M, Feng H, Kwun HJ, Rosen ST, Gjoerup O, Moore PS, Chang Y (2008) T antigen mutations are a human tumor-specific signature for Merkel cell polyomavirus. Proc Natl Acad Sci USA 105:16272–16277
13. Walboomers JMM, Jacobs MV, Manos MM, Bosch FZ, Kummer JA, Shah KV et al (1999) Human papillomavirus is a necessary cause of invasive cervical cancer worldwide. J Pathol 189:12–19
14. Weaver BA (2006) Epidemiology and natural history of genital human papillomavirus infection. J Am Osteopath Assoc 106:S2–S8
15. zur Hausen H (2002) Papillomaviruses and cancer: from basic studies to clinical application. Nat Rev Cancer 2:342–350
16. Steben M, Duarte-Franco E (2007) Human papillomavirus infection: epidemiology and pathophysiology. Gynecol Oncol 107:S2–S5
17. Hummel M, Hudson JB, Laimins LA (1992) Differentiation-induced and constitutive transcription of human papillomavirus type 31b in cell lines containing viral episomes. J Virol 66:6070–6080
18. Ruesch MN, Laimins LA (1998) Human papillomavirus oncoproteins alter differentiation-dependent cell cycle exit on suspension in semisolid medium. Virology 250:19–29
19. Barbosa MS, Lowy DR, Schiller JT (1989) Papillomavirus polypeptides E6 and E7 are zinc-binding proteins. J Virol 63:1404–1407
20. Cole ST, Danos O (1987) Nucleotide sequence and comparative analysis of the human papillomavirus type 18 genome. Phylogeny of papillomaviruses and repeated structure of the E6 and E7 gene products. J Mol Biol 193:599–608
21. Huibregtse JM, Scheffner M, Howley PM (1991) A cellular protein mediates association of p53 with the E6 oncoprotein of human papillomavirus types 16 or 18. EMBO J 10:4129–4135
22. Scheffner M, Huibregtse JM, Vierstra RD, Howley PM (1993) The HPV-16 E6 and E6-AP complex functions as a ubiquitin-protein ligase inthe ubiquination of p53. Cell 75:495–505
23. Scheffner M, Werness BA, Huibregtse JM, Levine AJ, Howley PM (1990) The E6 oncoprotein encoded by human papillomavirus types 16 and 18 promotes the degradation of p53. Cell 63:1129–1136
24. Werness BA, Levine AJ, Howley PM (1990) Association of human papillomavirus types 16 and 18 E6 proteins with p53. Science 248: 76–79
25. Klingelhutz AJ, Foster SA, McDougall JK (1996) Telomerase activation by the E6 gene product of human papillomavirus type 16. Nature 380:79–82
26. Meyerson M, Counter CM, Eaton EN, Ellisen LW, Steiner P, Caddle SD et al (1997) hEST2, the putative human telomerase catalytic subunit gene, is upregulated in tumor cells and during immortalization. Cell 90:785–795
27. Gewin L, Galloway DA (2001) E box-dependent activation of telomerase by human papillomavirus type 16 E6 does not require induction of c-myc. J Virol 75:7198–7201
28. Oh ST, Kyo S, Laimins LA (2001) Telomerase activation by human papillomavirus type 16 E6 protein: induction of human telomerase reverse transcriptase expression through Myc and GC-rich Sp1 binding sites. J Virol 75:5559–5566
29. Kiyono T, Hiraiwa A, Fujita M, Hayashi Y, Akiyama T, Ishibashi M (1997) Binding of high-risk human papillomavirus E6 oncoproteins to the human homologue of the drosophilia discs large tumor suppressor protein. Proc Natl Acad Sci USA 94:11612–11616
30. Nakagawa S, Huibregtse JM (2000) Human scribble (Vartul) is targeted for ubiquitin-mediated degradation by the high-risk papillomavirus E6 proteins and the E6AP ubiquitin-protein ligase. Mol Cell Biol 20:8244–8253
31. Lee SS, Glaunsinger B, Mantovani F, Banks L, Javier RT (2000) Multi-PDZ domain protein MUPP1 is a cellular target for both adenovirus E4-ORF1 and high-risk papillomavirus type 18 E6 oncoproteins. J Virol 74:9680–9693
32. Glaunsinger B, Lee SS, Thomas M, Banks L, Javier RT (2000) Interactions of the PDZ-protein MAGI-1 with adenovirus E4^ORF1 and high risk papillomavirus E6 oncoproteins. Oncogene 19:5270–5280
33. Thomas M, Laura R, Hepner K, Guccione E, Sawyers C, Lasky L, Banks L (2002) Oncogenic human papillomavirus E6 proteins target the MAGI-2 and MAGI-3 proteins for degradation. Oncogene 21:5088–5096
34. Lee SS, Weiss RS, Javier RT (1997) Binding of human virus oncoproteins to hDlg/SAP97, a mammalian homolog of the Drosophilia discs large tumor suppressor protein. Proc Natl Acad Sci USA 94:6670–6675
35. Bedell MA, Jones KH, Grossman SR, Laimins LA (1989) Identification of human papillomavirus type 18 transforming genes in immortalized and primary cells. J Virol 63:1247–1255
36. Halbert CL, Demers GW, Galloway DA (1991) The E7 gene of human papillomavirus type 1 is

sufficient for immortalization of human epithelial cells. J Virol 65:473–478

37. Dyson N, Howley PM, Munger K, Harlow E (1989) The human papillomavirus 16 E7 oncoprotein is able to bind to the retinoblastoma gene product. Science 243:934–937
38. Davies R, Hicks R, Crook T, Morris J, Vousden K (1993) Human papillomavirus type 16 E7 associates with a histone H1 kinase and with p107 through sequences necessary for transformation. J Virol 67:2521–2528
39. Cheng S, Schmidt-Grimminger DC, Murant T, Broker TR, Chow LT (1995) Differentiation-dependent up-regulation of the human papillomavirus E7 gene reactivates cellular DNA replication in suprabasal differentiated keratinocytes. Genes Dev 9:2335–2349
40. Martin LG, Demers GW, Galloway DA (1998) Disruption of the G1/S transition in human papillomavirus type 16 E7-expressing human cells is associated with altered regulation of cyclin E. J Virol 72:975–985
41. Munger K, Werness BA, Dyson N, Phelps WC, Harlow E, Howley PM (1989) Complex formation of human papillomavirus E7 protein with the retinoblastoma tumor suppressor gene product. EMBO J 8:4099–4105
42. Funk JO, Waga S, Harry JB, Espling E, Stillman B, Galloway DA (1997) Inhibition of CDK activity and PCNA-dependent DNA replication by p21 is blocked by interaction with the HPV-16 E7 oncoprotein. Genes Dev 11:2090–2100
43. Jones DL, Lani RM, Munger K (1997) The human papillomavirus E7 oncoprotein can uncouple cellular differentiation and proliferation in human keratinocytes by abrogating p21Cip1-mediated inhibition of cdk2. Genes Dev 11:2101–2111
44. Zerfass-Thome K, Zwerschke W, Mannhardt B, Tindle R, Botz JW, Jansen-Durr P (1996) Interaction of the cdk inhibitor p27KIP1 by the human papillomavirus type 16 E7 protein. Oncogene 13:2323–2330
45. Doorbar JSE, Sterling J, Mclean C, Crawford L (1991) Specific interation between HPV-16 E1^E4 and cytokeratins results in collapse of the epithelial cell intermediate filament network. Nature 352:824–827
46. Wang Q, Griffin H, Southern S, Jackson D, Martin A, McIntosh P et al (2004) Functional analysis of the human papillomavirus type 16 E1^E4 protein provides a mechanism for in vivo and in vitro keratin filament reorganization. J Virol 78:821–833
47. Davy CE, Jackson DJ, Wang Q, Raj K, Masterson PJ, Fenner MF et al (2002) Identification of a G2 arrest domain in the E1^E protein of human papillomavirus type 16. J Virol 76:9806–9818
48. Nakahara T, Peh WL, Doorbar J, Lee D, Lambert PF (2005) Human papillomavirus type 16 E1^E4 contributes to multiple facets of the papillomavirus life cycle. J Virol 79:13150–13165
49. Wilson R, Fehrmann F, Laimins LA (2005) Role of the E1^E4 protein in the differentiation-dependent life cycle of human papillomavirus type 31. J Virol 79:6732–6740
50. Wilson R, Ryan GB, Knight GL, Laimins LA, Roberts S (2007) The full-length E1^E4 protein of human papillomavirus type 18 modulates differentiation-dependent viral DNA amplification and late gene expression. Virology 362:453–460
51. Conrad M, Bubb VJ, Schlegel R (1993) The human papillomavirus type 6 and 16 E5 proteins are membrane-associated proteins which associate with the 16-kilodalton pore-forming protein. J Virol 67:6170–6178
52. Disbrow GL, Hanover JA, Schlegel R (2005) Endoplasmic reticulum-localized human papillomavirus type 16 E5 protein alters endosomal pH but not trans-Golgi pH. J Virol 79: 5839–5846
53. Disbrow GL, Sunitha I, Baker CC, Hanover J, Schlegel R (2003) Codon optimization of the HPV-16 E5 gene enhances protein expression. Virology 311:105–114
54. Bouvard VG, Matlashewski G, Gu ZM, Storey A, Banks L (1994) The human papillomavirus type 16 E5 gene cooperates with the E7 gene to stimulate proliferation of primary cells and increases viral gene expression. Virology 203:73–80
55. Valle GF, Banks L (1995) The human papillomavirus HPV-6 and HPV-16 E5 proteins cooperate with HPV-16 E7 in the transformation of primary rodent cells. J Gen Virol 96:1239–1245
56. Bergman P, Ustav M, Sedman J, Moreno-Lopez J, Vennstrom B, Pettersson U (1988) The E5 gene of bovine papillomavirus type 1 is sufficient for complete oncogenic transformation of mouse fibroblasts. Oncogene 2:453–459
57. Schiller JT, Vass WC, Vousden KH, Lowy DR (1986) E5 open reading frame of bovine papillomavirus type 1 encodes a transforming gene. J Virol 57:1–6
58. Schlegel R, Wade-Glass M, Rabson MS, Yang YC (1986) The E5 transforming gene of bovine papillomavirus encodes a small, hydrophobic polypeptide. Science 233:464–467
59. Petti L, Nilson L, DiMaio D (1991) Activation of the platelet-derived growth factor receptor

by the bovine papillomavirus E5 transforming protein. EMBO J 10:845–855

60. Leechanachai P, Banks L, Moreau F, Matlashewski G (1992) The E5 gene from human papillomavirus type 16 is an oncogene which enhances growth factor-mediated signal transduction to the nucleus. Oncogene 7:19–25
61. Straight SW, Hinkle PM, Jewers RJ, McCance DJ (1993) The E5 oncoprotein of human papillomavirus type 16 transforms fibroblasts and effects the downregulation of the epidermal growth factor receptor in keratinocytes. J Virol 67:4521–4532
62. Stoppler MC, Straight SW, Tsao G, Schlegel R, McCance DJ (1996) The E5 gene of HPV-16 enhances keratinocyte immortalization by full-length DNA. Virology 223:251–254
63. Conrad M, Goldstein D, Andresson T, Schlegel R (1994) The E5 protein of HPV-6, but not HPV-16, associates efficiently with cellular growth factor receptors. Virology 200:796–800
64. Hwang ES, Nottoli T, DiMaio D (1995) The HPV16 E5 protein: expression, detection, and stable complex formation with transmembrane proteins in COS cells. Virology 211:227–233
65. Goldstein DJ, Schlegel R (1991) Bovine papillomavirus E5 oncoproteins binds to the 16 K component of vacuolar H^+-ATPases. Nature 354:347–349
66. Goldstein D, Schlegel R (1990) The E5 oncoprotein of bovine papillomavirus binds to a 16 kd cellular protein. EMBO J 9:137–145
67. Goldstein DJ, Schlegel R (1992) A glutamine residue in the membrane-associating domain of the bovine papillomavirus type 1 E5 oncoprotein mediates its binding to a transmembrane component of the vacuolar H(+)-ATPase. J Virol 66:405–413
68. Regan JA, Laimins LA (2008) Bap31 is a novel target of the human papillomavirus E5 protein. J Virol 82:10042–10051
69. Fehrmann F, Klumpp DJ, Laimins LA (2003) Human papillomavirus type 31 E5 protein supports cell cycle progression and activates late viral functions upon epithelial differentiation. J Virol 77:2819–2831
70. Genther S, Stering S, Duensing S, Munger K, Sattler C, Lambert P (2003) Quantitative role of the human papillomavirus type 16 E5 gene during the productive stage of the viral life cycle. J Virol 77:2832–2842

INDEX

A

Age-related macular degeneration 45, 47
Airway epithelium 109–110, 116, 117, 123, 124, 133
Alveolar epithelium 145–158

B

Basal 20, 21, 25, 69, 83, 96, 111, 113, 115, 124, 125, 128, 134–138, 140, 162, 177, 178, 182, 184, 194–199, 202, 203, 211, 215, 227, 229, 300, 312, 321, 322, 324–326, 333, 366, 367, 369, 372, 375–378, 384, 449, 451, 452
Basement membrane 52, 58, 68, 148, 149, 205, 222, 223, 226, 233, 242, 312–314, 316–317, 320, 321, 366, 394, 452
Basoapical polarity 194, 195, 197, 198, 202, 209, 210, 214, 225
Biliary epithelial cells (BEC) 296, 303–317
Biliary tract 303, 304, 306, 307, 312
Bladder 383, 384, 387–392, 394–398
Breast epithelium 193–196

C

Cancer progression 162, 196, 198, 200, 206
Cell
  engineering 417
  isolation and cultivation 387
  line 1, 2, 4, 6–10, 14, 32, 37–38, 42, 48, 51, 96, 145–146, 162, 188, 197–202, 205, 206, 209, 211, 232–234, 238–240, 245, 295, 298, 308, 327, 329, 341, 344, 348, 355, 380, 387, 390, 417–446, 456, 458
Cerebrospinal fluid (CSF) 13–28, 367
Characterization of RPE 50–51, 59–60, 329–344
Cholangiopathy 303–317
Choroid plexus 13–28
Cilia 109, 128–130, 134–136, 139, 140, 304, 360
Clonal culture system 273–285
Coculture 39, 40, 287–300, 348, 351, 356, 360, 361, 384, 385, 392–393
Collagen gel 171, 172, 243, 293, 304, 305, 307–310, 313, 460, 461
Colony assay 274, 454
Confocal immunofluorescence 88, 102, 315, 316
Conjunctival epithelial cell culture 37
Conjunctival epithelium 34, 36–38
Corneal epithelial cell culture 39, 41
Corneal epithelium 31, 34, 35, 38
Culture 1, 13, 31, 45, 67, 81, 95, 109, 123, 145, 161, 177, 193, 221, 251, 273, 287, 303, 319, 329, 347, 365, 383, 401, 417, 449
3D culture 194–215, 224, 242, 332, 367, 393
Culture medium 3, 5, 9, 15, 16, 19, 20, 24, 25, 27, 40, 41, 72, 86, 88, 97, 100, 102, 110, 148, 153, 157, 162, 178, 180–181, 198, 204, 206–210, 212–215, 229, 231–232, 276, 280, 282–284, 290, 293, 296–298, 300, 313, 321, 322, 324–327, 333, 336, 337, 342–344, 350–355, 358–361, 375, 384–385, 387, 388, 391, 392, 398, 402, 406–410, 412, 429, 438

D

Dermal fibroblasts 162–168, 170–173
Differentiation 32, 34, 37–42, 67, 78, 82, 91, 96, 111, 123–142, 161, 162, 178, 184–186, 189, 195–198, 200, 201, 203, 208, 210, 211, 214, 215, 222, 224–232, 239, 240, 246, 274, 298, 320, 349, 365, 367, 376, 381, 384, 392, 395, 397, 401, 419, 426, 449, 451–453, 455–462
Drosophila 2, 6, 7, 9

E

Ectodermal organ 401–414
Endometrium 347–349, 354, 356–360
Epidermal growth factor (EGF) 15, 32–34, 36, 37, 42, 54, 70, 76, 112, 113, 115, 124, 126, 127, 141, 164, 170, 171, 199, 202, 204, 206, 208, 210, 276, 289, 290, 297, 300, 305, 309–313, 317, 320, 321, 333, 351, 353, 358, 367, 368, 384, 390, 454, 456, 461
Epidermis 1, 67, 162, 177, 178, 182

Scott H. Randell and M. Leslie Fulcher (eds.), *Epithelial Cell Culture Protocols: Second Edition*, Methods in Molecular Biology, vol. 945, DOI 10.1007/978-1-62703-125-7, © Springer Science+Business Media, LLC 2012

Epithelia 1, 15, 19, 26, 28, 31–42, 77, 162, 195, 198, 202, 351–354, 410, 449–450, 456–457
Epithelial cell culture 1, 13, 31, 45, 67, 81, 95, 109, 123, 145, 161, 177, 193, 221, 251, 273, 287, 303, 319, 329, 347, 365, 383, 401, 417, 449
Epithelial cell line 2, 4, 7–9, 14, 197–198, 202, 238–239, 419
Epithelial cells 1, 13, 31, 45, 67, 81, 95, 109, 123, 145, 161, 177, 193, 221, 251, 273, 287, 303, 319, 329, 347, 365, 383, 401, 417, 449
Epithelial differential in culture 365
Epithelial-like construct 68
Epithelial mesenchymal transition (EMT) 46–49, 62, 314, 317, 359

F

Filaggrin 177–179, 185, 186, 190, 452
Flow cytometry 130, 138, 154–155, 187, 252, 254–255, 262–263, 270, 277, 280, 282, 370, 378–379, 435

G

GAL4/UAS system 2, 4, 7
Genetic engineering 222, 417–446
Gene transfer 125, 137, 140, 421
Gingival 71, 77

H

Hepatocytes 225, 273–274, 276, 280, 282–285, 287–300
Human
  adult retinal pigment epithelium 45
  papillomavirus 450–452
  renal epithelial cells 329–344

I

Immortalization 202, 334, 337–339, 350, 354, 355, 418, 419, 425, 426, 429, 438–439, 443, 453
Immunofluorescence 38, 59, 60, 88, 102, 105, 129–130, 137–138, 155, 156, 185–187, 213, 244, 315, 316, 396, 397
In situ hybridization (ISH) 83, 405, 410–411, 414
Intestinal stem cells 320–323, 325–327
In vitro culture 27, 46, 307, 402

K

Keratin 32, 177, 178, 185, 187, 190, 269, 353, 404, 452, 454
Keratinocytes 32, 33, 67–78, 161–174, 177–190, 368, 384, 390, 449–452, 455, 456, 458–462

L

Lentivector 420–421, 423–429, 431, 433–440, 442–444
Lentivirus 129, 137, 188, 334, 338, 420
Lgr5 .... 320, 321
Liver 21–23, 28, 258, 260, 273–285, 287–288, 291–294, 296–299, 304, 311, 314, 317
Liver stem/progenitor cells 274, 277–278, 281–285
Loricrin 177–179, 185, 186, 190
Luminal 194, 195, 198, 202, 222–225, 238, 324, 348, 357, 366, 367, 369, 375–377, 381, 384, 397
Lung 110, 115, 116, 120, 123, 145–158, 298, 419

M

MAPK kinase 310, 311
MEK5 .... 309, 311–313
Metabolism 215, 287, 288, 297–299
Mouse 15, 16, 28, 32, 59, 123–142, 151, 162, 177–190, 200, 201, 221–246, 252, 257–259, 266, 268, 273–285, 296, 298, 319–327, 350, 354, 368, 369, 375, 377, 394, 396, 402–406, 409, 414, 417, 420, 454, 456, 458, 460, 462
Mouse mammary epithelium 221–246
Mucous 124, 129, 135, 136
Murine 3T3 fibroblasts 162, 165

N

Neuropsychiatric disorders 81
Neuroregeneration 81
Non-parenchymal cells 274, 296–298

O

Ocular surface epithelia 32, 38, 40
Odor 85, 89, 90
Olfactory 81–91
Oral mucosa 67, 68, 77
Organ culture 197, 256, 266–268, 401–414
Organotypic culture 69–71, 73–78, 171–173
Ovary 1, 256, 347, 348, 351
Oviduct 347–349, 353, 356
Oxygen 120, 288, 297–299, 395

P

Pathogenesis 178, 304, 455
Pharyngeal mucosa 77
Physiology 46, 47, 82, 200, 201, 329
Polarity 1, 2, 194–200, 202, 203, 209–211, 214, 215, 225, 226, 229, 295, 367, 453
Polycystic liver 303

Primary 1–10, 14–28, 32, 34–37, 48, 51, 52, 59, 63, 68–74, 76, 78, 81–91, 95–106, 109–111, 115–117, 119, 120, 123, 146, 155–157, 162, 174, 178, 182–189, 194, 198, 201–202, 224–228, 230–240, 244–246, 269, 287–300, 319–327, 329–337, 339–342, 352, 353, 359, 360, 365–381, 388–391, 397, 413, 420, 425, 426, 429, 434, 436, 438–440, 449, 450, 452–454
Primary human keratinocytes 68, 367
Prospective isolation 281
Prostate 365–381
Proximal tubule 329, 330, 334

Q

Quantitative real-time PCR (qPCR) 51, 52, 55, 60, 61, 256, 267, 271, 402, 405, 406, 408, 409, 412, 428, 429, 435–441, 443, 444, 446

R

Raft 460–461
Ras 1–2, 418, 419
Re-aggregate cultures 251–271
Reproduction 348
Respiratory tract 109, 123, 131
RPE culture 48, 50, 51, 54
RPE expansion 45

S

Scaffold 226, 320, 384, 394–396
Serum free medium 37, 76, 127, 236, 245, 384, 392
Skin 17–18, 67–68, 78, 127, 131, 162–164, 170–174, 177–178, 181–182, 189, 224, 252, 257–259, 269, 273, 291, 402, 413, 419, 449, 451
Skin equivalent 163, 171–173
Small intestinal crypts 321, 323–325
Smell 84
Stem cells 31, 67–68, 96, 273–285, 320–323, 325–327, 349, 356–359, 361, 366–368, 374, 392, 419

T

Three-dimensional organotypic cell culture 69–71, 76–78
Thymus 251–270
Tissue architecture 196, 197, 202, 203, 210
Tissue engineering 383, 384, 387, 394, 395
Tissue-specific signaling 195, 197, 200, 229, 231
Tooth 401–403, 406, 409, 412, 414
Trachea 109, 110, 116, 124, 125, 127, 131–133, 136, 140, 141, 148, 150
Transformation 78, 197, 198, 200, 307, 390, 417–419, 425, 449–462
Transthyretin 13–14
Tumor suppressor 1, 2, 4, 196, 354, 452
Type I cells 145
Type II cells 96, 145–158

U

Ureter 384, 398
Urinary reconstruction 383
Urothelial cells 383–398

V

Vector 3, 39, 55, 85, 88, 99, 130, 137, 178, 179, 255, 334, 338, 355, 386, 418–428, 430–443, 445, 446, 458
Virus 125, 129, 140, 188, 330, 338, 339, 348, 355, 360, 420–425, 433, 440, 449–452, 454

W

Wnt signal 320, 401

Z

Z310 cell 14–16, 22–25, 27–28

MIX
Papier aus verantwortungsvollen Quellen
Paper from responsible sources
FSC® C105338

If you have any concerns about our products,
you can contact us on
ProductSafety@springernature.com

In case Publisher is established outside the EU,
the EU authorized representative is:
Springer Nature Customer Service Center GmbH
Europaplatz 3, 69115 Heidelberg, Germany

Printed by Libri Plureos GmbH
in Hamburg, Germany